T0212870

Lecture Notes in Artificial Intelligence 9376

Subseries of Lecture Notes in Computer Science

LNAI Series Editors

Randy Goebel
University of Alberta, Edmonton, Canada
Yuzuru Tanaka
Hokkaido University, Sapporo, Japan
Wolfgang Wahlster
DFKI and Saarland University, Saarbrücken, Germany

LNAI Founding Series Editor

Joerg Siekmann
DFKI and Saarland University, Saarbrücken, Germany

More information about this series at http://www.springer.com/series/1244

Van-Nam Huynh · Masahiro Inuiguchi
Thierry Denoeux (Eds.)

Integrated Uncertainty in Knowledge Modelling and Decision Making

4th International Symposium, IUKM 2015
Nha Trang, Vietnam, October 15–17, 2015
Proceedings

 Springer

Editors
Van-Nam Huynh
Japan Advanced Institute of Science
    and Technology
Nomi
Japan

Thierry Denoeux
Université de Technologie de Compiègne
Compiègne
France

Masahiro Inuiguchi
Graduate School of Engineering Science
Osaka
Japan

ISSN 0302-9743   ISSN 1611-3349 (electronic)
Lecture Notes in Artificial Intelligence
ISBN 978-3-319-25134-9   ISBN 978-3-319-25135-6 (eBook)
DOI 10.1007/978-3-319-25135-6

Library of Congress Control Number: 2015951823

LNCS Sublibrary: SL7 – Artificial Intelligence

Springer Cham Heidelberg New York Dordrecht London
© Springer International Publishing Switzerland 2015

Printed on acid-free paper

Springer International Publishing AG Switzerland is part of Springer Science+Business Media
(www.springer.com)

Preface

The International Symposium on Integrated Uncertainty in Knowledge Modelling and Decision Making (IUKM) aims to provide a forum for the exchange of research results as well as ideas for and experience of applications among researchers and practitioners involved in all aspects of uncertainty management and application. This conference series started in 2010 at the Japan Advanced Institute of Science and Technology, Ishikawa, under the title of the International Symposium on Integrated Uncertainty Management and Applications (IUM 2010). The name of the conference series was then changed to its current title at the second edition held in October 2011 at Zhejiang University, Hangzhou, China, while the third edition was held in July 2013 at Beihang University, Beijing, China.

The fourth IUKM was held during October 15–17, 2015 in Nha Trang City, Vietnam, and was jointly organized by the Pacific Ocean University (POU), VNU-Hanoi University of Engineering and Technology (VNU-UET), Hanoi National University of Education (HNUE), and Japan Advanced Institute of Science and Technology (JAIST).

The organizers received 58 submissions. Each submission was peer reviewed by at least two members of the Program Committee. While 26 papers (45%) were accepted after the first round of reviews, 17 others were conditionally accepted and underwent a rebuttal stage in which authors were asked to revise their paper in accordance with the reviews, and prepare an extensive response addressing the reviewers' concerns. The final decision was made by the program chairs. Finally, a total of 40 papers (69%) were accepted for presentation at IUKM 2015 and publication in the proceedings. The keynote and invited talks presented at the symposium are also included in this volume.

As a follow-up of the symposium, a special volume of the *Annals of Operations Research* is anticipated to include a small number of extended papers selected from the symposium as well as other relevant contributions received in response to subsequent open calls. These journal submissions will go through a fresh round of reviews in accordance with the journal's guidelines.

The IUKM 2015 symposium was partially supported by The National Foundation for Science and Technology Development of Vietnam (NAFOSTED). We are very grateful to the local organizing team from Pacific Ocean University for their hard working, efficient services, and wonderful local arrangements.

We would like to express our appreciation to the members of the Program Committee for their support and cooperation in this publication. We are also thankful to Alfred Hofmann, Anna Kramer, and their colleagues at Springer for providing a meticulous service for the timely production of this volume. Last, but certainly not the least, our special thanks go to all the authors who submitted papers and all the attendees for their contributions and fruitful discussions that made this symposium a success.

We hope that you will find this volume helpful and motivating.

October 2015

Van-Nam Huynh
Masahiro Inuiguchi
Thierry Denoeux

Organization

General Co-chairs

Viet-Ha Nguyen   University of Engineering and Technology, VNU-Hanoi, Vietnam

Sadaaki Miyamoto   University of Tsukuba, Japan

Honorary Co-chairs

Michio Sugeno   European Center for Soft Computing, Spain

Hung T. Nguyen   New Mexico State University, USA; Chiang Mai University, Thailand

Dinh Lien Quach   Pacific Ocean University, Nha Trang, Vietnam

Program Co-chairs

Van-Nam Huynh   Japan Advanced Institute of Science and Technology, Japan

Masahiro Inuiguchi   University of Osaka, Japan

Thierry Denoeux   University of Technology of Compiegne, France

Local Arrangements Co-chairs

Van-Hai Pham   Pacific Ocean University, Nha Trang, Vietnam

Anh-Cuong Le   University of Engineering and Technology, VNU-Hanoi, Vietnam

Publication and Financial Chair

Dang-Hung Tran   Hanoi National University of Education, Vietnam

Secretary

Van-Doan Nguyen   Japan Advanced Institute of Science and Technology, Japan

Program Committee

Byeong-Seok Ahn   Chung-Ang University, Korea

Yaxin Bi   University of Ulster, UK

Bernadette Bouchon-Meunier   Université Pierre et Marie Curie, France

Lam Thu Bui   Le Quy Don Technical University, Vietnam

Tru Cao	Ho Chi Minh City University of Technology, Vietnam
Christer Carlsson	Åbo Akademi University, Finland
Fabio Cuzzolin	Oxford Brookes University, UK
Bernard De Baets	Ghent University, Belgium
Thierry Denoeux	University of Technologie of Compiegne, France
Zied Elouedi	LARODEC, ISG de Tunis, Tunisia
Tomoe Entani	University of Hyogo, Japan
Lluis Godo	IIIA - CSIC, Spain
Quang Ha	University of Technology, Sydney, Australia
Enrique Herrera-Viedma	University of Granada, Spain
Katsuhiro Honda	Osaka Prefecture University, Japan
Tzung-Pei Hong	National Univesity of Kaohsiung, Taiwan
Van-Nam Huynh	Japan Advanced Institute of Science and Technology, Japan
Masahiro Inuiguchi	Osaka University, Japan
Radim Jirousek	University of Economics, Czech Republic
Gabriele Kern-Isberner	Technische Universität Dortmund, Germany
Laszlo T. Koczy	Budapest University of Technology and Economics, Hungary
Vladik Kreinovich	University of Texas at El Paso, USA
Rudolf Kruse	University of Magdeburg, Germany
Yasuo Kudo	Muroran Institute of Technology, Japan
Yoshifumi Kusunoki	Osaka University, Japan
Jonathan Lawry	University of Bristol, UK
Hoai Bac Le	University of Science, VNU-Ho Chi Minh, Vietnam
Churn-Jung Liau	Academia Sinica, Taipei, Taiwan
Jun Liu	University of Ulster, UK
Luis Magdalena	European Centre for Soft Computing, Spain
Sadaaki Miyamoto	Tsukuba University, Japan
Tetsuya Murai	Hokkaido University, Japan
Xuan Hoai Nguyen	Hanoi University, Vietnam
Akira Notsu	Osaka Prefecture University, Japan
Vilem Novak	Ostrava Univeristy, Czech Republic
Irina Perfilieva	Ostrava Univcristy, Czech Republic
Zengchang Qin	Beihang University, China
Yasuo Sasaki	Japan Advanced Institute of Science and Technology, Japan
Hirosato Seki	Kwansei Gakuin University, Japan
Dominik Slezak	University of Warsaw and Infobright Inc., Poland
Songsak Sriboonchitta	Chiang Mai University, Thailand
Yongchuan Tang	Zhejiang University, China
Vicenc Torra	University of Skovde, Sweden
Seiki Ubukata	Osaka University, Japan
Hong-Bin Yan	East China University of Science and Technology, China

Local Organizing Committee

Van Ninh Do	Pacific Ocean University, Nha Trang, Vietnam
Dang Bac Nguyen	Pacific Ocean University, Nha Trang, Vietnam
Da Thao Nguyen	Pacific Ocean University, Nha Trang, Vietnam
Quoc Tien Nguyen	Pacific Ocean University, Nha Trang, Vietnam
Thi Hong Phuong Nguyen	Pacific Ocean University, Nha Trang, Vietnam
Thuc Tri Nguyen	Pacific Ocean University, Nha Trang, Vietnam
Huu Quy Nguyen	Pacific Ocean University, Nha Trang, Vietnam
Thi Cam Huyen Mai	Pacific Ocean University, Nha Trang, Vietnam
Thi Tuyet Nhung Trinh	Pacific Ocean University, Nha Trang, Vietnam

Sponsors

 The National Foundation for Science and Technology Development of Vietnam (NAFOSTED)

 Japan Advanced Institute of Science and Technology

 Pacific Ocean University, Nha Trang, Vietnam

Contents

Keynote and Invited Talks

Epistemic Uncertainty Modeling: The-state-of-the-art.................. 1
   Hung T. Nguyen

Fuzzy Sets, Multisets, and Rough Approximations 11
   Sadaaki Miyamoto

What Is Fuzzy Natural Logic: Abstract 15
   Vilém Novák

Contributed Papers

Combining Fuzziness and Context Sensitivity in Game Based
Models of Vague Quantification 19
   Christian G. Fermüller

A New Model of a Fuzzy Associative Memory 32
   Irina Perfilieva

Construction of Associative Functions for Several Fuzzy Logics via the
Ordinal Sum Theorem 43
   Mayuka F. Kawaguchi and Michiro Kondo

Cognitively Stable Generalized Nash Equilibrium in Static Games with
Unawareness 54
   Yasuo Sasaki

Maximum Lower Bound Estimation of Fuzzy Priority Weights from a Crisp
Comparison Matrix 65
   Tomoe Entani and Masahiro Inuiguchi

Logarithmic Conversion Approach to the Estimation of Interval Priority
Weights from a Pairwise Comparison Matrix 77
   Masahiro Inuiguchi and Shigeaki Innan

An Effective Method for Optimality Test Over Possible Reaction Set for
Maximin Solution of Bilevel Linear Programming with Ambiguous
Lower-Level Objective Function 89
   Puchit Sariddichainunta and Masahiro Inuiguchi

Proposal of Grid Area Search with UCB for Discrete Optimization
Problem.   102
Akira Notsu, Koki Saito, Yuhumi Nohara, Seiki Ubukata,
and Katsuhiro Honda

Why Copulas Have Been Successful in Many Practical Applications:
A Theoretical Explanation Based on Computational Efficiency.   112
Vladik Kreinovich, Hung T. Nguyen, Songsak Sriboonchitta,
and Olga Kosheleva

A New Measure of Monotone Dependence by Using Sobolev Norms
for Copula   126
Hien D. Tran, Uyen H. Pham, Sel Ly, and T. Vo-Duy

Why ARMAX-GARCH Linear Models Successfully Describe Complex
Nonlinear Phenomena: A Possible Explanation   138
Hung T. Nguyen, Vladik Kreinovich, Olga Kosheleva,
and Songsak Sriboonchitta

A Copula-Based Stochastic Frontier Model for Financial Pricing.   151
Phachongchit Tibprasorn, Kittawit Autchariyapanitkul,
Somsak Chaniam, and Songsak Sriboonchitta

Capital Asset Pricing Model with Interval Data.   163
Sutthiporn Piamsuwannakit, Kittawit Autchariyapanitkul,
Songsak Sriboonchitta, and Rujira Ouncharoen

Confidence Intervals for the Difference Between Normal Means
with Known Coefficients   171
Suparat Niwitpong and Sa-Aat Niwitpong

Approximate Confidence Interval for the Ratio of Normal Means
with a Known Coefficient of Variation   183
Wararit Panichkitkosolkul

Confidence Intervals for the Ratio of Coefficients of Variation of the
Gamma Distributions.   193
Patarawan Sangnawakij, Sa-Aat Niwitpong, and Suparat Niwitpong

A Deterministic Clustering Framework in MMMs-Induced Fuzzy
Co-Clustering.   204
Shunnya Oshio, Katsuhiro Honda, Seiki Ubukata, and Akira Notsu

FCM-type Co-clustering Transfer Reinforcement Learning for Non-Markov
Processes   214
Akira Notsu, Takanori Ueno, Yuichi Hattori, Seiki Ubukata,
and Katsuhiro Honda

MMMs-Induced Fuzzy Co-clustering with Exclusive Partition
Penalty on Selected Items 226
   Takaya Nakano, Katsuhiro Honda, Seiki Ubukata, and Akira Notsu

Clustering Data and Vague Concepts Using Prototype Theory Interpreted
Label Semantics 236
   Hanqing Zhao and Zengchang Qin

An Ensemble Learning Approach Based on Rough Set Preserving the
Qualities of Approximations................................ 247
   *Seiki Ubukata, Taro Miyazaki, Akira Notsu, Katsuhiro Honda,
   and Masahiro Inuiguchi*

Minimum Description Length Principle for Compositional Model
Learning 254
   Radim Jiroušek and Iva Krejčová

On the Property of SIC Fuzzy Inference Model with Compatibility
Functions.. 267
   Hirosato Seki

Applying Covering-Based Rough Set Theory to User-Based Collaborative
Filtering to Enhance the Quality of Recommendations............... 279
   Zhipeng Zhang, Yasuo Kudo, and Tetsuya Murai

Evidence Combination Focusing on Significant Focal Elements for
Recommender Systems 290
   Van-Doan Nguyen and Van-Nam Huynh

A Multifaceted Approach to Sentence Similarity.................... 303
   Hien T. Nguyen, Phuc H. Duong, and Tuan Q. Le

Improving Word Alignment Through Morphological Analysis 315
   *Vuong Van Bui, Thanh Trung Tran, Nhat Bich Thi Nguyen,
   Tai Dinh Pham, Anh Ngoc Le, and Cuong Anh Le*

Learning Word Alignment Models for Kazakh-English Machine
Translation... 326
   Amandyk Kartbayev

Application of Uncertainty Modeling Frameworks to Uncertain Isosurface
Extraction 336
   Mahsa Mirzargar, Yanyan He, and Robert M. Kirby

On Customer Satisfaction of Battery Electric Vehicles Based on Kano
Model: A Case Study in Shanghai............................ 350
   Yanping Yang, Hong-Bin Yan, and Tieju Ma

Co-movement and Dependency Between New York Stock Exchange,
London Stock Exchange, Tokyo Stock Exchange, Oil Price, and Gold
Price   362
  Pathairat Pastpipatkul, Woraphon Yamaka, and Songsak Sriboonchitta

Spillovers of Quantitative Easing on Financial Markets of Thailand,
Indonesia, and the Philippines.   374
  *Pathairat Pastpipatkul, Woraphon Yamaka, Aree Wiboonpongse,
  and Songsak Sriboonchitta*

Impacts of Quantitative Easing Policy of United States of America on Thai
Economy by MS-SFABVAR   389
  *Pathairat Pastpipatkul, Warawut Ruankham, Aree Wiboonpongse,
  and Songsak Sriboonchitta*

Volatility and Dependence for Systemic Risk Measurement of the
International Financial System   403
  *Jianxu Liu, Songsak Sriboonchitta, Panisara Phochanachan,
  and Jiechen Tang*

Business Cycle of International Tourism Demand in Thailand: A Markov-
Switching Bayesian Vector Error Correction Model.   415
  Woraphon Yamaka, Pathairat Pastpipatkul, and Songsak Sriboonchitta

Volatility Linkages Between Price Returns of Crude Oil and Crude Palm
Oil in the ASEAN Region: A Copula Based GARCH Approach.   428
  *Teera Kiatmanaroch, Ornanong Puarattanaarunkorn,
  Kittawit Autchariyapanitkul, and Songsak Sriboonchitta*

The Economic Evaluation of Volatility Timing on Commodity Futures
Using Periodic GARCH-Copula Model   440
  Xue Gong, Songsak Sriboonchitta, and Jianxu Liu

On the Estimation of Western Countries' Tourism Demand for Thailand
Taking into Account of Possible Structural Changes Leading to a Better
Prediction.   452
  Nyo Min and Songsak Sriboonchitta

Welfare Measurement on Thai Rice Market: A Markov Switching Bayesian
Seemingly Unrelated Regression   464
  Pathairat Pastpipatkul, Paravee Maneejuk, and Songsak Sriboonchitta

Modeling Daily Peak Electricity Demand in Thailand   478
  Jirakom Sirisrisakulchai and Songsak Sriboonchitta

Author Index   489

Epistemic Uncertainty Modeling:
The-state-of-the-art

Hung T. Nguyen[1,2](✉)

[1] New Mexico State University, Las Cruces, USA
[2] Chiang Mai University, Chiang Mai, Thailand
hunguyen@nmsu.edu

Abstract. This paper is about the state-of-the-art of epistemic uncertainty modeling from subjective probability (Thomas Bayes) to fuzzy measures (Michio Sugeno).

Keywords: Bayesian statistics · Epistemic uncertainty · Fuzzy measures · Random sets

1 Introduction

Epistemic uncertainty (uncertainty due to lack of knowledge) clearly is an important factor in decision-making problems. The Bayesian approach to modeling epistemic uncertainty, in a statistical context, say, is well known since Thomas Bayes [1] where uncertainty about "population parameters" is described as subjective probability measures. The fact that uncertainty can take multiple forms has been known since the beginning of probability theory. See, e.g., Fox and Ulkumen [2] for a clear distinction between stochastic and epistemic uncertainties, being the two main sources causing uncertainty.

A rationale for the Bayesian approach to modeling epistemic uncertainty is provided by De Finetti's theorem [3], and strongly defended by Dennis Lindley [4] where he showed that confidence intervals, belief functions and possibility measures are all inadmissible in a reasonable decision framework.

Belief functions (Shafer [5]) and possibility measures (Zadeh [6]), among other non-additive set-functions, were suggested as extensions of Bayesian probability measures to model epistemic uncertainty. It should be noted that it was Michio Sugeno [7] who considered non-additive set-functions, known as fuzzy measures, to model subjective evaluations, before Shafer and Zadeh, and Shafer's belief functions and Zadeh's possibility measures are special cases of fuzzy measures.

It turns out that the message of Dennis Lindley, namely "You cannot avoid probability" (when modeling epistemic uncertainty) is misleading, as discovered by Goodman et al. [8]. Specifically, Lindley's theorem says that, in a scoring rule framework, admissible measures of uncertainty are only those which are functions of probability measures. But then, a set-function which is a function of a probability measure needs not be additive! Whether or not this "discovery"

Dedicated to Michio Sugeno

V.-N. Huynh et al. (Eds.): IUKM 2015, LNAI 9376, pp. 1–10, 2015.
DOI: 10.1007/978-3-319-25135-6_1

provided confidence for researchers to "propose" new alternatives to Bayesian probability measures does not matter. Criticisms of the Bayesian approach to epistemic uncertainty modeling from many perspectives seem to be responsible for alternatives.

We focus here on epistemic uncertainty modeling in general contexts for decision-making. Typically, these are situations where modeling epistemic uncertainty is a must.

Even if we restrict ourself to the familiar practice of using statistical theory to handle uncertainty, we are still facing many embarassing cryptic issues, especially those concerning epistemic uncertainty modeling.

In 1900, David Hilbert posed 23 (unsolved at that time) mathematical problems to the mathematical community whose solutions (in his view) should advance mathematics. Almost 100 years later, at his (Johann)Bernoulli Lecture 1997, at Groningen University, The Netherlands, D.R. Cox handed out to the public a list of 14 cryptic (obscure/ambiguous) issues concerning the nature of statistical inference. See Kardaun et al. [9]. Among them, here are the cryptic issues which seem to be of our concern:

(a) How possible and fruitful is it to treat quantitatively uncertainty not derived from statistical variability?

(b) Are all sensitive probabilities ultimately frequency based?

(c) Is the formulation of personalistic probability by de Finetti and Savage the wrong way round? It puts betting behaviour first and belief to be determined from that?

(d) In a Bayesian formulation should priors constructed retrospectively, after seeing the data, be treated distinctively?

(e) What is the role in theory and practice of upper and lower probabilities? Clearly, unlike Hilbert's list, the 14 "questions" of Cox are issues and not problems (to be investigated for possible solutions).

2 Information Measures

Epistemic uncertainty is due to our lack of knowledge. For example, we do not know the exact value of the number π. Unlike stochastic uncertainty, we can reduce epistemic uncertainty by seeking information (here, locating information, e.g., $3.1 < \pi < 3.2$, $3.14 < \pi < 3.15$, ...).

The fact that bounds on unknown quantities provide non probabilistic information leads to the actual method of *interval analysis* for modeling epistemic uncertainty. See, e.g., Swiler et al. [10].

Thus, modeling epistemic uncertainty should start with information (since after all, information is more general than data as it can take on "non-conventional" forms such as linguistic values) and not probability! In fact, as Kolmogorov [11] put it "Information theory must precede probability and not be based on it". It should be noted that the Wiener-Shannon information measure $J(A) = c \log \frac{1}{P(A)}$, based upon probability, cannot be used for epistemic uncertainty modeling as, e.g., the probability that $\pi \in A$ does not make any sense.

For a *semantic* information theory which could be applied to modeling epistemic uncertainty, see Kampe' de Feriet [12]. Essentially, an information measure is a set function $J(.)$ such as $A \subseteq B \implies J(A) \geq J(B)$. This order relation $A \succcurlyeq B$ (more informative) when $A \subseteq B$ in information theory is essential for defining appropriate lattice-based topologies for random fuzzy sets [13], [14] which are useful in extending Bayesian prior probability measures to linguistic information gathering processes. Note that while the Wiener-Shannon information measure is decomposable as

$$J(A \cup B) = -c \log[e^{-\frac{J(A)}{c}} + e^{-\frac{J(B)}{c}}]$$

general information measures, not necessarily based on probability, could be decomposable differently, such as

$$J(A \cup B) = \min\{J(A), J(B)\}$$

see Nguyen [15].

3 Fuzzy Measures

Fuzziness is a type of uncertainty due to the very nature of fuzzy concepts in everyday natural language uses. A fuzzy subset A of a set U is a map, denoted also as $A(.) : U \to [0, 1]$ where for $u \in U$, the value $A(u) \in [0, 1]$ is interpreted as the degree to which u is a member of A. The "set" A is called "fuzzy" because its boundary is not sharply defined. For example, A is the label describing the set of low income people in a given population. Unless we define a threshold (a poverty line) to distinguish low income people with others, the fuzzy concept "low income" is fuzzy, by natural language. It is the set which is fuzzy, not membership degrees of elements. Defining a "set" by its membership function is just an extension of ordinary sets (using the equivalence with their indicator functions).

An interesting example is in the designs of *Regression Discontinuity* for causal inference in social sciences, where the use of membership functions with values in $[0, 1]$, rather than in $\{0, 1\}$, is necessary.

In 1974, M. Sugeno [7] developed a theory of subjective assignments (e.g., for pattern recognition problems) as follows. Let $\theta_o \in \Theta$ be known. We wish to know whether $\theta_o \in A$, for a given subset $A \subseteq \Theta$. Of course, if we know the indicator function $1_A(.)$, then we can tell!

Suppose we cannot assess $1_A(.)$. The available (observed) characteristics of θ_o are not sufficient to make a sure identification. There is uncertainty in the assignment of θ_o to any set A. So let $\mu(A) = \mu(\theta_o \in A) \in [0, 1]$ be the degree to which θ_o is in a (crisp) set A. Operating on, say, 2^Θ, the set-function $\mu(.)$ is called a fuzzy measure. Note that fuzzy measures differ from fuzzy sets since the sets in the arguments of μ are not fuzzy: what is "fuzzy" is in the degree to which we think A contains θ_o. Put it differently, with incomplete information, we are unable to know whether our θ_o is in A or not, i.e., the situation is "fuzzy" (not clear, ambiguous, imprecise), we are forced to use some assignment reflecting our uncertainty about our assignment. That was what Sugeno did: A fuzzy measure is an ordinary set function whose values are "fuzzy" in the sense they represent our uncertainty (not by probability necessarily), usually subjective, of another type. Thus, a measure (a procedure to "measure" something) is fuzzy because its values represent the "likelihoods" of membership, whereas, a fuzzy set is "fuzzy" since the "set" itself is fuzzy (not sharply specified).

Formally, a fuzzy measure on a set Θ is a (monotone) map $F(.) : 2^\Theta \to [0, 1]$ such that $F(\varnothing) = 0$, $F(\Theta) = 1$, and $A \subseteq B \implies F(A) \leq F(B)$.

In a sense, fuzzy measures provide information about the location of θ_0 in Θ, similar to the Bayesian probability, but no need to "view" θ_0 as a random variable, i.e., can "vary" in Θ.

When the fuzzy measure is (formally) a distribution of a random set on Θ, then while retaining θ_0 fixed (but unknown, as it should be), it induces a random set modeling of the possible location of θ_0, a reasonable way to express our prior knowledge (information) about θ_0, i.e., a "good" way to model epistemic uncertainty in statistics (lack of knowledge about "where" is θ_0). In other words, we do not need to "view" our unknown (but fixed) θ_0 as a random variable (like the Bayesians did) in order to put our prior information on it by a probability measure, we simply model our epistemic uncertainty about θ_0 by putting our prior information in the form of a bona fide random set on Θ whose "distribution function" is a special fuzzy measure. What is "moving around" is not θ_0 but our random set governed by the distribution we provide from our (prior) information about θ_0. Extensions to random fuzzy sets could be considered, especially when seeking experts' knowledge to reduce epistemic uncertainty.

Note that we leave aside the now well known epistemic uncertainty due to fuzziness in our natural language. We note however that there is a usefulconnection between fuzzy and interval methods in treating epistemic uncertainty (for processing fuzzy data), via, e.g., Nguyen [16], which has been classified, at the recent FUZZ-IEEE (Istanbul, August 3-5, 2015/celebrating the 50th anniversary of Zadeh's first paper on fuzzy sets [17]), as one among 100 most influential papers and books in research involving fuzziness.

4 Belief Functions

One year after Sugeno's thesis [7], G. Shafer [5] wrote a Ph.D. thesis in which he proposed to establish a mathematical theory of evidence to model epistemic uncertainty. This can be seen as an attempt to replace Bayesian priors by something more realistic from an application viewpoint (possibly in a more general area of decision-making than statistics). Since personal probabilities in the Bayesian approach are degrees of beliefs (of agents), Shafer called his new non-additive set-functions induced from "evidence" *belief functions*. A belief function $F(.)$ on a (finite) set Θ, is a map $F(.) : 2^\Theta \to [0, 1]$ defined by weakening Henri Poincare's equality, namely $F(\varnothing) = 0$, $F(\Theta) = 1$, and, for any $k \geq 2$, and

$A_1, ..., A_k$, subsets of Θ, with $|I|$ denoting the cardinality of the set I,

$$F(\cup_{i=1}^{k} A_i) \geq \sum_{\varnothing \neq I \subseteq \{1,2,...,k\}} (-1)^{|I|+1} F(\cap_{i \in I} A_i)$$

Clearly, belief functions are fuzzy measures in Sugeno's sense. In fact, they are special types of fuzzy measures as they correspond to distributions and not probability measures of random elements. This has been pointed out immediately by Nguyen [18] in 1976 (but appeared in 1978). Specifically, a given belief function $F(.)$ is precisely the distribution of a *random set* $S : (\Omega, \mathcal{A}, P) \to 2^{\Theta}$, i.e., $F(A) = P(\omega : S(\omega) \subseteq A)$. Note that, in the finite case, the fact that $F(.)$ characterizes the probability law of S is obvious. From an interpretation point of view, this remark says that, using evidence to derive quantitative modeling of epistemic uncertainty is somewhat similar to Bayesian approach by treating unknown (but fixed) quantities, not as random variables, but as random sets, i.e., as a more flexible ("higher") level. For an introduction to the mathematical theory of random sets, see Nguyen [19].

5 Possibility Measures

We are certainly familiar with categorical data analysis (e.g., via contingency tables in elementary statistics) where variables take values as categories labeled linguistically, such as "low", "medium", "high", etc....In using natural language to impart knowledge and information, there is a great deal of fuzziness in imprecise statements such as "John is young". But, given only the information that "John is young", we do not know his exact (true) age (it is an epistemic uncertainty), but we can "infer" subjectively a *possibility distribution* $\mu_A(.)$ with the interpretation that, for each $x \in [0, 120]$, $\mu_A(x)$ is the degree of possibility that John is x years old. Clearly, possibility distributions are different from probability distributions. A possibility distribution on a set U is simply a function $\varphi : U \to [0, 1]$.

According also to Zadeh, a *possibility measure* asociated with a possibility distribution φ on U is a set-function Ψ defined on the power set 2^U, with values in $[0, 1]$, defined simply as

$$\Psi(A) = \sup\{\varphi(u) : u \in A\}$$

Possibility measures are not additive. In fact, they are "maxitive":

$$\Psi(A \cup B) = \max\{\Psi(A), \Psi(B)\}$$

6 Imprecise Probabilities

Putting a (subjective) prior probability measure to describe an epistemic uncertainty, by Bayesians, is just proposing a "model" for it. But "All models are wrong, but some are useful" (Box, [20]). So, let's try some models! Note that we are not discussing how "not to use priors" in a statistical framework! Interested readers could read recent works of Liu and Martin [21], [22] on "prior-free probabilistic inference", using random sets.

While the Bayesian approach to statistics consists essentially of modeling the epistemic uncertainty of model parameters by prior (subjective) probability measures, its usefulness (or reliability) is often questionable because of specific choices of priors. From the above "*random set connection*", Dempster-Shafer's work can be viewed as an attempt to address this reliability problem by considering a special case of the so-called *imprecise probability* approach, which, at present, is still in full research activities.

The imprecise probability approach is this. In a robustness spirit, rather than specifying a precise (prior) probability measure P_o on Θ, it is "safer" to consider some family of probability measures $\mathcal{P}$, one of its (unknown) element could be used as prior. But then, without knowning P_o, we are force to work with the whole family $\mathcal{P}$, for example with its "lower probability"

$$F(A) = \inf\{P(A) : P \in \mathcal{P}\}$$

or its "upper probability" $G(A) = \sup\{P(A) : P \in \mathcal{P}\} = 1 - F(A^c)$.

The set-function $F(.)$ (lower envelop), viewed as a prior, is no longer additive. The problem is this. Having a prior (whatever you model it) is just the first step in the Bayesian program. The second and essential step is updating it in view of new (observed) data via Bayes' formula which is "doable" only with probability measure prior. Research in this direction consists of, among others, generalizing Bayes' updating process for non-additive set functions, using, say, Choquet integral concept in Potential Theory.

It might happen that, in some cases, the lower envelop F of a suitable $\mathcal{P}$ turns out to be a distribution function of some random set on Θ, for which we could use random set theory to proceed ahead. For example, let Θ be a finite set, and $A_1, A_2, ..., A_n$ be a partition of Θ. Let $\mathcal{P}$ be the set of probability measures on Θ such that $P \in \mathcal{P}$ iff $P(A_i) = \alpha_i$, $i = 1, 2, ..., n$, with given $\alpha_i > 0$, $i = 1, 2, ..., n$, $\sum_{i=1}^n \alpha_i = 1$. Let $F(A) = \inf\{P(A) : P \in \mathcal{P}\}$. Then $F(.)$ is a belief function.

Now, if we view a random set S as a coarse data, i.e., the variable of interest is $X : \Omega \to \Theta$, which is not directly observable, and S (observable) is its coarsening, i.e. $P(X \in S) = 1$ (the other way around, X is an almost sure selector of S). An event $A \subseteq \Theta$ is realized (occured) if $X(\omega) \in A$. But X is not observable, only $S(\omega)$ is, so that we are uncertain about the occurence of A. If $S(\omega) \subseteq A$, then we are sure that A occurs since $S(\omega) \ni X(\omega)$ and hence $X(\omega) \in A$. Thus, from a pessimistic viewpoint, we quantify our degree of belief in the occurence of A based on S by $P(S \subseteq A) = F(A)$ which is less than the actual probability that A occurs, namely $P(X \in A)$. On the other hand, if $S(\omega) \cap A \neq \varnothing$, then it is possible that A occurs since $\{\omega : X(\omega) \in A\} \subseteq \{\omega : S(\omega) \cap A \neq \varnothing\}$. Thus we quantify this "plausibility" by

$$T(A) = P(S \cap A \neq \varnothing) = 1 - F(A^c)$$

The set-function T is dual to F, and referred to as the *capacity functional* of the random set S. The capacity functionals play the role of distribution functions for random sets. In view of Choquet's work in Capacity Theory, it is more convenient to use T rather than F.

We turn now to the continuous case, where $\Theta = \mathbb{R}^d$. We restrict ourself to the case of *random closed sets*, i.e., random elements taking closed subsets of $\mathbb{R}^d$ as values. Let $\mathcal{F}$ denote the set of all closed sets of $\mathbb{R}^d$. A random closed set S is a measurable map from $(\Omega, \mathcal{A}, P)$ to $(\mathcal{F}, \sigma(\mathcal{F}))$ where $\sigma(\mathcal{F})$ is the $\sigma-$ field of subsets of $\mathcal{F}$ generated by Matheron's hit-or-miss topology (we skip technical details here).

The counter-part of Lebesgue-Stieltjes characterization theorem for probability measures on $\mathbb{R}^d$ is the following Choquet theorem which provides the axiomatic definition of capacity functional in the continuous case.

Let $\mathcal{K}$ denotes the set of all compact subsets of $\mathbb{R}^d$. A function $T : \mathcal{K} \to \mathbb{R}$ is called a *capacity functional* (of a random closed set) if it satisfies

(i) $0 \leq T(.) \leq 1, T(\varnothing) = 0$
(ii) T is alternating of infinite order, i.e., for any $n \geq 2$, and $K_1, K_2, ..., K_n$ in $\mathcal{K}$,

$$T(\cap_{i=1}^{n} K_i) \leq \sum_{\varnothing \neq I \subseteq \{1,2,...,n\}} (-1)^{|I|+1} T(\cup_{i \in I} K_i)$$

(iii) If $K_n \searrow K$ in $\mathcal{K}$, then $T(K_n) \searrow T(K)$ (right continuity)

Choquet Theorem: If $T : \mathcal{K} \to \mathbb{R}$ is a capacity functional, then there exists a unique probability measure Q on $\sigma(\mathcal{F})$ such that, for all $K \in \mathcal{K}$,

$$Q(\mathcal{F}_K) = T(K)$$

where

$$\mathcal{F}_K = \{A \in \mathcal{F} : K \cap A \neq \varnothing\}$$

Choquet theorem justifies the above definition of a capacity functional, in the sense that, like distribution functions of random vectors determine completely corresponding probability laws, capacity functionals as defined above determine completely (and uniquely) corresponding probability measures (laws) of random closed sets on euclidean spaces.

Since $T(K) = P(K \cap S \neq \varnothing)$, it follows, in particular, for signleton sets, that

$$T(\{x\}) = P(\{x\} \cap S \neq \varnothing) = P(x \in S)$$

The function $\pi_S : \mathbb{R}^d \to [0,1]$, defined by $\pi_S(x) = P(x \in S)$ is referred to as the (one-point) *coverage function* of the random set S. The coverage function is obtained from the capacity functional T_S of S by setting $\pi(x) = T(\{x\})$. Coverage functions are essential in sampling designs in finite populations in classical statistics (see Hajek [23]). The coverage function alone is not enough to conduct inference since it is much weaker than the associated capacity functional.

There is, however, a "canonical" case where coverage functions determine capacity functionals. Let $\phi : \mathbb{R}^d \to [0,1]$ be an upper- semi - continuous (usc) function, i.e., $\{x \in \mathbb{R}^d : \phi(x) \geq s\}$ is closed, for any $s \in \mathbb{R}$.

a) Let $\alpha : (\Omega, \mathcal{A}, P) \to [0,1]$ be a random variable, uniformly distributed on $[0,1]$. Then $S : \Omega \to \mathcal{F}(\mathbb{R}^d)$, defined by $S(\omega) = \{x \in \mathbb{R}^d : \phi(x) \geq \alpha(\omega)\}$, is clearly a random closed set on $\mathbb{R}^d$. What is its coverage function? Well,

$$\pi_S(x) = P(x \in S) = P(\phi(x) \geq \alpha(\omega)) = \phi(x)$$

What is its capacity functional? Well,

$$T_S(K) = P(\omega : K \cap S(\omega) \neq \varnothing) = \sup_{x \in K} \phi(x)$$

i.e, this canonical random set is completely determined by its coverage function.

Several remarks are in order:

(i) The set-function $K \in \mathcal{K} \to \sup_{x \in K} \phi(x)$ is a capacity functional since it is precisely of the form $P(K \cap S \neq \varnothing) = Q(\mathcal{F}_K)$,

(ii) If $\phi(.)$ is taken as a Zadeh's possibility distribution, then we got a neat relation between possibility theory and Shafer's (dual) belief functions (which are subsumed by random sets).

(iii) The canonical random set $S(\omega) = \{x \in \mathbb{R}^d : \phi(x) \geq \alpha(\omega)\}$ is **nested**, i.e., for any $\omega \neq \omega'$, either $S(\omega) \subseteq S(\omega')$, or $S(\omega) \supseteq S(\omega')$. It can be shown that, in general, if a random set is nested, then its coverage function determines its capacity functional.

b) The above canonical random closed set suggests a way to construct capacity functionals. For any $\phi : \mathbb{R}^d \to [0,1]$, an upper- semi - continuous (usc) function, the set function

$$K \in \mathcal{K} \to \sup_{x \in K} \phi(x) = T(K)$$

is a capacity functional of some random closed set.

7   Conclusions

We can say now that: There are many ways of representing epistemic uncertainty, such as probability theory, fuzzy sets, possibility theory, and imprecise probability, interval analysis, Dempster-Shafer evidence theory, and for mixed aleatory/epistemic uncertainties, second-order probability.

Note however that the problem of selecting an appropriate mathematical structure to represent epistemic uncertainty is challenging!

As econophysicists emphasized in the context of modeling uncertainty for decision-making, the difference of socio-economic phenomena with natural phenomena (e.g., physics, chemistry, cell biology) should be clearly noted. See e.g., McCauley [24]. While physics can be reduced to mathematical laws of nature (at any levels: quantum field, Newtonian mechanics, general relativity) in view of invariance principles, human invented laws can always be violated. Thus, one should be careful when tempting to make, say, economics look like an exercise in calculus, and financial econometrics as a subset of stochastic processes obeying the martingale representation theorem! As far as decision-making is concerned, as McCauley put it, "We have a law to understand and explain everything, at leat qualitatively, except phenomena involving decisions made by minds".

The buzz word is uncertainty. And the questions are, e.g., what is uncertainty? are there different types of uncertainty? why we run into uncertainty? what about "known unknowns? unknown unknowns?", and then, how to model various types of uncertainty? how to combine different types of uncertainty in a given problem? how to make decisions in the face of these uncertainties?

Can we answer the above questions by "solving" mathematical equations? Is it obvious that not everything can be mathematized?

While random uncertainty is modeled by frequentist probability (accepted universally), despite the objection of John Maynard Keynes (1883-1946): "In the long run, we are all dead", the problem of epistemic uncertainty modeling seems to be an art rather than a science (just like in everyday medical diagnosis!).

References

1. Bayes, T.: An essay towards solving a problem in the doctrine of chance. Philo. Trans. Royal Soc. London **58**, 370–418 (1763)
2. Fox, C.R., Ulkumen, G.: Distinguishing two dimensions of uncertainty. In: Brun, W., et al. (eds.) Perspectives on Thinking, Judging, and Decision Making, Universitetsforlaget, Oslo, Norway, pp. 21–35 (2011)
3. La De Finetti, B.: Prevision: ses lois logiques, ses spurces subjectives. Ann. H. Poincaré **7**, 1–68 (1937)
4. Lindley, D.V.: Scoring rules and the inevitability of probability. International Statist. Review **50**, 1–26 (1982)
5. Shafer, G.: A Mathematical Theory of Evidence. Princeton University Press (1976)
6. Zadeh, L.: Fuzzy sets as a basis for a theory of possibility. J. Fuzzy Sets and Systems **1**, 3–28 (1978)
7. Sugeno, M.: Theory of fuzzy integrals and its applications, Ph.D. Thesis, Tokyo Institute of Technology (1974)
8. Goodman, I.R., Nguyen, H.T., Rogers, G.S.: On the scoring approach to admissibility of uncertainty measures in expert systems. J. Math. Anal. and Appl. **159**, 550–594 (1991)
9. Kardaun, O.J.W.F., et al.: Reflections on Fourteen Cryptic Issues Concerning The Nature of Statistical Inference. International Statistical Review **17**(2), 277–318 (2003)
10. Swiler, L.P., Paez, T.L., Mayes, R.L.: Epistemic uncertainty quantification tutorial. In: Proceedings of the IMAC-XXVII (2009)
11. Kolmogorov, A.N.: On logical foundations of probability theory. In: Probability Theory and Mathematical Statistics (Tbilisi). Lecture Notes in Mathematics, vol. 1021, pp. 1–5. Springer (1983)
12. Kampé de Feriet, J.: La theorie generalisee de l'information et la mesure subjective de l'information. Lecture Notes in Mathematics, vol. 398, pp. 1–35. Springer (1974)
13. Gierz, G., et al.: A Compendium of Continuous Lattices. Springer (1980)
14. Nguyen, H.T., Tran, H.: On a continuous lattice approach to modeling of coarse data in systems analysis. J. Uncertain Systems **1**(1), 62–73 (2007)

15. Nguyen, H.T.: Sur les mesures d'information de type Inf. Lecture Notes in Mathematics, vol. 398, pp. 62–75. Springer (1974)
16. Nguyen, H.T.: A note on the extension principle for fuzzy sets. J. Math. Anal. and Appli. **64**, 369–380 (1978)
17. Zadeh, L.A.: Fuzzy sets. Inf. and Control. **8**, 338–353 (1965)
18. Nguyen, H.T.: On random sets and belief functions. J. Math. Anal. and Appli. **65**, 531–542 (1978)
19. Nguyen, H.T.: An Introduction to Random Sets. Chapman and Hall/CRC Press (2006)
20. Box, G.E.P., Draper, N.R.: Empirical Model-Building and Response Surfaces, p. 424. J. Wiley (1987)
21. Martin, R., Liu, C.: Inferential models: A framework for prior-free posterior probabilistic inference. J. Amer. Statist. Assoc. **108**, 301–313 (2013)
22. Martin, R.: Random sets and exact confidence regions. Sankhya A **76**(2), 288–304 (2014)
23. Hajek, K.: Sampling From A Finite Population. Marcel Dekker (1981)
24. McCauley, J.L.: Dynamics of Markets. Econophysics and Finance. Cambridge University Press (2010)

Fuzzy Sets, Multisets,
and Rough Approximations

Sadaaki Miyamoto$^{(\boxtimes)}$

Department of Risk Engineering, Faculty of Systems and Information Engineering,
University of Tsukuba, 1-1-1 Tennodai, Tsukuba, Ibaraki 305-8573, Japan
miyamoto@risk.tsukuba.ac.jp

Abstract. Multisets alias bags are similar to fuzzy sets but essentially different in basic concepts and operations. We overview multisets together with basics of fuzzy sets in order to observe differences between the two. We then introduce fuzzy multisets and the combination of the both concepts. There is another concept of real-valued multisets as a generalization of multisets. Rough approximations of multisets and fuzzy multisets are discussed which uses a natural projection of the universal set onto the set of equivalence classes.

1 Multisets

Let $X = \{x_1, \ldots, x_n\}$ is a universal set. As is well-known, a fuzzy set F of X is characterized by a membership function $\mu_A \colon X \to [0,1]$, where $\mu_A(x)$ is the degree of relevance of x to a concept represented by set symbol A. In contrast, a multiset M of X is characterized by a count function $C_M \colon X \to N$, where $N = \{0, 1, \ldots\}$. $C_M(x) = m$ implies that x exists m times in multiset M. the inclusion, equality, union, and intersection of multisets are defined by the same relations as those of fuzzy sets except that the relations and operations are on N instead of $[0,1]$. Multisets have the addition ($\oplus$) and minus ($\ominus$) operations which fuzzy sets do not have, while multisets do not have the complement while a fuzzy set has $\mu_{\bar{A}}(x) = 1 - \mu_A(x)$. Note that $C_{M_1 \oplus M_2}(x) = C_{M_1}(x) + C_{M_2}(x)$ and $C_{M_1 \ominus M_2}(x) = \max\{0, C_{M_1}(x) - C_{M_2}(x)\}$.

Note 1. Multisets [1] are also called bags [7]. The name of multisets are used throughout this paper. Crisp multisets are simply called multisets.

2 Fuzzy Multisets

An easy generalization of multisets is real-valued multiset which generalizes $C_M \colon X \to N$ to $C_M \colon X \to [0, +\infty]$ which includes the point of $+\infty$. We omit the details of real-valued multisets here but there are important properties in multi-relations and its algebra (see, e.g., [4]).

Another well-known generalization is fuzzy multisets which originally have been proposed by Yager [7]. Later the author [2] proposed another set of operations as the union and intersection of fuzzy multisets. A fuzzy multiset is a multiset of $X \times [0, 1]$ [7].

V.-N. Huynh et al. (Eds.): IUKM 2015, LNAI 9376, pp. 11–14, 2015.
DOI: 10.1007/978-3-319-25135-6_2

Example 1. Let $X = \{a, b, c, d\}$. A crisp multiset M is expressed as $\{a, b, d, a, a, b\} = \{a, a, a, b, b, d\}$, i.e., $C_M(a) = 3$, $C_M(b) = 2$, $C_M(c) = 0$, and $C_M(d) = 1$. An example of fuzzy multiset A of X is

$$A = \{(a, 0.1), (b, 0.5), (c, 0.6), (c, 0.6), (a, 0.2), (c, 0.8)\}.$$

A membership sequence is the collection of memberships for a particular element of X which is arranged into decreasing order. In the above example, the membership sequence form of the above A is

$$A = \{(0.2, 0.1)/a, (0.5)/b, (0.8, 0.6.0.6)/c\}.$$

The ith member of membership sequence is denoted by $\mu_A^i(x)$. Thus $\mu_A^2(a) = 0.1$, $\mu_A^3(c) = 0.6$, etc. The inclusion, union, and intersection are then defined as follows:

inclusion: $A \subseteq B \iff \mu_A^i(x) \leq \mu_B^i(x)$, $\forall x \in X$, $i = 1, 2, \ldots$,
union: $\mu_{A \cup B}^i(x) = \max\{\mu_A^i(x), \mu_B^i(x)\}$, $i = 1, 2, \ldots$,
intersection: $\mu_{A \cap B}^i(x) = \min\{\mu_A^i(x), \mu_B^i(x)\}$, $i = 1, 2, \ldots$,
minus: $\mu_{A \ominus B}^i(x) = \max\{0, \mu_A^i(x) - \mu_B^i(x)\}$, $i = 1, 2, \ldots$,

while addition is defined without the use of the membership sequence, by simply gathering members of the two fuzzy multisets.

Example 2. Let

$$B = \{(a, 0.1), (b, 0.5), (c, 0.6), (a, 0.1), (c, 0.7)\}$$
$$= \{(0.1, 0.1)/a, (0.5)/b, (0.7, 0.6)/c\}$$

and

$$C = \{(a, 0.3), (a, 0.1), (c, 0.5), (a, 0.2), (c, 0.9), (d, 0.7)\}$$
$$= \{(0.3, 0.2, 0.1)/a, (0.9, 0.5)/c, (0.7)/d\}.$$

Then we have $B \subseteq A$ and

$$A \cup C = \{(0.3, 0.2, 0.1)/a, (0.5)/b, (0.9, 0.6, 0.6)/c, (0.7)/d\}$$
$$A \cap C = \{(0.2, 0.1)/a, (0.8, 0.5)/c\}$$
$$A \oplus C = \{(a, 0.1), (b, 0.5), (c, 0.6), (c, 0.6), (a, 0.2), (c, 0.8),$$
$$(a, 0.3), (a, 0.1), (c, 0.5), (a, 0.2), (c, 0.9), (d, 0.7)\}.$$

Let X and $Y = \{y_1, \ldots, y_l\}$ be two universal sets and $f: X \to Y$. Suppose G is an ordinary set: since $f(G) = \bigcup_{x \in A} \{f(x)\}$, the extension principle of fuzzy sets is derived. Thus we have

$$C_{f(M)}(y) = \max_{x \in X, f(x)=y} C_M(x).$$

In the same way, we have

$$\mu_{f(A)}^i(y) = \max_{x \in X, f(x)=y} \mu_A^i(x),$$

as the extension principle of fuzzy multisets, which is used below.

3 Rough Approximations

Let us assume that an equivalence relation R is given on X, in order to consider rough approximations [5,6]. In other words, relation xRy classifies X into equivalence classes $U_1, \ldots, U_K$. For convenience, let $\mathcal{U} = \{U_1, \ldots, U_K\}$. A natural projection of X onto $\mathcal{U}$ is given by $g_R \colon X \to \mathcal{U}$:

$$g_R(x) = [x]_R = U_i \iff x \in U_i.$$

Let us recall that, given an ordinary set G of X, its upper approximation and lower approximation are respectively given by

$$R^*(G) = \bigcup \{U_i \colon U_i \cap G \neq \emptyset\},$$
$$R_*(G) = \bigcup \{U_i \colon U_i \subseteq G\}.$$

Moreover the rough boundary is $B(G) = R^*(G) - R_*(G)$.

The authors already defined rough approximations of fuzzy sets [3]:

$$\mu_{R^*(F)}(x) = \max_{x' \in U_i} \mu_F(x'), \iff x \in U_i,$$
$$\mu_{R_*(F)}(x) = \min_{x' \in U_i} \mu_F(x'), \iff x \in U_i.$$

We use the natural projection here to define the rough approximations of multisets and fuzzy multisets.

Note first that

$$R^*(G) = g_R^{-1}(g_R(G)) = (g_R^{-1} \circ g_R)(G),$$

where G is an arbitrary crisp set. In contrast, $R_*(G)$ is expressed as follows: Let

$$\hat{G} = g_R^{-1}(g_R(G)) - G.$$

Then

$$B(G) = g_R^{-1}(g_R(\hat{G}))$$

and

$$R_*(G) = g_R^{-1}(g_R(G)) - g_R^{-1}(g_R(\hat{G})).$$

Rough approximation of a multiset M is then straightforward:

$$\mu_{R^*(M)}(x) = \max_{x' \in U_i} C_M(x'), \iff x \in U_i,$$
$$\mu_{R_*(M)}(x) = \min_{x' \in U_i} C_M(x'), \iff x \in U_i.$$

by using the natural projection g_R. In a similar way, we can show that

$$\mu^i_{R^*(A)}(x) = \max_{x' \in U_i} \mu^i_A(x'), \iff x \in U_i,$$
$$\mu^i_{R_*(A)}(x) = \min_{x' \in U_i} \mu^i_A(x'), \iff x \in U_i.$$

for a fuzzy multiset A.

4 Conclusion

We briefly overviewed the theory of multisets and fuzzy multisets as well as their rough approximations. In this paper we focused upon the way how rough approximations are derived using the natural projection. In contrast, many fundamental topics of multisets have been omitted here, but more about basics of multisets and fuzzy multisets will be explained in the talk with illustrations so that readers with no sufficient background will understand their fundamental concepts and basic theory. Readers interested in this theory could refer to, e.g., [3, 4, 7]. Applications of fuzzy multisets will also be mentioned.

Acknowledgments. This work has partly been supported by the Grant-in-Aid for Scientific Research, Japan Society for the Promotion of Science, Project number 26330270.

References

1. Knuth, D.E.: Seminumerical Algorithms. Addison-Wesley (1969)
2. Miyamoto, S.: Fuzzy multisets with infinite collections of memberships. In: Proc. of the 7th International Fuzzy Systems Association World Congress (IFSA 1997), June 25–30, 1997, Prague, Chech, vol. 1, pp. 61–66 (1997)
3. Miyamoto, S.: Generalizations of multisets and rough approximations. International Journal of Intelligent Systems 19(7), 639–652 (2004)
4. Miyamoto, S.: Generalized bags and bag relations: toward an alternative model for fuzzy set applications. In: Proc. of CIMCA 2008, December 10–12, 2008, Vienna, pp. 380–385 (2008)
5. Pawlak, Z.: Rough sets. International Journal of Computer and Information Sciences 11, 341–356 (1982)
6. Pawlak, Z.: Rough Sets. Kluwer, Dordrecht (1991)
7. Yager, R.R.: On the theory of bags. Int. J. General Systems 13, 23–37 (1986)

What Is Fuzzy Natural Logic
Abstract

Vilém Novák[✉]

Institute for Research and Applications of Fuzzy Modeling, University of Ostrava,
NSC IT4Innovations, 30. Dubna 22, 701 03 Ostrava 1, Czech Republic
vilem.novak@osu.cz
http://irafm.osu.cz/

Natural Logic. In 1970, G. Lakoff published a paper [8] in which he introduced the concept of *natural logic* with the following goals:

– to express all concepts capable of being expressed in natural language,
– to characterize all the valid inferences that can be made in natural language,
– to mesh with adequate linguistic descriptions of all natural languages.

Natural logic is thus a collection of terms and rules that come with natural language and that allows us to reason and argue in it. According to G. Lakoff's hypothesis, natural language employs a relatively small finite number of atomic predicates that take sentential complements (sentential operators) and are related to each other by meaning-postulates that do not vary from language to language. The concept of natural logic has been further developed by several authors (see, e.g., [2, 9] and elsewhere).

In this paper, we will briefly overview a special extension of the mathematical fuzzy logic in narrow sense (FLn) that is called *Fuzzy Natural Logic* (FNL). Its goal stems from the above Lakoff's characterization and can be specified as follows: to develop a *formal theory of human reasoning that includes mathematical models of the semantics of certain classes of special expressions of natural language and generalized quantifiers with regard to presence of the vagueness phenomenon.* The main difference from the Lakoff's characterization is that FNL is a *mathematical* theory which, in addition, includes also the model of the vagueness phenomenon using tools of FLn. At the same time it must follow results of the logical analysis of natural language (see, e.g., [3]).

Fuzzy Logic in Narrow Sense. Recall that FLn is a generalization of classical mathematical logic (see [6, 14]) in the sense that it has formally established syntax and semantics. The syntax consists of precise definitions of a formula, proof, formal theory, provability, model, etc. It is extended by more connectives and more logical axioms. Semantics of this logic is many-valued. The completeness theorem says that a formula A is provable iff it is true in the degree 1 in all models.

There are many formal calculi in FLn, for example MTL (monidal t-norm-based logic), BL (basic logic), Łukasiewicz, product, LΠ and other logics. They differ from each other by the considered structure of truth values. For FNL,

© Springer International Publishing Switzerland 2015
V.-N. Huynh et al. (Eds.): IUKM 2015, LNAI 9376, pp. 15–18, 2015.
DOI: 10.1007/978-3-319-25135-6_3

the most important is higher-order fuzzy logic called Fuzzy Type Theory (FTT) because the experience indicates that first-order logical systems are not powerful enough for the proper formalization of linguistic semantics which is a necessary constituent of FNL.

Paradigm of FNL. The fuzzy natural logic consists of several formal theories developed on a unique formal basis:

(a) Formal theory of evaluative linguistic expressions [16]; see also [15].
(b) Formal theory of fuzzy/linguistic IF-THEN rules and linguistic descriptions; approximate reasoning based on them [5,13,20,21].
(c) Formal theory of intermediate and generalized quantifiers [4,7,10,11,17].

Fuzzy Type Theory. This is a higher-order fuzzy logic being generalization of classical type theory introduced by B. Russel, A. Church and L. Henkin (for extensive presentation of it see [1]). The generalization consists especially in replacement of the axiom stating "there are two truth values" by a sequence of axioms characterizing structure of the algebra of truth values.

The fuzzy type theory (FTT) is the basic formal tool for FNL. We work with a special case of the Łukasiewicz FTT, which is based on the algebra of truth values forming the standard Łukasiewicz MV_Δ-algebra $\mathcal{L} = \langle [0,1], \vee, \wedge, \otimes, \oplus, \Delta, \rightarrow, 0, 1 \rangle$. Important concept in FTT is that of a *fuzzy equality*, which is a reflexive, symmetric and $\otimes$-transitive binary fuzzy relation on some set M, i.e. it is a function $\doteq: M \times M \rightarrow L$.

Syntax of FTT is a generalization of the lambda-calculus constructed in a classical way, but differing from classical type theory by definition of additional special connectives and larger list of axioms. It has been proved that the fuzzy type theory is complete. The details can be found in [12,18].

Evaluative Linguistic Expressions. These are expressions of natural language, for example, *small, medium, big, roughly one hundred, very short, more or less deep, not tall, roughly warm or medium hot, quite roughly strong, roughly medium size,* etc. They form a small, syntactically simple, but very important part of natural language which is present in its everyday use any time. The reason is that people regularly need to evaluate phenomena around them and to make important decisions, learn how to behave, and realize various kinds of activities based on evaluation of the given situation. In FNL, a special formal theory of FTT has been constructed using which semantics of the evaluative expressions including their vagueness is modeled. The details can be found in [16].

Fuzzy/Linguistic IF-THEN Rules. In FNL, these are taken as genuine conditional clauses of natural language with the general form

$$\text{IF } X \text{ is } \mathcal{A} \text{ THEN } Y \text{ is } \mathcal{B}, \tag{1}$$

where "X is $\mathcal{A}$", "Y is $\mathcal{B}$" are the, so called, evaluative linguistic predications. A typical example is

IF *temperature* is *extremely small* THEN *the amount of gas* is *very big*.

A finite set of rules (1) is called *linguistic description* and it is construed as a special text. The method of *perception-based logical deduction* enables to derive conclusion from linguistic descriptions, thus simulating reasoning of people.

Intermediate and Fuzzy Quantifiers. These are natural language expressions such as *most, a lot of, many, a few, a great deal of, large part of, small part of.* In correspondence with the analysis given by P. Peterson in [22] they are in FNL modeled as special formulas of fuzzy type theory in a certain extension of the formal theory of evaluative linguistic expressions. Typical elaborated quantifiers are

"Most (Almost all, Few, Many) B are A".

The developed model stems from the assumption that intermediate quantifiers are classical general or existential quantifiers for which the universe of quantification is modified and the modification can be imprecise.

Intermediate quantifiers occur also in generalized Aristotle syllogisms, for example:

PPI-III:
Major premise P_1: Almost all employed people have a car
Minor premise P_2: Almost all employed people are well situated
Conclusion C: Some well situated people have a car

Formal validity of more than 120 such syllogisms was already proved. This means that the implication $P_1 \Rightarrow (P_2 \Rightarrow C)$ is true in the degree 1 in all models.

Modeling Human Reasoning. Human reasoning is typical by employing natural language. We argue that formalism of FNL is rich enough to be able to develop a sufficiently well working model of human reasoning. One such possibility was described in [19] where a model of non-monotonic reasoning based on a series of linguistic descriptions characterizing a criminal case faced by a detective Columbo was developed.

References

1. Andrews, P.: An Introduction to Mathematical Logic and Type Theory: To Truth Through Proof. Kluwer, Dordrecht (2002)
2. van Benthem, J.: A brief history of natural logic. In: Chakraborty, M., Löwe, B., Nath Mitra, M., Sarukkai, S. (eds.) Logic, Navya-Nyaya and Applications, Homage to Bimal Krishna Matilal. College Publications, London (2008)
3. Duží, M., Jespersen, B., Materna, P.: Procedural Semantics for Hyperintensional Logic. Springer, Dordrecht (2010)
4. Dvořák, A., Holčapek, M.: L-fuzzy quantifiers of the type $\langle 1 \rangle$ determined by measures. Fuzzy Sets and Systems **160**, 3425–3452 (2009)
5. Dvořák, A., Novák, V.: Formal theories and linguistic descriptions. Fuzzy Sets and Systems **143**, 169–188 (2004)

6. Hájek, P.: What is mathematical fuzzy logic. Fuzzy Sets and Systems **157**, 597–603 (2006)

7. Holčapek, M.: Monadic L-fuzzy quantifiers of the type $\langle 1^n, 1 \rangle$. Fuzzy Sets and Systems **159**, 1811–1835 (2008)

8. Lakoff, G.: Linguistics and natural logic. Synthese **22**, 151–271 (1970)

9. MacCartney, B., Manning, C.D.: An extended model of natural logic. In: IWCS-8 1909 Proc. Eighth Int. Conf. on Computational Semantics, pp. 140–156. Association for Computational Linguistics, Stroudsburg, PA, USA (2009)

10. Murinová, P., Novák, V.: A formal theory of generalized intermediate syllogisms. Fuzzy Sets and Systems **186**, 47–80 (2012)

11. Murinová, P., Novák, V.: The structure of generalized intermediate syllogisms. Fuzzy Sets and Systems **247**, 18–37 (2014)

12. Novák, V.: On fuzzy type theory. Fuzzy Sets and Systems **149**, 235–273 (2005)

13. Novák, V.: Perception-based logical deduction. In: Reusch, B. (ed.) Computational Intelligence, Theory and Applications, pp. 237–250. Springer, Berlin (2005)

14. Novák, V.: Which logic is the real fuzzy logic? Fuzzy Sets and Systems **157**, 635–641 (2006)

15. Novák, V.: Mathematical fuzzy logic in modeling of natural language semantics. In: Wang, P., Ruan, D., Kerre, E. (eds.) Fuzzy Logic - A Spectrum of Theoretical & Practical Issues, pp. 145–182. Elsevier, Berlin (2007)

16. Novák, V.: A comprehensive theory of trichotomous evaluative linguistic expressions. Fuzzy Sets and Systems **159**(22), 2939–2969 (2008)

17. Novák, V.: A formal theory of intermediate quantifiers. Fuzzy Sets and Systems **159**(10), 1229–1246 (2008)

18. Novák, V.: EQ-algebra-based fuzzy type theory and its extensions. Logic Journal of the IGPL **19**, 512–542 (2011)

19. Novák, V., Dvořák, A.: Formalization of commonsense reasoning in fuzzy logic in broader sense. Journal of Applied and Computational Mathematics **10**, 106–121 (2011)

20. Novák, V., Lehmke, S.: Logical structure of fuzzy IF-THEN rules. Fuzzy Sets and Systems **157**, 2003–2029 (2006)

21. Novák, V., Perfilieva, I.: On the semantics of perception-based fuzzy logic deduction. International Journal of Intelligent Systems **19**, 1007–1031 (2004)

22. Peterson, P.: Intermediate Quantifiers. Logic, linguistics, and Aristotelian semantics. Ashgate, Aldershot (2000)

Combining Fuzziness and Context Sensitivity in Game Based Models of Vague Quantification

Christian G. Fermüller$^{(\boxtimes)}$

Vienna University of Technology, Vienna, Austria
chrisf@logic.at

Abstract. We introduce a game semantic approach to fuzzy models of vague quantification that addresses a number of problems with previous frameworks. The main tool is the introduction of a new logical operator that supports context based evaluations of suitably quantified formulas.

1 Introduction

Vague quantifier expression, like few, many, about half are notoriously difficult to model adequately in a degree based setting. The problems already arise with precise quantifiers, like all, at most three, more than half, if range and scope predicates are vague. The literature summarized in the recent survey article [3] and the monograph [13] certainly documents research progress on this topic. Nevertheless there remain a number of challenges that call for variations and extensions of traditional frameworks for fuzzy quantifiers. After quickly reviewing relevant terminology and a very useful classification of fuzzy quantifiers by Liu and Kerre [18] in Section 2, we formulate in Section 3 four desiderata for linguistically adequate models of vague quantification that arguably have not yet been fully met in current proposals. We argue the Giles's game semantic approach to Łukasiewicz logic Ł, revisited in Section 4, provides a suitable basis for addressing the outlined challenges in principle. However, Giles did not consider generalized quantifiers, as needed for natural language semantics. Moreover, as explained in Section 5, the standard semantics for (unary) universal quantification in Ł cannot be straightforwardly extended to an adequate model of the corresponding (binary) natural language quantifier all. To meet the crucial desideratum of respecting a vagueness specific type of intensionality we adapt Giles's game to evaluation with respect to contexts modeled by sets of precisifications. For this purpose we introduce a new logical operator ∘ that signifies a random choice of some precisification. Random choices, not of precisifications, but of witness elements have already been introduced to obtain game based models of semi-fuzzy quantifiers in [7,9]. In Section 6 we indicate that the new operator ∘ allows one to lift these models to fully fuzzy quantification.

We emphasize that the focus of this paper is on conceptional analysis, rather than on generalizing mathematical results. The limited space forces us to delegate the systematic investigation of the suggested models, including corresponding adequateness theorems, to future work.

C.G. Fermüller—Supported by Austrian Science Fund (FWF) I1827-N25.

V.-N. Huynh et al. (Eds.): IUKM 2015, LNAI 9376, pp. 19–31, 2015.
DOI: 10.1007/978-3-319-25135-6_4

2 Classifying Vague and Fuzzy Quantifiers

The monograph [13] provides an extensive discussion of the achievements and drawbacks in the literature on fuzzy quantifiers, initiated by Zadeh's seminal paper [23]; a more recent survey paper on the topic is [3]. However, fully adequate models of vague quantifiers call for respecting certain features of vagueness that are usually neglected in fuzzy models, but are in the focus of contemporary linguistic research on natural language semantics (see, e.g., [10,1,22]). We refer the reader to those sources and just review some useful terminology here. We are interested in formal models of quantifier expressions like many, few, about a half, almost all, but also of statements like All athletes are tall or At least two rich men are unhappy, where the quantifier expression has a precise meaning, but at least some of the arguments to which it applies are vague. Natural language quantifiers are frequently *binary*; i.e., they have the logical form $Qx(R(x), S(x))$, where the unary argument predicate R is called the *range* (or *restriction*) and the unary argument predicate S the *scope* of the quantifier Qx. If the range is missing—i.e., implicitly assumed to coincide with the domain of discourse—then the quantifier is called *unary*. Occasionally, more than two arguments are present and a quantifier may bind more than one variable (like in There are more bankers who are rich than doctors who are healthy). But we will focus on unary and binary quantifiers here for sake of clarity.

In the fuzzy logic approach a unary predicate is interpreted by a *fuzzy set* $\widetilde{B}$, i.e., by a function $\widetilde{B} : U \to [0,1]$, where the real number $\widetilde{B}(a) \in [0,1]$ assigned to an element a of the universe U is called the *membership degree* of a in $\widetilde{B}$. Crisp (i.e., classical) sets are obtained as a special case, where the membership degree is either 0 or 1. A unary quantifier Q^1 is modeled by a function $\widetilde{Q}^1 : \mathcal{P}(\widetilde{B}) \to [0,1]$, where the real number in $[0,1]$ assigned to the fuzzy set $\widetilde{B}$ is called the *truth degree (truth value)* of the formula $Q^1x\, S(x)$, if S is interpreted by $\widetilde{B}$. Similarly, a binary quantifier is modeled by a function $\widetilde{Q}^2 : \mathcal{P}(\widetilde{A} \times \widetilde{B}) \to [0,1]$, where $\widetilde{A}$ interprets the range and $\widetilde{B}$ the scope of Q^2. A quantifier is called *precise* if resulting the truth degree is always either 0 or 1, otherwise it is called *fuzzy*. Following Liu and Kerre [18], we distinguish four types of quantified statements:

Type I: the quantifier is precise and its arguments are crisp;
Type II: the quantifier is precise, but the arguments may be fuzzy;
Type III: the quantifier is fuzzy, but its arguments are crisp;
Type IV: the quantifier as well as its arguments are fuzzy.

Statements of type III are also called *semi-fuzzy*. Strictly speaking, the above classification only applies to the fuzzy logic model of quantification; but it can easily be adapted to a more general setting, where vague predicates may not necessarily be modeled as fuzzy sets and vague quantifier expressions may be modeled differently as well. Thus we will, e.g., talk of type II quantification in All children are young and type III quantification in Many doors are locked, without insisting that these statements are formalized according to the traditional fuzzy logic approach.

3   Problems with Fuzzy Models of Vague Quantifiers

It has been pointed out, e.g. in [13], that traditional fuzzy models of vague quantification—based, e.g.,on Zadeh's Σ-count or FG-count or on more sophisticated methods like (variants of) OWA—are only applicable to certain types of quantifiers and are frequently linguistically inadequate when applied to type II or type IV quantification. For example, none of the above-mentioned methods yields a fuzzy unary quantifier **half** : $\tilde{B} \to [0,1]$, that would correctly predict that the statement **Half are tall**, interpreted with respect to a domain U of $2n$ persons, is usually classified by English speakers as definitely false if all $a \in U$ are of the same height h, independently of whether a person of height h is considered tall, borderline to tall, or not tall at all. On the other hand Glöckner [13] maintains that plausible results can usually be achieved with fuzzy models of semi-fuzzy (type III) quantification, i.e., quantification over crisp arguments, like in **About half of the candidates passed the exam**. For this reason Glöckner proposes an approach to fuzzy quantification that proceeds in two separate steps:

1. Pick linguistically adequate truth functions for type I and III quantifiers.
2. Lift the crisp and and semi-fuzzy quantifiers obtained in step 1 to type II and IV models by applying a so-called *quantifier fuzzification mechanism (QFM)* that respects certain general conditions, presented as *determiner fuzzification scheme (DFS) axioms*.

The term '*linguistic adequateness*' is used by fuzzy logicians mostly in a rather loose sense. Linguists usually apply stricter criteria when analyzing the meaning of utterances of declarative sentences in given contexts. While linguists certainly acknowledge the importance of graded adverbs, linguistic hedges (including hedges regarding truth), degrees of emphasis and similar linguistic devices (see, e.g., [16]) they prefer to model these phenomena with respect to a binary interface (accepted/rejected, received/ignored) at the sentence level. Consequently fuzzy models are often deemed incompatible with main stream paradigms in natural language semantics. Moreover, linguistic research on the semantics of quantifier expressions like **many** (see, e.g., [1,17,10]) focuses on subtle *intensional* aspects of word meaning that calls for the consideration of context specific expectations of speakers and hearers.

Some of the mentioned criticism of fuzzy models for natural phenomena relating to vagueness can certainly be defused. For example, one might combine fuzzy quantifier models with a de-fuzzification mechanism in order to restore the binary interface at the outermost level of language processing. More importantly, we suggest that the alleged incompatibility of the linguistic and the fuzzy approach can at least partly be resolved by making the different aims of the respective formal models more explicit. In particular, fuzzy approaches often aim at simple computational devices for automated information extraction or data summary that deliberately abstract from all kinds of subtleties of human communication. For such purposes, rough, quickly computable *approximations* of natural language meanings that are only adequate with respect to simplifying assumptions about context, expectations, pragmatic principles, etc, might well turn out to be adequate. Nevertheless, we will see that certain features of natural language meaning in contexts of vagueness call for variations and extensions of the traditional fuzzy logic approach to vague quantification. We agree with Glöckner that it is useful to disentangle semi-fuzzy and fully fuzzy quantification. However, we suggest a number of additional desiderata for vague quantifier models that have not (yet) been adequately addressed in the literature on fuzzy quantifiers, including Glöckner's own approach.

(D1) **Embeddability in t-norm based fuzzy logics:** Fuzzy quantifiers are often investigated in isolation from proof systems for propositional and first order fuzzy logics. We argue that formal models of fuzzy quantifiers are most useful if they can be directly embedded into at least some of the standard systems of t-norm based logics as presented in, e.g., [2,19]. The most prominent logics based on concrete continuous t-norms are Łukasiewicz logic **Ł**, Gödel logic **G**, and Product logic **P** (see [14,2]. Of these, we single out Łukasiewicz logic here, since it is the only fuzzy logic, where all logical connectives are modeled by *continuous* truth function, which amounts to a useful desideratum in its own right. Unfortunately, neither the original proposals by Zadeh [23], nor the later approaches to fuzzy quantifiers as reviewed in [13] and [3] are fully compatible with principles of (continuous) t-norm based deductive logic, as stated by Hàjek in [14].

(D2) **Interpretability of truth degrees:** In all previous approaches (including Glöckner's), the challenge to come up with an adequate semantics for vague quantifier expressions is separated from the challenge to find convincing interpretations of truth degrees assigned to formalizations of partially true statements. While we agree on the usefulness of separating these two issues, we argue that a well-justified choice of truth functions for logical connectives (see D3) cannot remain fully independent of particular models of the meaning of truth degrees.

(D3) **Guidance for the choice of truth functions:** In traditional fuzzy approaches the choice of truth functions for quantifier expressions like many or about a half usually remains ad-hoc and is not guided by a systematic method for determining optimal candidate functions. As indicated above, Glöckner [13] convincingly criticizes this fact and proposes an axiomatic approach that seeks to justify the choice of truth functions for fuzzy quantifiers *relative* to truth functions for semi-fuzzy quantifiers. While this certainly ameliorates the problem, it does not fully address the challenge to justify the choice of particular families of truth functions with respect to first principles about logical reasoning.

(D4) **Respecting vagueness specific intensionality:** As we will explain in Sections 5 and 6, all current fuzzy models of quantifiers run into troubles if range and scope predicates are vague (type II and type IV quantification). The main reason for the lack of linguistic adequateness in these cases is that semantic dependencies, that may well be ignored for semi-fuzzy quantification, become essential if range or scope predicates are vague. We argue this fact calls for an extension of the traditional fuzzy approach that allows one to model certain intensional aspects of vague predicates.

4 Giles's Game for Łukasiewicz Logic

As pointed out above, we focus on Łukasiewicz logic **L**, here. The standard semantics of propositional **L** is obtained by extending assignments $v_I(\cdot)$ of values $\in [0, 1]$ from atomic to complex **L**-formulas as follows:

$$v_I(F \wedge G) = \min(v_I(F), v_I(G)), \qquad v_I(F \vee G) = \max(v_I(F), v_I(G)),$$
$$v_I(F \,\&\, G) = \max(0, v_I(F){+}v_I(G){-}1), \quad v_I(\bot) = 0,$$
$$v_I(\neg F) = 1 - v_I(F), \qquad v_I(F \to G) = \min(1, 1{-}v_I(F){+}v_I(G)).$$

Note the **L** features two different forms of conjunction: weak ($\wedge$) and strong ($\&$), respectively. At the first order level an interpretation I consists of a *universe (domain)* U and a *signature interpretation* $\varPhi$ that assigns a fuzzy relation $\widetilde{P}$: $U^n \to [0, 1]$ to each n-ary predicate symbol P. (Clearly $\varPhi(P)$ is a fuzzy set for $n = 1$.) Moreover $\varPhi$ assigns a domain element to every constant symbol. Abusing notation, we identify each domain element with a constant symbol denoting it (via $\varPhi$). Thus, for atomic formulas we may write

$$v_I(P(a_1, \ldots, a_n)) = \varPhi(P)(a_1, \ldots, a_n).$$

The semantics of the standard universal and existential quantifier is given by

$$v_I(\forall x F(x)) = \inf_{a \in U}(v_I(F(a))), \qquad v_I(\exists x F(x)) = \sup_{a \in U}(v_I(F(a))).$$

Since we are interested in modeling natural language expressions we may follow the conventional assumption that the universe (i.e., the relevant domain of discourse) U is always finite. For later reference we point out that $I = \langle U, \varPhi \rangle$ is called *classical* if the range of the signature interpretation $\varPhi$ is restricted to $\{0, 1\}$. Thus evaluating statements with crisp predicates ('classical reasoning') amounts to a sub-case of evaluating **L**-formulas.

As demonstrated by Robin Giles in [12,11] the truth functions for **L** need not be imposed without further explanation, but rather can be extracted from principles of reasoning with vague (fuzzy) predicates, modeled by combining a *dialogue game* for reducing logically complex statements with a *betting scheme* for evaluating final game states, where all remaining statements are atomic.

The dialogue part of Giles's game is a two-player zero-sum game with perfect information. The players are called 'you' and 'me', respectively. At every state of the game each player maintains a multi-set of currently asserted statements (called the *tenet* of that player), represented by **L**-formulas. Accordingly a game state is denoted by $[F_1, \ldots, F_n \mid G_1, \ldots, G_m]$, where $[F_1, \ldots, F_n]$ is your tenet and $[G_1, \ldots, G_m]$ is my tenet, respectively. In each move of the game one of the players picks an occurrence of a formula from her opponent's tenet and either attacks or grants it explicitly. In both cases the picked formula is removed from the tenet and thus cannot be attacked again. The other player has to respond to the attack in accordance with the following rules.

$(R_\wedge)$ If I assert $F \wedge G$ then you attack by pointing either to F or to G; in response I have to assert either F or G, accordingly.

$(R_\&)$ If I assert $F \& G$ then, in response to your attack, I have to assert either F as well as G or, alternatively, $\perp$.

$(R_\vee)$ If I assert $F \vee G$ then, in response to your attack, I have to assert either F or G at my own choice.

$(R_\rightarrow)$ If I assert $G \rightarrow F$ then, if you choose to attack, you have to assert G and I have to assert F in reply.

$(R_\neg)$ If I assert $\neg F$ then you assert F and I have to assert $\perp$ in reply.

$(R_\forall)$ If I assert $\forall x F(x)$ then you attack by choosing some domain element (constant) a and I have to reply by asserting $F(a)$.

$(R_\exists)$ If I assert $\exists x F(x)$ then I have to choose a domain element (constant) a and to assert $F(a)$.

Perfectly dual rules apply for your assertions and my corresponding attacks.

At the final state of a game, when all formulas have been replaced by atomic subformulas, the players have to pay a fixed amount of money, say 1, for each atomic statement A in their tenet that is evaluated as 'false' (0) according to an associated experiment E_A. Except for the experiment $E_\perp$ which always yields 'false', these experiments may show dispersion, i.e., they may yield different answers upon repetition. However a fixed *risk* $\langle A \rangle$ specifies the probability that E_A results in a negative answer ('false'). My final risk, i.e., the total *expected* amount of money that I have to pay to you in the final state $[A_1, \ldots, A_n \mid B_1, \ldots, B_m]$ therefore is $\langle A_1, \ldots, A_n \mid B_1, \ldots, B_m \rangle = \sum_{i=1}^{m} \langle B_i \rangle - \sum_{j=1}^{n} \langle A_j \rangle$.

Theorem 1. ([**12,6**]) *For every atomic formula A let $\langle A \rangle$ be its risk and let I be the **L**-interpretation given by $v_I(A) = 1 - \langle A \rangle$. Then, if we both play rationally, any game starting in state $[\mid F]$ will end in a state where my final risk is $1 - v_I(F)$.*

Note that Giles's approach directly addresses desiderata D1 and D3, formulated in Section 3: it provides a semantic framework for Łukasiewicz logic (D1) that is based on a specific interpretation of truth degrees (D3). As far as the standard logical connectives are concerned, also desideratum D2 is satisfied: truth functions are extracted from reasoning principles presented by rules for attacking and defending statements of a particular logical form. Giles only considered the standard quantifiers $\forall$ and $\exists$. However, we suggest that a large range of precise quantifiers can be readily characterized by game rules as well. We do not have space for a systematic investigation here, but rather present just two examples of such rules, here.[1]

$(R_{\leq 2})$ If I assert At most two $x\, F(x)$ then you attack by choosing three different $a_1, a_2, a_3 \in U$ and I have to reply by asserting $\neg F(a_i)$ for some $i \in \{1, 2, 3\}$.

$(R_{\geq U/3})$ If I assert At least a third $x\, F(x)$ then, in reply to your attack, I have to choose different $a_1, \ldots a_n \in U$ where $n \geq |U|/3$ and assert $F(a_1), \ldots F(a_n)$.

---

[1] For sake of conciseness we only present rules for unary quantifiers. Similar rules can be formulated for binary and other quantifiers.

In principle, these rules are applicable to crisp as well as to fuzzy predicates, i.e., for risk values that are not just either 0 or 1. But, as we will see in the next section, the linguistic adequateness of the resulting model is questionable if we apply it to type II, rather than just to type I quantification.

5   From Type I to Type II Quantifiers Via Precisifications

For sake of clarity we will focus on the quantifier expression all in this section, but *mutatis mutandis* our discussion also applies to the transition from type I to type II quantification for quantifiers like some, at most 3, at least a half, etc.

Recall from Section 4 that in the unary case, i.e. if there is only a scope, but no range predicate, all is modeled in Lukasiewicz logic by[2] $v_I(\forall x F(x)) = \min_{a \in U}(v_I(F(a)))$. Moving to the more important binary case, i.e. models for sentences of the form All F are G, at least two options arise.

O1: We may reduce the binary universal quantifier $\forall^2$ to the unary one by setting $\forall^2 x(F(x), G(x)) := \forall x(F(x) \to G(x))$, which amounts to

$$v_I(\forall^2 x(F(x), G(x))) = \min_{a \in U}(\min(1, 1 - v_I(F(a)) + v_I(G(a)))).$$

O2: Alternatively, we may follow the 'fuzzy quantifier tradition' outlined in Section 2 (see, e.g., [13]) and set $v_I(\forall^2 x(F(x), G(x))) = \min_{a \in U}(1 - \widetilde{F}(a), \widetilde{G}(a))$, where $\widetilde{F}$ is the fuzzy set corresponding to the (possibly logically complex) predicate F and thus $\widetilde{F}(a)$ coincides with $v_I(F(a))$ (and likewise for $\widetilde{G}$). Accordingly, we obtain

$$v_I(\forall^2 x(F(x), G(x))) = \min_{a \in U}(\max(1 - v_I(F(a)), v_I(G(a)))),$$

which amounts to identifying $\forall^2 x(F(x), G(x))$ with $\forall x(\neg F(x) \lor G(x))$.

Clearly O1 and O2 coincide for crisp predicates (type I quantification). To assess their respective linguistic adequateness for type II quantification let us look at corresponding formalizations of All children are poor, evaluated in a context where the domain consists of persons that are all borderline cases of children as well as of poor persons; i.e., $v_I(F(a)) = v_I(G(a)) = 0.5$ for all $a \in U$. According to option O1 the sentence is evaluated as perfectly true (1), which (at least according to Glöckner [13] and other fuzzy logicians) hardly is adequate. Option O2 returns 0.5. The latter may seem more reasonable at a first glimpse; but when we evaluate All children are children instead, which yields exactly the same values in the given interpretation, O1 seems more plausible than O2.

In such a situation one might be tempted to talk about an inherent ambiguity of all. However this idea creates more problems than it solves, since it runs counter to established linguistic wisdom about the concept of ambiguity (see, e.g., [20]) and moreover would call for non-ad-hoc criteria for choosing the

---

[2] Since we restrict attention to finite domains we may replace inf by min.

'correct' indexical meaning of all. We claim that the presented example rather hints to the fact that any fully adequate model of the meaning of all has to respect *intensionality*. In other words: even if we accept the representation of the meaning of vague predicates F and G as fuzzy sets $\widetilde{F}$ and $\widetilde{G}$, respectively, no uniformly adequate degree of truth can be computed for All F are G from $\widetilde{F}$ and $\widetilde{G}$ alone. As already indicated in formulating desiderata D4 and D5 in Section 3, we need to extend the traditional framework for fuzzy quantifiers by incorporating *contexts* in the evaluation mechanism.

Semanticists routinely employ various different mechanisms for context representation. In particular Discourse Representation Theory (DRS) [15] has emerged as a scientific standard that is amenable to many application scenarios. While we maintain that linguistically adequate models of vague quantifiers will eventually have to make extensive use of the corresponding literature, we think that a straightforward representation of a given context as simply a (finite) set of classical interpretations (similar, e.g., to [1]) is sufficient to illustrate the essential features relevant for adequately evaluating vague quantified statements.

Definition 1. *A context C is a non-empty finite set of classical interpretations over the same signature that share the same universe U_C and assign the same domain elements to the constant symbols.*

Some remarks and clarifications are appropriate regarding Definition 1.

- A context is not intended to represent real or potential states of 'the world'. It just represents information deemed sufficient for evaluating a given statement in a given situation, independently of how this information is arrived at.
- Each element $I \in C$ is called a *precisification* and signifies an admissible manner to classify the elements of U_C as either satisfying or not satisfying a given predicate. One may imagine a 'forced march sorites' situation as, e.g., explained in [21], where competent speakers are forced to either accept or reject some currently asserted statement, subject to revision.
- The fact that precisifications are classical does not entail classical evaluations. A statement is considered vague according to this model precisely if no single classical truth value emerges in a given context. Intermediary truth degrees arise by taking the whole context of evaluation into account.
- The assumption of finiteness corresponds to the fact that cognition always entails some level of granularity. In many situations very coarse levels of granularity (in assessing, e.g., tallness, age, richness, etc.) will be sufficient to represent the relevant information conveyed by uttering a sentence like Peter is a tall child in a given context.

Given a context C with universe U_C, every classical formula $F(x)$ determines a fuzzy set $\widetilde{F}_C$, where the membership degrees are defined by

$$\widetilde{F}_C(a) = \frac{|\{I \in C : v_I(F(a)) = 1\}|}{|C|}.$$

One might want to refine this interpretation of membership degrees by endowing the context C with a measure ν, where $\nu(I)$ is intended to represent the plausibility of the precisification I relative to the other admissible precisifications that form C. Equivalently, one may view the context as a *fuzzy set* of plausible precisifications. However, here, we restrict attention to the basic case, where all precisification are considered equally plausible.

Example 1. Consider a context C with universe $U_C = \{\mathsf{Eve}, \mathsf{Joe}\}$ that consists of all possible precisifications that arise from assigning, independently of each other, either 0 or 1 to $\mathsf{child}(a)$ and $\mathsf{poor}(a)$ for $a \in U_C$. Then $\widetilde{\mathsf{child}}_C(\mathsf{Eve}) = \widetilde{\mathsf{child}}_C(\mathsf{Joe}) = 0.5$ and likewise $\widetilde{\mathsf{poor}}_C(\mathsf{Eve}) = \widetilde{\mathsf{poor}}_C(\mathsf{Joe}) = 0.5$. Returning to the two different options O1 and O2 for formalizing the binary all-quantifier in Lukasiewicz logic, we obtain the following values for the corresponding **L**-interpretation I^C:

O1: $v_{I^C}(\forall x(\mathsf{child}(x) \rightarrow \mathsf{child}(x))) = 1$ and $v_{I^C}(\forall x(\mathsf{child}(x) \rightarrow \mathsf{poor}(x))) = 1$
O2: $v_{I^C}(\forall x(\neg\mathsf{child}(x) \vee \mathsf{child}(x))) = 0.5$ and $v_{I^C}(\forall x(\neg\mathsf{child}(x) \vee \mathsf{poor}(x))) = 0.5$

As already pointed out, both options are problematic. In contrast, in classical logic, $\forall x(P(x) \rightarrow Q(x))$ is equivalent to $\forall x(\neg P(x) \vee Q(x))$. Obviously, for $P = Q = $ child all classical interpretations and therefore all precisifications evaluate both formulas as true (1), whereas only 9/16 of the precisifications in C evaluate the formulas as true for $P = $ child and $Q = $ poor. Clearly this latter style of context based evaluation yields more plausible results than either O1 or O2.

To model the latter 'context sensitive' style of evaluation in Giles's fashion we stipulate that the dispersive experiment E_A associated with an atomic formula A consists in a (uniformly) random choice of some precisification $I \in C$, returning $v_I(A)$ as result of E_A. The risk value associated with A then amounts to the fraction of precisifications in C that evaluate A as false. More formally, we have $\langle A \rangle_C = |\{I \in C : v_I(A) = 0\}|/|C|$. Note that this interpretation of E_A does not entail any change in Giles's game itself. However the implied shift of perspective suggests a mechanism that allows one to refer to precisifications not only at final game states, but already at earlier states of evaluating logically complex statements. To this aim we stipulate that each formula F occurring in either my or your tenet may carry a reference to some $I \in C$, denoted by $F{\uparrow}I$. Accordingly, we introduce an operator $\circ$ and specify its semantics by the following game rule:

$(R_\circ)$ If I assert $\circ F$ then, in reply to your attack, some precisification $I \in C$ is chosen randomly and $\circ F$ is replaced with $F{\uparrow}I$ in my tenet.

The other game rules remain unchanged, except for stipulating that any reference $\uparrow I$ is inherited from formulas to subformulas. A final state is now of the form $\Sigma = [A_1\rho_1, \ldots, A_n\rho_n \mid B_1\rho_1', \ldots, B_m\rho_m']$, where each ρ_i, ρ_j' is either a reference $\uparrow I$ for some $I \in C$ or else is empty. This yields a local risk $\langle \Sigma \rangle$ by setting my risk $\langle A \rangle_\Sigma$ associated with A in Σ to $1 - v_I(A)$ in the former case and to $\langle A \rangle_C$ in the latter case. To obtain the overall (expected) risk associated with a *non-final* state we have to take into account that all choices of precisifications are uniformly random and thus have to compute the average over corresponding local risks. Taking our clue from Theorem 1., let $\langle F \rangle_C$ denote my final overall risk with respect to context C in a game starting in $[| F]$, where we both play rationally. We set $v_{I^C}(F) = 1 - \langle F \rangle_C$. This in particular yields $v_{I^C}(\circ F) = |\{I \in C : v_I(F) = 1\}|/|C|$. Our model amounts to simply adding $\circ$ in front of any formula F_S that is intended to formalize a universally quantified natural language statement S according to either O1 or—now equivalently—O2. The original options emerge as special cases that provide bounds for the context sensitive evaluation.

Proposition 1. *For every context C and its corresponding* **L**-*interpretation I^C:*
$$v_{I^C}(\forall x(\neg F(x) \vee G(x))) \leq v_{I^C}(\circ \forall x(\neg F(x) \vee G(x))) = v_{I^C}(\circ \forall x(F(x) \to G(x))) \leq v_{I^C}(\forall x(F(x) \to G(x))).$$

Similar context sensitive models arise for precise quantifiers specified by game rules like $R_{\leq 2}$ and $R_{\geq U/3}$ (see Section 4) In any case, adding $\circ$ amounts to a simple mechanism for the shift from type I to type II quantification, addressing desideratum D4 of Section 3.[3]

6 From Type III to Type IV Quantifiers: Random Witnesses

In [7,9] we have introduced a concept for characterizing semi-fuzzy (type II) quantifiers game semantically, motivated by the challenges formulated as desiderata D1, D2, and D3 in Section 3. The central idea is to allow for a *sampling mechanism* that augments the choice of witness constants by player 'you' in rule $R_\forall$ and by player 'I' in $R_\exists$ with various types of random choices of witness constants. For example, we may add the following rule for a family of so-called blind choice proportional quantifiers L_m^k to Giles game for **L**.

$(R_{\mathsf{L}_m^k})$ If I assert $\mathsf{L}_m^k x \hat{F}(x)$ then you may attack by asserting k random instances of $\hat{F}(x)$, in reply to which I have to assert m random instances of $\neg \hat{F}(x)$.

The hat-symbol attached to the scope formula $\hat{F}(x)$ indicates that it has to be classical. This is established by working in a two-tiered language that syntactically separates classical from fuzzy predicates. We have argued in [9] that this leads to adequate models for proportional type III quantifiers like about a half or at least about a third in combination with other game based connectives.

Although rules like $R_{\mathsf{L}_m^k}$ are, in principle, also applicable if the scope formulas are fuzzy, we have deliberately refrained from generalizing type III to type IV quantifiers in this manner, since the resulting models are hardly linguistically

---

[3] Our model is reminiscent of the combination of supervaluation theoretic and fuzzy evaluation described in [5]. However while $\circ$ measures the fraction of precisifications satisfying the formula F in its scope, the supertruth operator **S** of [5] maps its scope into classical logic. Another attempt to combine fuzzy logic with context based evaluation can be found in [8], but only propositional logic is covered there.

adequate, in general. More precisely: desideratum D4 of Section 3 cannot be fulfilled in this manner, even for unary quantifiers.

Example 2. Consider the statement About half [of the people] are tall to be evaluated in a fixed universe of discourse. We argue that the degree of truth of this statement should reflect the fraction of admissible precisifications of the predicate tall that separate the universe into roughly equal numbers of people judged as tall and as not tall, respectively. Note that this entails *two levels of uncertainty*: (1) What precisifications of tall are plausible in the given context? (2) When are two subsets of people to be judged as roughly equal in size? There is nothing wrong with a graded approach in addressing these questions. However, the problem with traditional fuzzy models of about half is that these two quite different types of uncertainty are not clearly distinguished in the evaluation.

In analogy to Section 5, we lift type semi-fuzzy (type II) to type IV quantification by adding the o-operator in front of a corresponding formula of Ł. The combination of an appropriate game rule for the semi-fuzzy quantifier, as indicated above, and the rule R_o triggers an evaluation with respect to a given context C. Note that this model nicely separates the indicated different sources of uncertainty by referring to two different levels of randomization:

(1) uncertainty arising from the non-crispness of tall is modeled by the random choice of an admissible precisification in R_o;
(2) uncertainty regarding standards for about half is modeled by random choice of witness constants in the corresponding game rule for the quantifier.

7   Conclusion

We have singled out four desiderata for models of vague quantification: (D1) they should be embedded into deductive fuzzy logics, in particular Łukasiewicz logic; (D2) they should relate to some interpretation of truth degrees; (D3) truth functions should not be imposed ad-hoc, but rather should be derived from basic reasoning principles; and (D4)—central to this paper—certain intensional aspects arising in contexts of vagueness should be respected. We have argued that specific extensions of Giles's game based semantics for Łukasiewicz logic enable one to meet these desiderata. In particular, we have suggested to re-interpret the evaluation of atomic statements via dispersive experiments as evaluation in randomly chosen precisifications specifying a given context. This triggers the introduction of a new logical operator o, that allows one to refer to precisifications not only at the final stage of the semantic game, but already during the process of evaluation. This leads to new fuzzy as well as intensional (non-truth functional) models of quantifiers, that address concerns about the linguistic adequateness of fuzzy quantifiers with vague range and scope predicates.

We have outlined the main conceptual challenges and proposed solutions only by way of specific examples, here. Clearly a more systematic exploration of the scope and limits of context based game models for vague quantifiers than is possible in the limited space available here, is needed. In future work we plan to systematically investigate properties of the indicated models, including formal adequateness proofs in analogy to those for game based models of proportional semi-fuzzy quantifiers in [7,9] and of specific vagueness related connectives in [4].

References

1. Barker, C.: The dynamics of vagueness. Linguistics & Philosophy **25**(1), 1–36 (2002)
2. Cintula, P., Hájek, P., Noguera, C. (eds.): Handbook of Mathematical Fuzzy Logic. College Publications (2011)
3. Delgado, M., Ruiz, M.D., Sánchez, D., Vila, M.A.: Fuzzy quantification: a state of the art. Fuzzy Sets and Systems **242**, 1–30 (2014)
4. Fermüller, C.G.: Hintikka-style semantic games for fuzzy logics. In: Beierle, C., Meghini, C. (eds.) FoIKS 2014. LNCS, vol. 8367, pp. 193–210. Springer, Heidelberg (2014)
5. Fermüller, C.G., Kosik, R.: Combining supervaluation and degree based reasoning under vagueness. In: Hermann, M., Voronkov, A. (eds.) LPAR 2006. LNCS (LNAI), vol. 4246, pp. 212–226. Springer, Heidelberg (2006)
6. Fermüller, C.G., Metcalfe, G.: Giles's game and the proof theory of Lukasiewicz logic. Studia Logica **92**(1), 27–61 (2009)
7. Fermüller, C.G., Roschger, C.: Randomized game semantics for semi-fuzzy quantifiers. In: Greco, S., Bouchon-Meunier, B., Coletti, G., Fedrizzi, M., Matarazzo, B., Yager, R.R. (eds.) IPMU 2012, Part IV. CCIS, vol. 300, pp. 632–641. Springer, Heidelberg (2012)
8. Fermüller, C.G., Roschger, C.: Bridges between contextual linguistic models of vagueness and t-norm based fuzzy logic. In: Montagna, F. (ed.) Petr Hájek on Mathematical Fuzzy Logic, vol. 6, pp. 91–114. Springer (2014)
9. Fermüller, C.G., Roschger, C.: Randomized game semantics for semi-fuzzy quantifiers. Logic Journal of the IGPL **223**(3), 413–439 (2014)
10. Fernando, T., Kamp, H.: Expecting many. In: Proceedings of SALT, vol. 6, pp. 53–68 (2011)
11. Giles, R.: A non-classical logic for physics. Studia Logica **33**(4), 397–415 (1974)
12. Giles, R.: A non-classical logic for physics. In: Wojcicki, R., Malinkowski, G. (eds.) Selected Papers on Lukasiewicz Sentential Calculi, pp. 13–51. Polish Academy of Sciences (1977)
13. Glöckner, I.: Fuzzy Quantifiers. STUDFUZZ, vol. 193. Springer, Heidelberg (2006)
14. Hájek, P.: Metamathematics of Fuzzy Logic. Kluwer Academic Publishers (2001)
15. Kamp, H., Van Genabith, J., Reyle, U.: Discourse representation theory. In: Handbook of Philosophical Logic, vol. 15, pp. 125–394. Springer (2011)
16. Kennedy, C.: Vagueness and grammar: The semantics of relative and absolute gradable adjectives. Linguistics & Philosophy **30**(1), 1–45 (2007)
17. Lappin, S.: An intensional parametric semantics for vague quantifiers. Linguistics & Philosophy **23**(6), 599–620 (2000)
18. Liu, Y., Kerre, E.E.: An overview of fuzzy quantifiers (I) interpretations. Fuzzy Sets and Systems **95**(1), 1–21 (1998)

19. Metcalfe, G., Olivetti, N., Gabbay, D.: Proof Theory for Fuzzy Logics. Applied Logic, vol. 36. Springer (2008)
20. Sauerland, U.: Vagueness in language: The case against fuzzy logic revisited. In: Cintula, P., Fermüller, C., Godo, L., Hájek, P. (eds.) Understanding Vagueness - Logical, Philosophical and Linguistic Perspectives. College Publications (2011)
21. Shapiro, S.: Vagueness in Context. Oxford University Press, USA (2006)
22. Solt, S.: Vagueness in quantity: Two case studies from a linguistic perspective. In: Cintula, P., Fermüller, C., Godo, L., Hájek, P. (eds.) Understanding Vagueness - Logical, Philosophical and Linguistic Perspectives. College Publications (2011)
23. Zadeh, L.A.: A computational approach to fuzzy quantifiers in natural languages. Computers & Mathematics with Applications 9(1), 149–184 (1983)

A New Model of a Fuzzy Associative Memory

Irina Perfilieva[✉]

Centre of Excellence IT4Innovations, Institute for Research
and Applications of Fuzzy Modeling, University of Ostrava, 30. dubna 22,
701 03 Ostrava 1, Czech Republic
Irina.Perfilieva@osu.cz

Abstract. We propose a new theory of implicative fuzzy associative memory. This memory is modeled by a fuzzy preorder relation. We give a necessary and sufficient condition on input data that guarantees an effective composition of a fuzzy associative memory, which is moreover, insensitivity to a certain type of noise.

Keywords: Fuzzy associative memory · Fuzzy preorder · Upper set · Noise

1 Introduction

In this contribution, we are focused on knowledge integration in uncertain environments and especially on data storage in the form of fuzzy associative memory (FAM) and retrieval. The latter is considered even in the case of damaged, incomplete or noisy requests.

The first attempt to construct a *fuzzy associative memory* (FAM) has been made by Kosko - [4]. This approach presented FAM as a single-layer feedforward neural net containing nonlinear matrix-vector product. This approach was later extended with the purpose to increase the storage capacity (e.g. [2]). Significant progress was achieved by the introduction of the so called learning implication rules [1,3], that afterwards led to *implicative fuzzy associative memory* (IFAM) with *implicative fuzzy learning*. Theoretical background of IFAM were discussed in [12].

In our contribution, we give a new theoretical justification of IFAM that is based on the notion of a fuzzy preorder relation. This enables us to discover conditions on input data that guarantee that IFAM works properly. We constructively characterize all types of noise that do not influence the successful retrieval.

2 Preliminaries

2.1 Implicative Fuzzy Associative Memory

In this Section, we explain background of the theory of fuzzy associative memories and their implicative forms. We choose database $\{(\mathbf{x}^1, \mathbf{y}^1), \ldots, (\mathbf{x}^p, \mathbf{y}^p)\}$ of

© Springer International Publishing Switzerland 2015
V.-N. Huynh et al. (Eds.): IUKM 2015, LNAI 9376, pp. 32–42, 2015.
DOI: 10.1007/978-3-319-25135-6_5

input-output objects (images, patterns, signals, texts, etc.) and assume that they can be represented by couples of normal fuzzy sets so that a particular fuzzy set $\mathbf{x}^k$, $k = 1,\ldots,p$, is a mapping $\mathbf{x}^k : X \to [0,1]$, and similarly, $\mathbf{y}^k : Y \to [0,1]$ where $X = \{u_1,\ldots,u_n\}$, $Y = \{v_1,\ldots,v_m\}$.

A model of FAM is associated with a couple (W,θ), consisted of a fuzzy relation $W : X \times Y \to [0,1]$ and a bias vector $\theta \in [0,1]^m$. A model of FAM connects every input $\mathbf{x}^k$ of a corresponding database with the related to it output $\mathbf{y}^k$, $k = 1,\ldots,p$. The connection can be realized by a sup $-t$ composition[1] $\circ$, so that

$$\mathbf{y}^k = W \circ \mathbf{x}^k \vee \theta, \; k = 1,\ldots,p, \tag{1}$$

or by a one level fuzzy neural network endowed with Pedrycz's neurons. The first one is represented by the following expression

$$\mathbf{y}_i^k = \bigvee_{j=1}^{n} (w_{ij} \; t \; \mathbf{x}_j^k) \vee \theta_i, \; i = 1,\ldots,m, \tag{2}$$

where $\mathbf{x}_j^k = \mathbf{x}^k(u_j)$, $\mathbf{y}_i^k = \mathbf{y}(v_i)$ and $w_{ij} = W(u_i,v_j)$. The second one is a computation model which realizes (2). In the language of fuzzy neural networks, we say that $W = (w_{ij})$ is a synaptic weight matrix and p is a number of constituent input-output patterns.

In practice, the crisp equality in (1) changes to

$$\mathbf{y}^k \approx W \circ \mathbf{x}^k, k = 1,\ldots,p, \tag{3}$$

where the right-hand side is supposed to be close to $\mathbf{y}^k$. Moreover, FAM is supposed to be tolerant to a particular input noise.

In [12], a model (W,θ) of implicative fuzzy associative memory (IFAM) has been proposed where

$$w_{ij} = \bigwedge_{k=1}^{p} (\mathbf{x}_j^k \to \mathbf{y}_i^k), \tag{4}$$

$$\theta_i = \bigwedge_{k=1}^{p} \mathbf{x}_i^k,$$

and $\to$ is an adjoint implication with respect to the chosen continuous t-norm.

One important case of IFAM is specified by identical input-output patterns. This memory is called an *autoassociative fuzzy implicative memory* (AFIM), and it is aimed at memorizing patterns as well as error correction or removing of noise.

For a given input $\mathbf{x}$, AFIM returns output $\mathbf{y}$ in accordance with (2). If $\mathbf{x}$ is close to some pattern $\mathbf{x}^k$, then $\mathbf{y}$ is close to the same pattern $\mathbf{x}^k$. In the ideal case, patterns from $\{\mathbf{x}^1,\ldots,\mathbf{x}^p\}$ are eigen vectors of an AFIM model.

---

[1] t is a t-norm, i.e. a binary operation on $[0,1]$, which is commutative, associative, monotone and has 1 as a neutral element.

One of the main benefits of AFIM is its error correction ability. By this we mean that if an input $\mathbf{x}$ is close to some pattern $\mathbf{x}^k$, then the output $\mathbf{y}$ is equal to the same pattern $\mathbf{x}^k$.

In the proposed contribution, we analyze the AFIM retrieval mechanism with respect to two goals: (a) to have patterns from $\{\mathbf{x}^1, \ldots, \mathbf{x}^p\}$ as eigen vectors of that fuzzy relation W, which constitutes a model of AFIM; (b) to correct a certain type of noise. We find a necessary and sufficient condition on input patterns that guarantees that the goal (a) is fulfilled and moreover, we characterize a noise that can be successfully removed by the retrieval procedure.

Our technical platform is more general than that in [12]: we replace $[0, 1]$ by an arbitrary complete residuated lattice $\mathcal{L}$ and consider initial objects as fuzzy sets with values in $\mathcal{L}$. This allows us to utilize many known facts about fuzzy sets and fuzzy relations of particular types.

2.2 Algebraic Background

In this Section, we will step aside from the terminology of associative memories and introduce an algebraic background of the technique proposed below.

Let $\mathcal{L} = \langle L, \vee, \wedge, *, \rightarrow, 0, 1 \rangle$ be a fixed, complete, integral, residuated, commutative l-monoid (*a complete residuated lattice*). We remind the main characteristics of this structure: $\langle L, \vee, \wedge, 0, 1 \rangle$ is a complete bounded lattice, $\langle L, *, \rightarrow, 1 \rangle$ is a residuated, commutative monoid.

Let X be a non-empty set, L^X a class of *fuzzy sets* on X and $L^{X \times X}$ a class of *fuzzy relations* on X. Fuzzy sets and fuzzy relations are identified with their membership functions, i.e. elements from L^X and $L^{X \times X}$, respectively. A fuzzy set A is *normal* if there exists $x_A \in X$ such that $A(x_A) = 1$. The (ordinary) set $Core(A) = \{x \in X \mid A(x) = 1\}$ is the *core* of the normal fuzzy set A. Fuzzy sets $A \in L^X$ and $B \in L^X$ are *equal* ($A = B$), if for all $x \in X$, $A(x) = B(x)$. A fuzzy set $A \in L^X$ is *less than or equal* to a fuzzy set $B \in L^X$ ($A \leq B$), if for all $x \in X$, $A(x) \leq B(x)$.

The lattice operations $\vee$ and $\wedge$ induce the union and intersection of fuzzy sets, respectively. The binary operation $*$ of $\mathcal{L}$ is used below for set-relation composition of the type sup-$*$, which is usually denoted by $\circ$ so that

$$(A \circ R)(y) = \bigvee_{x \in X} (A(x) * R(x, y)).$$

Let us remind that the $\circ$ composition was introduced by L. Zadeh [15] in the form max $-$ min.

3   Fuzzy Preorders and Their Eigen Sets

In this Section, we introduce theoretical results which will be used below in the discussed application. The results are formulated in the language of residuated lattices. We will first recall basic facts about fuzzy preorder relations as they were presented in [5]. Then we will characterize eigen sets of fuzzy preorder relations and how they can be reconstructed.

Our interest to fuzzy preorder relations came from the analysis of the expression (4) – a representation of a fuzzy relation in a model of AFM. In the particular case of autoassociative fuzzy implicative memory, expression (4) changes to

$$w_{ij} = \bigwedge_{k=1}^{p} (\mathbf{x}_j^k \rightarrow \mathbf{x}_i^k).$$

This is a representation of the Valverde (fuzzy) preorder (see Remark 1 below).

3.1 Fuzzy Preorders and their Upper and Lower Sets

The text in this Section is an adapted version of [8].

A binary fuzzy relation on X is a $*$-*fuzzy preorder* of X, if it is reflexive and $*$-transitive. The fuzzy preorder $Q^* \in L^{X \times X}$, where

$$Q^*(x,y) = \bigwedge_{i \in I} (A_i(x) \rightarrow A_i(y)), \tag{5}$$

is *generated* by an arbitrary family of fuzzy sets $(A_i)_{i \in I}$ of X.

Remark 1. The fuzzy preorder Q^* (5) is often called the Valverde order on X determined by a family of fuzzy sets $(A_i)_{i \in I}$ of X (see [14] for details).

If Q is a fuzzy preorder on X, then the fuzzy set $A \in L^X$ such that

$$A(x) * Q(x,y) \leq A(y) \quad (A(y) * Q(x,y) \leq A(x)), \quad x, y \in X,$$

is called an *upper set* (a *lower set*) of Q (see [5]). Denote $Q^t(x) = Q(t,x)$ $(Q_t(x) = Q(x,t))$, $x \in X$, and see that Q^t (Q_t) is an upper set (lower set) of Q. The fuzzy set Q^t (Q_t) is called a *principal* upper set (lower set).

If Q is a fuzzy preorder on X, then $Q^{op} \in L^{X \times X}$ such that $Q^{op}(x,y) = Q(y,x)$ is a fuzzy preorder on X as well. It follows that an upper set of Q is a lower set of Q^{op} and vice versa. For this reason, our results will be formulated for upper sets of respective fuzzy preorders.

The necessary and sufficient condition that a family of fuzzy sets of X constitutes a family of upper sets of some fuzzy preorder on X has been proven in [5]. In Theorem 1 [8], given below, we characterize principal upper sets of a fuzzy preorder on X. Let us remark that assumptions of Theorem 1 are different from those in [5].

Theorem 1. *Let I be an index set, $(A_i)_{i \in I} \subseteq L^X$ a family of normal fuzzy sets of X and $(x_i)_{i \in I} \subseteq X$ a family of pairwise different core elements such that for all $i \in I$, $A_i(x_i) = 1$. Then the following statements are equivalent:*

(i) There exists a fuzzy preorder Q on X such that for all $i \in I$, $x \in X$, $A_i(x) = Q(x_i, x)$ (A_i is a principal upper set of Q).

(ii) For all $i \in I$, $x \in X$, $A_i(x) = Q^*(x_i, x)$ *(A_i is a principal upper set of Q^*)* where Q^* is given by (5).

(iii) For all $i, j \in I$,

$$A_i(x_j) \leq \bigwedge_{x \in X} (A_j(x) \to A_i(x)). \tag{6}$$

Corollary 1. *Let $(A_i)_{i \in I} \subseteq L^X$ be a family of normal fuzzy sets of X and $(x_i)_{i \in I} \subseteq X$ a family of pairwise different core elements such that for all $i, j \in I$, (6) holds true. Then Q^* is the coarsest fuzzy preorder on X such that every fuzzy set A_i, $i \in I$, is a principal upper set of Q^*.*

Remark 2. On the basis of Theorem 1 and its Corollary 1, we conclude that a family of normal fuzzy sets $(A_i)_{i \in I} \subseteq L^X$ with pairwise different core elements $(x_i)_{i \in I} \subseteq X$, such that (6) is fulfilled, generates the coarsest fuzzy preorder Q^* on X such that every family element A_i is a principal upper set of Q^* that corresponds to its core element x_i.

3.2   Eigen Sets of Fuzzy Preorders and their "Skeletons"

In this Section, we show that if the assumptions of Theorem 1 are fulfilled, and if the fuzzy preorder Q^* on X is generated (in the sense of (5)) by normal fuzzy sets $(A_i)_{i \in I} \subseteq L^X$ with pairwise different core elements $(x_i)_{i \in I} \subseteq X$, then these fuzzy sets are the eigen (fuzzy) sets of Q^* (see [10]), i.e. they fulfill

$$A_i \circ Q^* = A_i, \ i \in I. \tag{7}$$

Proposition 1. *Let family $(A_i)_{i \in I} \subseteq L^X$, $i \in I$, of normal fuzzy sets of X with pairwise different core elements $(x_i)_{i \in I} \subseteq X$ fulfill (6) and generate fuzzy preorder Q^* in the sense of (5). Then every A_i, $i \in I$, is an eigen set of Q^*.*

Proof. Let us choose and fix A_i, $i \in I$. By Theorem 1, $A_i(x) = Q^*(x_i, x)$. Then

$$A_i(x) \circ Q^*(x, y) = \bigvee_{x \in X} (Q^*(x_i, x) * Q^*(x, y)) \leq$$

$$Q^*(x_i, y) = A_i(y).$$

On the other hand,

$$A_i(x) \circ Q^*(x, y) = \bigvee_{x \in X} (Q^*(x_i, x) * Q^*(x, y)) \geq$$

$$Q^*(x_i, y) * Q^*(y, y) = Q^*(x_i, y) = A_i(y).$$

Corollary 2. *Let the assumptions of Propositions 1 be fulfilled and fuzzy set $\bar{A}_i \in L^X$, $i \in I$, be a "skeleton" of A_i, where*

$$\bar{A}_i(x) = \begin{cases} 1, & \text{if } x \in Core(A_i), \\ 0, & \text{otherwise.} \end{cases} \tag{8}$$

Then A_i can be reconstructed from $\bar{A}_i$, i.e.

$$\bar{A}_i \circ Q^* = A_i. \tag{9}$$

Proof. By Proposition 1, and the inequality $\bar{A}_i \leq A_i$, we have $\bar{A}_i \circ Q^* \leq A_i \circ Q^* = A_i$. On the other hand,

$$(\bar{A}_i \circ Q^*)(y) = \bigvee_{x \in X} (\bar{A}_i(x) * Q^*(x,y)) \geq Q^*(x_i, y) = A_i(y).$$

Corollary 3. *Let the assumptions of Propositions 1 be fulfilled and fuzzy set $\tilde{A}_i \in L^X$ be "in between" $\bar{A}_i$ and A_i, i.e.*

$$\bar{A}_i \leq \tilde{A}_i \leq A_i,$$

where $i \in I$. Then A_i can be reconstructed from $\tilde{A}_i$, i.e.

$$\tilde{A}_i \circ Q^* = A_i. \tag{10}$$

Proof. The proof follows from the following chain of inequalities:

$$A_i = A_i \circ Q^* \geq \tilde{A}_i \circ Q^* \geq \bar{A}_i \circ Q^* = A_i.$$

The following proposition is important for the below considered applications. It shows that under the assumptions of Propositions 1, every A_i, $i \in I$, is an eigen set of another fuzzy preorder Q^r, which is composed from all these constituent fuzzy sets. By saying "composed", we mean that opposite to Q^*, Q^r does not require any computation.

Proposition 2. *Let family $(A_i)_{i \in I} \subseteq L^X$, $i \in I$, of normal fuzzy sets of X with pairwise different core elements $(x_i)_{i \in I} \subseteq X$ fulfill (6). Then every A_i, $i \in I$, is an eigen set of the following fuzzy preorder*

$$Q^r(x,y) = \begin{cases} A_i(y), & \text{if } x = x_i, \\ 1, & \text{if } x = y, \\ 0, & \text{otherwise .} \end{cases} \tag{11}$$

4   Fuzzy Preorders and AFIM

In this Section, we will put a bridge between the theory, presented in Section 3, and the theory of autoassociative fuzzy implicative memories (AFIM), presented in Section 2. We will see that in the proposed below model of AFIM, a connecting fuzzy relation (denoted above by W) is a fuzzy preorder relation.

In details, we choose a residuated lattice with the support $L = [0,1]$ and a database $\{\mathbf{x}^1, \ldots, \mathbf{x}^p\}$ of initial objects that are represented by fuzzy sets or fuzzy relations. In the first case, we have a database of signals, while in the second one, we have 2D gray-scaled images. The second case can be easily reduced to the first one - it is enough to represent an image as a sequence of rows. Below, we assume that our objects are normal fuzzy sets identified with their membership functions, i.e. they are elements of $[0,1]^X$, where X is a finite universe. The assumption of normality does not put any restriction, because any given finite collection of fuzzy sets on a finite universe can be normalized. Because we illustrate the proposed technique by images, we refer to the initial objects as to images.

In accordance with (5), we construct the fuzzy preorder relation Q^*, such that

$$Q^*(i,j) = \bigwedge_{k=1}^{p} (\mathbf{x}^k(i) \rightarrow \mathbf{x}^k(j)).$$

We remark that this is the reverse fuzzy preorder with respect to that given by (4). In the terminology of the theory of autoassociative memories, the results from Section 3 show that under condition (6),

- each constituent input image $\mathbf{x}^k$, $k = 1, \ldots, p$, can be retrieved, if the weight matrix W is equal to Q^* and the computation of the output is based on the simpler version of (2), i.e.

$$\mathbf{y}_i = \bigvee_{j=1}^{n} (\mathbf{x}_j^k \iota w_{ij}), \, l = 1, \ldots, m, \tag{12}$$

which does not involve bias θ;
- each constituent input image $\mathbf{x}^k$, $k = 1, \ldots, p$, can be retrieved, if the weight matrix W is equal to Q^r (see (11)) with the subsequent computation of the output by (12);
- each constituent input image $\mathbf{x}^k$, $k = 1, \ldots, p$, can be fully reconstructed from its binary "skeleton" (see (8) in Corollary 2).

Let us remark that from the second result, listed above, it follows that there exists a weight matrix W which can be assembled from constituent input images in accordance with (11) and by this, no computation is needed. This fact leads to a tremendous saving of computational complexity.

Moreover, from Corollary 3 we deduce a complete characterization of a noise N_k that can be "added to" (actually, subtracted from) a constituent input image $\mathbf{x}^k$ without any corruption of the output. In details,

$$N_k(t) = \begin{cases} n_k(t), & \text{if } t \notin Core\mathbf{x}^k, \\ 0, & \text{otherwise,} \end{cases} \tag{13}$$

where for $t \notin Core\mathbf{x}^k$, the value $n_k(t)$ fulfills the requirement $0 \leq n_k(t) \leq \mathbf{x}^k(t)$, $k = 1, \ldots, p$.

Below, we demonstrate how the presented above theory works in the case of some benchmark input images.

5 Illustration

The aim of this Section is to give illustrations to the theoretical results of this paper. We used gray scaled images with the range $[0, 1]$, where 0 (1) represents the black (white) color. We chose two different sets of images, both were artificially created from available databases. The sets contain 2D images of 20×20 and 32×32 pixels, respectively. All images are represented by vectors that are comprised by successive rows. Each image corresponds to a fuzzy set on $\{1, \ldots, 20\} \times \{1, \ldots, 20\}$ or $\{1, \ldots, 32\} \times \{1, \ldots, 32\}$ with values in $[0, 1]$.

5.1 Experiments with Abstract Images

We have created three databases A, B, and C of 2D images of 20×20 pixels, where condition (6) is/is not fulfilled, details are below.

- **Database A** - contains three images (see figure 1) such that (6) is fulfilled.
- **Database B** - contains four images (see figure 2) such that (6) is not fulfilled with the non-separability degree as follows:

$$D_B = \bigwedge_{x \in X} (A_j(x) \to A_i(x)) \to A_i(x_j). \tag{14}$$

- **Database C** - contains eight images (see figure 3) such that (6) is not fulfilled, and the corresponding non-separability degree D_C is less than D_B.

Fig. 1. Database A contains three images such that (6) is fulfilled.

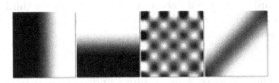

Fig. 2. Database B contains four images such that (6) is not fulfilled with the degree D_B.

For each database of images, we computed he corresponding fuzzy preorder Q^* and its reduction Q^r.

In Fig. 4, we demonstrate the influence of condition (6) on the quality of retrieval by the AFIM mechanism, where the computation of the output is based on (12) and the weight matrix W is equal to Q^* or Q^r. For this purpose we choose

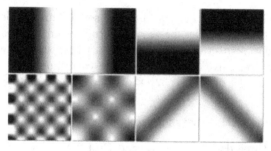

Fig. 3. Database C contains eight images such that (6) is not fulfilled with the degree D_C such that $D_C \leq D_B$.

Fig. 4. *Top left* Third image from database A. *Top right* Binary "skeleton" of the third image from database A. *Bottom left* Output image retrieved from the binary "skeleton" of the third image and database A - identical with the third image. *Bottom right* Output image retrieved from the third image and database C - different from the third input image.

the third image from database A as an input and retrieve it from each of three databases A, B and C.

In Fig. 4, we see that if the weight matrix W is computed (comprised) from database A as fuzzy preorder Q^* (Q^r), then the output coincides with the identical to it input. Moreover, any image from database A can be retrieved from its binary "skeleton". If W is computed from database C as fuzzy preorder Q^*, then the output differs from the input, i.e. images from database C cannot be retrieved precisely.

Fig. 5. *Left* Third image from database A with 70% density of eroded noise. *Right* The output, retrieved by AFIM with the weight matrix Q^r. Eroded noise has been completely removed by the AFIM retrieval.

In Fig. 5, we demonstrate how eroded noise (13) can be removed by the AFIM retrieval. We added 70% dense erosion to the third image from database A and process the obtained eroded image by the AFIM with the weight matrix W that corresponds to fuzzy preorder Q^r, computed from database A. In the right-hand side of Fig. 5, we show the output, retrieved by IFAM with the weight matrix Q^r. This output coincides with the original (non-eroded) third image.

6 Conclusion

A new theory of implicative fuzzy associative memory has been proposed. We showed that

1. every database pattern can be successfully retrieved,
   - if all database patterns are well separated, the weight matrix W corresponds to a certain fuzzy preorder relation and the computation of the output is based on a composition with W, which does not involve bias θ;
   - if additionally to the above conditions, the weight matrix W corresponds to a certain reduction of the fuzzy preorder relation.
2. the weight matrix W does not require computation, if the above mentioned conditions are fulfilled.

We discovered a necessary and sufficient condition that guarantees insensitivity to a certain type of noise. The latter is precisely characterized.

Acknowledgment. This paper was supported by the European Regional Development Fund in the IT4Innovations Centre of Excellence project (CZ.1.05/ 1.1.00/02.0070) and Ministry of Science.

The author expresses sincere thanks to Dr. Marek Vajgl for experimental tests in Section 5.

References

1. Cheng, Q., Fan, Z.-T.: The stability problem for fuzzy bidirectional associative memories. Fuzzy Sets Syst. **132**, 83–90 (2002)
2. Chung, F.-L., Lee, T.: On fuzzy associative memory with multiple-rule storage capacity. IEEE Trans. Fuzzy Syst. **4**, 375–384 (1996)
3. Junbo, F., Fan, J., Yan, S.: A learning rule for fuzzy associative memories. In: Proc. IEEE Int. Conf. Neural Networks, IEEE World Congr. Computational Intelligence, vol. 7, pp. 4273–4277 (1994)
4. Kosko, B.: Neural Networks and Fuzzy Systems: A Dynamical Systems Approach to Machine Intelligence. Prentice-Hall (1992)
5. Lai, H., Zhang, D.: Fuzzy preorder and fuzzy topology. Fuzzy Sets and Syst. **157**, 1865–1885 (2006)
6. Perfilieva, I.: Fuzzy relation equations in semilinear spaces. In: Hüllermeier, E., Kruse, R., Hoffmann, F. (eds.) IPMU 2010. CCIS, vol. 80, pp. 545–552. Springer, Heidelberg (2010)
7. Perfilieva, I. Systems of fuzzy relation equations in a space with fuzzy preorder. In: Proc. of IFSA-EUSFLAT 2009 Joint Conf., Lisbon, pp. 1601–1605 (2009)
8. Perfilieva, I.: Finitary solvability conditions for systems of fuzzy relation equations. Inf. Sci. **234**, 29–43 (2013)
9. Ruspini, E.H.: A new approach to clustering. Inf. and Cont. **15**, 22–32 (1969)
10. Sanchez, E.: Resolution of Eigen fuzzy sets equations. Fuzzy Sets and Syst. **1**, 69–74 (1978)
11. Steinbuch, K.: Die lernmatrix. Kybernetik **1**, 36–45 (1961)
12. Sussner, P., Valle, M.-E.: Implicative Fuzzy Associative Memories. IEEE Trans. Fuzzy Systems **14**, 793–807 (2006)
13. Taylor, W.: Eletrical simulation on some nervous system functional activities. Inf. Theory **15**, 166–177 (2004)
14. Valverde, L.: On the structure of F-indistinguishability operators. Fuzzy Sets and Syst. **17**, 313–328 (1985)
15. Zadeh, L.A.: The concept of a linguistic variable and its application to approximate reasoning I, II, III. Inf. Sci. **8–9**, 199–257, 301–357, 43–80 (1975)

Construction of Associative Functions for Several Fuzzy Logics via the Ordinal Sum Theorem

Mayuka F. Kawaguchi[1](✉) and Michiro Kondo[2]

[1] Hokkaido University, Sapporo 060-0814, Japan
mayuka@ist.hokudai.ac.jp
[2] Tokyo Denki University, Inzai, Chiba 270-1382, Japan
mkondo@mail.dendai.ac.jp

Abstract. In this report, the ordinal sum theorem of semigroups is applied to construct logical operations for several fuzzy logics. The generalized form of ordinal sum for fuzzy logics on [0, 1] is defined in order to uniformly express several families of logical operations. Then, the conditions in ordinal sums for various properties of logical operations are presented: for examples, the monotonicity, the location of the unit element, the left/right-continuity, or and/or-likeness. Finally, some examples to construct pseudo-uninorms by the proposed method are illustrated.

Keywords: Ordinal sum · Pseudo-t-norms · Pseudo t-conorms · Pseudo-uninorms

1 Introduction

The concept of ordinal sums has been originated by Climescu [3], and then has been generalized by Clifford [1], [2] to a method for constructing a new semigroup from a given linearly-ordered system of semigroups. In the history of the research on fuzzy logical connectives as t-norms, t-conorms, and uninorms, the ordinal sum has been often appeared as representations of such operations [4]-[14].

In this paper, the authors challenge to reform the ordinal sum on [0,1] to a more general scheme as a common platform to construct fuzzy logical connectives in the broader sense: including non-commutative ones besides t-norms, t-conorms and uninorms. The results of this work would be useful for obtaining an associative operation suitable for human thinking/evaluation, in several applications such as information aggregation in diagnoses systems, construction of metrics based on fuzzy relation, constraint satisfaction in multicriteria decision making, and so on.

2 Origin of Ordinal Sum Theorem

Climescu [3] has introduced the original concept of ordinal sums which is a method to construct a new semigroup from a family of semigroups. According to Schweizer et al. [13], his definition of an ordinal sum is expressed as follows.

© Springer International Publishing Switzerland 2015
V.-N. Huynh et al. (Eds.): IUKM 2015, LNAI 9376, pp. 43–53, 2015.
DOI: 10.1007/978-3-319-25135-6_6

Ordinal Sum Theorem by Climescu [3] and Schweizer et al. [13]

Let (A, F) and (B, F) be semigroups. If the sets A and B are disjoint and if H is the mapping defined on $(A \cup B) \times (A \cup B)$ by

$$H(x, y) = \begin{cases} F(x, y), & x \in A, \ y \in A, \\ x, & x \in A, \ y \in B, \\ y, & x \in B, \ y \in A, \\ G(x, y), & x \in B, \ y \in B, \end{cases} \tag{1}$$

then $(A \cup B, H)$ is a semigroup.

On the other hand, Clifford [1], [2] has introduced a more generalized definition of the same concept, and has named it an ordinal sum. The following theorem is the reformatted version by Klement et al.

Ordinal Sum Theorem by Clifford [1], [2], and Klement et al. [11]

Let $(A, \leq)$ with $A \neq \varnothing$ be a linearly ordered set and $(G_\alpha)_{\alpha \in A}$ with $G_\alpha = (X_\alpha, *_\alpha)$ be a family of semigroups. Assume that for all $\alpha, \beta \in A$ with $\alpha < \beta$ the sets X_α and X_β are either disjoint or that $X_\alpha \cap X_\beta = \{x_{\alpha\beta}\}$, where $x_{\alpha\beta}$ is both the unit element of G_α and the annihilator of G_β, and where for each $\gamma \in A$ with $\alpha < \gamma < \beta$ we have $X_\gamma = \{x_{\alpha\beta}\}$. Put $X = \bigcup_{\alpha \in A} X_\alpha$ and define the binary operation $*$ on X by

$$x * y = \begin{cases} x *_\alpha y & \text{if } (x, y) \in X_\alpha \times X_\alpha, \\ x & \text{if } (x, y) \in X_\alpha \times X_\beta \text{ and } \alpha < \beta, \\ y & \text{if } (x, y) \in X_\alpha \times X_\beta \text{ and } \beta < \alpha. \end{cases} \tag{2}$$

Then $G = (X, *)$ is a semigroup. The semigroup G is commutative if and only if for each $\alpha \in A$ the semigroup G_α is commutative.

Here, G is called the ordinal sum of $(G_\alpha)_{\alpha \in A}$, and each G_α is called a summand.

3   A Generalization of Ordinal Sums on the Unit Interval [0, 1]

In this research work, let us restrict the linearly ordered set A, mentioned in Section 2, to be finite. One of the main ideas proposed here is to give an indexing independently from ordering to the set of summands $(G_\alpha)_{\alpha \in A}$ by introducing a bijection as a correspondence between them.

Definition 1. *Consider a permutation* σ *on* $A=\{1,2,..,n\}$, *i.e. a bijection* $\sigma:\{1,2,..n\}\to\{1,2,..n\}$, *then define a linear order* $\leq$ *in the family of sets* $\{X_i\}_{i\in\{1,2,...,n\}}$ *as follows:*

$$X_i \leq X_j \overset{def.}{\Leftrightarrow} \sigma(i) \leq \sigma(j) \quad for \ \forall i,j \in \{1,2,...,n\}.\tag{3}$$

Example 1. If $n=6$, and a permutation σ is given as

$$\sigma = \begin{pmatrix} 1 & 2 & 3 & 4 & 5 & 6 \\ 1 & 3 & 5 & 6 & 4 & 2 \end{pmatrix} = \begin{pmatrix} 1 & 6 & 2 & 5 & 3 & 4 \\ 1 & 2 & 3 & 4 & 5 & 6 \end{pmatrix},$$

then we get the linear order in $\{X_i\}_{i\in\{1,2,...,6\}}$ as $X_1 \leq X_6 \leq X_2 \leq X_5 \leq X_3 \leq X_4$. This permutation σ works to locate an element X_i with index i at the $\sigma(i)$ -th position.

Hereafter, we treat the case that $X = \bigcup_{i=1}^n X_i = [0,1]$ and $X_i \ (i=1,2,...n)$ are disjoint each another, in order to apply the ordinal sum theorem for constructing various logical connectives defined in $[0,1]$.

Definition 2. *Let* $I = \{I_i\}_{i=1,2,...,n}$ *be a partition by a finite number of non-empty subintervals of* $[0,1]$, *i.e.* $\bigcup_{i=1}^n I_i = [0,1]$ *and* $I_i \cap I_j = \varnothing \ (i \neq j)$ *hold. Also, we denote* $a_i = \inf I_i, \ b_i = \sup I_i$.

There exists the linear order relation among pairwise disjoint real subintervals according to the real number order. Thus, the permutation σ in Definition 1 gives the indexing as $I_i \leq I_j \Rightarrow \sigma(i) \leq \sigma(j)$. In other words, the subinterval at k-th position is indexed as $I_{\sigma^{-1}(k)}$.

Example 2. If $n=4$, and a partition I and a permutation σ are given as

$$I = \{[0,0.25], \ ]0.25,0.5], \ ]0.5,0.75], \ ]0.75,1]\} \quad and$$

$$\sigma = \begin{pmatrix} 1 & 2 & 3 & 4 \\ 2 & 4 & 1 & 3 \end{pmatrix} = \begin{pmatrix} 3 & 1 & 4 & 2 \\ 1 & 2 & 3 & 4 \end{pmatrix},$$

respectively, then we have the following indexing for subintervals :

$$I_3 = [0,0.25], I_1 =]0.25,0.5], I_4 =]0.5,0.75], I_2 =]0.75,1];$$

$$I_3 \leq I_1 \leq I_4 \leq I_2.$$

Definition 3. *Let I' be a subset of $I = \{I_i\}_{i=1,2,...,n}$. I' is called to be ascending ordered if $I_i \leq I_j$ holds for $\forall I_i, I_j \in I'$ $(i \leq j)$. Similarly, I' is called to be descending ordered if $I_i \leq I_j$ holds for $\forall I_i, I_j \in I'$ $(j \leq i)$.*

Definition 4. *Two subsets $\underline{I}$ and $\tilde{I}$ of $I = \{I_i\}_{i=1,2,...,n}$ are defined as follows:*

$$\underline{I} \stackrel{def.}{=} \left\{ I_i \middle| I_i \leq I_n \right\}, \quad \tilde{I} \stackrel{def.}{=} \left\{ I_i \middle| I_n \leq I_i \right\}.$$

Example 3. When $n = 6$ and a permutation σ is given as

$$\sigma = \begin{pmatrix} 1 & 2 & 3 & 4 & 5 & 6 \\ 1 & 6 & 2 & 5 & 3 & 4 \end{pmatrix} = \begin{pmatrix} 1 & 3 & 5 & 6 & 4 & 2 \\ 1 & 2 & 3 & 4 & 5 & 6 \end{pmatrix},$$

we obtain the indexing of $I = \{I_i\}_{i=1,2,...,6}$ as $I_1 \leq I_3 \leq I_5 \leq I_6 \leq I_4 \leq I_2$, and we have $\underline{I} = \{I_1, I_3, I_5, I_6\}$ and $\tilde{I} = \{I_6, I_4, I_2\}$. Here, $\underline{I}$ is ascending ordered, and $\tilde{I}$ descending ordered.

Definition 5. *Assign each binary operation $H_i : [0,1]^2 \to [0,1]$ to each direct product $I_i \times I_i$ of a subinterval $I_i \in I$. Then, we define the binary operation $H : [0,1]^2 \to [0,1]$ as the following ordinal sum:*

$$H(x, y) = \begin{cases} a_i + (b_i - a_i)H_i\left(\dfrac{x - a_i}{b_i - a_i}, \dfrac{y - a_i}{b_i - a_i} \right) & \text{if } (x, y) \in I_i \times I_i \\ x & \text{if } (x, y) \in I_i \times I_j \text{ and } i < j \quad (4) \\ y & \text{if } (x, y) \in I_i \times I_j \text{ and } j < i. \end{cases}$$

4   Construction of Logical Connectives on [0, 1]

4.1   Properties Required for Fuzzy Logical Connectives

Conjunctive/disjunctive operations on [0,1] used in various fuzzy logic are defined by combining some of the following properties.

Associativity:
(A) $a * (b * c) = (a * b) * c$

Commutativity:
(C) $a * b = b * a$

Existence of the unit element:
(U1) $a * 1 = 1 * a = a$
(U0) $a * 0 = 0 * a = a$
(UE) $a * e = e * a = a$ $(e \in]0,1[)$

Boundary conditions:
(Bmin) $a * b \leq \min(a,b)$
(Bmax) $a * b \geq \max(a,b)$

Monotonicity:
(M) $a \leq b \;\Rightarrow\; a * c \leq b * c,\; c * a \leq c * b$

Left-continuity, right-continuity:
(LC) $\lim\limits_{x \to b-0} a * x = a * b,\quad \lim\limits_{x \to a-0} x * b = a * b$
(RC) $\lim\limits_{x \to b+0} a * x = a * b,\quad \lim\limits_{x \to a+0} x * b = a * b$

And-like, or-like [6]:
(AL) $0 * 1 = 1 * 0 = 0$
(OL) $0 * 1 = 1 * 0 = 1$

The definitions of already-known logical operations are expressed by the combinations of the above-mentioned properties as follows.

- t-Norms: (A), (C), (U1), (M)
- t-Conorms: (A), (C), (U0), (M)
- Uninorms: (A), (C), (UE), (M)

- Pseudo-t-norms [5]:   (A), (U1), (M)
- Pseudo-t-conorms: (A), (U0), (M)
- Pseudo-uninorms [10]: (A), (UE), (M)

- t-Subnorms [8], [9]: (A), (C), (Bmin), (M)
- t-Subconorms: (A), (C), (Bmax), (M)

Here, the authors introduce the notion of pseudo-t-sub(co)norms as follows.

- Pseudo-t-subnorms: (A), (Bmin), (M)
- Pseudo-t-subconorms: (A), (Bmax), (M)

4.2   Realizations of the Properties in the Framework of Ordinal Sum

We obtain the following theorems regarding to a binary operation $H : [0,1]^2 \to [0,1]$ defined in Definition 5.

Theorem 1. (Clifford [1], [2])

(i) H *is associative if and only if all summands* H_i ($i = 1,...,n$) *are associative.*

(ii) H *is commutative if and only if all summands* H_i ($i = 1,...,n$) *are commutative.*

Theorem 2

If $\underline{I}$ *is ascending ordered,* $\tilde{I}$ *is descending ordered,* H_i *for* $\sigma(i) < \sigma(n)$ *satisfy the boundary condition (Bmin),* H_i *for* $\sigma(n) < \sigma(i)$ *satisfy the boundary condition (Bmax), and all summands including* H_n *are monotone-increasing (M), then* H *is monotone-increasing (M).*

See Appendix for the detailed proof of Theorem 2.

Theorem 3

(i)   *If* I_n *is right-closed (i.e. there exists* $\max I_n$ *) and* H_n *satisfies (U1) (i.e. it has the unit element* 1 *), then* $e = \max I_n = b_n$ *is the unit element of* H .

(ii)   *If* I_n *is left-closed (i.e. there exists* $\min I_n$ *) and* H_n *satisfies (U0) (i.e. it has the unit element* 0 *), then* $e = \min I_n = a_n$ *is the unit element of* H .

(iii)   *If* H_n *satisfies (UE) (i.e. it has the unit element* $e' \in]0,1[$ *), then* $e = a_n + e'(b_n - a_n)$ *is the unit element of* H .

Corollary of Theorem 3

(i)   *If* $\sigma(n) = n$ *and* H_n *satisfies (U1) (i.e. it has the unit element* 1 *), then* H *satisfies (U1).*

(ii)   *If* $\sigma(n) = 1$ *and* H_n *satisfies (U0) (i.e. it has the unit element* 0 *), then* H *satisfies (U0).*

Theorem 4

(i)   *Let the subinterval including* 0 *be closed, and the other subintervals be left-open and right-closed as* $I_i =]a_i, b_i]$ *. Then,* H *is left-continuous (LC) if and only if all summands* H_i ($i = 1,...,n$) *are left-continuous (LC).*

(ii)   *Let the subinterval including* 1 *be closed, and the other subintervals be left-closed and right-open as* $I_i = [a_i, b_i[$ *. Then,* H *is right-continuous (RC) if and only if all summands* H_i ($i = 1,...,n$) *are right-continuous (RC).*

Theorem 5

Suppose that the indices $i, j \in \{1, 2, ..., n\}$ *satisfy* $\sigma(i) = 1$ *and* $\sigma(j) = n$.

(i) *If* $i < j$, *then* H *is and-like* (AL).

(ii) *If* $j < i$, *then* H *is or-like* (OL).

5   Applications

Example 4. Let us consider the case to construct "a left-continuous t-norm." We can obtain it by applying the following conditions to eq. (4):

Theorem 1 (i), (ii)	for associativity and commutativity,
Theorem 2	for monotonicity,
Corollary of Theorem 3 (i)	for unit element $e = 1$, and
Theorem 4 (i)	for left-continuity.

The above result is a finite version of Jenei's method [8], [9], to construct a left-continuous t-norm.

Example 5. Also, we can construct various kinds of pseudo-uninorms through applying the following conditions to eq. (4):

Theorem 1 (i)	for associativity,
Theorem 2	for monotonicity,
Theorem 3	for unit element $e \in [0, 1]$,
Theorem 4	for left-continuity/right-continuity, and
Theorem 5	for and-likeness/or-likeness.

Fig.1 (a) illustrates a case of left-continuous and-like pseudo-uninorms, where $n = 3$, $\sigma = \begin{pmatrix} 1 & 2 & 3 \\ 1 & 3 & 2 \end{pmatrix} = \begin{pmatrix} 1 & 3 & 2 \\ 1 & 2 & 3 \end{pmatrix}$ and $I = \{[0, a],]a, e],]e, 1]\}$. All summands $\tilde{T}_1$, $\hat{T}_3$ and $\tilde{S}_2$ are associative, monotone increasing and left-continuous. Since $\sigma(1) = 1$ and $\sigma(2) = 3$, Th.5(i) is applicable. Also, Th.3(i) is applicable because $\hat{T}_3$ is a pseudo-t-norm and $I_3 = (a, e]$ is right-closed, thus $e = \max I_3$ is the unit element.

Fig.1 (b) illustrates a case of right-continuous and-like pseudo-uninorms, where $n = 3$, $\sigma = \begin{pmatrix} 1 & 2 & 3 \\ 1 & 2 & 3 \end{pmatrix}$ and $I = \{[0, a[, [a, e[, [e, 1]\}$. All summands $\tilde{T}_1$, $\tilde{T}_2$ and $\hat{S}_3$ are associative, monotone increasing and right-continuous. Since $\sigma(1) = 1$ and $\sigma(3) = 3$, Th.5(i) is applicable. Also, Th.3(ii) is applicable because $\hat{S}_3$ is a pseudo-t-conorm and $I_3 = [a, e)$ is left-closed, thus $e = \min I_3$ is the unit element.

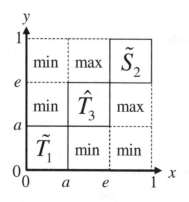

(a) left-continuous (LC) case

$I = \{[0,a],]a,e],]e,1]\}$

$\tilde{T}_1$: LC pseudo-t-subnorm

$\tilde{S}_2$: LC pseudo-t-subconorm

$\hat{T}_3$: LC pseudo-t-norm

(b) right-continuous (RC) case

$I = \{[0,a[, [a,e[, [e,1]\}$

$\tilde{T}_1$: RC pseudo-t-subnorm

$\tilde{T}_2$: RC pseudo-t-subnorm

$\hat{S}_3$: RC pseudo-t-conorm

Fig. 1. Examples of and-like pseudo-uninorms $H(x,y)$ ($n = 3$, e: unit element)

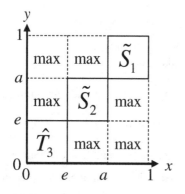

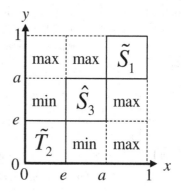

(a) left-continuous (LC) case

$I = \{[0,e],]e,a],]a,1]\}$

$\tilde{S}_1$: LC pseudo-t-subconorm

$\tilde{S}_2$: LC pseudo-t-subconorm

$\hat{T}_3$: LC pseudo-t-norm

(b) right-continuous (RC) case

$I = \{[0,e[, [e,a[, [a,1]\}$

$\tilde{S}_1$: RC pseudo-t-subconorm

$\tilde{T}_2$: RC pseudo-t-subnorm

$\hat{S}_3$: RC pseudo-t-conorm

Fig. 2. Examples of or-like pseudo-uninorms $H(x,y)$ ($n = 3$, e: unit element)

Fig. 2 (a) illustrates a case of left-continuous or-like pseudo-uninorms, where $n = 3$, $\sigma = \begin{pmatrix} 1 & 2 & 3 \\ 3 & 2 & 1 \end{pmatrix} = \begin{pmatrix} 3 & 2 & 1 \\ 1 & 2 & 3 \end{pmatrix}$ and $I = \{[0,e], \,]e,a], \,]a,1]\}$. All summands $\hat{T}_3$, $\tilde{S}_2$ and $\tilde{S}_1$ are associative, monotone increasing and left-continuous. Since $\sigma(3) = 1$ and $\sigma(1) = 3$, Th. 5(ii) is applicable. Also, Th. 3(i) is applicable because $\hat{T}_3$ is a pseudo-t-norm and $I_3 = [0,e]$ is right-closed, thus $e = \max I_3$ is the unit element.

Fig. 2 (b) illustrates a case of right-continuous or-like pseudo-uninorms, where $n = 3$, $\sigma = \begin{pmatrix} 1 & 2 & 3 \\ 3 & 1 & 2 \end{pmatrix} = \begin{pmatrix} 2 & 3 & 1 \\ 1 & 2 & 3 \end{pmatrix}$ and $I = \{[0,e[, \, [e,a[, \, [a,1]\}$. All summands $\tilde{T}_2$, $\hat{S}_3$ and $\tilde{S}_1$ are associative, monotone increasing and right-continuous. Since $\sigma(2) = 1$ and $\sigma(1) = 3$, Th. 5(ii) is applicable. Also, Th. 3(ii) is applicable because $\hat{S}_3$ is a pseudo-t-conorm and $I_3 = [e,a)$ is left-closed, thus $e = \min I_3$ is the unit element.

6   Concluding Remarks

In this paper, the authors proposed a general method to construct various fuzzy logical connectives on $[0,1]$ by the ordinal sum scheme which generates a new semigroup from a system of semigroups. Through our proposed method, we can generate various fuzzy logical connectives as t-norms, t-conorms, uninorms, and also non- commutative ones such as pseudo-t-norms, pseudo-t-conorms, pseudo-uninorms, by combining the conditions corresponding to the required properties, and by choosing adequate summands from already-known operations.

References

1. Clifford, A.: Naturally totally ordered commutative semigroups. Amer. J. Math. **76**, 631–646 (1954)
2. Clifford, A.: Totally ordered commutative semigroups. Bull. Amer. Math. Soc. **64**, 305–316 (1958)
3. Climescu, A.C.: Sur l'équation fonctionelle de l'associativité. Bull. École. Polytech. Jassy. **1**, 1–16 (1946)
4. De Baets, B., Mesiar, R.: Ordinal sums of aggregation operators. In: Bouchon-Meunier, B., Gutierrez-Rios, J., Magdalena, L., Yager, R. (eds.) Technologies for Constructing Intelligent Systems: Tools, pp. 137–147. Springer-Verlag, Heidelberg (2002)
5. Flondor, P., Georgescu, G., Iorgulescu, A.: Pseudo-t-norms and pseudo-BL algebras. Soft Computing **5**, 355–371 (2001)
6. Fodor, J.C., Yager, R.R., Rybalov, A.: Structure of uninorms. Int. J. of Uncertainty, Fuzziness and Knowledge-Based Systems **5**, 411–427 (1997)

7. Frank, M.J.: On the simultaneous associativity of $F(x, y)$ and $x + y - F(x, y)$. Aequationes Math. **19**, 194–224 (1979)
8. Jenei, S.: Generalized ordinal sum theorem and its consequence to the construction of triangular norms. BUSEFAL **80**, 52–56 (1999)
9. Jenei, S.: A note of the ordinal sum theorem and its consequence for construction of triangular norms. Fuzzy Sets and Systems **126**, 199–205 (2002)
10. Kawaguchi, M.F., Watari, O., Miyakoshi, M.: Fuzzy logics and substructural logics without exchange. In: Proc. of FUSFLAT-LFA 2005, Barcelona, pp. 973–978 (2005)
11. Klement, E.P., Mesiar, R., Pap, E.: Triangular Norms. Kluwer Academic Pub. (2000)
12. Ling, C.-H.: Representation of associative functions. Pub. Math. Debrecen **12**, 189–212 (1965)
13. Schweizer, B., Sklar, A.: Associative functions and abstract semigroups. Pub. Math. Debrecen **10**, 69–81 (1963)
14. Yager, R.R., Rybalov, A.: Uninorm aggregation operators. Fuzzy Sets and Systems **80**, 111–120 (1996)

Appendix

Proof of Theorem 2

(a) $x \in I_i$ $(\sigma(i) < \sigma(n))$

Let us consider a partition of $[0,1]$ as follows:

$$\bigcup_{1 \le \sigma(j) < \sigma(i)} I_j < I_i < \bigcup_{\sigma(i) < \sigma(j) \le \sigma(n)} I_j < \bigcup_{\substack{\sigma(n) < \sigma(j) \le n, \\ i<j}} I_j < \bigcup_{\substack{\sigma(n) < \sigma(j) \le n, \\ j<i}} I_j .$$

Then we have

$$H(x, y) = \begin{cases} y = \min(x, y) & \text{if } y \in \bigcup_{1 \le \sigma(j) < \sigma(i)} I_j \\[2mm] a_i + (b_i - a_i)H_i\left(\dfrac{x - a_i}{b_i - a_i}, \dfrac{y - a_i}{b_i - a_i}\right) & \text{if } y \in I_i \\[2mm] x = \min(x, y) & \text{if } y \in \bigcup_{\sigma(i) < \sigma(j) \le \sigma(n)} I_j \\[2mm] x = \min(x, y) & \text{if } y \in \bigcup_{\sigma(n) < \sigma(j) \le n, i<j} I_j \\[2mm] y = \max(x, y) & \text{if } y \in \bigcup_{\sigma(n) < \sigma(j) \le n, j<i} I_j . \end{cases}$$

Since H_i satisfies (Bmin) and (M), $\min(x, a_i) \le a_i \le H(x, y) \le \min(x, b_i)$ for any $y \in I_i$. Besides min, max and H_i satisfy (M). Thus, $H(x, y)$ is monotone increasing w.r.t. $y \in [0,1]$.

(b) $x \in I_n$

Let us consider a partition of $[0,1]$ as follows:

$$\bigcup_{1 \le \sigma(j) < \sigma(n)} I_j < I_n < \bigcup_{\sigma(n) < \sigma(j) \le n} I_j .$$

Then we have

$$H(x,y) = \begin{cases} y = \min(x,y) & \text{if } y \in \bigcup_{1 \le \sigma(j) < \sigma(n)} I_j \\[2ex] a_n + (b_n - a_n) H_n \left(\dfrac{x - a_n}{b_n - a_n}, \dfrac{y - a_n}{b_n - a_n} \right) & \text{if } y \in I_n \\[2ex] y = \max(x,y) & \text{if } y \in \bigcup_{\sigma(n) < \sigma(j) \le n} I_j . \end{cases}$$

Since $\min(x,a_n) \le a_n \le H(x,y) \le b_n \le \max(x,b_n)$ for any $y \in I_n$ and min, max and H_n satisfy (M), $H(x,y)$ is monotone increasing w.r.t. $y \in [0,1]$.

(c) $x \in I_i \quad (\sigma(n) < \sigma(i))$

Let us consider a partition of $[0,1]$ as follows:

$$\bigcup_{\substack{1 \le \sigma(j) < \sigma(n), \\ j < i}} I_j < \bigcup_{\substack{1 \le \sigma(j) < \sigma(n), \\ i < j}} I_j < \bigcup_{\sigma(n) \le \sigma(j) < \sigma(i)} I_j < I_i < \bigcup_{\sigma(i) < \sigma(j) \le n} I_j .$$

Then we have

$$H(x,y) = \begin{cases} y = \min(x,y) & \text{if } y \in \bigcup_{\substack{1 \le \sigma(j) < \sigma(n), \\ j < i}} I_j \\[2ex] x = \max(x,y) & \text{if } y \in \bigcup_{\substack{1 \le \sigma(j) < \sigma(n), \\ i < j}} I_j \\[2ex] x = \max(x,y) & \text{if } y \in \bigcup_{\sigma(n) \le \sigma(j) < \sigma(i)} I_j \\[2ex] a_i + (b_i - a_i) H_i \left(\dfrac{x - a_i}{b_i - a_i}, \dfrac{y - a_i}{b_i - a_i} \right) & \text{if } y \in I_i \\[2ex] y = \max(x,y) & \text{if } y \in \bigcup_{\sigma(i) < \sigma(j) \le n} I_j . \end{cases}$$

Since H_i satisfies (Bmax) and (M), $\max(x,a_i) \le H(x,y) \le b_i \le \max(x,b_i)$ for any $y \in I_i$. Besides min, max and H_i satisfy (M). Thus, $H(x,y)$ is monotone increasing w.r.t. $y \in [0,1]$.

From (a), (b) and (c), for any $x \in [0,1]$, $H(x,y)$ is monotone increasing w.r.t. $y \in [0,1]$. The similar discussion is valid for the case w.r.t. x. Therefore, H is monotone increasing. □

Cognitively Stable Generalized Nash Equilibrium in Static Games with Unawareness

Yasuo Sasaki[✉]

School of Knowledge Science, Japan Advanced Institute of Science and Technology,
1-1, Asahidai, Nomi, Ishikawa, Japan
sasaki@jaist.ac.jp

Abstract. In game theory, recently models and solution concepts of games with unawareness have been developed. This paper focuses on static games with unawareness and points out a conceptual problem of an existing equilibrium concept called generalized Nash equilibrium. Some generalized Nash equilibria can be cognitively unstable in the sense that, once such an equilibrium is played, some agent may feel that the outcome is unexpected one at some level of someone's perception hierarchy. This may lead to change in the agent's perception and thus her behavior. Based on the observation, we characterize a class of generalized Nash equilibrium that satisfies cognitive stability so that it can avoid such a problem. Then we discuss relationships between cognitively sable generalized Nash equilibrium and Nash equilibrium of the objective game, that is, how unawareness can or cannot change the equilibrium convention.

Keywords: Game theory · Unawareness · Generalized Nash equilibrium · Perception hierarchy · Cognitive stability

1 Introduction

Standard game theory assumes that every agent (decision maker) is aware of all the components of the game such as agents, actions and possible states of the nature. On the other hand, recently models and solution concepts of games with unawareness have been developed. (For a historical survey, see [11].) Focusing on static (normal-form) games with unawareness, this paper characterizes a particular class of existing solution concept called generalized Nash equilibrium based on cognitive aspects of the agents and examines its properties.

Unawareness is distinguished from uncertainty (in the sense of standard incomplete information in game theory [4], or traditional Knight's definition of risk) in that under uncertainty the agent conceives all the relevant states to

Y. Sasaki—The author thanks two anonymous referees for helpful comments on this paper. This research is supported by KAKENHI Grant-in-Aid for Young Scientists (B) 15K16292.

V.-N. Huynh et al. (Eds.): IUKM 2015, LNAI 9376, pp. 54–64, 2015.
DOI: 10.1007/978-3-319-25135-6_7

which she can assign probabilities, while under unawareness she cannot even conceive all the states [1,10]. In strategic situations called games, the problem becomes more complex because of asymmetric unawareness of the agents. This includes not only the possibility that an agent may be unaware of something but also the possibility that an agent believes another agent may be unaware of something and so on. Thus we need to take into account perception hierarchies of the agents in order to deal with these situations, i.e. an agent's view about the game, an agent's view about the others' views about the game, and so on. Several models and solution concepts of games with unawareness have been proposed to deal with such interactive situations [2,3,5,8].

Among these, [3] defined generalized Nash equilibrium for games with unawareness as a generalization of Nash equilibrium, the central solution concept of game theory. Roughly speaking, in a generalized Nash equilibrium, in any games someone believes to be the true game in someone's perception hierarchy, the agent who perceives the game takes her best response there given the others' actions chosen in the game according to the equilibrium (precisely, see 2.2).

In this paper, we point out a conceptual problem of the equilibrium concept, that is, some generalized Nash equilibria can be cognitively unstable in the sense that, once such an equilibrium is played, some agent may feel that the outcome is unexpected one at some level of someone's perception hierarchy. This may lead to change in the agent's perception and thus her behavior. (We will state the problem more clearly in 3.1.) If this is the case, it becomes hard to be motivated with the standard argument about interpretation of equilibrium concepts: Nash equilibrium is usually motivated as a convention[1]. Based on the observation, we characterize a class of generalized Nash equilibrium that satisfies cognitive stability so that it can avoid such a problem. Then we examine its properties particularly in terms of how unawareness can or cannot change the equilibrium convention by discussing relationships between cognitively sable generalized Nash equilibrium and Nash equilibrium of the objective game (which is interpreted as the "true" game that would have been played if there had not existed any kind of unawareness).

Following this introduction, Section 2 introduces static games with unawareness and generalized Nash equilibrium. Then, in Section 3, we define cognitive stability of the equilibrium concept and examine its properties. Finally we add conclusion in Section 4. Proofs are presented in the appendix.

---

[1] Therefore [11] pointed out, "equilibrium notions in strategic situations with unawareness may make sense only in special situations such as when players' awareness along the equilibrium path never changes, or when becoming aware also implies magically also mutual knowledge of the new equilibrium convention." Indeed [3] also wrote, "While we think that an understanding of these generalized Nash equilibrium will be critical to understanding solution concepts in games with awareness ... we are not convinced that (generalized) Nash equilibrium is necessarily the "right" solution concept ... this still leaves open the question of what is the "right" solution concept."

2 Model

2.1 Static Games with Unawareness

We define static games with unawareness based on [9], which is essentially a restriction of dynamic (extensive-form) games with unawareness in [3] to static cases. Unlike [9], we do not consider unawareness of agents for simplicity. In other words, the agent set is assumed to be common knowledge among the agents[2]. In addition, we do not take into account the possibility that an agent may be uncertain about the others' awareness levels. An example of a game with unawareness will be illustrated in the end of 2.2.

Definition 1. $\Gamma^U = (\mathcal{G}, \mathcal{F})$ is a *static game with unawareness*, where:

- $\mathcal{G} = \{G^k\}_{k=0,\ldots,n}$ is a set of static games, where G^0 is the objective game while $G^1, \ldots, G^n$ are subjective games. For every k, $G^k = (N, A^k, u^k)$, where:
  - N is a finite set of the agents, which is common in every $G^k \in \mathcal{G}$.
  - $A^k = \times_{i \in N} A_i^k$, where A_i^k is a finite set of agent i's actions.
  - $u^k = (u_i^k)_{i \in N}$, where $u_i^k : A^k \to \Re$ is agent i's utility function.
- $\mathcal{F} = (f_i)_{i \in N}$ is a collection of awareness correspondences for each agent. For any $i \in N$, $f_i : \mathcal{G} \to \mathcal{G}$.

To define a static game with unawareness, we need a collection of static games, $G^0, \ldots, G^n$. Among them G^0 is the objective game, that is, the game perceived by the modeler, or the "true" game. Since we assume that, by the nature of unawareness, the agents may perceive different games, they may believe the others may perceive different games, and so on, we need other games, $G^1, \ldots, G^n$, in order to describe such agents' views, which are called subjective games. (n should be finite in order to keep the model tractable, but we allow it to be infinite.) The collection of these games, $\mathcal{G}$, may contain two or more games which has the same structure. To avoid confusion, in this paper, for any $G^k, G^l \in \mathcal{G}$, we write $G^k = G^l$ if and only if $k = l$. On the other hand, we write $G^k \simeq G^l$ if and only if the two games has the same components, that is, $A^k = A^l$ and $u_i^k(a) = u_i^l(a)$ for any agent i and outcome a. If this is the case, we say G^k and G^l are equivalent.

The agents' perceptions are described with awareness correspondences. Agent i's awareness correspondence, f_i, is a mapping from a static game to another (or possibly the same) game. When $f_i(G^k) = G^l$, this is interpreted as, "An agent who perceives G^k (or the modeler if $k = 0$) believes agent i perceives G^l". We say G^l is reachable from G^k if $G^k = G^l$ or there exist compositions of awareness correspondences that can map G^k to G^l, formally $f_i \circ f_j \circ \cdots \circ f_x(G^k) = G^l$ for some $i, j, \ldots, x \in N$. We denote $G^k \succeq G^l$ when G^l is reachable from G^k.

---

[2] This is because of the difficulty in determining utility functions under unawareness of agents. That is, the condition about relations of the agents' utility functions among static games, which we will define as C2 below, becomes not straightforward in such cases.

In a game with unawareness, an agent may not know all the games in $\mathcal{G}$ and all the mappings of the awareness correspondences in $\mathcal{F}$. In general, we assume that, when $f_i(G^k) = G^l$, every relevant perception of agent i there is all the games reachable from G^l and all the mappings by the awareness correspondences defined on such games. This can be captured equivalently by an infinite hierarchy of static games, $\mathcal{V}_i(G^l) = (v_i^1(G^l), v_i^2(G^l), ...)$, with the interpretation that $v_i^n(G^l)$ is i's n-th order perception at G^l. Then we call $\mathcal{V}_i(G^l)$ the whole view of agent i at G^l. It is defined for any positive integer n inductively as follows: $v_i^1(G^l) = G^l$, $v_i^2(G^l) = (f_j(G^l))_{j \in N_{-i}}$, $v_i^3(G^l) = (v_j^2(f_j(G^l)))_{j \in N_{-i}}$, ..., $v_i^n(G^l) = (v_j^{n-1}(f_j(G^l)))_{j \in N_{-i}}$, ... In this way, awareness correspondences can deal with an agent's any higher-order perceptions. Note that, in G^l, i's whole view can be defined only when there exists $G^k \in \mathcal{G}$ such that $f_i(G^k) = G^l$. That is, for agent i, let $\mathcal{G}_i = \{G^l \in \mathcal{G}|$ for some $G^k \in \mathcal{G}, f_i(G^k) = G^l\}$, and then it can be defined at each game in $\mathcal{G}_i$. Intuitively, $\mathcal{G}_i$ consists of static games such that i views as the true game at some point in the model. We say $\mathcal{V}_i(G^k)$ and $\mathcal{V}_i(G^l)$ are equivalent when $\mathcal{V}_i(G^k)$ can be reformulated as $\mathcal{V}_i(G^l)$ by replacing appropriately each game in $\mathcal{V}_i(G^k)$ with some equivalent game, and vice versa. If this is the case, we also denote $\mathcal{V}_i(G^k) \simeq \mathcal{V}_i(G^l)$. We say i believes G^k is common knowledge if and only if every game in $\mathcal{V}_i(G^k)$ is equivalent to G^k. If this is the case, we also simply say G^k is believed to be common knowledge

In this study, we assume the following conditions C1-C4 with respect to the structure of static games with unawareness so that the model works and makes sense.

C1. For any $G^k, G^l \in \mathcal{G}$, if $G^k \succeq G^l$, then $A_i^k \supseteq A_i^l$ for any $i \in N$.
C2. For any $a \in A^0$, in any $G^k \in \mathcal{G}$, if $a \in A^k$, then $u_i^k(a) = u_i^0(a)$ for any $i \in N$.
C3. For any $G^k \in \mathcal{G}$ and $i \in N$, if $f_i(G^k) = G^l$, then $f_i(G^l) = G^l$.
C4. For any $i \in N$ and $G^k, G^l \in \mathcal{G}$, if $\mathcal{V}_i(G^k) \simeq \mathcal{V}_i(G^l)$, then $G^k = G^l$.

C1 refers to a natural property of unawareness of actions. It states that if G^l is reachable from G^k, then an agent who perceives G^k is aware of all the actions perceived by some agent who believes the true game is G^l, while the converse may not hold true. Thus, for example, agent i's view about agent j's view cannot contain any action of which agent i is unaware. C2 states that whenever an agent is aware of an outcome, every agent's utility there is same as that in the objective game. C3 requires that, in a game reached by a mapping of an agent's awareness corresponding, it maps the game into the game itself. This is assumed to avoid having two different games to describe agent i's view and i's view about i's view. The condition also makes it reasonable to define an agent's first order perception in the way above. C4 requires that an agent cannot possess equivalent whole views at any two different games. Without this condition, equilibria may depend on how to formulate the situation even if the modeler would like to analyze the same situation [9].

2.2   Generalized Nash Equilibrium

We introduce generalized Nash equilibrium [3] as follows. The formulation here follows [9].

We first define local actions to describe each agent's choice at each game in $\mathcal{G}$. For each $i \in N$, they are defined only in games in $\mathcal{G}_i$. (Recall that these are the games i believes to be the true game at some point in the model.) In order to deal with mixed actions, denote the set of the probability distributions on A_i^k by ΔA_i^k, and let $\Delta A^k = \times_{i \in N} \Delta A_i^k$. Thus $\delta_i \in \Delta A_i^k$ is i's mixed action in G^k. Then i's local action $\sigma_{(i,k)}$ describes her (possibly stochastic) choice in $G^k \in \mathcal{G}_i$.

Definition 2. In a static game with unawareness $\Gamma^U = (\mathcal{G}, \mathcal{F})$, for every $i \in N$ and $G^k \in \mathcal{G}_i$, $\sigma_{(i,k)} \in \Delta A_i^k$ is called agent i's *local action* in G^k.

Let σ_i be a combination of agent i's local actions in all the games in $\mathcal{G}_i$ and denote the set of such combinations by Σ_i, i.e. $\sigma_i \in \Sigma_i$. Although σ_i looks like a contingency plan of agent i which specifies her choices for every possible game, this should be interpreted in a different way. Rather, $\sigma_{(i,k)}$ is considered as her choice in G^k when at some point in someone's whole view she is viewed as believing G^k is the true game. (Thus agent i herself may not conceive of such a local action.) σ_i is just the collection of such her choices. Let us denote $\Sigma = \times_{i \in N} \Sigma_i$ and call $\sigma \in \Sigma$ a generalized action profile. By the nature of unawareness, Σ may not be common knowledge. For convenience, when $\sigma_{(i,k)}$ is a pure action that assigns probability one to $a_i \in A_i^k$, we may simply write $\sigma_{(i,k)} = a_i$.

Then let $Eu_i^k(\sigma)$ be agent i's expected utility in $G^k \in \mathcal{G}_i$ when $\sigma \in \Sigma$ is used. Only the actions (in σ) actually taken in G^k are needed to calculate this. Here, i takes $\sigma_{(i,k)}$ but another agent j's choice may not be given as $\sigma_{(j,k)}$ if $f_j(G^k) \neq G^k$. Generally, each agent $j (\neq i)$ uses $\sigma_{(j,l)}$ when $f_j(G^k) = G^l$. (Note that this makes sense due to C1.) Then $Eu_i^k(\sigma)$ can be calculated based on u_i^k given that each agent uses such an action in G^k. Also let $\sigma_{-(i,k)}$ be all the local actions in $\sigma \in \Sigma$ other than $\sigma_{(i,k)}$. Then the equilibrium concept is defined as follows.

Definition 3. In a static game with unawareness $\Gamma^U = (\mathcal{G}, \mathcal{F})$, $\sigma^* \in \Sigma$ is a *generalized Nash equilibrium* if and only if for every $i \in N$, $G^k \in \mathcal{G}_i$ and $\sigma_{(i,k)} \in \Delta A_i^k$, $Eu_i^k(\sigma^*) \geq Eu_i^k(\sigma_{(i,k)}, \sigma_{-(i,k)}^*)$.

A generalized Nash equilibrium is such a generalized action profile σ^* that, for every agent i and game G^k, if i believes G^k is the true game, then, in G^k, her local action is a best response to the others' choices specified by σ^*. (For more detail about its interpretation, see [3] and [9].)

For agent i's action sets in two different games, A_i^k and A_i^l, when $\delta_i \in \Delta A_i^k$ and $\delta_i' \in \Delta A_i^l$, let us denote $\delta_i \equiv \delta_i'$ if and only if δ_i and δ_i' have common supports and moreover probabilities on them are all same. Usually the modeler's interest is in analyzing which outcome can be observed according to the equilibrium concept. To capture this, for a generalized Nash equilibrium σ^*, we call $(\delta_i^*)_{i \in N} \in \Delta A^0$ such that, for every $i \in N$, $\delta_i^* \equiv \sigma_{(i,k)}^*$ when $f_i(G^0) = G^k$ its objective outcome. This is interpreted as each agent's choice in the objective game.

Example: (This example is based on [2,9]). Consider a two-agent game played by Alice and Bob as shown in (a) in Fig. 1. Now Alice perceives this game, while she believes that Bob is unaware of one of her actions, a_3, and therefore views another game (b). She also believes Bob believes the game (b) is common knowledge. On the other hand, Bob actually is aware of a_3 and views the game (a). Furthermore he believes such a belief of Alice, that is, he believes that she believes that he believes the game (b) is common knowledge.

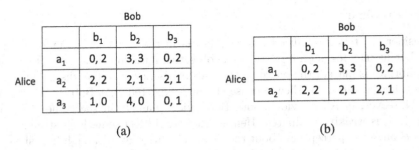

Fig. 1. Example static game with unawareness

In our framework, the situation can be formulated as depicted in Fig. 2. $\mathcal{G} = \{G^0, G^1, G^2\}$. Among them, G^0 and G^1 are same as the game (a) in Fig. 1, while G^2 is same as the game (b). f_A and f_B describe the awareness correspondences of Alice and Bob, respectively. (Henceforth A and B in mathematical symbols indicate Alice and Bob, respectively.) For instance, the arrow from G^0 to G^1 labeled as "f_A" means $f_A(G^0) = G^1$.

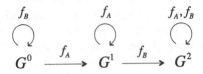

Fig. 2. Formulation of the example

In this game, $\mathcal{G}_A = \{G^1, G^2\}$ and $\mathcal{G}_B = \{G^0, G^2\}$, and there are two pure-action generalized Nash equilibria. In the first equilibrium, which we denote σ_1^*, $(\sigma_{(A,1)}, \sigma_{(A,2)}) = (a_2, a_2)$ and $(\sigma_{(B,0)}, \sigma_{(B,2)}) = (b_1, b_1)$. Thus its objective outcome is (a_2, b_1), which is a Nash equilibrium of G^0. On the other hand, in the second equilibrium, which we denote σ_2^*, $(\sigma_{(A,1)}, \sigma_{(A,2)}) = (a_3, a_1)$ and $(\sigma_{(B,0)}, \sigma_{(B,2)}) = (b_3, b_2)$. Thus its objective outcome is (a_3, b_3), which is not a

Nash equilibrium of G^0. The basic idea to derive a generalized Nash equilibrium is similar to backward induction as follows: Alice and Bob play some Nash equilibrium in G^2, then Alice takes a best response in G^1 to the Bob's choice in the Nash equilibrium, and finally Bob takes a best response in G^0 to the Alice's choice in G^1. Hence if they are expected to play the Pareto-dominant Nash equilibrium in G^2, (a_1, b_2), then σ_2^* is achieved, and as a result the objective outcome, (a_3, b_3), is Pareto-dominated by the objective outcome of σ_1^*, namely (a_2, b_1). Since (a_2, b_1) is the only Nash equilibrium in the objective game, this example illustrates how unawareness can change the equilibrium set.

3 Cognitive Stability

3.1 Problem

Consider the two generalized Nash equilibria in the previous example, σ_1^* and σ_2^*. Let us first focus on σ_2^*, where the objective outcome is (a_3, b_3). Once this equilibrium is played, the outcome gives a surprise, or cognitive dissonance in a psychological term, to Alice because she had expected that Bob would choose b_2. Moreover, in Alice's view about Bob's view, namely in the game of (b) in Fig. 1, b_3 is weakly dominated. Hence, Alice would feel something strange and may change her perception about the game in some way. Although how she can change her view is not clear[3], the update may lead to her behavioral change if they play the game again. Likewise, since Bob had expected that Alice had expected that Bob would take b_2, he would notice that his choice of b_3 may bring such a surprise to her. Therefore he would change his view about her perception after the play, which may also lead to his behavioral change next time. In standard game theory, Nash equilibrium is often motivated with the notion of convention: if a game is played repeatedly and the agents can learn somehow the others' choices, then their choices will be an equilibrium. But, according to the discussion above, this interpretation cannot be applied to this generalized Nash equilibrium.

On the other hand, σ_1^*, where the objective outcome is (a_2, b_1), never brings such cognitive dissonance to the both agents because, in the equilibrium, not only that each agent considers the opponent's choice is just as expected, but also that each agent believes the opponent considers the agent's choice is just as expected and so on. Therefore, in contrast to σ_2^*, it can be motivated with convention like the standard argument above. That is, if they somehow come to play this outcome in their repetitive interactions, they would naturally continue to play it even under the existence of unawareness because the outcome never reveals such misperceptions to the agents.

Hence there is a crucial conceptual difference between the two equilibria in the example game. In general, generalized Nash equilibrium can be categorized

---

[3] This would be related to "reverse Bayesianism," i.e., in the belief update, an initially null event will receive positive weight [6]. It would also be related to belief revision studied in artificial intelligence.

into two classes, one that can give cognitive dissonance in the sense above to some agent at some level of someone's perception hierarchy and the other that can never bring such a problem. Thus only the latter can persist as an equilibrium convention. Given the standard argument to motivate an equilibrium concept, we need to characterize a class of generalized Nash equilibrium which excludes the former and includes only the latter in order to deal with the intuition. Next we define cognitive stability of generalized Nash equilibrium based on this motivation.

3.2 Definition

Precisely, cognitive stability of generalized Nash equilibrium requires that all of every agent's expectation about the others' choices, every agent's expectation about the others' expectations about the others' choices, every agent's expectation about the others' expectations about the others' expectations about the others' choices, and so on are same as the actions taken in an outcome that will be realized, i.e. the objective outcome of the generalized Nash equilibrium. In other words, for every agent i, anyone's expectation about i's choice, anyone's expectation about anyone's expectation about i's choice, and so on are same as i's choice in the objective outcome[4]. Hence we define cognitively stable of generalized Nash equilibrium as follows.

Definition 4. In a static game with unawareness $\Gamma^U = (\mathcal{G}, \mathcal{F})$, let $\sigma^* \in \Sigma$ be a generalized Nash equilibrium. Then it is *cognitively stable* if and only if for every $i \in N$ and $G^k, G^l \in \mathcal{G}_i$, $\sigma^*_{(i,k)} \equiv \sigma^*_{(i,l)}$.

That is, cognitively stable generalized Nash equilibrium is a class of generalized Nash equilibrium in which, for every agent i, the local actions in every game in $\mathcal{G}_i$ are all identical. In the previous example, σ^*_1 is a cognitively stable generalized Nash equilibrium while σ^*_2 is not.

3.3 Properties

We investigate some properties of cognitively stable generalized Nash equilibrium. In particular, when we consider it as such a class of equilibrium that can be motivated as an equilibrium convention, let us examine how the existence of unawareness can or cannot change the set of possible equilibrium conventions of the game. We here consider relationships of the following three outcomes given by different equilibrium concepts in a static game with unawareness:

- $E \subseteq \Delta A^0$: the set of Nash equilibria in G^0.
- $E_g \subseteq \Delta A^0$: the set of objective outcomes of generalized Nash equilibria.
- $E_c \subseteq \Delta A^0$: the set of objective outcomes of cognitively stable generalized Nash equilibria.

---

[4] This is same as simple belief hierarchies discussed in [7].

Based on the standard interpretation of Nash equilibrium, E can be considered as the set of outcomes that can be an equilibrium convention under non-existence of unawareness. In contrast, E_c is interpreted as such a set under existence of unawareness. To see how cognitive stability can provide unique insights, we also compare it to E_g. In general, their inclusive relations can be depicted as an Euler diagram of Fig. 3.

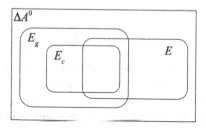

Fig. 3. Relationships among the equilibria (general)

We here present two propositions that shows, under specific condition, we can have unique inclusive relations.

Proposition 1. In a static game with unawareness $\Gamma^U = (\mathcal{G}, \mathcal{F})$ in which, for every $i \in N$, $A_i^k = A_i^0$ when $f_i(G^0) = G^k$, $E_c \subseteq E$.

Proposition 1 claims that, when every agent is aware of all of her own actions (i.e., her action set in the objective game), an objective outcome of cognitively stable generalized Nash equilibrium is always a Nash equilibrium of the objective game. Thus this can be interpreted as a sufficient condition that existence of unawareness cannot make a new equilibrium convention. The inclusive relation of this case is illustrated in (a) of Fig. 4. (Compare this to Fig. 3.)

Then, to describe the next proposition, we introduce a particular class of E. Let $E' = \{\delta \in E|$ for every $i \in N$ and $G^k \in \mathcal{G}_i$, $\mathrm{supp}(\delta) \subseteq A^k$ in $G^k\}$, where $\mathrm{supp}(\delta) \subseteq A^k$ is the set of combinations of pure actions in G^k that can be played with positive probability under a given mixed outcome $\delta \in \Delta A^k$. That is, $\delta \in E'$ is a Nash equilibrium of the objective game such that every agent is aware of existence of such an outcome, every agent believes every agent is aware of it, and so on. In other words, it is a Nash equilibrium of the objective game such that the existence of the outcome itself is common knowledge.

Proposition 2. In a static game with unawareness $\Gamma^U = (\mathcal{G}, \mathcal{F})$, $E_c \supseteq E'$.

Proposition 2 claims that if the objective game has a Nash equilibrium included in E', then it is an objective outcome of cognitively stable generalized Nash equilibrium. Therefore such an outcome can always be a candidate of

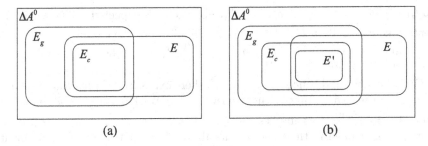

Fig. 4. Relationships among the equilibria (a: Proposition 1; b: Proposition 2)

an equilibrium convention under unawareness. This does not hold generally for any outcome in E. See (b) in Fig. 4 for the inclusive relation.

4 Conclusion

In this paper, we have defined cognitive stability of generalized Nash equilibrium in static games with unawareness based on the observation that some generalized Nash equilibria can bring cognitive dissonance and, in such cases, they cannot be motivated as equilibrium conventions. Furthermore, we have shown some results about its property from a viewpoint of how unawareness can or cannot change the equilibrium convention.

Cognitive stability adds unique insights to the notion of generalized Nash equilibrium by taking into account cognitive aspects of the agents, i.e., how they can feel when they observe a particular outcome is played. We view the analysis in 3.3 is just the first step and we need to characterize the concept in a more rigorous way. Furthermore, we admit that our definition of cognitive stability might not be the "right" definition of it but just its sufficient condition, that is, it may be too strong to characterize an equilibrium that will never bring cognitive dissonance in our sense. We still need to investigate this kind of stability.

Moreover, while we have focused on static games in this paper for simplicity, it is possible to apply the similar discussion to dynamic games with unawareness. This would be another future topic.

Appendix

Proof of Proposition 1

In Γ^U in which, for every $i \in N$, $A_i^k = A_i^0$ when $f_i(G^0) = G^k$, suppose there exists $\delta^* = (\delta_i^*, \delta_{-i}^*) \in E_c \setminus E$. For every $i \in N$, let $f_i(0) \in \{0, ..., n\}$ such that $G^{f_i(0)} = f_i(G^0)$. Since $\delta^* \in E_c$, there exists a cognitively stable generalized Nash equilibrium $\sigma^* \in \Sigma$, where, for any $i \in N$, $\sigma^*_{(i,f_i(0))} \equiv \delta_i^*$, and thus, for any $j \in N_{-i}$, $\sigma^*_{(j,f_j \circ f_i(0))} \equiv \delta_j^*$, where $f_j \circ f_i(0) \in \{0, ..., n\}$ such that $G^{f_j \circ f_i(0)} =$

$f_j \circ f_i(G^0)$, since $\sigma^*_{(j,f_j(0))} \equiv \delta^*_j$. Let $\hat{u}^k_i(\delta)$ be agent i's expected utility when the agents take $\delta \in \Delta A^k$ in G^k. Then, by the definition of the equilibrium, for any $i \in N$ and $\delta_i \in \Delta A^{f_i(0)}_i$, $\hat{u}^{f_i(0)}_i(\delta^*_i, \delta'_{-i}) \geq \hat{u}^{f_i(0)}_i(\delta_i, \delta'_{-i})$, where $\delta'_{-i} = (\delta'_j)_{j \in N_{-i}}$ such that $\delta'_j \equiv \delta^*_j$ for any $j \in N_{-i}$.

On the other hand, since $\delta^* \notin E$, in G^0, there exists $k \in N$ such that δ^*_k is not a best response to δ^*_{-k}, that is, there exists $\delta_k \in \Delta A^0_k$, $\hat{u}^0_k(\delta_k, \delta^*_{-k}) > \hat{u}^0_k(\delta^*_k, \delta^*_{-k})$. Since $A^0_k = A^{f_k(0)}_k$, this implies, in $f_k(G^0)$, $\hat{u}^{f_k(0)}_k(\delta_k, \delta'_{-k}) > \hat{u}^{f_k(0)}_k(\delta^*_k, \delta'_{-k})$, where δ'_{-k} is defined in the same way as above. This contradicts the statement above. Hence $E_c \setminus E = \phi$, that is, $E_c \subseteq E$. $\square$

Proof of Proposition 2

Consider $\delta^* = (\delta^*_i, \delta^*_{-i}) \in E'$. Then, in every $G^k \in \mathcal{G}_i$ for any $i \in N$, $(\delta'_i, \delta'_{-i}) \in \Delta A^k$ is a Nash equilibrium, where $\delta'_i \equiv \delta^*_i$ for any $i \in N$, that is, for any $i \in N$ and $\delta_i \in \Delta A^k_i$, $\hat{u}^k_i(\delta'_i, \delta'_{-i}) \geq \hat{u}^k_i(\delta_i, \delta'_{-i})$, where $\hat{u}^k_i$ is defined in the same way as in the proof of Proposition 1.

Now consider a generalized action profile $\sigma^* \in \Sigma$ such that, for every $i \in N$ and $G^k \in \mathcal{G}_i$, $\sigma^*_{(i,k)} \equiv \delta^*_i$. (We can always define such a generalized action profile based on δ^* by the assumption of $\delta^* \in E'$.) Then, based on the discussion above, for every $i \in N$, $G^k \in \mathcal{G}_i$ and $\sigma_{(i,k)} \in \Delta A^k_i$, $Eu^k_i(\sigma^*) \geq Eu^k_i(\sigma_{(i,k)}, \sigma^*_{-(i,k)})$. This means that σ^* is a cognitively stable generalized Nash equilibrium. Therefore, for every $i \in N$, $\sigma^*_{(i,k)} \equiv \delta^*_i$ when $f_i(G^0) = G^k$, which means δ^* is the objective outcome of σ^*. Hence $E_c \supseteq E'$. $\square$

References

1. Dekel, E., Lipman, B.L., Rustichini, A.: Standard state-space models preclude unawareness. Econometrica **66**(1), 159–173 (1998)
2. Feinberg, Y.: Games with unawareness. Stanford University, Graduate School of Business discussion papers (2012)
3. Halpern, J.Y., Rêgo, L.C.: Extensive games with possibly unaware players. Mathematical Social Sciences **70**, 42–58 (2014)
4. Harsanyi, J.C.: Games with incomplete information played by Bayesian players. Management Science **14**, 159–182, 320–334, 486–502 (1967)
5. Heifetz, A., Meier, M., Schipper, B.C.: Dynamic unawareness and rationalizable behavior. Games and Economic Behavior **81**, 50–68 (2013)
6. Karni, E., Vierø, M.L.: "Reverse Bayesianism": A choice-based theory of growing awareness. The American Economic Review **103**(7), 2790–2810 (2013)
7. Perea, A.: Epistemic Game Theory: Reasoning and Choice. Cambridge University Press (2012)
8. Sasaki, Y.: Subjective rationalizability in hypergames. Advances in Decision Sciences (2014). doi:10.1155/2014/263615
9. Sasaki, Y.: Bayesian representation of static games with unawareness. unpublished manuscript (2015)
10. Schipper, B.C.: Awareness-dependent subjective expected utility. International Journal of Game Theory **42**(3), 725–753 (2013)
11. Schipper, B.C.: Unawareness–a gentle introduction to both the literature and the special issue. Mathematical Social Sciences **70**, 1–9 (2014)

Maximum Lower Bound Estimation of Fuzzy Priority Weights from a Crisp Comparison Matrix

Tomoe Entani[1]([✉]) and Masahiro Inuiguchi[2]

[1] Graduate School of Applied Informatics, University of Hyogo,
Kobe, Hyogo 650-0047, Japan
entani@ai.u-hyogo.ac.jp
[2] Graduate School of Engineering Science, Osaka University,
Toyonaka, Osaka 560-8531, Japan

Abstract. In Interval AHP, our uncertain judgments are denoted as interval weights by assuming a comparison as a ratio of the real values in the corresponding interval weights. Based on the same concept as Interval AHP, this study denotes uncertain judgments as fuzzy weights which are the extensions of the interval weights. In order to obtain the interval weight for estimating a fuzzy weight, Interval AHP is modified by focusing on the lower bounds of the interval weights similarly to the viewpoint of belief function in evidence theory. It is reasonable to maximize the lower bound since it represents the weight surely assigned to one of the alternatives. The sum of the lower bounds of all alternatives is considered as a membership value and then the fuzzy weight is estimated. The more consistent comparisons are given as a result of the higher-level sets of fuzzy weights in a decision maker's mind.

Keywords: Interval AHP · Fuzzy weight · Interval weight · Membership function · Uncertainty

1 Introduction

AHP (Analytic Hierarchy Process) is a useful tool to extract a decision maker's preference from his/her intuitive judgments [9]. When a decision maker gives the comparisons of all pairs of alternatives, his/her preferences are obtained as the weights of alternatives. It is easy for a decision maker to give his/her intuitive judgments as pairwise comparisons since s/he focuses on comparing a pair of alternatives without caring for the other alternatives. As a result, an alternative is compared several times and the given comparisons are seldom consistent each other.

The inconsistency of the given comparisons is well-known and discussed a lot in AHP. One of the ways to treat inconsistency is to introduce the consistency index and distinguish whether the given comparisons are too inconsistent [1,8]. On the other hand, Interval AHP [10,11] takes the comparisons possibly into consideration, instead of distinguishing them. It is based on the idea that a

© Springer International Publishing Switzerland 2015
V.-N. Huynh et al. (Eds.): IUKM 2015, LNAI 9376, pp. 65–76, 2015.
DOI: 10.1007/978-3-319-25135-6_8

decision maker does not perceive a precise weight of an alternative but a range of its weight in his/her mind. In Interval AHP, a comparison is considered as a part of the rational decision so that the interval weights are obtained so as to include the given comparisons. In short, AHP and Interval AHP induce plausible and possible preferences from the given judgments, respectively.

Based on the same idea as Interval AHP, this paper assumes that our judgments are uncertain and obtains the possible preferences reflecting the uncertainty. Instead of the interval weights, such uncertain judgments are denoted as the fuzzy weights. Therefore, the method to estimate the fuzzy weights from the given crisp comparisons is proposed. In some fuzzy approaches in AHP [2,4,7], a fuzzy or interval comparison matrix is used. From the viewpoint of uncertainty, several models of Interval AHP have been proposed [6]. This paper modifies Interval AHP by focusing on the lower bound of an interval weight since it represents the weight surely assigned to an alternative similarly to the viewpoint of belief function in evidence theory [5]. The left weight is considered as ignorance since it is common of some alternatives and possible to be assigned to more than two alternatives. The upper bound of an interval weight includes such a weight as possibly assigned to an alternative. It is reasonable to assign the weight to one of the alternatives as much as possible so that the lower bounds are maximized. Then, the sum of the lower bounds of all alternatives is considered as a membership value of a fuzzy weight in a decision maker's mind. The more consistent the given comparisons are, the higher the membership value becomes. The decision maker gives the comparisons based on this certain level sets of the fuzzy weights of alternatives.

The given comparison is represented as the ratio of the weights of the corresponding alternatives. In Interval AHP, the inclusion relation between a comparison and the interval ration of its corresponding interval weights are used to obtain the interval weights. In the proposed model, the relation between a comparison and a ratio of weights is reconsidered. It assumes that the weight of an alternative is estimated by the corresponding comparisons and the weights of the other alternatives. The interval weight of an alternative is estimated by the other alternatives and the relation between the weight and its estimations are used to obtain the interval weights. Since the proposed model is based on the lower bound, the estimations by the lower bounds of the others are used. Such lower bounds of estimations are compared to the upper bound of an interval weight so that the interval weight and its estimations are common.

In order to estimate fuzzy weights from their certain level sets which are interval weights focusing on their lower bounds, it assumes some membership values, i.e., the sums of the lower bounds from 0 to 1, in addition. The relation between an interval weight and its estimations are modified depending on whether the membership value is higher than the certain level or not. In the higher case, the weight is forced to be assigned to one of the alternatives so that a weight and its estimations may not be common and the deficiency is minimized. While, in the lower case, they are always common so that their differences can be minimized. In each case, the interval weights are obtained based on the proposed model as the respective level sets of the fuzzy weight. Then, the fuzzy weight is estimated such interval weights as its representative level sets.

This paper is organized as follows. In the next section, as the modification of Interval AHP, the lower bound based Interval AHP which focuses on the lower bounds is proposed. Then, in section 3, the fuzzy weights are estimated from some representative interval weights by the proposed model as their level sets. Section 4 shows two numerical examples and discusses the results. The last section is the conclusion.

2 Lower Bound Based Interval AHP

A decision maker gives the pairwise comparisons on n alternatives as follows.

$$A = \begin{bmatrix} 1 & \cdots & a_{1n} \\ \vdots & a_{ij} & \vdots \\ a_{n1} & \cdots & 1 \end{bmatrix}, \tag{1}$$

where a_{ij} is his/her intuitive judgment on the importance ratio of alternative i to that of alternative j. The comparisons are identical and reciprocal as $a_{ii} = 1$ and $a_{ij} = 1/a_{ji}$. The comparisons are consistent if and only if

$$a_{ij} = a_{il}a_{lj}, \forall i, j, l. \tag{2}$$

However, (2) is seldom satisfied since an alternative is compared to the other $(n-1)$ alternatives.

In AHP, the weights of alternatives are obtained from (1) by eigenvector method $Aw = \lambda w$, where $w = (w_1, \ldots, w_n)^T$ is the eigenvector corresponding to principal eigenvalue. The weights are normalized such that $\sum_i w_i = 1$. The weight is assigned to one of the alternatives without ignorance so that the plausible preferences are obtained by AHP.

In Interval AHP, the interval weight $W_i = [w_i^L, w_i^R]$ which includes the given comparisons is obtained by the following LP problem [10,11]. It is assumed that the given comparisons are inconsistent since the weights of alternatives in a decision maker's mind are uncertain. A decision maker may use a real value in interval weight W_i in giving comparison a_{ij}, where $j > i$.

$$\begin{aligned} \min \quad & \sum_i (w_i^R - w_i^L), \\ \text{s.t.} \quad & \sum_{i \neq j} w_i^R + w_j^L \geq 1, \forall j, \\ & \sum_{i \neq j} w_i^L + w_j^R \leq 1, \forall j, \\ & \frac{w_i^L}{w_j^R} \leq a_{ij} \leq \frac{w_i^R}{w_j^L}, \forall i, j, j > i, \\ & w_i^L \geq \varepsilon, \forall i, \end{aligned} \tag{3}$$

where the first two kinds of constraints are for the normalization of intervals based on interval probability [3,12]. They are the interval counterparts of the ordinal crisp probability. When the weights are real values as $w_i^R = w_i^L = w_i, \forall i$, two inequalities are replaced into $\sum_i w_i = 1$. The redundancy of the intervals to make the sum of any real values in the intervals be 1 is excluded. For instance, the 1st inequality for j requires w_j^L not to be too small. The next inequalities for a_{ij} are the inclusion constraints. They require the obtained interval weights to include the given comparisons as

$$a_{ij} \in \frac{W_i}{W_j} = \frac{[w_i^L, w_i^R]}{[w_j^L, w_j^R]} = \left[\frac{w_i^L}{w_j^R}, \frac{w_i^R}{w_j^L} \right], \tag{4}$$

where the fraction of intervals is defined as its maximum range. By minimizing the widths of the interval weights, both bounds in the right side of (4) become the closest to the give comparison in the left side. In other words, the primal objective is to minimize uncertainty of the interval weight. If the comparisons are perfectly consistent as in (2), the weights are obtained as real values $w_i^R = w_i^L, \forall i$ and they equal to those by eigenvector or geometric mean method in AHP. On the other hand, the more inconsistent the given comparisons are, the wider the obtained interval weights become. The lower bound of the interval weight is considered as the weight surely assigned to one of the alternatives. While, its upper bound includes the possibly assigned weight in addition and such weight is a common weight of some alternatives. As the surely assigned weight decreases, the ignorance which is the possibly assigned weight increases.

In the same concept as Interval AHP, this paper assumes that our judgments are often uncertain and then they are represented as fuzzy weights of alternatives, instead of interval weights. The fuzzy weight can be considered as a set of interval weights. In order to estimate a fuzzy weight, some representative interval weights are used. The core interval weights, based on which a decision maker gives the comparisons, are obtained as follows. We revisit Interval AHP by focusing on the lower bound of interval weight. It is reasonable to focus more on the weight surely assigned to one of the alternatives than the weight assigned to more than two alternatives. Interval AHP by (3) is modified so as to be suitable for fuzzy weight estimation and we name the proposed model lower bound based Interval AHP. In Interval AHP, the inclusion relation as in (4) is considered based on the relation between the given comparison and the corresponding weights as $a_{ij} = \frac{w_i}{w_j}$. In the proposed lower bound based Interval AHP, based on the same relation, the weight of an alternative is estimated by the weight of the other alternative as $w_i = a_{ij} w_j$. In case of the crisp weights, there are $n-1$ estimations of the weight of alternative i as $w_i' = a_{ij} w_j, \forall j \neq i$. When the weights are extended to interval $W_i = [w_i^L, w_i^R]$, its estimations are intervals $W_i' = [a_{ij} w_j^L, a_{ij} w_j^R], \forall j \neq i$. Since the proposed model is based on the lower bound of the interval weight, the estimations by the lower bounds $a_{ij} w_j^L, \forall j \neq i$ are used. The relation between the interval weight of alternative i, W_i, and its estimation, W_i', by the other alternative $j \neq i$ is as follows.

$$a_{ij} w_j^L \leq w_i^R, \forall j \neq i, \tag{5}$$

which are satisfied if the interval weight and its estimation have at least a value in common.

By replacing the inclusion relation (4) in (3) into the estimation (5) and maximizing the lower bound w_i^L, the lower bound based Interval AHP is formulated as the following LP problem.

$$\alpha = \max \sum_i w_i^L,$$
$$s.t. \ \sum_{i \neq j} w_i^R + w_j^L \geq 1, \forall j,$$
$$\sum_{i \neq j} w_i^L + w_j^R \leq 1, \forall j, \qquad (6)$$
$$a_{ij} w_j^L \leq w_i^R, \forall i, j, j \neq i,$$

where the optimal objective function value α represents the weight surely assigned to one of the alternatives and the left weight $1 - \alpha$ may be assigned to some alternatives. The lower and upper bounds of the interval weight, w_i^L and w_i^R, represent the weights surely and possibly assigned to alternative i, respectively. In this way, the possible preferences are obtained by lower bound based Interval AHP.

Let us denote the maximum surplus of the upper bound of alternative i from its estimations by the other alternatives $\forall j \neq i$ in (5) as p_i. The surplus should be minimized and (6) is rewritten as follows.

$$\max \left(\sum_i w_i^L - \varepsilon \sum_i p_i \right),$$
$$s.t. \ \sum_{i \neq j} w_i^R + w_j^L \geq 1, \forall j,$$
$$\sum_{i \neq j} w_i^L + w_j^R \leq 1, \forall j, \qquad (7)$$
$$0 \leq w_i^R - a_{ij} w_j^L \leq p_i, \forall i, j, j \neq i,$$

where ε is a small positive value so that the surplus of the interval weight from its estimations is minimized secondarily.

When the given comparisons are perfectly consistent as in (2), the optimal solutions of (7) are $w_i = w_i^R = w_i^L, \forall i$ and then $\alpha = 1$. The more inconsistent the comparisons are, the less α becomes.

3   Estimating Fuzzy Weight

The weight of alternative i in a decision maker's mind is denoted as fuzzy weight $\tilde{W}_i$. Let us denote the membership function of fuzzy weight $\tilde{W}_i$ as $\mu_{\tilde{W}_i}$. In this section, the fuzzy weight is estimated by its α-level set obtained by (7) in the previous section. A decision maker gives the comparisons based on α-level sets of the fuzzy weights in his/her mind. The sum of the weights of all alternatives surely assigned to one of the alternatives represents a membership value of a fuzzy weight. It means that $\alpha = \mu_{\tilde{W}_i}(w_i^L) = \mu_{\tilde{W}_i}(w_i^R), \forall i$, i.e., α-level sets of fuzzy weights $\tilde{W}_i, \forall i$ are intervals $[w_i^L, w_i^R], \forall i$ by (7). They are core interval weights to estimate the fuzzy weights. The fuzzy weight consists of some representative interval weights which are their level sets. As far as $\mu_{\tilde{W}_i}$ is less than $\alpha = \sum_i w_i^L$ by (7), the relation between an interval weight and its estimations satisfy (5), however, in the other cases the relation is not satisfied. Therefore, we assume β_k

and γ_k, where $\alpha = \beta_0 < \beta_1 < \ldots < \beta_m = 1$ and $\alpha = \gamma_0 > \gamma_1 > \ldots > \gamma_l \geq 0$, respectively, for given m and l. As m and l increase, the more precise estimations can be done.

First, let assume $\mu_{\tilde{W}_i} = \beta_k$, which requires that the weight surely assigned to an alternative is more than β_k. Because of $\alpha \leq \beta_k$, they cannot satisfy (5). The maximum deficiency of interval weight W_i from its estimations, q_i, should be minimized.

$$\min \sum_i q_i,$$
$$s.t. \ \beta_k \leq \sum_i w_i^{L\beta_k},$$
$$w_i^{L\beta_{k-1}} \leq w_i^{L\beta_k}, w_i^{R\beta_k} \leq w_i^{R\beta_{k-1}}, \forall i$$
$$\sum_{i \neq j} w_i^{R\beta_k} + w_j^{L\beta_k} \geq 1, \forall j, \qquad (8)$$
$$\sum_{i \neq j} w_i^{L\beta_k} + w_j^{R\beta_k} \leq 1, \forall j,$$
$$-q_i \leq w_i^{R\beta_k} - w_j^{L\beta_k} a_{ij}, \forall i, j, j \neq i,$$
$$q_i \geq 0, \forall i,$$

where $[w_i^{L\beta_k}, w_i^{R\beta_k}], \forall i$ are the variables and $[w_i^{L\beta_{k-1}}, w_i^{R\beta_{k-1}}], \forall i$ are obtained previously. By repeating solving (8) from $k = 1$ to m, sequentially m representative interval weights are obtained. They are higher-level sets of the fuzzy weight than α-level one by (7).

In case of $m = 1$, (8) is reduced to the following problem where $\beta_1 = \mu_{\tilde{W}_i} = 1$ indicates crisp weights as $w_i = w_i^{L\beta_1} = w_i^{R\beta_1}, \forall i$. The weight is surely assigned to one of the alternatives so that there is no ignorance which is possibly assigned weight to an alternative.

$$\min \sum_i q_i,$$
$$s.t. \ \sum_i w_i = 1,$$
$$w_i^L \leq w_i \leq w_i^R, \forall i, \qquad (9)$$
$$-q_i \leq w_i - a_{ij} w_j, \forall i, j, j \neq i,$$
$$q_i \geq 0, \forall i,$$

where $w_i^L = w_i^{L\beta_0}, \forall i$ and $w_i^R = w_i^{R\beta_0}, \forall i$ are the optimal solutions of (7). The weight is forced to be assigned to one of the alternatives without ignorance, instead of allowing some estimations of an alternative to be more than its crisp weight. The crisp weights $w_i, \forall i$ make their estimations be the closest to them and be included in α-level set of the fuzzy weights denoted as interval weights $W_i, \forall i$ by (7).

Next, let assume $\mu_{\tilde{W}_i} = \gamma_k$, which requires that the weight surely assigned to one of the alternatives is at most γ_k. Because of $\alpha \geq \gamma_k$, (5) is satisfied and an interval weight and its estimations are always common. Since the upper bound of the interval weight of an alternative is always more than the estimations by the lower bounds of the others, the constraint of the sum of the lower bounds is added into (7). The problem to obtained interval weights $[w_i^{L\gamma_k}, w_i^{R\gamma_k}], \forall i$, is formulated as follows.

$$\max \left(\sum_i w_i^{L\gamma_k} - \varepsilon \sum_i p_i\right),$$

$$s.t. \ \sum_i w_i^{L\gamma_k} \leq \gamma_k,$$

$$\varepsilon \leq w_i^{L\gamma_k} \leq w_i^{L\gamma_{k-1}}, w_i^{R\gamma_{k-1}} \leq w_i^{R\gamma_k}, \forall i$$

$$\sum_{i \neq j} w_i^{R\gamma_k} + w_j^{L\gamma_k} \geq 1, \forall j, \tag{10}$$

$$\sum_{i \neq j} w_i^{L\gamma_k} + w_j^{R\gamma_k} \leq 1, \forall j,$$

$$0 \leq w_i^{R\gamma_k} - w_j^{L\gamma_k} a_{ij} \leq p_i, \forall i, j, j \neq i,$$

where $p_i, \forall i$ are the surpluses of the weights from their estimations and in the same way as (8), $[w_i^{L\gamma_{k-1}}, w_i^{R\gamma_{k-1}}], \forall i$ are obtained previously. By repeating solving (10) from $k = 1$ to l, sequentially l representative interval weights are obtained and they are lower-level sets of the fuzzy weight than its α-level set.

In case of $\gamma_l = 0$, the lower bounds of 0-level sets of the fuzzy weights are 0 as $w_i^{L\gamma_l} = w_i^{L0} = 0, \forall i$. As a result, the estimations are also 0, as $a_{ji}w_i^{L0} = 0, \forall i$, whatever a_{ij} is. In order not to ignore and to reflect the given comparison a_{ij} to the weight of alternative i, its lower bound is assumed as $w_i^{L\gamma_l} = w_i^{L0} = \varepsilon, \forall i$, where ε is a small positive number so that $\gamma_l = n\varepsilon$. Then, in case of $l = 1$, (10) is reduced to the following problem to obtain the upper bounds $= w_i^{R\gamma_1} = w_i^{R0}, \forall i$.

$$\min \sum_i w_i^{R0} - \varepsilon \sum_i r_i,$$

$$s.t. \ w_i^R \leq w_i^{R0}, \forall i$$

$$\sum_{i \neq j} w_i^{R0} + \varepsilon \geq 1, \forall j, \tag{11}$$

$$\sum_{i \neq j} \varepsilon + w_j^{R0} \leq 1, \forall j,$$

$$0 \leq w_i^{R0} - \varepsilon a_{ij} \leq p_i, \forall i, j, j \neq i,$$

where $w_i^{L\gamma_1} = w_i^{L0}, \forall i$ are replaced into ε and the obtained interval weights $[\varepsilon, w_i^{R0}], \forall i$ should include α-level sets of the fuzzy weights $[w_i^L, w_i^R], \forall i$ by (7). The weight surely assigned to an alternative is reduced to $n\varepsilon \leq \alpha$ and the ignorance is increased to $1 - n\varepsilon$.

For simplicity, assuming $m = l = 1$, the membership function of fuzzy weight $\tilde{W}_i$ is illustrated as in Figure 1. It is estimated by three representative interval weights such as its 0-level set $[\varepsilon, w_i^{R0}]$, its α-level set $[w_i^L, w_i^R]$ as a core interval weight, and its 1-level set w_i as follows.

$$\mu_{\tilde{W}_i}(x) = \begin{cases} \dfrac{\alpha}{w_i^L}x, & 0 < x \leq w_i^L \\[2mm] \dfrac{1-\alpha}{w_i - w_i^L}(x - w_i^L) + \alpha, & w_i^L \leq x \leq w_i \\[2mm] \dfrac{\alpha - 1}{w_i^R - w_i}(x - w_i) + 1, & w_i \leq x \leq w_i^R \\[2mm] \dfrac{-\alpha}{w_i^{R0} - w_i^R}(x - w_i^R) + \alpha, & w_i^R \leq x \leq w_i^{R0} \\[2mm] 0, & w_i^{R0} \leq x \end{cases} \tag{12}$$

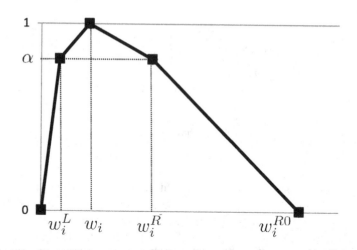

Fig. 1. Membership function of fuzzy weight of alternative i

The interval weights by any level-sets of the fuzzy weights by (12) are normalized so that they are interval probabilities and satisfy the 1st and 2nd constraints in (3). For instance, the 1st constraint in case of $\alpha \leq \beta \leq 1$ is verified as follows. Assume interval weights $[w_i^{L\beta}, w_i^{R\beta}], \forall i$, where $\mu_{\tilde{W}_i}(w_i^{L\beta}) = \mu_{\tilde{W}_i}(w_i^{R\beta}) = \beta$. Their bounds are denoted by the 2nd and 3rd functions in (12) as follows.

$$\beta = \frac{1-\alpha}{w_i - w_i^L}(w_i^{L\beta} - w_i^L) + \alpha \leftrightarrow w_i^{L\beta} = \frac{1-\beta}{1-\alpha}w_i^L + \frac{\beta-\alpha}{1-\alpha}w_i, \forall i,$$

$$\beta = \frac{\alpha-1}{w_i^R - w_i}(w_i^{R\beta} - w_i) + 1 \leftrightarrow w_i^{R\beta} = \frac{1-\beta}{1-\alpha}w_i^R + \frac{\beta-\alpha}{1-\alpha}w_i, \forall i,$$

where $[w_i^L, w_i^R], \forall i$ by (7) satisfy the 1st constraint in (3) so that $\sum_{i \neq j} w_i^R + w_j^L \geq 1, \forall j$ and $w_i, \forall i$ by (9) satisfy $\sum_i w_i = 1$. The 1st constraint in (3) for β-level sets of the fuzzy weights $[w_i^{L\beta}, w_i^{R\beta}], \forall i$ is verified as follows.

$$\sum_{i \neq j} w_i^{R\beta} + w_j^{L\beta}$$
$$= \sum_{i \neq j} \left(\frac{1-\beta}{1-\alpha}w_i^R + \frac{\beta-\alpha}{1-\alpha}w_i \right) + \left(\frac{1-\beta}{1-\alpha}w_j^L + \frac{\beta-\alpha}{1-\alpha}w_j \right) \qquad (13)$$
$$= \frac{1-\beta}{1-\alpha}(\sum_{i \neq j} w_i^R + w_j^L) + \frac{\beta-\alpha}{1-\alpha}(\sum_i w_i) \geq 1, \forall j.$$

Similarly, the 2nd constraint in (3) for β-level sets is verified. In case of $0 < \gamma \leq \alpha$, in the same way, it is verified that γ-level sets of the fuzzy weights satisfy the 1st and 2nd constraints in (3). Therefore, any-level sets of the fuzzy weights which are denoted as interval weights are normalized from the viewpoint of interval probability.

In general, the membership function of fuzzy weight $\tilde{W}_i$ is denoted by the representative interval weights by (8) and (10) as follows.

$$
\mu_{\tilde{W}_i}(x) =
\begin{cases}
\dfrac{\gamma_k - \gamma_{k-1}}{w_i^{L\gamma_{k-1}} - w_i^{L\gamma_k}}(x - w_i^{L\gamma_k}) + \gamma_k, & w_i^{L\gamma_k} \leq x \leq w_i^{L\gamma_{k-1}},\ k = l, \dots, 1 \\[3ex]
\dfrac{\beta_k - \beta_{k-1}}{w_i^{L\beta_k} - w_i^{L\beta_{k-1}}}(x - w_i^{L\beta_{k-1}}) + \beta_{k-1}, & w_i^{L\beta_{k-1}} \leq x \leq w_i^{L\beta_k},\ k = 1, \dots, m \\[3ex]
\dfrac{\beta_{k-1} - \beta_k}{w_i^{R\beta_{k-1}} - w_i^{R\beta_k}}(x - w_i^{R\beta_k}) + \beta_k, & w_i^{R\beta_k} \leq x \leq w_i^{R\beta_{k-1}},\ k = m, \dots, 1 \\[3ex]
\dfrac{\gamma_k - \gamma_{k-1}}{w_i^{R\gamma_k} - w_i^{R\gamma_{k-1}}}(x - w_i^{R\gamma_{k-1}}) + \gamma_{k-1}, & w_i^{R\gamma_{k-1}} \leq x \leq w_i^{R\gamma_k},\ k = 1, \dots, l \\[3ex]
0, & w_i^{R\gamma_l} \leq x
\end{cases}
\tag{14}
$$

The fuzzy weights denoted as in Figure 1 are obtained from a crisp comparison matrix A in (1) reflecting the uncertainty in A. They reflect the possibilities of the given information with membership values. When a rigid order of alternatives is needed, the interval weight by a high-level set of a fuzzy weight is used and a crisp weight is found as a focal point in case of 1-level set. While, when the possibility of an alternative is a concern, the low-level set is useful.

4 Numerical Examples

Two decision makers A and B give the pairwise comparison matrices on 4 alternatives as shown in Table 1. By (7), the core interval weights for fuzzy weight estimation are obtained and they are shown next to each matrix. The weight is assigned to each alternative as much as possible. The sum of lower bounds of the interval weights by decision maker B, $\alpha_B = 0.929$, is more than that by decision maker A, $\alpha_A = 0.813$, so that B gives the comparisons based on higher membership value than A. These interval weights are assumed as 0.813-level sets and 0.929-level sets of their fuzzy weights, respectively. For comparison, the interval weights by (3), where the widths are minimized, are shown at the right column of Table 1. Their sums of the lower bounds are shown at the 1st rows and they are less than those by (7). It mentions that the more weight is surely assigned to one of the alternatives by the proposed (7) than (3). Since the lower bound is more suitable to represent a membership value than the width, Interval AHP is modified by focusing on the lower bounds of the interval weights and the proposed model is used to estimate a fuzzy weight.

In addition, the representative interval weights by $\beta_{A1} = 0.9$, $\beta_{A2} = 1$, $\gamma_{A1} = 0.5$, and $\gamma_{A2} = n\varepsilon = 0.004$ for decision maker A and those by $\beta_{B1} = 1$, $\gamma_{B1} = 0.5$, and $\gamma_{B2} = 0.004$ for decision maker B are obtained by (8) and (10). Then, the fuzzy weights by decision makers A and B are estimated by (14) and illustrated as in Figures 2 and 3, respectively. It is noted that the numbers of representatives, m and l, are arbitrary and the more they are, the more precisely a fuzzy weight is estimated. For instance, we can obtain 11 representative interval weights assuming membership values, β or γ, as 0, 0.1, ..., 0.9, and 1.0.

Table 1. Two comparison matrices

	A1	A2	A3	A4	α_A=0.813 by (7)	0.785 by (3)
A1	1	1/3	7	1/7	[0.037, 0.224]	[0.025, 0.230]
A2	3	1	6	4	[0.671, 0.671]	[0.689, 0.689]
A3	1/7	1/6	1	3	[0.032, 0.219]	[0.033, 0.115]
A4	7	1/4	1/3	1	[0.073, 0.260]	[0.038, 0.172]

	B1	B2	B3	B4	α_B=0.929 by (7)	0.914 by (3)
B1	1	7	1	9	[0.463, 0.534]	[0.444, 0.444]
B2	1/7	1	1/7	1/9	[0.015, 0.066]	[0.015, 0.063]
B3	1	7	1	3	[0.392, 0.463]	[0.406, 0.444]
B4	1/9	9	1/3	1	[0.059, 0.131]	[0.049, 0.135]

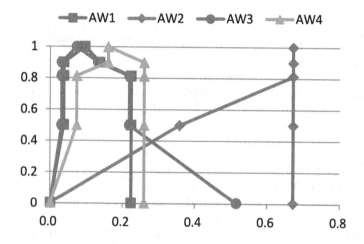

Fig. 2. Fuzzy weights by decision maker A

As for the fuzzy weights from the crisp comparisons by decision maker A, alternative A2 is apparently better than the other alternatives regardless of levels. Among the other alternatives, alternative A4 is a little better than alternatives A1 and A3 at 0.813-level. As higher the level becomes to 0.9 or 1, it becomes more apparent that A4 is better than them. The common weight of some alternatives is assigned more to alternative A4 than to alternatives A1 and A3. It may be because decision maker A potentially evaluates alternative A4 better but s/he is not very sure of it. While, at 0-level, where the weight is assigned to more than two alternatives, the possible weight of alternatives A3 increases, instead of the decrease of the weight surely assigned to alternative A2, because alternatives A2 and A3 are similar and substitutes in a sense.

As for the fuzzy weights from the crisp comparisons by decision maker B, at 0.929-level alternative B1 is better than B3, however, at 1-level their weights are the same. The weight surely assigned to B1 does not increase when the

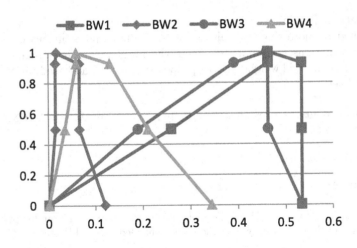

Fig. 3. Fuzzy weights by decision maker B

membership value is higher than 0.929. Most of the left weight, 0.069 of 0.071, is assigned to B3 at 1-level since decision maker B may not know alternative B3 well. S/he is sure of the weight of alternatives B1, B2 and B4 so that they are assigned little weight at 1-level sets of their fuzzy weights.

5 Conclusion

Based on the idea that the decision maker's judgments are uncertain, the fuzzy weights have been obtained from the given crisp comparisons. This paper proposed the lower bound based Interval AHP which obtains the interval weights of alternatives from the given crisp comparisons for fuzzy weight estimation. In the proposed model, the lower bound of an interval weight is focused on. It is reasonable to maximize the lower bound since it represents the weight surely assigned to one of the alternatives and an alternative is assigned the weight as much as possible. The left weight is considered as ignorance and it is common of some alternatives. The upper bound includes such a weight as possibly assigned to the alternative, in addition to the lower bound. For fuzzy weight estimation, the sum of the lower bounds of all alternatives is considered as a membership value of a fuzzy weight in a decision maker's mind. In other words, it is considered that the decision maker gives comparisons based on the certain level set of the fuzzy weight in his/her mind. The comparisons are more consistent each other when a decision maker gives them based on the higher membership values of his/her fuzzy weights. The sums of the lower bounds are assumed to some values from 0 to 1 so as to estimate a fuzzy weight. The relation between an interval weight and its estimations are modified depending on whether the value is higher than the certain level or not. For the higher-level set of the fuzzy weight, whose extreme case is a crisp weight, the maximum deficiency of a weight from its estimations is minimized, while for its lower-level set, where some weights are assigned to more than two alternatives, the maximum surplus of a weight from its estimations is minimized. Then, the fuzzy weight is estimated by the representative interval weights by the proposed lower bound based Interval AHP as its level sets.

Acknowledgments. This work was partially supported by JSPS KAKENHI Grant Number 26350423.

References

1. Alonso, J.A., Lamata, M.T.: Consistency in the analytic hierarchy process: A new approach. Uncertainty, Fuzziness and Knowledge-based Systems **14**, 445–459 (2006)
2. Buckley, J.J.: Fuzzy hierarchical analysis. Fuzzy Sets and Systems **17**(3), 175–189 (1985)
3. de Campos, L.M., Huete, J.F., Moral, S.: Probability intervals: a tool for uncertain reasoning. International Journal of Uncertainty **2**(2), 167–196 (1994)
4. Csutora, R., Buckley, J.J.: Fuzzy hierarchical analysis: the lambda-max method. Fuzzy Sets and Systems **120**(2), 181–195 (2001)
5. Dempster, A.P.: Upper and lower probabilities induced by a multivalued mapping. The Annals of Mathematical Statistics **38**(2), 325–339 (1967)
6. Entani, T., Sugihara, K.: Uncertainty index based interval assignment by interval AHP. European Journal of Operational Research **219**(2), 379–385 (2012)
7. van Laarhoven, P., Pedrycz, W.: A fuzzy extension of Saaty's priority theory. Fuzzy Sets and Systems **11**, 199–227 (1983)
8. Pelaez, J., Lamata, M.: A new measure of consistency for positive reciprocal matrices. Computers and Mathematics with Applications **46**, 1839–1845 (2003)
9. Saaty, T.L.: The Analytic Hierarchy Process. McGraw-Hill, New York (1980)
10. Sugihara, K., Ishii, H., Tanaka, H.: Interval priorities in AHP by interval regression analysis. European Journal of Operational Research **158**(3), 745–754 (2004)
11. Sugihara, K., Tanaka, H.: Interval evaluations in the Analytic Hierarchy Process by possibilistic analysis. Computational Intelligence **17**(3), 567–579 (2001)
12. Tanaka, H., Sugihara, K., Maeda, Y.: Non-additive measures by interval probability functions. Information Sciences **164**, 209–227 (2004)

Logarithmic Conversion Approach to the Estimation of Interval Priority Weights from a Pairwise Comparison Matrix

Masahiro Inuiguchi$^{(\boxtimes)}$ and Shigeaki Innan

Graduate School of Engineering Science, Osaka University,
Toyonaka, Osaka 560-8531, Japan
inuiguti@sys.es.osaka-u.ac.jp, innan@inulab.sys.es.osaka-u.ac.jp

Abstract. An alternative method for the estimation of interval priority weights in Interval AHP is proposed. The proposed method applies the logarithmic conversion to components of the pairwise comparison matrix and obtains logarithmically-converted interval priority weights so as to minimize the sum of their widths. The logarithmically-converted interval priority weights are estimated as an optimal solution of a linear programming problem. The interval priority weights are obtained by the antilogarithmic conversion of the optimal solution and by the normalization. Numerical experiments are conducted to demonstrate the advantages of the proposed estimation over the conventional estimation.

Keywords: Interval AHP · Interval priority weight estimation · Logarithmic conversion · Linear programming · Normalization

1 Introduction

AHP (Analytic Hierarchy Process) [1] is a useful tool to evaluate the priority weights of alternatives and criteria in multiple criteria decision problems. In the method, a decision maker gives a pairwise comparison matrix whose component shows the relative importance between the alternatives/criteria corresponding to the row and the column. Based on the given pairwise comparison matrix, the priority weights are estimated by maximum eigenvalue method [1] or by geometric mean method [2]. However, the obtained pairwise comparison matrix is seldom consistent each other because of the vagueness of human judgements. The degree of inconsistency is measured by the consistency index and the estimated priority weights are adopted for decision analysis when the consistency index is in the empirically-defined allowable range. To cope with the inaccuracy, imprecision and vagueness of human judgements, fuzzy and interval techniques are applied to AHP. In many of fuzzy approaches, a pairwise comparison matrix with fuzzy components is used to represent the vagueness of human judgment on the relative importance between criteria. Then a fuzzy priority weight vector has been estimated so as to approximate the fuzzy pairwise comparison matrix [3–5]. Similarly, a pairwise comparison matrix with interval components is also used

© Springer International Publishing Switzerland 2015
V.-N. Huynh et al. (Eds.): IUKM 2015, LNAI 9376, pp. 77–88, 2015.
DOI: 10.1007/978-3-319-25135-6_9

to represent the vagueness of human judgment. However, a crisp priority weight vector consistent or nearly consistent to a given interval pairwise comparison matrix is estimated (see [6]).

On the other hand, in Interval AHP [7], an interval priority weight vector is estimated from a given crisp pairwise comparison matrix. This approach is based on the idea that the decision maker's evaluation is vague so that the weights of criteria are expressed by intervals and that the pairwise comparison matrix is obtained by judgments with arbitrarily selected values from the intervals. An interval priority weight vector covering the given pairwise comparison matrix is estimated so as to minimize its total spreads. Later Interval AHP is extended to treat interval pairwise comparison matrices [8] but we treat only a crisp pairwise comparison matrix in this paper.

Although Interval AHP is developing, the validity of estimation method for interval priority weights from a given pairwise comparison matrix has not yet investigated considerably. In this paper, we propose an alternative estimation method for interval priority weights and compare the validity of the conventional and proposed estimation methods by numerical experiments. The proposed estimation method, all components of the given pairwise comparison matrix are logarithmically-converted and logarithmically-converted interval priorities are estimated so as to minimize the sum of their widths. The interval priority weights are obtained by the antilogarithmic conversion of the optimal solution and by the normalization. We compare their validities in the similarities to the true interval priority weights and in the adequateness in the alternative ranking.

This paper is organized as follows. In Section 2, AHP and Interval AHP are briefly reviewed. The proposed approach is described in Section 3. In Section 4, numerical experiments for comparison between the conventional estimation of interval priority weights and the proposed one are described. The results of the experiments are described also in Section 4. The concluding remarks are given in Section 5.

2   Interval AHP

We introduce Interval AHP for multiple criteria decision problem with n criteria. For simplicity and the consistency of notation with subsequent sections, we define $N = \{1, 2, \ldots, n\}$ and $N \backslash j = N \backslash \{j\} = \{1, 2, \ldots, j - 1, j + 1, \ldots, n\}$ for $j \in N$.

In AHP, the decision problem is structured hierarchically as criteria and alternatives. At each node except leaf nodes of the hierarchical tree, a priority weight vector for criteria or for alternatives is obtained from a pairwise comparison matrix A.

We first describe the estimation of a priority weight vector $\boldsymbol{w} = (w_1, w_2, \ldots, w_n)^{\mathrm{T}}$ from a given pairwise comparison matrix,

$$A = \begin{bmatrix} 1 & \cdots & a_{1n} \\ \vdots & a_{ij} & \vdots \\ a_{n1} & \cdots & 1 \end{bmatrix}, \tag{1}$$

where we assume the reciprocity, i.e., $a_{ij} = 1/a_{ij}$, $i,j \in N$. Because the (i,j) component a_{ij} of A shows the relative importance of the i-th criterion over the j-th criterion, theoretically, we have $a_{ij} = w_i/w_j$, $i,j \in N$. If a_{ij}, $i,j \in N$ are obtained exactly, the strong transitivity $a_{ij} = a_{il}a_{lj}$, $i,j,l \in N$ should be satisfied. However, human evaluation is not very accurate so that the strong transitivity is not satisfied. In the conventional approach [1], a_{ij}, $i,j \in N$ are assumed to be approximations of w_i/w_j. w is estimated as the normalized eigenvector corresponding to the maximal eigenvalue, because the nonnegativity of the eigenvector corresponding to the maximal eigenvalue is guaranteed by Perron-Frobenius theorem. There is the other popular way to estimate w. It is the geometric mean method [4]. To evaluate the consistency of the given pairwise comparison matrix, the following consistency index is used:

$$C.I. = \frac{\lambda_{\max} - n}{n - 1}, \tag{2}$$

where $\lambda_{\max}$ is the maximal eigenvalue of A. If $C.I.$ is not greater than 0.1, it is often considered that the obtained vector w is acceptable.

To this estimation problem, the idea of interval regression analysis [9] was applied. In this approach, we assume that decision maker's evaluation is not very accurate to be expressed by a unique priority weight vector w but intrinsically vague so that the priority weight vector has a range. Therefore, the non-fulfillment of the strong transitivity is not regarded as inconsistency but due to the intrinsic variety of possible priority weight vectors. Accordingly, we consider an interval priority weight vector $W = (W_1, W_2, \ldots, W_n)^{\mathrm{T}}$ instead of a priority weight vector w, where $W_i = [w_i^{\mathrm{L}}, w_i^{\mathrm{R}}]$, $i \in N$ and $w_i^{\mathrm{L}} \leq w_i^{\mathrm{R}}$, $i \in N$. To fit the given pairwise comparison matrix, we require the interval priority weight vector W to satisfy

$$\frac{w_i^{\mathrm{L}}}{w_j^{\mathrm{R}}} \leq a_{ij} \leq \frac{w_i^{\mathrm{R}}}{w_j^{\mathrm{L}}}, \ i,j \in N, \ i < j. \tag{3}$$

We note that by the reciprocity, $a_{ij} = 1/a_{ji}$, $i,j \in N$, we only consider $i,j \in N$ such that $i < j$. The set of interval priority weight vectors W satisfying (3) is denoted by $\mathcal{W}(A)$. Moreover, corresponding to the normalization condition of w in the conventional AHP, we require the interval priority weight vector W to satisfy

$$\sum_{j \in N \setminus i} w_j^{\mathrm{R}} + w_i^{\mathrm{L}} \geq 1, \ i \in N, \tag{4}$$

$$\sum_{j \in N \setminus i} w_j^{\mathrm{L}} + w_i^{\mathrm{R}} \leq 1, \ i \in N. \tag{5}$$

(4) and (5) ensure that, for any $w_i^{\circ} \in W_i$, there exist $w_j \in W_j$, $j \in N \setminus i$ such that $\sum_{j \in N \setminus i} w_j + w_i^{\circ} = 1$ (see [10]). Namely, any values in W_i, $i \in N$ are meaningful (there is no ineffective subarea in W). The set of interval priority weight vectors W satisfying (4) and (5) is denoted by $\mathcal{W}^{\mathrm{N}}$.

Under conditions (3), (4) and (5) as well as $\epsilon < w_i^{\mathrm{L}} \leq w_i^{\mathrm{R}}$, $i \in N$, we calculate a suitable $\boldsymbol{W}$, where ϵ is a very small positive number. The wider each W_i is, the easier W_i, $i \in N$ satisfy (3), (4) and (5). The narrower interval priority weights give clearer preferences in the comparison of alternatives. Then we minimize the following total widths of interval priority weights W_i, $i \in N$:

$$d(\boldsymbol{W}) = \sum_{i \in N} (w_i^{\mathrm{R}} - w_i^{\mathrm{L}}). \tag{6}$$

Consequently, the interval priority weight vector $\boldsymbol{W}$ is estimated by solving the following linear programming problem:

$$\underset{\boldsymbol{W}}{\mathrm{minimize}} \{ d(\boldsymbol{W}) \mid \boldsymbol{W} \in \mathcal{W}(A) \cap \mathcal{W}^{\mathrm{N}}, \epsilon \leq w_i^{\mathrm{L}} \leq w_i^{\mathrm{R}}, \ i \in N \}. \tag{7}$$

The set of optimal solutions and the optimal value to this problem is denoted by $\mathcal{W}^{\mathrm{DM}}$ and $\hat{d}$, respectively.

Once an interval priority weight vector $\boldsymbol{W}$ is obtained, we define a dominance relation between alternatives under the assumption that utility scores $u_i(o_p)$ of alternatives o_p in view of each criterion are given. Sugihara et. al [7] and Guo and Tanaka [11] proposed a dominance relation based on the overall interval scores. We use the dominance relation proposed by Entani and Inuiguchi [10] because it considers all possible priority weights suitable for the given matrix A. We use two dominance relations defined by

$$o_p \succ^\pi o_q \Leftrightarrow \exists \boldsymbol{W} \in \mathcal{W}^{\mathrm{DM}}, \ \exists \boldsymbol{w} \in \boldsymbol{W}, \mathbf{e}^{\mathrm{T}} \boldsymbol{w} = 1;$$
$$\sum_{i \in N} w_i (u_i(o_p) - u_i(o_q)) > 0, \tag{8}$$

$$o_p \succ^\nu o_q \Leftrightarrow \exists \boldsymbol{W} \in \mathcal{W}^{\mathrm{DM}}, \ \forall \boldsymbol{w} \in \boldsymbol{W}, \mathbf{e}^{\mathrm{T}} \boldsymbol{w} = 1;$$
$$\sum_{i \in N} w_i (u_i(o_p) - u_i(o_q)) > 0, \tag{9}$$

where $\mathbf{e} = (1, 1, \ldots, 1) \in \mathbf{R}^n$. $o_p \succ^\pi o_q$ implies that o_p possibly dominates o_q and $o_p \succ^\nu o_q$ implies that o_p certainly dominates o_q.

3   Logarithmic Conversion Approach

The interval priority weight vector estimated by solving (7) has a tendency that the width of an interval priority weight vanishes if its center value is larger than others. Indeed it has been observed often by numerical examples [7,10]. This tendency can be understood by the following fact: increasing the width of $W_j = [w_j^{\mathrm{L}}, w_j^{\mathrm{R}}]$ by $\epsilon > 0$ is more effective than increasing the width of $W_i = [w_i^{\mathrm{L}}, w_i^{\mathrm{R}}]$ by $\epsilon > 0$ to satisfy (3) for some (i, j) such that $a_{ij} > 1$ and $i < j$. Indeed if we increase w_i^{R} by $\epsilon > 0$, the first inequality of (3) becomes (a) $w_i^{\mathrm{L}} \leq a_{ij} w_j^{\mathrm{R}} + a_{ij}\varepsilon$. On the other hand, if we decrease w_i^{L} by $\epsilon > 0$, the first inequality of (3) becomes (b) $w_i^{\mathrm{L}} \leq a_{ij} w_j^{\mathrm{R}} + \varepsilon$. Because $a_{ij} > 1$, (a) is weaker than (b).

Therefore, increasing w_j^{R} by $\epsilon > 0$ is more effective than decreasing w_i^{L} by $\epsilon > 0$. Same explanation is effective also for the second inequality of (3). However, by the normality condition (4) and (5), the explanation above does not work conclusively but suggestively to an assertion "the width of an interval priority weight vanishes if its center value is larger than others".

This tendency can deteriorate the quality of the interval priority weight estimation. As we observed in the explanation above, the influence of the width increment of an interval priority weight for satisfaction of (3) under objective function $d(\mathbf{W})$ is different by a_{ij}. We resolve this difference by considering the sum of widths in the logarithmic priority weight space.

Let $c_{ij} = \log a_{ij}$, $u_i^{\mathrm{L}} = \log w_i^{\mathrm{L}}$ and $u_i^{\mathrm{R}} = \log w_i^{\mathrm{R}}$, $i,j \in N$, $i \neq j$. Namely, instead of interval priority weights W_i, $i \in N$, we consider logarithmically converted interval priority weights $U_i = [u_i^{\mathrm{L}}, u_i^{\mathrm{R}}]$, $i \in N$. Then (3) is rewritten by

$$u_i^{\mathrm{L}} - u_j^{\mathrm{R}} \leq c_{ij} \leq u_i^{\mathrm{R}} - u_j^{\mathrm{L}}, \ i,j \in N, \ i < j. \tag{10}$$

Let $\mathcal{U}(A)$ be a set of U_i, $i \in N$ satisfying (10). Corresponding to $d(\mathbf{W})$ defined by (6), we consider

$$d_{\mathrm{L}}(U) = \sum_{i \in N} (u_i^{\mathrm{R}} - u_i^{\mathrm{L}}). \tag{11}$$

Increasing the width of $U_j = [u_j^{\mathrm{L}}, u_j^{\mathrm{R}}]$ by $\epsilon > 0$ has a same effect as increasing the width of $U_i = [u_i^{\mathrm{L}}, u_i^{\mathrm{R}}]$ by $\epsilon > 0$ to satisfy (10) for any (i,j). Therefore, the tendency of the conventional method can be resolved by those replacements.

However, the normality conditions (4) and (5) cannot be expressed by linear inequalities of u_i $i \in N$. Then we first obtain an optimal solution, $U_i = [u_i^{\mathrm{L}}, u_i^{\mathrm{R}}]$, $i \in N$ by solving linear programming problem,

$$\underset{U}{\text{minimize}}\{d(U) \mid U \in \mathcal{U}(A), u_i^{\mathrm{L}} \leq u_i^{\mathrm{R}}, \ i \in N\}. \tag{12}$$

Then we calculate $W_i = [w_i^{\mathrm{L}}, w_i^{\mathrm{R}}]$, $i \in N$ by the antilogarithmic conversion and normalization of $U_i = [u_i^{\mathrm{L}}, u_i^{\mathrm{R}}]$, $i \in N$.

Theorem 1. *Linear programming problem (12) has always infinitely many optimal solutions but a unique optimal value.*

When interval priority weights $U_i = [u_i^{\mathrm{L}}, u_i^{\mathrm{R}}]$, $i \in N$ in the logarithmic priority are given, a normalized interval priority weight vector $W_i = [w_i^{\mathrm{L}}, w_i^{\mathrm{R}}]$, $i \in N$ is obtained by

$$w_i^{\mathrm{L}} = \frac{\exp(u_i^{\mathrm{L}})}{\sum_{j \in N \setminus i} \exp(u_j^{\mathrm{R}}) + \exp(u_i^{\mathrm{L}})}, \ w_i^{\mathrm{R}} = \frac{\exp(u_i^{\mathrm{R}})}{\sum_{j \in N \setminus i} \exp(u_j^{\mathrm{L}}) + \exp(u_i^{\mathrm{R}})}, \ i \in N.$$

$$\tag{13}$$

We note that, for every $i \in N$, we have

$$[w_i^L, w_i^R] = \left\{ \left. \frac{\exp(u_i)}{\sum\limits_{j \in N \setminus i} \exp(u_j) + \exp(u_i)} \right| u_i \in U_i, \ u_j \in U_j, \ j \in N \right\}. \qquad (14)$$

Intervals $[\log(w_i^L), \log(w_i^R)]$, $i \in N$ are not always optimal to Problem (12). However, because of its simplicity, we adopt $[w_i^L, w_i^R]$, $i \in N$ in our numerical experiments. Because (14) implies that a weight vector $([w_1^L, w_2^R], \ldots, [w_n^L, w_n^R])^T$ is a collection of normalized weight vectors, $[w_i^L, w_i^R]$, $i \in N$ satisfy (4) and (5). Moreover, because $w_i^L / w_j^R \leq \exp(u_i^L) / \exp(u_i^R)$ and $w_i^R / w_j^L \geq \exp(u_i^R) / \exp(u_i^L)$, $[w_i^L, w_i^R]$, $i \in N$ satisfy (3), too.

4 Numerical Experiments

4.1 Outline

In this section, we compare the proposed estimation method to the conventional one. To this end, we assume two possible cases: one is a case where the true crisp priority weights exist but the decision maker's perception is vague so that s/he perceives the weights with errors obeying a logarithmic normal distribution when s/he makes pairwise comparisons of criteria, and the other is a case where true interval priority weights exist but the decision maker perceives values randomly selected from the true intervals when s/he makes pairwise comparisons. For both cases, we evaluate to what extent the estimated interval priority weights are similar to the true ones and to what extent the dominance relations between alternatives obtained by the estimated interval priority weights coincide with those by the true ones.

4.2 Experiments with a Logarithmic Normal Distribution

We assume that there are true crisp priority weights but the decision maker's perception is vague so that s/he perceives the weights with errors obeying a logarithmic normal distribution when s/he makes pairwise comparisons of criteria. Then we consider five criteria $X_1, X_2, \ldots, X_5$ and prepare a set of true crisp priority weights, i.e., $w_1 = 0.3$, $w_2 = 0.25$, $w_3 = 0.2$, $w_4 = 0.15$ and $w_5 = 0.1$ in this paper. We generate a_{ij} by $a_{ij} = \exp(\log(w_i) - \log(w_j) + \omega_{ij})$ for $i > j$ and $i, j \in N$, where ω_{ij} is a random variable obeying a normal distribution with average 0 and variance 0.01 for $i \in N$, $j \in N \cap [i+1, \infty)$. Then $a_{ji} = 1/a_{ij}$, $i > j$ and $i, j \in N$ and $a_{ii} = 1$, $i \in N$. By this way, we generate 1,000 matrices A and for each of them, we applied the conventional and proposed methods for estimating interval priority weights $W_i = [w_i^L, w_i^R]$, $i \in N$.

Table 1. Ratios satisfying $w_i \in W_i$ for each criterion X_i

criterion	true w_i	Conventional	Quadratic	Proposed	Widened
X_1	0.3	10.8 %	36.7 %	71.1 %	39.5 %
X_2	0.25	35.0 %	45.6 %	70.0 %	60.5 %
X_3	0.2	51.1 %	56.5 %	70.6 %	75.5 %
X_4	0.15	59.9 %	67.9 %	67.4 %	81.1 %
X_5	0.1	70.8 %	72.5 %	80.5 %	92.5 %

Table 2. Ratios satisfying at least k conditions among $w_i \in W_i$, $i \in N$, simultaneously

k	Conventional	Quadratic	Proposed	Widened
all 5	0 %	3.2 %	24.4 %	15.4 %
at least 4	5.10 %	26.8 %	55.8 %	51.8 %
at least 3	40.6 %	59.7 %	84.1 %	85.0 %
at least 2	84.0 %	88.4 %	95.5 %	97.0 %
at least 1	97.9 %	98.6 %	99.8 %	99.9 %

Inclusion of True Priority Weights in the Estimated Intervals. For each estimated interval priority weights W_i, $i \in N$, we check whether the true priority weights are belonging to the estimated interval priority weights, i.e., we check $w_i \in W_i$ for each $i \in N$. The results of these numerical experiments are shown in Tables 1 and 2. In those tables, "Conventional" and "Proposed" stand for the conventional estimation and the proposed estimation. "Quadratic" implies an estimation method based on a quadratic programming problem obtained by replacing $d(\boldsymbol{W})$ with $d_2(\boldsymbol{W}) = \sum_{i \in N}(w_i^{\mathrm{R}} - w_i^{\mathrm{L}})^2$ in problem (7). "Widened" implies a modified conventional estimation method where the widths of the estimated interval priority weights are widened by dif/n without changing the center values. dif is the difference between the sum of widths of interval priority weights estimated by the conventional method and that estimated by the proposed method. Namely, the widened estimated interval priority weights $\hat{W}_i = [\hat{w}_i^{\mathrm{L}}, \hat{w}_i^{\mathrm{R}}]$, $i \in N$ are obtained by $\hat{w}_i^{\mathrm{L}} = w_i^{\mathrm{L}} - dif/(2n)$ and $\hat{w}_i^{\mathrm{R}} = w_i^{\mathrm{R}} + dif/(2n)$, $i \in N$. The widened estimated interval priority weights are considered because ratios in Tables 1 and 2 increase as the widths increase.

As shown in Table 1, the interval priority weights estimated by the proposed method include the true priority weights with high probability and the ratio distribution over criteria is most well-balanced among all estimated interval priority weights. For the other estimated interval priority weights, the ratio increases as true priority weight w_i decreases. The ratio for criterion X_4 of the proposed method is worse than that of the estimation by "Quadratic". Ratios for criteria X_3, X_4 and X_5 are worse than those of the estimation by "Widened".

Moreover, from Table 2, we observe that the proposed estimation takes larger ratios than others for $k =$ 'all 5' and 'at least 4'. Therefore, the interval priority weights by the proposed method include many true priority weights simultaneously. However, for the other k's, ratios in the proposed estimation are smaller than those in "Widened". The proposed method is better in all cases than other estimation methods. However, the differences decrease as k decreases.

Table 3. Utility scores of five alternatives

alternative	criteria				
	X_1	X_2	X_3	X_4	X_5
o_1	0.25	0.30	0.10	0.15	0.20
o_2	0.20	0.25	0.30	0.10	0.15
o_3	0.15	0.20	0.25	0.30	0.10
o_4	0.10	0.15	0.20	0.25	0.30
o_5	0.30	0.10	0.15	0.20	0.25

Table 4. Dominance ratios for alternative pairs

(o_p, o_q)	Conventional			Proposed		
	$o_p \succ^\nu o_q$	$o_q \succ^\nu o_p$	unknown	$o_p \succ^\nu o_q$	$o_q \succ^\nu o_p$	unknown
(o_1, o_2)	26.2 %	22.7 %	51.1 %	13.3 %	16.1 %	70.6 %
(o_1, o_3)	99.6 %	0 %	0.4 %	96.9 %	0 %	3.1 %
(o_1, o_4)	100 %	0 %	0 %	100 %	0 %	0 %
(o_1, o_5)	99.6 %	0 %	0.4 %	99.2 %	0 %	0.8 %
(o_2, o_3)	99.9 %	0 %	0.1 %	100 %	0 %	0 %
(o_2, o_4)	100 %	0 %	0 %	100 %	0 %	0 %
(o_2, o_5)	99.4 %	0 %	0.6 %	94.8 %	0 %	5.2 %
(o_3, o_4)	100 %	0 %	0 %	100 %	0 %	0 %
(o_3, o_5)	31.0 %	26.0 %	43.0 %	4.3 %	6.3 %	89.4 %
(o_4, o_5)	0 %	100 %	0 %	0 %	100 %	0 %

To sum up, the proposed estimation performs better than the conventional estimation and "Quadratic" estimation. However, there is not big difference between the proposed estimation and "Widened" estimation. In the sense of the balance over the criteria, the proposed estimation is the best. We note that "Widened" estimation cannot work without the execution of the proposed estimation because the modified widths are given by the widths of the interval priority weights estimated by the proposed method.

As we mentioned earlier, the evaluation used in this subsubsection is improved as the widths increase. Therefore, we cannot conclude the advantages of the estimation method only by these experiments. Then we executed another experiments for evaluating the correctness in ranking the alternatives.

Correctness in Ranking Alternatives. In order to evaluate the correctness in ranking alternatives by the estimated interval priority weights, we evaluate the coincidences of dominance relations between alternatives. To this end, we consider five alternatives whose utility scores $u_i(o_p)$ in each criterion are given in Table 3. Using the true weights, the order of five alternatives is obtained as $o_1 \sim o_2 \succ o_3 \sim o_5 \succ o_4$, where $o_p \succ o_q$ implies that o_p dominates o_q and $o_p \sim o_q$ implies that o_p and o_q are indifferent.

Applying (9) to 1,000 estimated interval priority weights in each of conventional and proposed estimation methods, we calculated dominance ratios for each pair (o_p, o_q) of alternatives. The results are shown in Table 4. As shown in Table 4, dominance relations for all pairs except (o_1, o_2) and (o_3, o_5) are almost correctly evaluated by the interval priority weights estimated by both conventional and proposed methods. Pairs (o_1, o_2) and (o_3, o_5) are indifferent alternative pairs in the true ranking order. For those pairs, the correct answer would be 'unknown' because it is hard to obtain the indifference between alternatives under interval priority weights. Comparing the columns of 'unknown' of both estimation methods at rows of (o_1, o_2) and (o_3, o_5), we find that the proposed method is better. However, for pairs (o_1, o_3) and (o_2, o_5), the ratios of 'unknown' in the proposed method are a little larger than those in the conventional method. We think that the disadvantage of the proposed method is much smaller than its advantage. Then, by this experiment, we conclude that the proposed method is better than the conventional method.

4.3   Experiments with True Interval Priority Weights

Different from the previous experiments, we assume that the decision makers have interval priority weights $T_i = [t_i^{\mathrm{L}}, t_i^{\mathrm{R}}]$, $i \in N$. When s/he make pairwise comparisons, priority weights $v_i \in T_i$ and $v_j \in T_j$ are selected uniformly from T_i and T_j and a_{ij} is evaluated by v_i/v_j for $i, j \in N$ such that $i < j$. We assume that $a_{ji} = 1/a_{ij}$ and $a_{ii} = 1$ for $i, j \in N$. Under this setting, we execute similar experiments to the previous subsection. Then we consider five criteria $X_1, X_2, \ldots, X_5$ and prepare a set of true interval priority weights, i.e., $T_1 = [0.21, 0.39]$, $T_2 = [0.16, 0.34]$, $T_3 = [0.11, 0.29]$, $T_4 = [0.06, 0.24]$ and $T_5 = [0.01, 0.19]$ in this paper. Value t_i is selected by a random number obeying uniform distribution on T_i. By this way, we generate 1,000 matrices A and for each of them, we applied the conventional and proposed methods for estimating interval priority weights $W_i = [w_i^{\mathrm{L}}, w_i^{\mathrm{R}}]$, $i \in N$.

Coincidence of True Interval Priority Weights with Estimated Intervals. To measure the degree of coincidence of true interval priority weights T_i, $i \in N$ and the estimated interval priority weights W_i, $i \in N$, we observe values of the following two indices:

$$p_i = \frac{d(T_i \cap W_i)}{d(T_i) + d(W_i) - d(T_i \cap W_i)}, \quad q_i = \frac{d(T_i \cap W_i)}{d(T_i)}, \tag{15}$$

where we define $d([x^{\mathrm{L}}, x^{\mathrm{R}}]) = x^{\mathrm{R}} - x^{\mathrm{L}}$ and $d(\emptyset) = 0$ in the same concept of (6). We note that $p_i = 1$ if and only if $T_i = W_i$ and $q_i = 1$ if and only if $T_i \subseteq W_i$. Therefore, p_i shows a coincidence degree of the true interval priority weight with the estimated interval priority weight while q_i evaluates to what extent the true interval priority weight is belonging to the estimated interval priority weight. Moreover, from those, we obtain

$$r_i = \frac{d(T_i \cap W_i)}{d(W_i)} = \frac{p_i q_i}{q_i - p_i + p_i q_i}, \quad F_i = \frac{2 q_i r_i}{q_i + r_i}, \tag{16}$$

where r_i evaluates to what extent the estimated interval priority weight includes the true interval priority weight. F_i is a harmonic mean of q_i and r_i, and is

Table 5. Coincidence of true interval priority weights with estimated intervals

criterion	Conventional				Proposed			
	p_i	q_i	r_i	F_i	p_i	q_i	r_i	F_i
X_1	0.09622	0.09762	0.87043	0.17555	0.52373	0.59120	0.82108	0.68743
X_2	0.19958	0.20297	0.92505	0.33289	0.49721	0.55141	0.83493	0.66418
X_3	0.31542	0.32058	0.95145	0.47958	0.47120	0.49993	0.89130	0.64057
X_4	0.42141	0.42721	0.96879	0.59295	0.47039	0.48534	0.93853	0.63982
X_5	0.47149	0.47728	0.97491	0.64083	0.51727	0.53332	0.94504	0.68184

known as F-measure in data mining. In data mining terminology, q_i and r_i are called recall and precision. For all indices p_i, q_i, r_i and F_i, the larger values, the better estimation.

The obtained results are shown in Table 5. As sown in Table 5, values of p_i and F_i of the proposed estimation are better than those of the conventional estimation for every $i \in N$. Therefore, we understand that the proposed estimation is advantageous over the conventional estimation. Values of p_i, q_i, r_i and F_i decrease as the index i of X_i increases in the conventional estimation. the center values of true interval priority weights decreases as the index i of X_i increases. Namely, the quality of estimation increases as the center values of true interval priority weights decreases in the conventional estimation. Such tendency is not remarkably observed in the proposed estimation (although the value of r_i has such tendency).

From values of q_i and r_i, the interval priority weights estimated by the conventional method are almost included in the true interval priority weights. The interval priority weights estimated by the proposed method are better in values of q_i but worse in values of r_i than those estimated by the conventional method. This would be caused by the fact that the interval priority weights estimated by the conventional method are very narrow. Considering values of r_i are very large and values of q_i are around 0.5 in the proposed estimation, the estimated interval priority weights are narrow, too, but sufficiently wider than the interval priority weights estimated by the conventional method.

Correctness in Ranking Alternatives. In order to evaluate the correctness in dominance relation between alternatives by the estimated interval priority weights, we evaluate the coincidences of dominance relations between alternatives. To this end, we consider again the five alternatives given in Table 3.

Because we have true interval priority weights, we consider the strong dominance relation $\succ^\nu_T$ obtained by (9) with substitution of true interval priority weights T_i to W_i as the true strong dominance relation. The value of $ud(o_p, o_q) = \min_{t_i \in T_i,\ i \in N} \sum_{i \in N} t_i(u_i(o_p) - u_i(o_q))$ under $t_i \in T_i$, $i \in N$ are obtained as shown in Table 6. Here we note that $\succ^\nu_T$ shows the true strong dominance relation based on T_i, $i \in N$ while $\succ^\nu_W$ shows the estimated strong dominance relation based on W_i, $i \in N$. From (9), we obtain $o_3 \succ^\nu_T o_4$ and $o_5 \succ^\nu_T o_4$ and no dominance relations hold for other pairs of alternatives.

Table 6. The evaluation of strong dominance relation $\succ_T^\nu$

alternative	o_1	o_2	o_3	o_4	o_5
o_1	—	-0.0225	-0.0325	-0.0075	-0.01 %
o_2	-0.0225	—	-0.01	-0.0075	-0.0325
o_3	-0.0575	-0.035	—	0.0225	-0.045
o_4	-0.0825	-0.0825	-0.0475	—	-0.0475
o_5	-0.035	-0.0575	-0.045	0.0225	—

Table 7. Dominance ratios for alternative pairs

	Conventional			Proposed		
(o_p, o_q)	$o_p \succ_W^\nu o_q$	$o_q \succ_W^\nu o_p$	unknown	$o_p \succ_W^\nu o_q$	$o_q \succ_W^\nu o_p$	unknown
(o_1, o_2)	28.2 %	19.3 %	52.5 %	13.8 %	10.2 %	76.0 %
(o_1, o_3)	50.8 %	1.7 %	47.5 %	23.6 %	0.2 %	76.2 %
(o_1, o_4)	98.8 %	0 %	1.2 %	77.5 %	0 %	22.5 %
(o_1, o_5)	68.3 %	6.0 %	25.7 %	48.6 %	0.4 %	51.0 %
(o_2, o_3)	66.2 %	2.3 %	31.5 %	56.0 %	0.9 %	43.1 %
(o_2, o_4)	98.8 %	0 %	1.2 %	78.9 %	0 %	21.1 %
(o_2, o_5)	51.0 %	3.6 %	45.5 %	18.2 %	0.3 %	81.5 %
(o_3, o_4)	94.0 %	0 %	6.0 %	86.0 %	0 %	14.0 %
(o_3, o_5)	21.2 %	15.4 %	63.4 %	1.7 %	3.3 %	95.0 %
(o_4, o_5)	0.1 %	97.8 %	2.1 %	0 %	79.7 %	20.3 %

Applying (9) to 1,000 estimated interval priority weights W_i in each of conventional and proposed estimation methods, we calculated dominance ratios for each pair (o_p, o_q) of alternatives. The results are shown in Table 7. As shown in rows of (o_3, o_4) and (o_4, o_5) in Table 7, the interval priority weights estimated by the conventional method estimate correct dominance relations $o_3 \succ_W^\nu o_4$ and $o_5 \succ_W^\nu o_4$ more frequently than those by the proposed method. From this sense, the conventional estimation looks better than the proposed estimation. However, as shown in rows of (o_1, o_4) and (o_2, o_4) in Table 7, the interval priority weights estimated by the conventional method estimate wrong dominance relations $o_1 \succ_W^\nu o_4$ and $o_2 \succ_W^\nu o_4$ more frequently than those by the proposed method. Surprisingly, frequencies of $o_1 \succ_W^\nu o_4$ and $o_2 \succ_W^\nu o_4$ are bigger than those of $o_3 \succ_W^\nu o_4$ and $o_5 \succ_W^\nu o_4$ in the conventional estimation. This may be caused by the fact that differences $ud(o_1, o_4) - ud(o_4, o_1) = 0.075$ and $ud(o_2, o_4) - ud(o_4, o_2) = 0.075$ are bigger than differences $ud(o_3, o_4) - ud(o_4, o_3) = 0.07$ and $ud(o_5, o_4) - ud(o_4, o_5) = 0.07$. On the other hand, although the frequencies of $o_3 \succ_W^\nu o_4$ and $o_5 \succ_W^\nu o_4$ are not very big, frequencies of $o_1 \succ_W^\nu o_4$ and $o_2 \succ_W^\nu o_4$ are smaller than those of $o_3 \succ_W^\nu o_4$ and $o_5 \succ_W^\nu o_4$ in the proposed estimation. Non-dominance relations for pairs (o_1, o_2), (o_1, o_3) and (o_2, o_5) are estimated better by the proposed method than by the conventional method.

From those observation, we can conclude that the interval priority weights estimated by the conventional method is too narrow to have a safe estimation. The proposed estimation is better than the conventional one but, nevertheless, the widths of estimated intervals are still narrow so that it fails to estimate correct dominance relations.

5   Concluding Remarks

In this paper, we proposed a logarithmic conversion approach to the estimation of interval priority weights in the setting of Interval AHP. The proposed approach estimates interval priority weights by solving a single linear programming problem but the normalization is necessary for the optimal solution. By numerical experiments, we demonstrated the advantages of the proposed estimation over the conventional estimation. By the experiments, we found that the estimated interval priority weights are too narrow in the conventional method and the distribution of widths to priority weights of criteria is unbalanced. On the other hand, the proposed estimation is better than the conventional one because the widths of the estimated interval priority weights are larger and the distribution of the widths to priority weights of criteria is more balanced. Nevertheless, the widths of the interval priority weights estimated by the proposed method are still small for obtaining better estimations. Further studies on the estimation of interval priority weights are necessary without loss of the tractability of the estimation problem.

Acknowledgments. This work was supported by JSPS KAKENHI Grant Number 26350423.

References

1. Saaty, T.L.: The Analytic Hierarchy Process. McGraw-Hill, New York (1980)
2. Saaty, T.L., Vargas, C.G.: Comparison of eigenvalue, logarithmic least squares and least squares methods in estimating ratios. Mathematical Modelling **5**, 309–324 (1984)
3. van Laarhoven, P.J.M., Pedrycz, W.: A fuzzy extension of Saaty's priority theory. Fuzzy Sets and Systems **11**, 199–227 (1983)
4. Buckley, J.J.: Fuzzy hierarchical analysis. Fuzzy Sets and Systems **17**, 233–247 (1985)
5. Wang, Y.-M., Elhag, T.M.S., Hua, Z.: A modified fuzzy logarithmic least squares method for fuzzy analytic hierarchy process. Fuzzy Sets and Systems **157**, 3055–3071 (2006)
6. Arbel, A.: Approximate articulation of preference and priority derivation. European Journal of Operational Research **43m**, 317–326 (1989)
7. Sugihara, K., Tanaka, H.: Interval evaluations in the analytic hierarchy process by possibilistic analysis. Computational Intelligence **17**, 567–579 (2001)
8. Sugihara, K., Ishii, H., Tanaka, H.: Interval priorities in AHP by interval regression analysis. European Journal of Operational Research **158**, 745–754 (2004)
9. Tanaka, H., Nagasaka, K., Hayashi, I.: Interval regression analysis by possibilistic measures (in Japanese). The Japanese Journal of Behaviormetrics **16**, 1–7 (1988)
10. Entani, T., Inuiguchi, M.: Pairwise comparison based interval analysis for group decision aiding with multiple criteria. Fuzzy Sets and Systems **271**, 79–96 (2015)
11. Guo, P., Tanaka, H.: Decision making with interval probabilities. European Journal of Operational Research **203**, 444–454 (2010)

An Effective Method for Optimality Test Over Possible Reaction Set for Maximin Solution of Bilevel Linear Programming with Ambiguous Lower-Level Objective Function

Puchit Sariddichainunta[✉] and Masahiro Inuiguchi

Osaka University, 1-3 Kanemachiyama, Toyonaka, Osaka 560-8531, Japan
puchit@inulab.sys.es.osaka-u.ac.jp, inuiguchi@sys.es.osaka-u.ac.jp

Abstract. A bilevel linear optimization problem with ambiguous lower-level objective requires a decision making under uncertainty of rational reaction. With the assumption that the ambiguous coefficient vector of the follower lies in a convex polytope, we apply the maximin solution approach and formulate it as a special kind of three-level programming problem. According to its property that the optimal solution locates on an extreme point, we adopt k-th best method to search the optimal solution equipped with tests for possible optimality, local optimality and global optimality of a solution. In this study, we propose an effective method to verify the rational reaction of the follower which is essential to all steps of optimality test. Our approach uses a relatively small memory to avoid repetition of possible optimality tests. The numerical experiments demonstrate our proposed method significantly accelerates the optimality verification process and eventually computes an optimal solution more efficiently.

Keywords: Bilevel linear optimization · Possibly optimal decision making · Maximin solution

1 Introduction

Bilevel linear programming problem (BLP) is an extension of the linear programming problem that consists of two levels of decision making stage [2]. It is known as a sequential game in non-cooperative game theory or a so-called Stackelberg game. In such game, the leader at the upper level chooses his strategy first, and then the follower at the lower level makes his own decision taking the leader's decision into consideration as given. Its applications are found useful in many research areas where the model used is a hierarchical optimization problem [6]; e.g., principal-agency problem in economics, optimal chemical equilibria, and irrigation water resource management.

In the conventional BLP [2,3,6], each decision maker is assumed to have complete information about the game. The leader and the follower can exactly observe each other's payoff function and strategy space. However, in realistic scenarios, the accurate information of the counterpart is not easily observed and

© Springer International Publishing Switzerland 2015
V.-N. Huynh et al. (Eds.): IUKM 2015, LNAI 9376, pp. 89–101, 2015.
DOI: 10.1007/978-3-319-25135-6_10

often contains ambiguity. For instance, the principal-agent problem between the regulator and the electricity generating company in the monopoly market is the case. The regulator as a policy maker wants to impose an optimal measure for the environmental issue and the energy security policy. While the regulator cannot have the complete information that describes the company motivation for its profit maximizing, the regulator has to develop their policy based on the indeterminate information of the company such as fuel price, demand for electrical power and so on. For the regulator, these information are imprecise and ambiguous due to the uncertainty, but are necessary to foresee the profit which motivates the rational action of the company. This class of principal-agent problem is much discussed in the area of contract theory [4]. Hence, the BLP with ambiguous coefficients has attracted high attentions and has been recently studied [1,5,16].

In our research [11], we propose another approach to solve the BLP problem with an ambiguous follower's objective function based on a maximin decision principle under uncertainty, and a solution algorithm based on k-th best method. We assume that the coefficients in the leader's objective function are precise as well as the coefficients in the constraints are crisp number. Although some research [5,16] set the coefficient vector of leader as imprecise parameters, their models constructed best-case and worst-case scenario of the leader's objective function which eventually fixed the coefficient vector to develop the solution method. Thus, this scenario construction is equivalent to assume that the leader's objective function is precise. On the contrary, the coefficients in the follower's objective function are the most concerned. We assume it to be ambiguous and can be represented by a convex polytope. Since those coefficients often depend on the setting of follower's problem that are not known well by the leader and the objective function reflecting the follower's decision may be unclear to the leader, we think that the BLP problem with an ambiguous follower's objective function coefficients is one of the crucial parts of the BLP problem with ambiguous coefficients.

In this paper, we provide an effective method to possibly conduct the optimality test that helps verifying the possible rational action of the follower. This optimality verification process is essential for both local and global optimality test by the definition. It turns out that the acceleration of our k-th best method is very promising. We have employed the numerical experiments in order to observe the run time efficiency of our proposed methods.

The organization of this paper is as follows. In Section 2, we describe the formulation of the problem. We also discuss some properties related to the formulation. Our proposed solution methods and theoretical backgrounds are provided in Section 3. In Section 4, we show the numerical experiments. Finally, conclusion and potential future works are discussed in Section 5.

2   Problem Formulation

Briefly we introduce our model setting for BLP with ambiguous objective function of the follower as well as the definitions and properties in this model. Our model focus on the maximin decision criteria for the leader's decision when he cannot observe the follower's objective function precisely. We assume the leader just ambiguously knows follower's objective coefficient vector to some extent. This uncertainty is represented by the convex polytope Γ. Moreover, the strategic effect from the sequential decision is incorporated to the model as a two-step rational decision in the constraints. This problem can be written as follows:

$$
\begin{aligned}
&\underset{x}{\text{maximize}}\ c_1^\mathrm{T} x + c_2^\mathrm{T} y,\\
&\text{subject to } x \geq 0,\\
&\qquad y \text{ is determined by the follower so as to}\\
&\qquad \underset{y}{\text{maximize}}\ \ \tilde{c}_3^\mathrm{T} y,\ \exists \tilde{c}_3 \in \Gamma = \{c_3 \mid Gc_3 \leq g\},\\
&\qquad \text{subject to } A_1 x + A_2 y \leq b,\\
&\qquad\qquad y \geq 0,
\end{aligned}
\tag{1}
$$

where $x \in \mathbf{R}^p$ and $y \in \mathbf{R}^q$ are the decision variable vectors of the leader and the follower, respectively, $c_1 \in \mathbf{R}^p$, $c_2 \in \mathbf{R}^q$, $b \in \mathbf{R}^m$, $A_1 \in \mathbf{R}^{m \times p}$, and $A_2 \in \mathbf{R}^{m \times q}$ are constant vectors and matrices while $\tilde{c}_3$ is the ambiguous coefficient vector, and $\Gamma \subseteq \mathbf{R}^q$ is a polytope defined by a matrix $G \in \mathbf{R}^{l \times q}$ and a vector $g \in \mathbf{R}^l$.

We assume the feasible solution set $S = \{(x, y) \mid A_1 x + A_2 y \leq b,\ x \geq 0,\ y \geq 0\}$ is bounded and nonempty. The strategy set of the leader and the follower in the feasible region S are defined respectively by $X(S) = \{x \geq 0 \mid \exists y \geq 0;\ A_1 x + A_2 y \leq b\}$, and $S(x) = \{y \geq 0 \mid A_2 y \leq b - A_1 x\}$. Moreover, we assume that the possible range of ambiguous coefficient vector $\tilde{c}_3$ is known in a bounded convex polyhedron defined by $\Gamma = \{c \in \mathbf{R}^q \mid Gc \leq g\}$ where $G \in \mathbf{R}^{r \times q}$ and $g \in \mathbf{R}^r$. When the leader knows exactly the coefficient vector of the follower's objective function, he can understand the follower's rational reaction set, $\mathrm{Opt}(c_3, x) = \{y \in S(x) \mid c_3^\mathrm{T} y = \max_{z \in S(x)} c_3^\mathrm{T} z\}$. In the case of ambiguous coefficient vector $\tilde{c}_3$, the leader cannot know the follower's rational response exactly, but he can explore the follower's rational response in a larger region. Under the assumption that $\tilde{c}_3 \in \Gamma$, we define the follower's possible reaction set by $\mathit{\Pi} S(x) = \bigcup_{c_3 \in \Gamma} \mathrm{Opt}(c_3, x)$.

According to the above situation, we further assume that the leader will consider the worst effect of the follower's response to his strategy, and rationalize his solution by adopting maximin criteria. Immediately, linear programming problem (1) is formulated as

$$
(OP)\left\{
\begin{aligned}
&\underset{x}{\text{maximize}}\ c_1^\mathrm{T} x + c_2^\mathrm{T} y\\
&\text{subject to } x \geq 0\\
&\quad y,\ c_3 \text{ solves,}\\
&(SP(x))\left\{
\begin{aligned}
&\underset{y, c_3}{\text{minimize}}\ c_2^\mathrm{T} y\\
&\text{subject to}\\
&c_3^\mathrm{T} y = \max_{z \geq 0}\{c_3^\mathrm{T} z \mid A_2 z \leq b - A_1 x\}\\
&Gc_3 \leq g\\
&A_1 x + A_2 y \leq b\\
&y \geq 0
\end{aligned}
\right.
\end{aligned}
\right.
$$

The upper level problem is a maximization problem for the leader to decide his decision variable x. The lower level is a minimization problem to obtain the optimal y and c_3 which affect on the leader's objective function value at worst. We note that a maximization problem is included in the lower level problem which represents rationality of the follower. Thus, the problem (OP) can be seen as a three-level programming problem. Let us define the inducible region set as $IR = \{(x, y) \mid (x, y) \in S, \ y \in \Pi S(x)\}$. The problem (OP) is rewritten as follows:

$$\begin{cases} \underset{x,y}{\text{maximize}} \ c_1^T x + c_2^T y, \\ \text{subject to } (x, y) \in IR. \end{cases} \tag{2}$$

3 Proposed Solution Methods

In this section we introduce a solution framework based on k-th best method. It consists of the vertex enumeration process and the optimality test process for possible optimality, local optimality and global optimality. The vertex enumeration is done partly and terminates when the global optimal solution is verified. In addition, we revisit some fundamental ideas of optimality and propose an effective method to reduce computation time for optimality test which consequently improve the overall algorithm efficiency. The basic idea of k-th best method, optimality definitions and test processes are explored in the following subsections.

3.1 K-th Best Method

The k-th best method is a vertex enumeration algorithm. It is a search algorithm starting from the first best solution and sequentially suggests the second best and so on, if the former solution does not satisfy to the optimality conditions. This method is useful when the optimal solution occurs at the vertex; i.e., the extreme point of the feasible region. The following theorem enhances k-th best method applicable to solve our model.

Theorem 1. *[11] The optimal solution of problem (OP) is located on a vertex of feasible region, S.*

Thus we proposed a solution procedure based on the k-th best method by checking the feasibility in the descending sequence of leader's objective function value. The following is the linear problem which ignores the lower-level problem.

$$\max\{c_1^T x + c_2^T y \mid (x, y) \in S\} \tag{3}$$

$(x^1, y^1), \cdots, (x^N, y^N)$ denote the N ordered basic feasible solutions satisfying

$$c_1^T x^k + c_2^T y^k \geq c_1^T x^{k+1} + c_2^T y^{k+1}, \ k = 1, \cdots, N - 1 \tag{4}$$

According to the fact that there exists an optimal solution at an extreme point, solving problem (OP) is to find k^* which is the smallest index of extreme points in set IR.

$$k^* = \min\{k \in \{1, \cdots, N\} \mid (\boldsymbol{x}^k, \boldsymbol{y}^k) \in IR\} \tag{5}$$

Based on this method, we fix $\boldsymbol{x} = \boldsymbol{x}^k$ and then justify whether $\boldsymbol{y}^k$ is the optimal solution for problem $(SP(x))$. If so, we can conclude $(\boldsymbol{x}^k, \boldsymbol{y}^k) \in IR$. Because we assume S is bounded, we have a finite number of basic feasible solutions (BFS). Therefore, this procedure stops in a finite number of iterations. The procedure is written as follows:

[**Algorithm 1**]. The solution procedure for the problem (OP) based on k-th best algorithm

Step 1. $k = 1$. Solve problem (3) by simplex method, and get the solution $(\boldsymbol{x}^1, \boldsymbol{y}^1)$. Set $M = \{(\boldsymbol{x}^1, \boldsymbol{y}^1)\}$, $T = \emptyset$.

Step 2. Fix $\boldsymbol{x} = \boldsymbol{x}^k$. Justify whether $\boldsymbol{y}^k$ is the optimal solution of $(SP(\boldsymbol{x}^k))$ or not. If it is the optimal solution of $(SP(\boldsymbol{x}^k))$, the solution is also the optimal solution of problem (OP) and terminate the procedure with the optimal solution $(\boldsymbol{x}^k, \boldsymbol{y}^k)$. Otherwise, go to step 3.

Step 3. Generate M^k, a set of adjacent extreme points of $(\boldsymbol{x}^k, \boldsymbol{y}^k)$, which satisfies $\boldsymbol{c}_1^T \boldsymbol{x} + \boldsymbol{c}_2^T \boldsymbol{y} \leq \boldsymbol{c}_1^T \boldsymbol{x}^k + \boldsymbol{c}_2^T \boldsymbol{y}^k$. Then, $T = T \cup \{(\boldsymbol{x}^k, \boldsymbol{y}^k)\}$, $M = (M \cup M^k) \setminus T$.

Step 4. $k = k + 1$. Choose $(\boldsymbol{x}^k, \boldsymbol{y}^k) \in \arg\max_{\boldsymbol{x}, \boldsymbol{y}}\{\boldsymbol{c}_1^T \boldsymbol{x} + \boldsymbol{c}_2^T \boldsymbol{y} \mid (\boldsymbol{x}, \boldsymbol{y}) \in M\}$, and go to step 2.

Step 1, 3 and 4 of Algorithm 1 is the usual vertex enumeration process. On the other hand, step 2 is the optimality verification process of $\boldsymbol{y}^k$ to problem $(SP(\boldsymbol{x}^k))$, feasibility, local and global optimality. We set up these tests to avoid computational loads for the global optimality test and to aim at detecting inconsistent solutions to each optimality definition:

Definition 1. *Possible, local and global optimality*
   Possible: $\boldsymbol{y} \in \Pi S(\boldsymbol{x}^k)$.
   Local: $\boldsymbol{y}^k$ *s.t.* $\not\exists \, \boldsymbol{y} \in N(\boldsymbol{y}^k) \cap \Pi S(\boldsymbol{x}^k)$ *and* $\boldsymbol{c}_2^T \boldsymbol{y} < \boldsymbol{c}_2^T \boldsymbol{y}^k$,
        where $N(\boldsymbol{y}^k)$ *is a set of all adjacent basic feasible solutions of* $\boldsymbol{y}^k$.
   Global: $\boldsymbol{y}^k$ *s.t.* $\not\exists \, \boldsymbol{y} \in \Pi S(\boldsymbol{x}^k)$ *and* $\boldsymbol{c}_2^T \boldsymbol{y} < \boldsymbol{c}_2^T \boldsymbol{y}^k$.

Those specific details of each optimality definition and test procedures are discussed in the next subsections.

3.2   Rational Reactions and Possible Optimality Test

As the leader needs to speculate the act of follower $\boldsymbol{y}^k$, it is necessary to check whether that $\boldsymbol{y}^k$ is a rational choice and a possible reactions of the follower, so-called, possible optimality or the feasibility of rational response. The feasibility of $\boldsymbol{y}^k$ is verified if there is a vector $c_3 \in G$ to advocate it as a rational reaction of the follower in problem $(SP(x^k))$. This is equivalent to that y^k is the member of possibly optimal solution set, noted by $y^k \in \Pi S(x)$.

As shown by [9], the possibly optimal solution set equals to a weakly efficient solution set of a multiple objective linear programming problem. Since a weakly efficient solution set is connected and polyhedral [15], $\Pi S(x)$ is then connected and polyhedral. However, $\Pi S(x)$ is not a convex set, problem $(SP(x))$ cannot be solved easily even when x is determined. Given $(x, y) \in S$, the feasibility of a solution y can be tested by the possibly optimal test proposed by [8]. The characterization of possible optimality test is described in the next theorem.

Theorem 2. *[11] The necessity and sufficient condition for $y \in S(x)$ to be a possible reaction, i.e., $y \in \Pi S(x)$, is given by the consistency of the following system of linear inequalities:*

$$\sum_{i=1}^{\tilde{m}} u_i \bar{a}_i \in \Gamma, \quad u_i \geq 0, \quad \forall i = 1, 2, \ldots, \tilde{m}, \tag{6}$$

where $\bar{a}_i^{\mathrm{T}}$ is a row vector of $\begin{bmatrix} A_2 \\ -E \end{bmatrix}$ corresponding to the i-th active constraint of $A_2 y \leq b - A_1 x$ and $y \geq 0$, where E is an identity matrix and $\tilde{m}$ is the total number of active constraints.

The test result of the method following to theorem 2 can be reused for other BFSs that have the same set of basic variables. Intuitively, the information required for (6) are just the active normal vectors of binding constraints and $Gc_3 \leq g$. The former can be represented by a set of non-basic variables, and the latter is a necessary condition. To all the other iterations, any extreme points with the same set of basic/non-basic variables correspond to the same active constraints. We thus can identify whether the current BFS belongs to the possibly optimal solution set by considering the set of basic/non-basic variables of each solution.

Corollary 1. *Let I be a set of basic variables for a BFS. Every extreme point in the lower-level problem $(SP(x))$ represented by this set I retain the same result of possibly optimal test.*

Furthermore, some partial information from the test method by theorem 2 can be reused to avoid the repetition of solving the same linear programming. The solution of positive u_i preserves the feasibility of other y of which BFS contains all non-basic variables corresponding to those positive u_i. In other words, if one BFS has been tested and become a member of $\Pi S(x)$, the solution of positive u_i generates a subset of non-basic variables that correspond to normal vectors of a cone for c_3. Another BFS with such a subset included in its non-basic variable set is also possibly optimal since the same c_3 maintains the possible optimality for this BFS.

Corollary 2. *Let J be a set of non-basic variables such that it has a corresponding vector $u > 0$ satisfies (6). The BFS that has all of those non-basic variables corresponding to positive u_i is also a possibly optimal solution.*

Subsequently, we found a simple method to test a single non-basic variables that always generate its corresponding positive u_i. The idea is based on the fact that u_i is a positive coefficient extending the normal vector of active constraint to relocate in the convex polyhedron Γ. The requirement is nothing but to find an appropriate scalar value of that normal vector enhancing its feasibility in Γ.

Corollary 3. *There is a positive coefficient u_i for a normal vector $\bar{a}_i$ in matrix A_2 such that $u_i G \bar{a}_i = u_i \hat{g} \leq g$ if and only if there is no such $g_t < 0 < \hat{g}_t$ for $t = 1, \ldots, q$ in g and $\hat{g}$, and there is u_i such that $r' \leq u_i \leq r''$ where*

$$r' = \max\left\{\frac{g_t}{\hat{g}_t} > 0 \mid g_t, \hat{g}_t < 0\right\} \text{ and } r'' = \min\left\{\frac{g_t}{\hat{g}_t} > 0 \mid g_t, \hat{g}_t > 0\right\}.$$

According to corollary 2 and 3, we can adopt the results of non-basic variable with respect to u_i to immediately verify possible optimality of other BFS which have those corresponding non-basic variables. The repetitive computations of solving LP for (6) to check possible optimality are avoidable. Indeed, we reuse the test results that passed the possibly optimal test according to Theorem 2. It requires some memory space only linear order with the upper bound of all possible full-rank combinations in the lower-level problem. The usage of memory storage provide a quick guide for evaluation of unchecked BFSs, and accelerate possible optimality verification in total.

3.3 Local Optimality Test

Local optimal solution is a solution with the property that there is no neighbor BFS which is possibly optimal and improves the objective value of the lower-level problem $(SP(x))$. We define it on the BFS so that the implementation is readily to operate the simplex algorithm. The procedure of local optimality tests is written as follows:

[Algorithm 2]. Local optimality for problem $(SP(x^k))$.

Step 2-(a). Apply simplex method to confirm the existence of a solution satisfying equation (6). If it exists, we know $y^k \in \Pi S(x^k)$ and go to step 2-(b). If it does not exist, go to step 3.

Step 2-(b). For each adjacent basic solution y of the current basic solution corresponding to y^k, we check $c_2^T y < c_2^T y^k$ and $y \in \Pi S(x^k)$ by the same way as in step 2-(a). If such an adjacent solution is founded, y^k is not locally optimal and go to step 3. Otherwise, proceed to the global optimality test, i.e., step 2-(c).

Step 2-(c). Test the global optimality of y^k to problem $(SP(x^k))$. If the global optimality is verified, we found that (x^k, y^k) is the optimal solution to problem (OP). Otherwise, go to step 3.

The degenerate BFS causes a serious concern in the verification process since it deters the pivoting process to move to another BFS which improves the objective value in the preferable direction, $-c_2$. It usually occurs when the columns are reduced; in this model the leader's decision variables are eliminated in the lower-level problem. To reach a BFS in the preferable direction, we effectively operate the depth-first search in the BFS search space for local as well as global optimality test and expect for fast detection of non-optimal y^k, [14].

3.4 Global Optimality Test

Following to the definition of global optimality, it is necessary to check all feasible solutions of problem $(SP(x^k))$ to assure the maximin decision criteria. In this section, we adopt three algorithms for the global optimality test: the two step k-th best method [13], the enumeration of adjacent possibly optimal solutions [11] and the inner approximation method [7]. The first method is intuitive from the definition and the second method is developed following to the fact that possibly optimal solutions are connected [10]. On the other hand, the third method is an algorithm to approximate the global optimality which somewhat generates unstable results due to zero-rounding errors would probably occur.

First, the two step k-th best method for the global optimality test is a direct implementation, according to the global optimality definition, to find out that no other better solutions than y^k is a possible reaction of the follower. It sequentially enumerates all extreme points in $S(x)$. The computation could be costly if there are many extreme points to visit. The Algorithm 3A is a basis of comparison in the numerical experiments.

[**Algorithm 3A**]. Two step k-th best method

Step (a). $l = 1$ and $y^{[1]} = \underset{y \in S(x^k)}{\text{argmin}} \ c_2^T y. \ \mathcal{E} = \emptyset, \mathcal{T} = \{y^{[1]}\}$.

Step (b). Check $y^{[l]} \in \Pi S(x^k)$. If so, evaluate $c_2^T y^{[l]} = c_2^T y^k$. y^k is global optimal solution for $(SP(x^k))$ when the equality is true, and terminate algorithm. If it is not equal, we conclude y^k is not global optimal solution, and go back to Algorithm 1.

Step (c). Let set $\mathcal{A}$ contain all adjacent BFS $\hat{y}$ such that $c_2^T \hat{y} \geq c_2^T y^{[l]}$. $\mathcal{E} = \mathcal{E} \cup \{y^{[l]}\}, \mathcal{T} = \mathcal{T} \cup (\mathcal{A} \setminus \mathcal{E})$.

Step (d). $l = l + 1$, $y^{[l]} = \text{argmin}\{c_2^T y | y \in \mathcal{A}\}$, and go to step (b).

Next, the adjacent enumeration of possibly optimal solution is implemented by using a valid inequality to restrict the feasible region of problem $(SP(x^k))$ which we really concern. It eliminates the non-interesting area where $c_2^T y \geq c_2^T y^k$ holds. We then obtain $S^k(x) = S(x) \cap \{y \mid c_2^T y < c_2^T y^k\}$, and the global optimality condition of y^k becomes $S^k(x) \cap \Pi S(x) = \emptyset$. We set $\varepsilon = 10^{-4}$ for $S_\varepsilon^k(x) = S(x) \cap \{y \mid c_2^T y \leq c_2^T y^k - \varepsilon\}$, instead of $S^k(x)$.

[**Algorithm 3B**]. Enumeration of adjacent possibly optimal solutions

Step (a). Initialize the unchecked basic feasible solution set $\mathcal{U}$ and checked basic feasible solution set is $\mathcal{E}$ as empty sets. Choose a vector $\hat{c}_3 \in \Gamma$, and solve

$\max_{y \in S_\varepsilon^k(x)} \hat{c}_3^T y$. If it is infeasible, we terminate the algorithm concluding that y^k is a global optimal solution of problem $(SP(x^k))$. Otherwise, let $\hat{y}$ be the obtained solution and go to step (b).

Step (b). Check the membership of $\hat{y}$ to $\Pi S(x)$ by testing the consistency of (6) with respect to $\hat{y}$. If $\hat{y} \in \Pi S(x)$, we terminate the algorithm concluding that y^k is not a global optimal solution of problem $(SP(x^k))$. Otherwise, update $\mathcal{E} = \mathcal{E} \cup \{\hat{y}\}$ and go to step (c).

Step (c). Generate all adjacent basic feasible solutions which are members of $\Pi S_\varepsilon^k(x)$. Let $\mathcal{D}$ be the set of those solutions. Update $\mathcal{U} = \mathcal{U} \cup (\mathcal{D} \setminus \mathcal{E})$. If $\mathcal{U} = \emptyset$, we terminate the algorithm concluding that y^k is a global optimal solution of problem $(SP(x^k))$.

Step (d). Select a basic feasible solution $\hat{y}$ from $\mathcal{U}$ and update $\mathcal{U} = \mathcal{U} \setminus \{\hat{y}\}$. Go back to step (b).

Finally, the inner approximation method exerts the polyhedral annexation technique [7]. It has been developed for the reverse convex optimization, and then applied to BLP which possesses the similar structure. This method transforms BLP into the quotient space which has lower dimensions, depending mainly on the number of independence constraints of the lower-level problem. It does not directly evolve all the problem $S^k(x)$ at once, but gradually approximate the solution from partially a necessary interior set.

The global optimality is to verify whether $\bar{S}^k(x) \subseteq \text{int}\bar{W}(x)$, where $\bar{S}^k(x) = \{u \mid \bar{A}^N u \leq \bar{b}, \ u \geq 0\} \cap \{u \mid c_2^T(y^0 + \xi u) < c_2^T y^k\}$ and $\bar{W}(x) = \{u \mid c_3^T(y^0 + \xi u) \leq \max\limits_{z \in S(x)} c_3^T z, \ \forall c_3 \in \Gamma\}$. In practice, we construct $P_j = \{u \geq 0 \mid t^T u \leq 1, \ \forall t \in V_j\}$ for $V_j \subset \mathbb{R}^q$, and observe whether $P_1 \subset P_2 \subset \cdots \subset \bar{W}(x)$. t^j is the normal vector of simplex P_j. It provides the recession direction u^j that updates P_{j+1} when there is $t \in V_j$ such that $\nu(t) \geq 1$. The initial simplex $P_1 = \text{co}\{0, \mu^1 e_1, \mu^2 e_2, \cdots, \mu^q e_q\}$ is constructed by μ^i the solution of the following problem:

$$\begin{cases} \text{maximize} \ \tau \\ \quad\;\; v, \tau \\ \text{subject to } A_2(-G^T v + y^0 + \xi \tau u^j) \leq b - A_1 x \\ \qquad\quad -G^T v + y^0 + \xi \tau u^j \geq 0 \\ \qquad\quad g^T v \leq 0 \\ \qquad\quad v \geq 0, \ \tau \geq 0 \end{cases} \tag{7}$$

[Algorithm 3C]. Inner approximation method

Initialization: Solve $y^0 = \arg\min\limits_{y \in S(x)} c_2^T y$ and apply possible optimality test.

If feasible, y^k is not global optimal. Go back to algorithm 1.

If not, we construct $\bar{S}_\varepsilon^k(x)$, $\bar{W}(x)$ and solve for $\mu^i, i = 1, \cdots, q$.

If $\mu^i e_i \cap \bar{S}_\varepsilon^k(x) \neq \emptyset$, then y^k is global optimal and go back to algorithm 1.

If pass the above requirements, we find t^0 from μ^i. Set $j \leftarrow 1$ to initialize $N_1 = \{0\}$, $V_1 = \emptyset$, $\Lambda_1 = \mathbf{R}_+^q$.

Step (a). For every $\lambda \in N_j$, compute $\nu(\lambda)$. $\nu(\lambda) = \max\limits_\lambda \{(t^0 - \lambda)^T u \mid u \in \bar{S}_\varepsilon^k(x)\}$. Remove λ from N_j such that $\nu(\lambda) < 1$. If V_j and N_j are both empty, then y^k is global optimal. Terminate the algorithm.

Step (b). Set $V_j \leftarrow V_j \cup N_j$. Find $\boldsymbol{\lambda}^j \in \arg\max_{\boldsymbol{\lambda}}\{\nu(\boldsymbol{\lambda}) \mid \boldsymbol{\lambda} \in V_j\}$ and $\boldsymbol{u}^j \in$
$\arg\max_{\boldsymbol{u}}\{(\boldsymbol{t}^0 - \boldsymbol{\lambda}^j)^{\mathrm{T}}\boldsymbol{u} \mid \boldsymbol{u} \in \bar{S}_\varepsilon^k(\boldsymbol{x})\}$

Step (c). Check $\boldsymbol{u}^j \in \partial\bar{W}(\boldsymbol{x})$. If $\boldsymbol{u}^j \in \partial\bar{W}(\boldsymbol{x})$, $\boldsymbol{y}^k$ is not global optimal. Go back to algorithm 1. If $\boldsymbol{u}^j \notin \partial\bar{W}(\boldsymbol{x})$, solve θ^j which is the optimal value of (7).

Step (d). Use a cutting hyperplane, $\Lambda_{j+1} = \Lambda_j \cap \{\boldsymbol{\lambda} \mid (\boldsymbol{t}^0 - \boldsymbol{\lambda})^{\mathrm{T}}\boldsymbol{u}^j \leq \frac{1}{\theta^j}\}$. For all extreme points in Λ_{j+1}, there are some new extreme points generated by the cut. Put them into N_{j+1}, and separate the others to V_{j+1}. Update $j \leftarrow j + 1$ and go back to Step (a).

4   Numerical Experiments

We evaluate the performance of our proposed method on randomly generated problems in the following way. m-p-q-r is the parameter of our generated problems. m is the number of constraints of feasible set S. p and q are the numbers of decision variables of the leader and of the follower respectively. r is the number of constraints in Γ.

4.1   Problem Generation

To generate a set of feasible region S, we derive m tangent hyperplanes from the surface of a unit hypersphere in the positive coordinate plane [12]. $\boldsymbol{r}_1 \in \mathbf{R}^{p+q}$ is a uniform random vector for the hypersphere equation, $\frac{\boldsymbol{r}_1^{\mathrm{T}}}{|\boldsymbol{r}_1|}\begin{pmatrix}\boldsymbol{x}\\\boldsymbol{y}\end{pmatrix} \leq 1$. $|\boldsymbol{r}_1|$ means the norm of uniformly random vector $\boldsymbol{r}$. After that, we perturb these hyperplane by using a random vector $\boldsymbol{r}_2 \in \mathbf{R}^{p+q}$ of which each element randomizes between $[1, 3]$. The coordinate is converted as a direct multiplication, $(\hat{\boldsymbol{x}}, \hat{\boldsymbol{y}}) = \boldsymbol{r}_2 \otimes (\boldsymbol{x}, \boldsymbol{y})$. Now we have a convex set S as a feasible region defined by m tangent hyperplanes of a random ellipsoid.

The coefficient vectors $\boldsymbol{c}_1 \in \mathbf{R}^p$, $\boldsymbol{c}_2 \in \mathbf{R}^q$ is a random vector whose elements are uniformly random in $[1, 4]$. The set Γ defined by q constraints is also generated similarly to S. The further process is that we move the center of a hypersphere out of the origin and add another q parallel tangent hyperplanes to create a new convex polytope without the axes.

4.2   Numerical Results

The numerical experiments are set up to compare the effect before and after implementing the reuse of possible optimality test results for each global optimality test method. The base line algorithms are described in subsection global optimality: Algorithm (3A), (3B) and (3C). The programs are developed in Microsoft Visual C/C++ 2013 and run the experiments in a desktop computer (OS: Windows 8.1, CPU:3.4 GHz, RAM:8GB). The epsilon value 10^{-4} is used.

The numerical experiments aim to compare the improvement of efficiency in each method. We measure the CPU run time used in achieving the global optimal

Table 1. Results of experiment (CPU time (s))

Case	5-5-8-20			10-5-8-20			15-5-8-20		
Method	A	B	C	A	B	C	A	B	C
1st	22%	19%	35%	8%	12%	13%	2%	3%	10%
Mean	0.090	0.097	0.067	2.450	2.261	1.866	29.442	23.472	20.438
S.D.	0.105	0.118	0.072	2.735	3.141	3.208	33.671	29.527	26.081
Min	0	0	0	0	0	0	0	0	0
Max	0.562	0.609	0.375	10.125	13.954	27.877	136.775	128.009	113.227
Median	0.054	0.062	0.046	1.062	0.843	0.656	15.423	11.329	10.079
Method	A'	B'	C'	A'	B'	C'	A'	B'	C'
1st	63%	41%	73%	31%	50%	49%	14%	48%	32%
Mean	0.050	0.059	0.050	1.571	1.583	1.673	23.967	21.785	20.235
S.D.	0.049	0.058	0.053	1.753	2.058	3.028	28.876	28.503	26.332
Min	0	0	0	0	0	0	0	0.015	0
Max	0.250	0.250	0.328	6.672	8.891	26.798	113.742	124.228	113.899
Median	0.031	0.046	0.031	0.758	0.625	0.586	10.782	9.774	9.649
Ratio	55%	44%	43%	67%	50%	45%	73%	52%	45%

solution. 100 problems are generated for 3 cases: 5-5-8-20, 10-5-8-20, and 15-5-8-20 to observe the efficiency varied by the number of constraints which represents the size of problem. Algorithm (3A), (3B), and (3C) without the reuse strategy are represented by A, B, and C, otherwise denoted by A', B' and C'.

Table 1 reports the statistics of CPU run time in seconds. The row of 1st is the proportion of problem sets which the method can finish at the lowest CPU run time among 6 methods. It is derived from the number of problems it can finish first over the number of all problems. The sum of those percentage of each method in the same problem set dose not equal to one because there are some problems that more than one method finish at the same time. The ratio at the bottom of the table is a proxy represents the average rate of efficiency improvement in each problem set. It depicts how our proposed method for possible optimality test has saved time from the duplication of possible optimality test. It is calculated by the frequency of reusing test results over the frequency of test inquiries. The higher scale of ratio, the more efficiency gained.

The result in Table 1 illustrates an empirical evidence for the validity of our proposed method which save some computational time in the possible optimality test process. When the problem size increases, the ratio in column A (the two step k-th best method) also positively changes as the larger problem size encounters with more inquiries of the possible optimality test, especially for the two step k-th best method. Although the ratio for method B and C are not very large because these methods are designed to check fewer BFSs at the lower-level problem. However, the proportion is still notable.

Considering before and after implement our reuse strategy, method C (the inner approximation) seems to be the fastest finisher following by B and A according to the mean run time. In general, the computational time of method A' and B' come closed to the computational time of method C' when deploy the reuse strategy. Only the case 10-5-8-20, it seems that method B' win over method C', but the difference is not statistically significant as the dependent sample t-test is applied with the result: t-value 0.412 and two-sided probability 0.367. Following to the percentage of the 1st row, method B' has beaten method C' for the problem set 10-5-8-20 and 15-5-8-20. The result implies that the reuse strategy effectively improve computational time of method B in many types of problems. The further investigation in problem structure is required.

5 Conclusion

After we introduced the minimax decision model and solution method for bilivel linear programming with ambiguous lower-level objective function, we explained some theoretical background to support our effective method for the possible optimality test. The efficiency improvement in each setting in the numerical result is because the utilization of memory follows by our theoretical observations in possible optimality test.

The future work could be in several directions. For example, the improvement for local search in both local and global optimality tests are at our concern. The comparison with the commercial package that can solve BLP as a complementary optimization problem is also of interest.

References

1. Abass, S.A.: An Interval Number Programming Approach for Bilevel Linear Programming Problem. International Journal of Management Science and Engineering Management 5(6), 461–464 (2010)
2. Bard, J.F.: Practical Bilevel Optimization: Algorithms and Applications. Kluwer Academic Publishers, Dordrecht (1998)
3. Bialas, W., Karwan, M.: On Two-level Optimization. IEEE Transactions on Automatic Control 27, 211–214 (1982)
4. Bolton, P., Dewatripont, M.: Contract Theory. MIT Press, Cambridge (2005)
5. Calvete, H.I., Galé, C.: Linear Bilevel Programming with Interval Coefficients. Journal of Computational and Applied Mathematics, 3751–3762 (2012)
6. Dempe, S.: Foundations of Bilevel Programming. Kluwer Academic Publishers, Dordrecht (2002)
7. Horst, R., Tuy, H.: Global Optimization: Deterministic Approaches, 3rd edn. Springer-Verlag, Berlin (1995)
8. Inuiguchi, M., Sakawa, M.: Possible and Necessary Optimality Tests in Possibilistic Linear Programming Problems. Fuzzy Sets and Systems 67, 29–46 (1994)
9. Inuiguchi, M., Kume, Y.: Minimax Regret in Linear Programming Problems with an Interval Objective Function. In: Tzeng, G.H., Wang, H.F., Wen, U.P., Yu, P.L. (eds.) Multiple Criteria Decision Making, pp. 65–74. Springer, New York (1994)
10. Inuiguchi, M., Tanino, T.: Enumeration of All Possibly Optimal Vertices with Possible Optimality Degrees in Linear Programming Problems with A Possibilistic Objective Function. Fuzzy Optimization and Decision Making 3, 311–326 (2004)

11. Inuiguchi, M., Sariddichainunta, P., Kawase, Y.: Bilevel linear programming with ambiguous objective function of the follower - formulation and algorithm. In: Proceeding of the 8th International Conference on Nonlinear Analysis and Convex Analysis, pp. 207–217. Yokohama Publishers, Yokohama (2013)
12. Muller, M.E.: A Note on A Method for Generating Points Uniformly on N-Dimensional Sheres. Communications of the ACM **2**(4), 19–20 (1959)
13. Nishizaki, I., Sakawa, M.: Solution concepts and their computational methods in multiobjective two-level linear programming problems. In: Proceeding of 1999 IEEE International Conference on Systems, Man and Cybernetics, vol. 3, pp. 985–990. IEEE, Tokyo (1999)
14. Sariddichainunta, P., Inuiguchi, M.: The Improvement of Optimality Test over Possible Reaction Reaction Set in Bilevel Linear Optimization with Ambiguous Objective Function of the Follower. Journal of Advanced Computational Intelligence and Intelligent Informatics **19**(5) (2015) (forthcoming)
15. Steuer, R.E.: Multiple Criteria Optimization: Theory, Computation, and Application. John Wiley and Sons, New York (1986)
16. Ren, A., Wang, Y.: A Cutting Plane Method for Bilevel Linear Programming with Interval Coefficients. Annals of Operations Research, online publication (2014)

Proposal of Grid Area Search with UCB for Discrete Optimization Problem

Akira Notsu$^{(\boxtimes)}$, Koki Saito, Yuhumi Nohara, Seiki Ubukata,
and Katsuhiro Honda

Graduate School of Engineering, Osaka Prefecture University, Gakuen 1-1, Naka,
Sakai, Osaka 599-8531, Japan
notsu@cs.osakafu-u.ac.jp
http://www.cs.osakafu-u.ac.jp/hi/

Abstract. In this paper, a novel method for the discrete optimization problem is proposed based on the UCB algorithm. Definition of the neighborhood in the search space of the problem easily affects the performance of the existing algorithms because they do not well take into account the dilemma of exploitation and exploration. To optimize the balance of exploitation and exploration, we divide the search space into several grids to reconsider the discrete optimization problem as a Multi-Armed Bandit Problem, and therefore the UCB algorithm is directly introduced for the balancing. We proposed a UCB-grid area search and conducted numerical experiments on the 0-1 Knapsack Problem. Our method showed stable results in different environments.

Keywords: Discrete optimization problem · Multi-armed bandit problem · GA · UCB

1 Introduction

An optimization problem is finding the best solution in a search space, which can be continuous or discrete. This problem becomes very difficult in ordinary circumstances because the search space is too large to find the best solution in practice [1]. However, an approximate solution often satisfies a designer's requirements without certified optimality. Thus, several approximate optimization methods have been proposed to find the approximate solution quickly.

The trade-off between "exploitation" and "exploration" is a dilemma in the problem [2]. Exploitation is to use the known data for a better solution. On the other hand, exploration is to search the new area for an unexpected discovery. It is very important to balance between exploitation and exploration in the search process. There are well-known methods like GA [3], PSO [4], annealing [5], and so on. However, the trade-off between exploitation and exploration has not been argued very well.

Also for a discrete optimization problem, like the knapsack problem, the trade-off is an issue. Although GA is a standard method, its trade-off is considerable, and therefore GA performance is easily affected by the problem parameters, especially in the case where a gene is an ineffective expression for local search in the problem structure.

© Springer International Publishing Switzerland 2015
V.-N. Huynh et al. (Eds.): IUKM 2015, LNAI 9376, pp. 102–111, 2015.
DOI: 10.1007/978-3-319-25135-6_11

We introduce the UCB algorithm [6] to the discrete optimization problem in order to balance the exploitation and exploration. UCB is a well-publicized and studied algorithm for solving the dilemma in the n-armed bandit problem [7]-[10]. This algorithm balances between exploitation and exploration, and break regret [11] becomes $O(\log n)$[12]. Thus, we propose a modified random search based on UCB for the optimization problem and perform a numerical experiment on the 0-1 knapsack problem to test for effectiveness.

2 Fundamental Theory

2.1 Multi-Armed Bandit Problem

A multi-armed bandit problem is for finding the best method to maximize the sum of rewards earned through a sequence of lever pulls of slot machines [13]. Each lever provides a random reward from a distribution specific to that lever. Agents don't know the distribution at first; then they positively seek information about levers. However, they have to maximize their rewards from pulling the levers. Therefore, they are in a dilemma between exploitation and exploration, and must optimize the balance between them. A lot of methods have been proposed for this problem, and we introduce three typical ones [14] here.

ε-greedy. ε-greedy is the method in which choosing the action having the best estimated value $\bar{X}_i$. An agent chooses the action randomly with probability ε ($0 \leq \varepsilon \leq 1$) and chooses greedy action with probability $1 - \varepsilon$. This method is easy to implement, and if we play infinite times, this method ensures that the estimated value converges on the true value because all actions are tried infinitely. However, it also has a weak point in that there is a possibility to choose the worst action forever because the action is chosen randomly in exploring.

Softmax. Softmax is a method for improving the ε-greedy's shortcoming that all actions are chosen with equal probability in exploring. Actions are chosen with probability considering the estimated value in this method. An agent is able to choose the action comparing the large/small relation of the estimated value, but this method has the problem that it is difficult to determine the definition of that reference point. The Boltzmann distribution is often-used and its definition is as follows:

$$P_i = \frac{e^{\bar{X}_i/T}}{\sum_i e^{\bar{X}_i/T}}.$$

P_i is the probability of choosing the i-th slot machine, and T is the temperature parameter. The higher T is, the more equally all actions are chosen, and the lower T is, the higher the probability of actions having a large estimated value is.

UCB. UCB is a method for choosing the highest value of UCB instead of the estimated value $\bar{X}_i$. In this method, an agent chooses the action to compare the upper bound of the confidence interval of the estimated value [6]. The feature is that a non-selected lever is inevitably selected and an agent is ready to obtain an unknown estimated value. The definition of UCB is as follows:

$$UCB_i = \bar{X}_i + C\sqrt{\frac{\ln n}{n_i}}.$$

i is the number of levers, n is the number of total played times, and n_i is the number of times played on i-th lever ($\sum_i n_i = n$). $\bar{X}_i$ is the expectation of value of i-th lever. C is the constant to determine a tendency of exploration.

UCB algorithm is characterized by the regret optimization. Regret is considered as one of the most important criteria of bandit algorithm. It is defined as follows:

$$Regret_n = n \max_i \bar{X}_i - \sum_i n_i \bar{X}_i.$$

Typically, it is proved that the lower bound of the number of choosing the lever whose expectation is not maximum is $O(\log n)$, when the number of play time n is infinite. Therefore, $O(\log n)$ regret at least arise when using the best algorithm. ε-greedy's regret is $O(n)$, while UCB algorithm's regret is $O(\log n)$.

2.2 Optimization Problem

When moving to a destination, many people choose the shortest route. Such a shortest path problem is a type of closed optimization problem. Not only this problem, but optimization problems in general are studied in various fields such as engineering and economics.

Generally, they are expressed as follows [15].

$$\text{minimize or maximize } f(x)$$
$$\text{subject to } x \in F,$$

where f is the objective function, F is the feasible region, and x is the feasible solution. The optimization problems are divided into some classes by the kinds of objective functions and constraints.

2.3 0-1 Knapsack Problem

The 0-1 knapsack problem is a type of combinatorial optimization problem [16]. It evaluates combinations that maximize summation of value in a combination that satisfies weight limits W when knapsack which can accommodate items of the constant capacity W. N kinds of items $x_i (i = 1, \cdots, N)$ that determines value p_i and weight w_i are given. It has many applications: freight transport systems, personnel management based on the ability of employees, purchase of supplies within a budget, decision making in financial transactions. Summation of value P, which is the objective function, is as follows:

$$P = \begin{cases} \sum_{i=1}^{n} p_i x_i & (\sum_{i=1}^{n} w_i x_i \leq W) \\ 0 & (\sum_{i=1}^{n} w_i x_i > W), \end{cases} \tag{1}$$

where $x \in 0, 1$.

The simplest method of solving this problem is using all combinations, but it costs much calculation time by the combination of 2^N. Thus, this study proposes a method that improves random search using the UCB strategy.

3   Proposed Method

It's important in optimization algorithms to maintain a desirable ratio of exploitation, the choice to obtain unknown information, to exploration, the choice to maximize profit among known information. We propose a method to maintain the ratio with the UCB algorithm in the optimization by dividing the area into several grids that are considered as slots.

3.1   Grid Area Search (GAS) for 0-1 Knapsack Problem

In our method, an exploration area is divided into smaller areas the same size; we choose an area (grid area) by the bandit algorithm and search in the area at random. Methods for choosing the grid area are random search, ε-greedy method, UCB algorithm, and so on.

3.2   UCB - Grid Area Search

We propose a method that applies the UCB algorithm to the choice of grid area in GAS. The procedure is as follows.

1. Divide a searching space into sections of the same size and search in all grid areas once at random.
2. Decide the gird area to search with UCB algorithm.
3. Randomly search in the area.
4. Repeat a trial some times, and the maximum solution is the solution.

The definition of UCB is as follows:

$$UCB_i = \max V_t / N + C \sqrt{\frac{\log n}{n_i}}. \tag{2}$$

i is the number of grid areas, n_i is the number of times that A was chosen, and n is the number of total search times ($\sum_i n_i = n$). $\max V_t$ is the expectation of value of the i-th grid area, and N is the number of items. It derives $0 \leq \max V_t / N \leq 1$, which is the same domain of definition as the estimated value in the multi-armed bandit problem. C is the constant to determine a tendency of exploration.

In figure 1, assumed UCB-GAS features and other algorithm features are shown. We intend to develop the versatile method for the optimization.

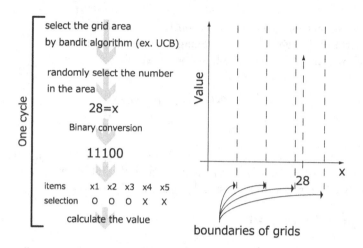

Fig. 1. Grid Area Search

Table 1. Algorithm features

algorithms	neighborhood	number of parameters	suitable space, set
random search	none	none	unknown space, set
PSO	near the particles	a few	continuous space
GA	similar combinations (gene)	several	combination set
UCB-GAS	in the same grid	a few	unknown space

4   Experiment

4.1   Experiment Summary

In this study, we conducted the performance comparison experiment with the 0-1 knapsack problem. This time, we used five kinds of optimization algorithms: Random Search, EG-Grid Area Search, UCB-Grid Area Search, SGA and GA. EG-Grid Area Search is the method that uses the ε-greedy method for the choice of the grid in the grid area search. SGA is a simple genetic algorithm that uses only roulette wheel selection, one-point crossover and the mutation in which a value of one place of heredity stroma is reversed. GA is the method that applies elite strategy to SGA.

4.2   Simulation Result

UCB-GAS Parameters. The effect of constant C and the number of grids are shown as follows.

We simulated each situation where constant C is 0.01, 0.1, 1 and 10, and weight limits $W = 6$. The number of grids in UCB-GAS is 100. The number of

Table 2. Simulation parameters (0-1 knapsack problem)

parameters	value
number of items N	20
weight limits W	2, 4, 6, 8, 10
item value p_i	randomly set $(0.0 < p_i < 1.0)$
item weight w_i	randomly set $(0.0 < w_i < 1.0)$

Table 3. Simulation parameters (GAS)

parameters	value
number of grid areas N_g	50, 100, 150, 200
ε	0.2
C	0.01, 0.1, 1, 10

Table 4. Simulation parameters (GA)

parameters	value
mutational rate	0.1
crossover rate	0.8
population	15, 20, 25

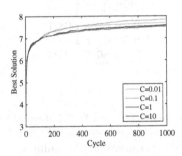

Fig. 2. Mean best solution $(C = 0.01, 0.1, 1$ or $10)$

explorations is 1,000, and we simulated 100 times in this situation. We show the graph of the average of the best solution, and show the table of the average and standard deviation of the final solutions as follows.

The vertical axis is for the number of cycles (exploration), and the horizontal one is for the mean of the best solution.

We also simulated each situation where the number of grids is 50, 100, 150 and 200 fixing C as 0.1.

Table 5. Best solution at 1,000th exploration

C	mean $\pm$ SD
0.01	7.6982 $\pm$ 0.9983
0.1	7.8494 $\pm$ 1.0275
1	7.5916 $\pm$ 0.9816
10	7.5290 $\pm$ 0.9268

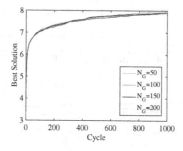

Fig. 3. Mean best solution (N_G = 50, 100, 150 or 200)

Table 6. Best solution at 1,000th exploration

N_G	mean $\pm$ SD
50	7.8980 $\pm$ 1.0480
100	7.9459 $\pm$ 1.0831
150	7.9037 $\pm$ 1.1033
200	7.8986 $\pm$ 1.0933

These results show that the sensitivity of parameter C and the number of grids have a low impact on search performance. Figure 3 shows GAS started to converge after searching all grid areas. If the number of grids becomes larger, GAS needs large computational times and its convergence becomes slower. However, they have little or no impact on convergence values.

Given that GAS can perform early search around an optimal solution, it is reasonable to set the number of grids to 100 in this problem.

Comparative Experiment. We similarly simulated with each method and show the graph of the average of the best solution, and show the table of the average and standard deviation of the final solution as follows. We also plotted the best solution every generation in SGA and GA. Their results are plotted every 20 cycles because they search the space 20 times in a generation. We set C as 0.1 and the number of grids as 100, referring to the above results.

Table 7. Best solution at 1,000th exploration

weight limits	$W = 2$	$W = 4$	$W = 6$	$W = 8$	$W = 10$
	mean $\pm$ SD	mean $\pm$ SD	mean $\pm$ SD	mean $\pm$ SD	mean $\pm$ SD
random search	3.52 $\pm$ 1.21	6.00 $\pm$ 0.97	7.55 $\pm$ 1.12	8.63 $\pm$ 1.01	9.00 $\pm$ 0.94
EG-GAS(N_G=100)	3.82 $\pm$ 0.98	6.18 $\pm$ 1.00	7.68 $\pm$ 1.16	8.91 $\pm$ 1.07	8.63 $\pm$ 1.01
UCB-GAS(N_G=50)	3.66 $\pm$ 1.07	6.30 $\pm$ 1.08	7.72 $\pm$ 1.09	9.00 $\pm$ 1.08	9.34 $\pm$ 1.19
UCB-GAS(N_G=100)	3.84 $\pm$ 1.07	6.27 $\pm$ 0.98	7.87 $\pm$ 1.21	9.00 $\pm$ 1.08	9.58 $\pm$ 1.00
UCB-GAS(N_G=200)	4.02 $\pm$ 1.22	6.34 $\pm$ 1.08	7.83 $\pm$ 1.29	9.06 $\pm$ 1.11	9.50 $\pm$ 1.11
GA(pop=15)	2.19 $\pm$ 2.06	5.83 $\pm$ 1.28	7.63 $\pm$ 1.24	8.93 $\pm$ 1.29	9.50 $\pm$ 1.30
GA(pop=20)	2.40 $\pm$ 2.07	5.86 $\pm$ 1.28	7.83 $\pm$ 1.25	9.10 $\pm$ 1.18	9.82 $\pm$ 1.04
GA(pop=25)	2.77 $\pm$ 1.99	6.18 $\pm$ 1.38	7.82 $\pm$ 1.28	9.21 $\pm$ 1.12	9.79 $\pm$ 1.09
SGA(pop=20)	2.42 $\pm$ 1.88	5.74 $\pm$ 1.14	7.52 $\pm$ 1.13	8.60 $\pm$ 1.03	8.63 $\pm$ 1.01

4.3 Discussion

Figures 4, 5 and 6 and Table 7($W < 6$) show that UCB-GAS offered a better search around an optimal solution than other methods. This is because the UCB strategy maintains the balance between exploitation and exploration. Computation time (s) was not too different without parallel computing.

On the other hand, Fig. 7 and Table 7($W > 6$) show that the result of GA is the best in all methods in aspects of convergence speed and averages of best

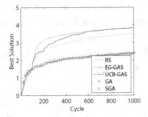

Fig. 4. Best solution ($W = 2$, N_G=100, pop=20)

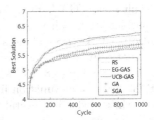

Fig. 5. Best solution ($W = 4$, N_G=100, pop=20)

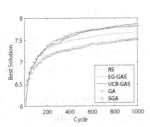

Fig. 6. Best solution ($W = 6$, N_G=100, pop=20)

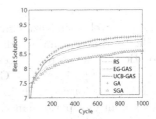

Fig. 7. Best solution ($W = 8$, N_G=100, pop=20)

Table 8. Computation time (s) without parallel computing (MATLAB2015a, Windows8.1, CPU:Xeon 2GHz, RAM: 16GB)

	mean ± SD
random search	0.0646 ± 0.0028
UCB-GAS (N_G=100)	0.0849 ± 0.0021
GA (pop=20)	0.0851 ± 0.0011

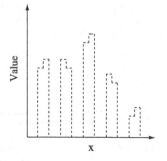

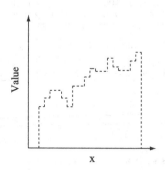

Fig. 8. Value function image (W=2)   **Fig. 9.** Value function image (W=8)

answers. However, most items are selected at best solution where $W = 8, 10$ ($N = 20$ and average weight of the item is 0.5, which means that sum of weight of all items is 10), and this result is in a particular situation.

This result means GA befits a local search to improve a solution for a optimization problem in which the true optimum solution exists near a local optimum solution. UCB-GAS balances trade-off between exploitation and exploration automatically, and therefore is unaffected by multimodal function. In value function like a figure 8, UCB-GAS is better then GA. While GA is better then UCB-GAS for value function like a figure 9.

Just as we intended, with several parameters, UCB-GAS developed a "stable" performance for the discrete optimization problem. Its convergence speed was worse than we expected but similar to GA.

5   Conclusion

We proposed a novel optimization method for the discrete optimization problem based on the UCB algorithm, which enables us to balance between exploitation and exploration in the search process. A numerical experiment was conducted on the 0-1 knapsack problem with GA.

Our method performed well in the problem. On the whole, taking into account its stability of performance, easiness and simplicity of implementation and parameter setting, it has better search ability than other methods.

Like a 0-1 knapsack problem, the search space and the grid area in it are too complex to define for solving the problem. Even though our algorithm used binary space, gridded it at regular intervals and showed better performance than the others, it leaves much room for discussion about better definition of the space, which is needed for further performance improvement.

Acknowledgment. This work was supported by JSPS KAKENHI Grant Number 15K00344.

References

1. Wolpert, D.H., Macready, W.G.: No free lunch theorems for optimization. IEEE Transactions on Evolutionary Computation 1(1), 67–82 (1997)
2. Cesa-Bianchi, N.: Prediction, learning, and games. Cambridge University Press (2006)
3. Davis, L.: Handbook of Genetic Algorithm. Van Nostrand Renhold (1990)
4. Kennedy, J., Eberhart, R.: Particle Swarm Optimization. IEEE Int. Conf. on Neural Networks 4, 1942–1948 (1995)
5. Voss, S., Osman, I.H., Roucairol, C.: Meta-Heuristics: Advances and Trends in Local Search Paradigms for Optimization (1999)
6. Auer, P., Cesa-Bianchi, N., Fischer, P., Informatik, L.: Finite-time analysis of the multi-armed bandit problem. Machine Learning 47, 235–256 (2002)
7. Audibert, J.Y., Munos, R., Szepesvari, C.: Exploration-exploitation trade-off using variance estimates in multi-armed bandits. Theoretical Computer Science 410, 1876–1902 (2009)
8. Audibert, J.Y., Bubeck, S.: Regret bounds and minimax policies under partial monitoring. Journal of Machine Learning Research 11, 2785–2836 (2010)
9. Kocsis, L., Szepesvari, C.: Discounted UCB. In: 2nd PASCAL Challenges Workshop (2006)
10. Radlinski, F., Kleinberg, R., Joachims, T.: Learning diverse rankings with multi-armed bandits. In: The 25th International Conference on Machine Learning, pp. 784–791 (2008)
11. Lai, T.L., Robbins, H.: Asymptotically efficient adaptive allocation rules. Advances in Applied Mathematics 6(1), 4–22 (1985)
12. Agrawal, R.: Sample mean based index policies with o(log n) regret for the multi-armed bandit problem. Advances in Applied Mathematics 27, 1054–1078 (1995)
13. Auer, P., Cesa-Bianchi, N., Freund, Y., Schapire, R.E.: Gambling in a rigged casino: the adversarial multi-armed bandit problem. In: Proc. of the 36th Annual Symposium on Foundations of Computer Science, pp. 322–331 (1995)
14. Sutton, R.S., Bart, A.G.: Generalization in Reinforcement Learning-An Introduction-. The MIT Press (1998)
15. Boyd, S., Vandenberghe, L.: Convex Optimization. Cambridge University Press (2004). https://web.stanford.edu/~boyd/cvxbook/bv_cvxbook.pdf
16. Kellerer, H., Pferschy, U., Pisinger, D.: Knapsack Problems. Springer (2004)

Why Copulas Have Been Successful in Many Practical Applications: A Theoretical Explanation Based on Computational Efficiency

Vladik Kreinovich[1]([✉]), Hung T. Nguyen[2,3], Songsak Sriboonchitta[3], and Olga Kosheleva[4]

[1] Department of Computer Science, University of Texas at El Paso, 500 W. University, El Paso, TX 79968, USA
vladik@utep.edu

[2] Department of Mathematical Sciences, New Mexico State University, Las Cruces, NM 88003, USA
hunguyen@nmsu.edu

[3] Faculty of Economics, Chiang Mai University, Chiang Mai, Thailand
songsakecon@gmail.com

[4] University of Texas at El Paso, 500 W. University, El Paso, TX 79968, USA
olgak@utep.edu

Abstract. A natural way to represent a 1-D probability distribution is to store its cumulative distribution function (cdf) $F(x) = \mathrm{Prob}(X \leq x)$. When several random variables $X_1, \ldots, X_n$ are independent, the corresponding cdfs $F_1(x_1), \ldots, F_n(x_n)$ provide a complete description of their joint distribution. In practice, there is usually some dependence between the variables, so, in addition to the marginals $F_i(x_i)$, we also need to provide an additional information about the joint distribution of the given variables. It is possible to represent this joint distribution by a multi-D cdf $F(x_1, \ldots, x_n) = \mathrm{Prob}(X_1 \leq x_1 \& \ldots \& X_n \leq x_n)$, but this will lead to duplication – since marginals can be reconstructed from the joint cdf – and duplication is a waste of computer space. It is therefore desirable to come up with a duplication-free representation which would still allow us to easily reconstruct $F(x_1, \ldots, x_n)$. In this paper, we prove that among all duplication-free representations, the most computationally efficient one is a representation in which marginals are supplements by a copula.

This result explains why copulas have been successfully used in many applications of statistics: since the copula representation is, in some reasonable sense, the most computationally efficient way of representing multi-D probability distributions.

1 Introduction

In many practical problems, we need to deal with joint distributions of several quantities, i.e., with multi-D probability distributions. There are many different ways to represent such a distribution in a computer. In many practical applications, it turns out to be beneficial to use a representation in which we store the marginal distributions (that describe the distribution of each quantity) and

© Springer International Publishing Switzerland 2015
V.-N. Huynh et al. (Eds.): IUKM 2015, LNAI 9376, pp. 112–125, 2015.
DOI: 10.1007/978-3-319-25135-6_12

a copula (that describe the relation between different quantities; definitions are given below). While this representation is, in many cases, empirically successful, this empirical success is largely a mystery.

In this paper, we provide a theoretical explanation of this empirical success, by showing that the copula representation is, in some reasonable sense, the most computationally efficient.

The structure of this paper is as follows: In Section 2, we explain why representing probability distributions is important for decision making: (consistent) decision making requires computing expected utility, and to perform this computation, we need to have the corresponding probability distribution in the computer represented in a computer. In Section 3, with this objective in mind, we consider the usual representations of probability distributions: what are the advantages and limitations of these representations. Section 3 ends with the main problem that we consider in this paper: what is the best representation of multi-D distributions? The formulation of this problem in Section 3 is informal. The problem is formalized in Section 4. This formalization enables us to prove that copulas are indeed the most effective computer representation of multi-D distributions.

2 Why It Is Important to Represent Probability Distributions: Since This Is Necessary For Decision Making

In this section, we explain why it is important to represent probability distributions in a computer.

Probability Distributions are Ubiquitous. To understand why representing probability distributions is important, let us recall that one of the main objectives of science and engineering is to predict the future state of the world – and to come up with decisions which lead to the most preferable future state.

These predictions are based on our knowledge of the current state of the world, and on our knowledge of how the state of the world changes with time. Our knowledge is usually approximate and incomplete. As a result, based on our current knowledge, we cannot predict the *exact* future state of the world, *several* future states are possible based on this knowledge. What we can predict is the set of possible states, and the *frequencies* with which, in similar situations, different future states will occur. In other words, what we can product is a *probability distribution* on the set of all possible future states.

This is how many predictions are made: weather predictions give us, e.g., a 60% chance of rain; economic predictions estimate the probability of different stock prices, etc.

Need to Consider Random Variables. Information about the world comes from measurements. As a result of each measurement, we get the values of the corresponding physical quantities. Thus, a natural way to describe the state of the world is to list the values of the corresponding quantities $X_1, \ldots, X_n$.

From this viewpoint, the probability distribution on the set of all possible states means a probability distribution on the set of the corresponding tuples $X = (X_1, \ldots, X_n)$.

How to Represent Probability Distributions: An Important Question. Due to ubiquity of probability distributions, it is important to select an appropriate computer representation of these distributions, a representation that would allow us to effectively come up with related decisions.

Thus, to come up with the best ways to represent a probability distribution, it is important to take into account how decisions are made.

How Decisions are made: A Reminder. In the idealized case, when we are able to exactly predict the consequences of each possible decision, decision making is straightforward: we select a decision for which the consequences are the best possible. For example:

- an investor should select the investment that results in the largest return,
- a medical doctor should select a medicine which leads to the fastest recovery of the patient, etc.

In reality, as we have mentioned, we can rarely predict the exact consequence of different decisions; we can, at best, predict the probabilities of different consequences of each decision. In such real-life setting, it is no longer easy to select an appropriate decision. For example:

- if we invest money in the US government bonds, we get a small guaranteed return;
- alternatively, if we invest into stocks, we may get much higher returns, but we may also end up with a loss.

Similarly:

- if we prescribe a well-established medicine, a patient will slowly recover;
- if instead we prescribe a stronger experimental medicine, the patient will probably recover much faster, but there is also a chance of negative side effects which may drastically delay the patient's recovery.

Researchers have analyzed such situations. The main result of the corresponding decision theory is that a consistent decision making under such probability uncertainty can be described as follows (see, e.g., [1,3,6]):

- we assign a numerical value u (called *utility*) to each possible consequence, and then
- we select a decision for which the expected value $E[u]$ of utility is the largest possible.

Since we want a representation of a probability distribution that would make decision making as efficient as possible, we thus need to select a representation that would allow us to compute the expected values of different utility functions as efficiently as possible.

In the next section, we use this motivation for representing probability distributions to explain which computer representations are most adequate.

Comment. From the strictly mathematical viewpoint, every decision making is based on comparing expected values of the utility functions. However, in many practical situations, the corresponding problem becomes much simpler, because the corresponding utility functions are simple.

For example, in risk analysis, e.g., when considering whether the bridge will collapse during a hurricane, we often do not differentiate between different *positive* situations (i.e., situations in which the bridge remains standing), and we also do not differentiate between different *negative* situations (i.e., situations in which the bridge collapses). In terms of the utility function, this is equivalent to considering a "binary" utility in which we only have two utility levels $u^+ > u^-$. In such situations, the expected utility is equal to $u^+ \cdot (1 - p^-) + u^- \cdot p^- = u^+ - (u^+ - u^-) \cdot p^-$, where p^- is the probability of the undesired scenario.

In such cases, comparing different values of expected utility is equivalent to comparing the corresponding probabilities p^-. Thus, instead of computing expected values of the utility function, it is sufficient to simply compute the corresponding probabilities.

This observation will be actively used in the following sections.

3   Different Computer Representations of Probability Distributions: Analysis from the Viewpoint of Decision-Making Applications

Case of 1-D Probability Distribution: What is Needed? In the previous section, we argued that since ultimately, our problem is to make a decision, and consistent decision making means comparing expected values of the utility function, it is reasonable to select such computer representations of probability distributions that would be the most efficient in computing the corresponding expected values.

Let us start with the simplest case of 1-D probability distributions. In view of the above argument, to understand which representation is the most appropriate, we need to describe possible utility functions. To describe such functions, let us start by considering a simple example: the problem of getting from point A to point B.

In general, all else being equal, we would like to get from A to B as fast as possible. So, in the idealized case, if we knew the exact driving time, we should select the route that takes the shortest time. In practice, random delays are possible, so we need to take into account the cost of different delays.

In some cases – e.g., if we drive home after a long flight – a small increase of driving time leads to a small decrease in utility. However, in other cases – e.g., if we are driving to the airport to take a flight – a similar small delay can make us miss a flight and thus, the corresponding decrease in utility will be huge. In our analysis, we need to take into account both types of situations.

In the situations of the *first type*, utility $u(x)$ is a smooth function of the corresponding variable x. Usually, we can predict x with some accuracy, so all possible values x are located in a small vicinity of the predicted value x_0. In this vicinity, we can expand the dependence $u(x)$ in Taylor series and safely ignore higher order terms in this expansion:

$$u(x) = u(x_0) + u'(x_0) \cdot (x - x_0) + \frac{1}{2} \cdot u''(x_0) \cdot (x - x_0)^2 + \dots$$

The expected value of this expression can be thus computed as the linear combination of the corresponding moments:

$$E[u] = u(x_0) + u'(x_0) \cdot E[x - x_0] + \frac{1}{2} \cdot u''(x_0) \cdot E[(x - x_0)^2] + \dots$$

Thus, to deal with situations of this type, it is sufficient to know the first few moments of the corresponding probability distribution.

In situations of the *second type*, we have a threshold x_t such that the utility is high for $x \le x_t$ and low for $x > x_t$. In comparison with the difference between high and low utilities, the differences between two high utility values (or, correspondingly, between two low utility values) can be safely ignored. Thus, we can simply say that $u = u^+$ for $x \le x_t$ and $u = u^- < u^+$ for $x > x_t$. In this case, the expected value of utility is equal to $E[u] = u^- + (u^+ - u^-) \cdot F(x_t)$, where $F(x_t) = \text{Prob}(x \le x_t)$ is the probability of not exceeding the threshold. So, to deal with situations of this type, we need to know the cdf $F(x)$.

1-D Case: What are the most Appropriate Computer Representations? Our analysis shows that in the 1-D case, to compute the expected utilities, we need to know the cdf *and* the moments.

Since the moments can be computed based on cdf, as

$$E[(x - x_0)^k] = \int (x - x_0)^k \, dF(x),$$

it is thus sufficient to have a cdf. From this viewpoint, the most appropriate way to represent a 1-D probability distribution in the computer is to store the values of its cumulative distribution function $F(x)$.

Multi-D Case. In the multi-D cases, we similarly have two types of situations. For situations of the first type, when small changes in the values x_i lead to small changes in utility, it is sufficient to know the first few moments.

In the situations of the second type, we want all the values not to exceed appropriate thresholds. For example, we want a route in which the travel time does not exceed a certain pre-set quantity, and the overall cost of all the tolls does not exceed a certain value. To handle such situations, it is desirable to know the following probabilities – that form the corresponding multi-D cdf:

$$F(x_1, \dots, x_n) \stackrel{\text{def}}{=} \text{Prob}(X_1 \le x_1 \,\&\, \dots \,\&\, X_n \le x_n).$$

So, in the multi-D case too, computing expected values of utility functions means that we need to compute both the moments and the multi-D cdf. Since the moments can be computed based on the cdf, it is thus sufficient to represent a cdf.

Situations when we go from 1-D to Multi-D Case. The above analysis of the multi-D case is appropriate for situations in which we acquire all our knowledge about the probabilities in one step:

- we start "from scratch", with no knowledge at all,
- then we gain the information about the joint probability distribution.

In such 1-step situations, as we have just shown, the ideal representation of the corresponding probability distribution is by its cdf $F(x_1, \ldots, x_n)$.

In many practical situations, however, knowledge comes *gradually*. Usually, first, we are interested in the values of the first quantity, someone else may be interested in the values of the second quantity, etc. The resulting information is provided by the corresponding marginal distributions $F_i(x_i)$.

After that, we may get interested in the relation between these quantities $X_1, \ldots, X_n$. Thus, we would like to supplement the marginal distributions with an additional information that would enable us to reconstruct the multi-D cdf $F(x_1, \ldots, x_n)$.

In principle, we can store this multi-D cdf as the additional information. However, this is not the most efficient approach. Indeed, it is well known that each marginal distribution $F_i(x_i)$ can be reconstructed from the multi-D cdf, as

$$F_i(x_i) = F(+\infty, \ldots, +\infty, x_i, +\infty, \ldots, \infty) = \lim_{T \to \infty} F(T, \ldots, T, x_i, T, \ldots, T).$$

So, if we supplement the original marginals with the multi-D cdf, we thus store duplicate information, and duplication is a waste of computer memory.

Situations when we go from 1-D to Multi-D Case: Resulting Problem. In the general multi-D case, we have shown that storing a cdf is an appropriate way of representing a multi-D distribution in a computer. However, in situations when we go from 1-D to multi-D case, this representation is no longer optimal: it involves duplication and is, thus, a waste of computer memory.

Copula-based Computer Representations: A Possible way to Solve this Problem. To avoid duplication, some researchers and practitioners use copula-based representations of multi-D distributions.

A copula corresponding to a multi-D distribution with cdf $F(x_1, \ldots, x_n)$ is a function $C(x_1, \ldots, x_n)$ for which

$$F(x_1, \ldots, x_n) = C(F_1(x_1), \ldots, F_n(x_n)),$$

where $F_i(x_i)$ are the corresponding marginal distributions; see, e.g., [2,4,5]. A copula-related way to represent a multi-D distribution is to supplement the marginals $F_i(x_i)$ with the copula $C(x_1, \ldots, x_n)$.

The above formula then enables us to reconstruct the multi-D cdf $F(x_1, \ldots, x_n)$. This representation has no duplication, since for the same copula, we can have many different marginals.

Remaining Problem. A copula-based representation avoids duplication and is, in this sense, better than storing the cdf. But *is the copula-based representation optimal* (in some reasonable sense), or is an even better representation possible? And if the copula-based representation *is* optimal, is it *the only optimal* one, or are there other representations which are equally good?

These are the questions that we will answer in the next section. Of course, in order to answer them, we need to first formulate them in precise terms.

4   Formalization of the Problem and the Main Result

Analysis of the Problem. Let us start by formalizing the above problem. We want a computer representation that will be duplicate-free and computationally efficient. What do we mean by computationally efficient?

As we have argued, for making decisions, we need to know the values of the multi-D cdf $F(x_1, \ldots, x_n)$. Thus, whatever representation we come up with, we need to be able to reconstruct the cdf based on this representation. Thus, to make the representation computationally efficient, we need to make sure that the algorithm for reconstructing the cdf is as fast as possible (i.e., that this algorithm consists of as few computational steps as possible), and that this representation uses as little computer memory as possible.

To find such an optimal representation, we need to have precise definitions of what is an algorithm, what is a computational step, and when is an algorithm computationally efficient (in terms of both computation time and computer space). Let us start with providing an exact definition of an algorithm.

Towards a Precise Description of what is an Algorithm. We want to be able, given the marginals and the additional function(s) used for representing the distribution, to reconstruct the multi-D cdf $F(x_1, \ldots, x_n)$. This reconstruction has to be done by a computer *algorithm*.

An algorithm is a sequence of steps, in each of which we either apply some operation $(+, -, \sin,$ given function) to previously computed values, or decide where to go further, or stop.

In our computations, we can use inputs, we can use auxiliary variables, and we can use constants. In accordance with the IEEE 754 standard describing computations with real numbers, infinite values $-\infty$ and $+\infty$ can be used as well.

It is also possible to declare a variable as "undefined" (in IEEE 754 this is called "not a number", NaN for short). For each function or operation, if at least one of the inputs is undefined, the result is also undefined.

We thus arrive at the following formal definition of an algorithm:

Definition 1.

- *Let F be a finite list of functions $f_i(z_1, \ldots, z_{n_i})$.*
- *Let $v_1, \ldots, v_m$ be a finite list of real-valued variable called* inputs.
- *Let $a_1, \ldots, a_p$ be a finite list of real-valued variables called* auxiliary variables.
- *Let $r_1, \ldots, r_q$ be real-valued variables; they will be called the* results *of the computations.*

An algorithm $\mathcal{A}$ *is a finite sequence of instructions $I_1, \ldots, I_N$ each of which has one of the following forms:*

- *an* assignment *instruction "$y \leftarrow y_1$" or "$y \leftarrow f_i(y_1, \ldots, y_{n_i})$", where:*
  - *y is one of the auxiliary variables or a result variable,*
  - *$f_i \in F$, and*
  - *each y_i is either an input, or an auxiliary variable, or a result, or a real number (including $-\infty$, $+\infty$, and NaN);*
- *an* unconditional branching *instruction "go to I_i;"*
- *a* conditional branching *instruction "if $y_1 \odot y_2$, then to I_i else go to I_j", where:*
  - *each y_i is either an input, or an auxiliary variable, or the result, or a real number (including $-\infty$ and $+\infty$); and*
  - *$\odot$ is one of the symbols $=$, $\neq$, $<$, $>$, $\leq$, and $\geq$;*
- *or a* stopping *instruction "stop".*

Definition 2. *The* result *of applying an algorithm $\mathcal{A}$ to the inputs $a_1, \ldots, a_m$ is defined as follows:*

- *in the beginning, we start with the given values of the inputs, all other variables are undefined;*
- *we then start with instruction I_1;*
- *on each instruction:*
  - *if this is an assignment instruction $y \leftarrow y_1$ or $y \leftarrow f_i(y_1, \ldots, y_{n_i})$, we assign, to the variable y, the new value y_1 or $f_i(y_1, \ldots, y_{n_i})$ and go to the next instruction;*
  - *if this is an unconditional branching instruction, we go to instruction I_i;*
  - *if this is a conditional branching instruction and both values y_1 and y_2 are defined, we check the condition $y_1 \odot y_2$ and, depending of whether this condition is satisfied, go to instruction I_i or to instruction I_j;*
  - *if this a conditional branching instruction, and at least one of the values y_i is undefined, we stop;*
  - *if this a stopping instruction, we stop.*

The values $r_1, \ldots, r_q$ at the moment when the algorithm stops are called the result *of applying the algorithm.*

Toward a Formal Definition of the Number of Computational Steps.
The above definition of an algorithm as a step-by-step procedure leads to the following natural definition of the number of computational steps:

Definition 3. *For every algorithm $\mathcal{A}$ and for each tuple of inputs $v_1, \ldots, v_m$, the number of instructions that the algorithm goes through before stopping is called the* running time *of $\mathcal{A}$ on $v_1, \ldots, v_m$.*

Examples. To illustrate the above definition, let us start with simple algorithms.

$1°$. The standard algorithm for computing the value $r_1 = v_1 \cdot (1 - v_1)$ requires the use of two arithmetic operations: subtraction $f_1(z_1, z_2) = z_1 - z_2$ and multiplication $f_2(z_1, z_2) = z_1 \cdot z_2$. Here, we can use a single auxiliary variable a_1. The corresponding instructions have the following form:

I_1: $a_1 \leftarrow f_1(1, v_1)$; this instruction computes $a_1 = 1 - v_1$;
I_2: $r_1 \leftarrow f_2(v_1, a_1)$; this instruction computes $r_1 = v_1 \cdot a_1 = v_1 \cdot (1 - v_1)$;
I_3: stop.

For all the inputs, this algorithm goes through two instructions before stopping, so its running time is 2.

$2°$. Computation of the absolute value $|v_1|$, i.e., v_1 if $v_1 \geq 0$ and $-v_1$ otherwise, requires that we use a unary minus operation $f_1(z_1) = -z_1$. The corresponding instructions have the following form:

I_1: if $v_1 \geq 0$, then go to I_2 else go to I_4;
I_2: $r_1 \leftarrow v_1$;
I_3: stop;
I_4: $r_1 \leftarrow f_1(v_1)$;
I_5: stop.

This algorithm also goes through two instructions before stopping, so its running time is also 2.

$3°$. Computation of $n! = 1 \cdot 2 \cdot \ldots \cdot n$ for a given natural number n requires:

- two arithmetic operations: addition $f_1(z_1, z_2) = z_1 + z_2$ and multiplication $f_2(z_1, z_2) = z_1 \cdot z_2$; and
- a loop, with an additional variable a_1 that takes the values $1, 2, \ldots, n$.

The corresponding instructions have the following form:

I_1: $r_1 \leftarrow 1$;
I_2: $a_1 \leftarrow 1$;
I_3: if $a_1 \leq v_1$, then go to I_4 else go to I_7;
I_4: $r_1 \leftarrow f_2(r_1, a_1)$;
I_5: $a_1 \leftarrow f_1(a_1, 1)$;
I_6: go to I_3;
I_7: stop.

The running time of this algorithm depends on the input v_1.

- When $v_1 = 0$, we go through three instructions I_1, I_2, and I_3 before stopping, so the running time is 3.

– When $v_1 = 2$, we go through I_1, I_2, I_3, I_5, I_6, then again I_3, I_4, I_5, and I_6, and finally I_3 and stop. In this case, the running time is 11.

4°. If we already have the multi-D cdf as one of the basic functions $f_i(z_1, \ldots, z_n) = F(z_1, \ldots, z_n)$, then computing cdf for given inputs requires a single computational step:

I_1: $r_1 \leftarrow f_1(v_1, \ldots, v_n)$;
I_2: stop.

The running time of this algorithm is 1.

5°. Similarly, if we have a copula $f_1(z_1, \ldots, z_n) = C(z_1, \ldots, z_n)$, and we can use the values $v_{n+i} = F_i(x_i)$ as additional inputs, the corresponding algorithm for computing the cdf also has a running time of 1:

I_1: $r_1 \leftarrow f_1(v_{n+1}, \ldots, v_{2n})$;
I_2: stop.

What is a Computer Representation of a Multi-D Distribution: Towards a Formal Definition. Now, we are ready to provide a formal definition of a computer representation: it is a representation in which, in addition to the marginals, we have one or more functions that enable us to algorithmically reconstruct the cdf.

Definition 3. *By a representation of an n-dimensional probability distribution, we mean a tuple consisting of:*

- *finitely many fixed functions $G_i(z_1, \ldots, z_{n_i})$, same for all distributions (such as $+$, $\cdot$, etc.);*
- *finitely many functions $H_i(z_1, \ldots, z_{m_i})$ which may differ for different distributions; and*
- *an algorithm (same for all distributions), that, using the above functions and $2n$ inputs x_1, ..., x_n, $F_1(x_1)$, ..., $F_n(x_n)$, computes the values of the cdf $F(x_1, \ldots, x_n)$.*

Examples.

- In the original representation by a cdf, we have $H_1(z_1, \ldots, z_n) = F(z_1, \ldots, z_n)$.
- In the copula representation, we have $H_1(z_1, \ldots, z_n) = C(z_1, \ldots, z_n)$.

The corresponding algorithms for computing the cdf are described in the previous text.

What is Duplication-free: Towards a Precise Definition. In the previous section, we argued that if we represent a distribution by storing both its marginals and its cdf, then this representation contains duplicate information: indeed, based on the cdf, we can reconstruct the marginals.

In precise terms, the original representation by a cdf, when we have $H_1(z_1, \ldots, z_n) = F(z_1, \ldots, z_n)$, is not duplication-free, since we can compute, e.g., the marginal $F_1(v_1)$ by applying the following algorithm:

I_1: $r_1 \leftarrow H_1(v_1, +\infty, \ldots, +\infty)$;

I_2: stop.

It is therefore reasonable to call a representation duplication-free if such a reconstruction is impossible:

Definition 4. We say that a representation is duplication-free *if no algorithm is possible that, given the functions H_i representing the distribution and the inputs $x_1, \ldots, x_n$, computes one of the marginals.*

Example. The copula representation is duplication-free: indeed, for the same copula, we can have different marginals, and thus, it is not possible to compute the marginals based on the copula.

A Representation must be Computationally Efficient: Towards Precise Definitions. First, we want the reconstruction of the cdf to be as fast as possible:

Definition 5. We say that a duplication-free representation is *time-efficient if for each combination of inputs, the running time of the corresponding algorithm does not exceed the running time of any other duplication-free algorithm.*

As we have mentioned earlier, in addition to an efficient use of computation *time*, it is also important to make sure that computer *memory* is used efficiently: this is why it makes sense to consider only duplication-free representations.

In general, we store the values of one of several functions of different number of variables. To store a function of m variables, we need to store, e.g., its values on the corresponding grid. If we use g different values of each of the coordinates, then we need to store the values of this function at g^m points, i.e., we need to store g^m real numbers. Thus, the smaller m, the more efficient we are. This leads to the following definition.

Definition 6.

- *We say that a representation $H_1(z_1, \ldots, z_{m_1}), \ldots, H_k(z_1, \ldots, z_{m_k})$ is more space-efficient than a representation $H_1'(z_1, \ldots, z_{m_1'}), \ldots, H_{k'}'(z_1, \ldots, z_{m_{k'}'})$ if $k \leq k'$ and we can sort the value m_i and m_i' in such as way that $m_i \leq m_i'$ for all $i \leq k$.*

- *We say that a time-efficient duplication-free representation is computationally efficient if it is more space-efficient than any other time-efficient duplication-free representation.*

Main Result. *The only computationally efficient duplication-free representation of multi-D probability distributions is the copula representation.*

Discussion. Thus, copulas are indeed the most efficient way of representing additional information about the multi-D distributions for which we already know the marginals. This theoretical result explains why copulas have been efficiently used in many applications.

Proof.

1°. By definition, a computationally efficient representation should be time-efficient. By definition of time efficiency, this means that for each combination of inputs, the running time of the corresponding algorithm should not exceed the running time of any other duplication-free algorithm.

We know that the copula representation is duplication-free and that its running time is 1 for all the inputs. Thus, for all the inputs, the running time of the computationally efficient algorithm should not exceed 1. Thus, this algorithm can have exactly one non-stop instruction.

2°. This instruction is our only chance to change the value of the output variable r_1, so this instruction must be of assignment type $r_1 \leftarrow f_1(y_1, \ldots, y_{n_1})$. Since we did not have time to compute the values of any auxiliary variables – this is our first and only instruction – the values $y_1, \ldots, y_{n_1}$ must be the original inputs.

3°. The function f_1 cannot be from the list of fixed functions, since otherwise

- we would get the same result for all possible probability distributions, and thus,
- we would not be able to compute the corresponding values of the cdf $F(x_1, \ldots, x_n)$, which are different for different distributions.

Thus, the function f_1 must be one of the functions H_i characterizing a distribution.

4°. This function $f_1 = H_i$ cannot have fewer than n inputs, because otherwise, some variable x_j will not be used in this computation. Thus, the list of functions H_i used to describe a probability distribution must include at least one function of n variables.

5°. We are interested in a computationally efficient duplication-free representation. By definition, this means that this representation must be more space-efficient than any other time-efficient duplication-free representation. We know one time-efficient duplication-free representation – it is the copula representation, in which we use a single function H_1 of n variables.

The fact that our representation is more space-efficient than this one means that it uses only one function, and this must be a function of n or fewer variables. We have already shown that we cannot have a function of fewer than n variables, so we must have a function of exactly n variables.

6°. The result $F(x_1, \ldots, x_n)$ of applying this function of n variables must depend on all n variables $x_1, \ldots, x_n$. Thus, for each of these variables x_i, either this same value x_i or the value $F_i(x_i)$ must be among its inputs.

7°. If one of the inputs is x_i, i.e., if the corresponding instruction has the form

$I_1: r_1 \leftarrow H_1(y_1, \ldots, y_{i-1}, x_i, y_{i+1}, \ldots, y_n);$

where each y_i is either x_i or $F_i(x_i)$, then we will be able to compute the corresponding marginal by using the instruction

I_1: $r_1 \leftarrow H_1(Y_1, \ldots, Y_{i-1}, x_i, Y_{i+1}, \ldots, Y_n)$;

where $Y_i = +\infty$ when $y_i = x_i$ and $Y_i = 1$ when $y_i = F_i(x_i)$. Since we assumed that our scheme is duplication-free, this means that such a case is not possible, and thus, all the inputs to the function H_1 are not the values x_i, but the values of the marginals. Thus, the corresponding instruction has the form

I_1: $r_1 \leftarrow H_1(F_1(x_1), \ldots, F_n(x_n))$;

The result of this computation should be the multi-D cdf, so we should have

$$F(x_1, \ldots, x_n) = H_1(F_1(x_1), \ldots, F_n(x_n))$$

for all possible values $x_1, \ldots, x_n$.

This is exactly the definition of the copula, so we indeed conclude that every computationally efficient representation of a multi-D probability distribution is the copula representation. The main result is proven.

5   Conclusions and Future Work

Conclusions. The need for representing multi-D distributions in a computer comes from the fact that to make decisions, we need to be able to compute (and compare) the expected values of different utility functions. So, from all possible computer representations of multi-D distributions, we should select the ones for which the corresponding computations are the most efficient.

In this paper, we have shown that in situations where we already know the marginals, copulas are indeed the most computationally efficient way of representing additional information about the multi-D distributions.

Possible Future Work. In this paper, we have concentrated on computing the cumulative distribution function (cdf). This computation corresponds to *binary* utility functions – i.e., utility functions that take only two values $u^+ > u^-$. Such binary functions provide a good first approximation to the user's utilities and user's preferences, but to obtain a more accurate description of user's preferences, we need to use utility functions from a wider class.

It is therefore desirable to find out, for wider classes of utility functions, which computer representations are the most computationally efficient for computing the corresponding expected values. The empirical success of copulas leads us to a natural conjecture that for many such classes, the copula-based computer representations will still be the most computationally efficient.

Acknowledgments. We acknowledge the partial support of the Center of Excellence in Econometrics, Faculty of Economics, Chiang Mai University, Thailand. This work was also supported in part by the National Science Foundation grants HRD-0734825 and HRD-1242122 (Cyber-ShARE Center of Excellence) and DUE-0926721.

The authors are thankful to the anonymous referees for valuable suggestions.

References

1. Fishburn, P.C.: Utility Theory for Decision Making. John Wiley & Sons Inc., New York (1969)
2. Jaworski, P., Durante, F., Härdle, W.K., Ruchlik, T. (eds.): Copula Theory and Its Applications. Springer Verlag, Heidelberg (2010)
3. Luce, R.D., Raiffa, R.: Games and Decisions: Introduction and Critical Survey. Dover, New York (1989)
4. McNeil, A.J., Frey, R., Embrechts, P.: Quantitative Risk Management: Concepts, Techniques Tools. Princeton University Press, Princeton (2005)
5. Nelsen, R.B.: An Introduction to Copulas. Springer Verlag, Heidelberg (1999)
6. Raiffa, H.: Decision Analysis. Addison-Wesley, Reading (1970)

A New Measure of Monotone Dependence by Using Sobolev Norms for Copula

Hien D. Tran[1]($\boxtimes$), Uyen H. Pham[2], Sel Ly[3], and T. Vo-Duy[4,5]

[1] Tan Tao University, Duc Hoa, Long An, Vietnam
hein.tran@ttu.ed.vn
[2] University of Economics and Law, Ho Chi Minh City, Vietnam
uyenph@uel.edu.vn
[3] Faculty of Mathematics and Statistics, Ton Duc Thang University,
Ho Chi Minh City, Vietnam
lysel@tdt.edu.vn
[4] Division of Computational Mathematics and Engineering (CME),
Institute for Computational Science (INCOS), Ton Duc Thang University,
Ho Chi Minh City, Vietnam
[5] Faculty of Civil Engineering, Ton Duc Thang University,
Ho Chi Minh City, Vietnam
voduytrung@tdt.edu.vn

Abstract. Dependence structure, e.g. measures of dependence, is one of the main studies in correlation analysis. In [10], B. Schweizer and E.F. Wolff used L^p-metric $d_{L^p}(C, P)$ to obtain a measure of monotone dependence where P is the product copula or independent copula, and in [11] P. A. Stoimenov defined Sobolev metric $d_S(C, P)$ to construct the measure $\omega(C)$ for a class of Mutual Complete Dependences (MCDs). Due to the fact that the class of monotone dependence is contained in the class of MCDs, we constructed a new measure of monotone dependence, $\lambda(C)$, based on Sobolev metric which can be used to characterize comonotonic, countermonotonic and independence.

Keywords: Copulas · Monotone dependence · Measures of dependence · Sobolev metric

1 Introduction

Let X and Y be random variables with continuous marginal distribution functions F and G, respectively, and a joint distribution function H, then by Sklar's theorem [7], there exists a unique copula C such that

$$H(x, y) = C(F(x), G(y)) . \tag{1}$$

This copula C captures the dependent structure of X and Y. In particular, X and Y are independent if and only if $C(u, v) = P(u, v) = uv$; While X and Y are comonotonic (i.e. $Y = f(X)$ a.s., where f is strictly increasing) if and

© Springer International Publishing Switzerland 2015
V.-N. Huynh et al. (Eds.): IUKM 2015, LNAI 9376, pp. 126–137, 2015.
DOI: 10.1007/978-3-319-25135-6_13

only if $C(u,v) = M(u,v) = \min(u,v)$ and X and Y are countermonotonic (i.e. $Y = f(X)$ a.s., where f is strictly decreasing) if and only if $C(u,v) = W(u,v) = \max(u+v-1,0)$.

L^p-metric $d_{L^p}(C,P)$ due to B. Schweizer and E.F. Wolff, e.g. L^1 distance,

$$\sigma(C) = d_{L^1}(C,P) = 12 \iint_{I^2} |C(u,v) - uv| \, du dv, \tag{2}$$

where $I = [0;1]$, could be used as a measure of monotone dependence because it attains its maximum of 1 if and only if X and Y are monotone dependence. However, monotone property of Y on X (or vice versa) is not specified, increasing or decreasing. Moreover, if X and Y are MCD (i.e. $Y = f(X)$ a.s., where f is a Borel measurable bijection), $\sigma(C)$ can attain any value in $(0;1]$; And the set of all copulas linking MCDs random variables is dense in the set of all copulas with respect to any L^p-distance, $p \geq 1$, see [11]. This implies that L^p-distances are limited to detect dependences.

In [11], instead of using the L^p norm, K.F. Siburg and P.A. Stoimenov proposed to use a modified Sobolev norm, given by

$$\|C\|_S = \left(\iint_{I^2} \left(\partial_1 C^2 + \partial_2 C^2 \right) du dv \right)^{1/2}, \tag{3}$$

where $\partial_i C's, i = 1,2$ are the partial derivatives of $C(u,v)$ with respect to the i-th variable,

$$\partial_1 C = \frac{\partial}{\partial u} C(u,v) \text{ and } \partial_2 C = \frac{\partial}{\partial v} C(u,v) \ .$$

Then, a new non-parametric measure of dependence for two continuous random variables X and Y with copula C, is defined by

$$\omega(C) = \sqrt{3} \|C - P\|_S = \left(3\|C\|_S^2 - 2 \right)^{1/2} \ . \tag{4}$$

The measures $\sigma(C)$ and $\omega(C)$ are defined via the independent copula P. By the advantages of Sobolev metrics, we use a new approach to define a different measure of monotone dependence, $\lambda(C)$, in terms of distance between copula C and the comonotonic copula M. In this paper, some examples are also motivated to compare the two measures $\lambda(C)$ and $\sigma(C)$.

2   Copula and Measures of Dependence

Let $I = [0;1]$ be the closed unit interval and $I^2 = [0;1] \times [0;1]$ be the closed unit square.

Definition 1. *A 2-copula (two dimensional copula) is a function* $C: I^2 \to I$ *satisfying the conditions:*
i) $C(u,0) = C(v,0) = 0$, *for any* $u,v \in I$.
ii) $C(u,1) = u$ *and* $C(1,v) = v$, *for any* $u,v \in I$.
iii) For any $u_1, u_2, v_1, v_2 \in I$ *such that* $u_1 \le u_2$ *and* $v_1 \le v_2$,

$$C(u_2, v_2) - C(u_2, v_1) - C(u_1, v_2) + C(u_1, v_1) \ge 0 .$$

Note (see [7]): for any copula C and for any $(u,v) \in I^2$, we have,

$$W(u,v) \le C(u,v) \le M(u,v) . \tag{5}$$

For monotone transformations of random variables, copulas are invariant or change in a predictable way. It makes copulas become useful in the study of non-parametric statistics.

Theorem 1. *X and Y be continuous random variables with copula* $C_{X,Y}$. *Let f and g be strictly monotone functions on RanX and RanY. Then, for any* $u,v \in I$, *the following statements· are true:*

i) If f and g are strictly increasing, then

$$C_{f(X),g(Y)}(u,v) = C_{X,Y}(u,v) .$$

ii) If f is strictly increasing and g is strictly decreasing, then

$$C_{f(X),g(Y)}(u,v) = u - C_{X,Y}(u, 1-v) .$$

iii) If f is strictly decreasing and g is strictly increasing, then

$$C_{f(X),g(Y)}(u,v) = v - C_{X,Y}(1-u, v) .$$

iv) If f and g are strictly decreasing, then

$$C_{f(X),g(Y)}(u,v) = u + v - 1 + C_{X,Y}(1-u, 1-v) .$$

Since a copula captures the dependent structure of random variables, one can construct measures of dependence using copulas with suitable metrics. In fact, some well-known measures can be written in terms of copula [7], for example, The Pearson's correlation coefficient $r(X,Y)$,

$$r(X,Y) = \frac{1}{\sigma_X \sigma_Y} \iint\limits_{I^2} [C(u,v) - uv] \, dF^{-1}(u) dG^{-1}(v) ; \tag{6}$$

The Kendall's $\tau(X,Y)$,

$$\tau(X,Y) = \tau(C,C) = 4 \iint\limits_{I^2} C(u,v) dC(u,v) - 1 ; \tag{7}$$

The Spearman's $\rho(X, Y)$,

$$\rho(X, Y) = \rho(C) = 12 \iint_{I^2} C(u, v) du dv - 3 \; ; \tag{8}$$

The Gini's $\gamma(X, Y)$,

$$\gamma(X, Y) = \gamma(C) = 2 \iint_{I^2} (|u + v - 1| - |u - v|) \, dC(u, v) \; . \tag{9}$$

For any copula A and B, we now denote the Sobolev scalar product for copulas by $\langle A, B \rangle_S$

$$\langle A, B \rangle_S = \iint_{I^2} [\partial_1 A (u, v) \, \partial_1 B (u, v) + \partial_2 A (u, v) \, \partial_2 B (u, v)] \, du dv \; .$$

Then, the Sobolev norm and its metric are given, respectively, by

$$\|A\|_S^2 = \langle A, A \rangle_S = \iint_{I^2} \left(\partial_1 A^2 + \partial_2 A^2 \right) du dv,$$

$$d_S^2 (A, B) = \langle A - B, A - B \rangle_S$$
$$= \|A - B\|_S^2 = \|A\|_S^2 + \|B\|_S^2 - 2\langle A, B \rangle_S \; .$$

In [2], William F. Darsow et.al. defined a ∗-product for copulas,

$$(A * B) (u, v) = \int_0^1 \partial_2 A (u, t) \partial_1 B (t, v) \, dt \; .$$

The ∗-product of two copulas is also a copula, see [[2],[11]]. In addition, the Sobolev scalar product can be reformed through the ∗-product.

Theorem 2. *Let A and B be copulas. Then the ∗-product satisfies:*

i) $\frac{1}{2} \leqslant \langle A, B \rangle_S \leqslant 1$;
ii)

$$\langle A, B \rangle_S = \int_0^1 \left(A^T * B + A * B^T \right) (u, u) \, du$$

$$= \int_0^1 \left(A^T * B + B * A^T \right) (u, u) \, du,$$

where $A^T (u, v) = A (v, u)$.

Finally, we present the key theorem leading to the measure of MCDs which is given by (4).

Theorem 3. *For any copula C, the following hold:*

i) $\langle C, P \rangle_S = \frac{2}{3}$;
ii) $\|C - P\|_S^2 = \|C\|_S^2 - \frac{2}{3}$;
iii) $\frac{2}{3} \leqslant \|C\|_S^2 \leqslant 1$.

3   A New Measure of Monotone Dependence, $\lambda(C)$

We now turn to construct a new measure of monotone dependence between two continuous random variables X and Y with copula C.

Definition 2. *Let C be a copula of (X, Y). We defined $\lambda(X, Y)$, or $\lambda(C)$, by*

$$\lambda(X, Y) = \lambda(C) = \|C\|_S^2 - 2\|C - M\|_S^2, \tag{10}$$

where, $\|\cdot\|_S$ is the Sobolev norm for copulas and M is the Frechet-Hoeffding upper-bound.

Here are other forms of $\lambda(C)$.

Theorem 4. *The measure $\lambda(C)$ can be represented under the following forms,*

i) $\lambda(C) = 8 \int_0^1 C(u, u)\, du - \int_0^1 \int_0^1 \left[\partial_1 C^2(u, v) + \partial_2 C^2(u, v)\right] du\, dv - 2$.

ii) $\lambda(C) = \|C - P\|_S^2 - 2\left[\|C - M\|_S^2 - \|W - P\|_S^2\right]$.

Proof. For *i)*, we have $\|M\|_S = 1$ (see [11]), and

$$\|C - M\|_S^2 = \|C\|_S^2 + \|M\|_S^2 - 2\langle C, M\rangle_S \ .$$

Then, by substituting into (10), we obtain

$$\lambda(C) = 4\langle C, M\rangle_S - \|C\|_S^2 - 2 \ . \tag{11}$$

Next, by applying *ii)* of the Theorem 2 and the fact that copula M is an unit element with respect to $* - product$, see [[11]], we get

$$\langle C, M\rangle_S = \int_0^1 \left(C^T * M + C * M^T\right)(u, u)\, du$$

$$= \int_0^1 \left(C^T + C\right)(u, u) = 2\int_0^1 C(u, u)\, du \ .$$

Thus, the first part *i)* is immediately satisfied from (11).

For *ii)*, by theorem 3, the Sobolev norm for copula C,

$$\|C\|_S^2 = \|C - P\|_S^2 + \frac{2}{3} \ .$$

Replacing this in the definition (10), the measure $\lambda(C)$ has the form,

$$\lambda(C) = \|C - P\|_S^2 - 2\|C - M\|_S^2 + \frac{2}{3} \ .$$

Then, the result follows due to the fact $\|W - P\|_S^2 = \frac{1}{3}$. $\square$

Remark 1.

1. The $\lambda(C)$ is actually free of M.
2. The measure $\lambda(C)$ also contains a metric $d(C, P)$, the same as the other measures of dependence. In addition, it contains a metric $d(C, M)$ and $d(W, P)$. Therefore, the measure $\lambda(C)$ is capable of detecting comonotonic, countermonotonic and independence.

The main theorem which makes $\lambda(C)$ becomes a measure of dependence (satisfying the Renyi's axioms 1959; B. Schweizer and E.F. Wolff 1981).

Theorem 5. *The measure $\lambda(C)$ or $\lambda(X, Y)$ satisfies following properties*

i) $\lambda(X, Y)$ *is defined for all* X, Y;
ii) $\lambda(X, Y) = \lambda(Y, X)$;
iii) $-1 \leq \lambda(X, Y) \leq 1$;
iv) $\lambda(X, Y) = 0$ *if* X *and* Y *are independent*;
v) $\lambda(X, Y) = 1$ *if and only if* $Y = f(X)$ *a.s. where* f *is strictly increasing*;
vi) $\lambda(X, Y) = -1$ *if and only if* $Y = f(X)$ *a.s. where* f *is strictly decreasing*;
vii) *If* f *and* g *are strictly increasing (or decreasing) a.s., then*

$$\lambda(f(X), g(Y)) = \lambda(X, Y) \ ;$$

viii) *If* (X_n, Y_n) *converges to* (X, Y) *in distribution, then* $\lambda(X_n, Y_n) \to \lambda(X, Y)$.

Proof. The part *i)* and *ii)* are obviously.

For *iii)*,*v)*, and *vi)*, we evaluate by the inequalities,

$$\|C\|_S^2 \leq 1 \quad \text{and} \quad \|C - M\|_S^2 \geq 0 \ .$$

Therefore,

$$\lambda(C) = \|C\|_S^2 - 2\|C - M\|_S^2 \leq 1 \ .$$

It is clear that when copula C is *"closer"* to copula M in the sense of Sobolev distance, the measure $\lambda(C)$ tends to 1. The equality is fully identified iff $C = M$. This proves (i).

Similarly, for the lower bound, by using part i) of Theorem 2,

$$\|C\|_S^2 \leq 1 \text{ and } \langle C, M \rangle_S \geq \frac{1}{2},$$

$$\lambda(C) = 4\langle C, M \rangle_S - \|C\|_S^2 - 2 \geq 4\frac{1}{2} - 1 - 2 = -1 \ .$$

Hence, the measure $\lambda(C)$ attains its minimum of -1 iff $C = W$. This proves (vi).

Part *iv)* follows from computing with copula $C = P$,

$$\|P\|_S^2 = \frac{2}{3}, \ \|P - M\|_S^2 = \frac{1}{3} \ .$$

To prove *vii)*, we divide it into two cases.

Firstly, when f and g are strictly increasing, we immediately have the identity from Theorem 1,

$$\lambda\left(f\left(X\right),g\left(Y\right)\right) = \lambda\left(C_{f(X),g(Y)}\right) = \lambda\left(C_{X,Y}\right) = \lambda\left(X,Y\right) \ .$$

Secondly, when f and g are strictly decreasing, due to the invariant of Sobolev norm for copula, we only need to show

$$\int_0^1 C_{f(X),g(Y)}\left(u,u\right)du = \int_0^1 C_{X,Y}\left(u,u\right)du \ .$$

Indeed, copula has a form as in iv) of Theorem 1, we obtain

$$\int_0^1 C_{f(X),g(Y)}\left(u,u\right)du = \int_0^1 \left[2u - 1 + C_{X,Y}\left(1 - u, 1 - u\right)\right]du$$

$$= \int_0^1 C_{X,Y}\left(1 - u, 1 - u\right)du$$

$$= \int_0^1 C_{X,Y}\left(t,t\right)dt \ .$$

For the last part *viii)*, let H_n, C_n, H and C be joint distributions, copulas of (X_n, Y_n) and (X, Y), respectively. In fact, if H_n converges to H, then C_n pointwise converges to C, where we used $C\left(u,v\right) = H\left(F^{-1}\left(u\right), G^{-1}\left(v\right)\right)$. Moreover, all copulas are Lipschizt, see [7]. Therefore, C_n uniformly converges to C. This implies that $\lim_{n \to \infty} \lambda\left(C_n\right) = \lambda\left(C\right)$.   □

In addition, If the joint distribution of X and Y is bivariate normal, i.e. the copula of them is Gaussian copula C_r

$$C_r\left(u,v\right) = \frac{1}{2\pi\sqrt{1 - r^2}} \int_{-\infty}^{\Phi^{-1}(u)} \int_{-\infty}^{\Phi^{-1}(v)} \exp\left(-\frac{s^2 - 2rst + t^2}{2\left(1 - r^2\right)}\right)dsdt,$$

where, r is the Pearson's correlation coefficient between X and Y, $|r| < 1$. Then, as the measure $\omega(C_r)$ is an increasing function of r (see [11]), so is $\lambda(C_r)$. In this paper, the $\lambda(C_r)$ is computed by using Gauss quadrature as shown in Fig. 1.

It maybe required that if the measure equals 0, then the independence is implied. Moreover, it might be invariant under all monotone transformations. Actually, conditions *iv)* and *vi)* could be modified to obtain these desired results.

Theorem 6. *X and Y are independent if and only if copula C satisfies*

$$\lambda\left(C\right) = 0, \text{ and } \int_0^1 C\left(u,u\right)du = \frac{1}{3} \ . \tag{12}$$

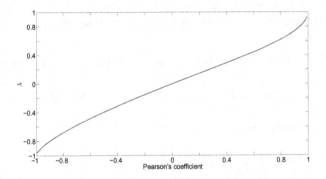

Fig. 1. The $\lambda(C_r)$ of Gaussian copula is an actually increasing function of r

Proof. The copula $C(u,v) = P(u,v) = uv$, captures the independence of two random variables. Thus, equalities in (12) hold.

Conversely, when (12) holds, then we easily have $\|C\|_S^2 = \frac{2}{3}$. Therefore, according to part ii) of Theorem 3, the independence is concluded.   □

Theorem 7. *The measure $\lambda(C)$ is invariant via f strictly increasing and g strictly decreasing, (or vice versa) if and only if copula C satisfies*

$$\int_0^1 [C(u,u) + C(u,1-u)]\, du = \frac{1}{2} \ . \tag{13}$$

Proof. First, for f strictly increasing and g strictly decreasing, copula of $f(X)$ and $g(Y)$ is given by,

$$C_{f(X),g(Y)}(u,v) = u - C_{X,Y}(u,1-v) \ .$$

Then, $\lambda(f(X),g(Y)) = \lambda(X,Y)$ iff

$$\int_0^1 [u - C_{X,Y}(u,1-u)]du = \int_0^1 C_{X,Y}(u,u)du$$

$$\int_0^1 [C_{X,Y}(u,u) + C_{X,Y}(u,1-u)]\, du = \frac{1}{2} \ .$$

Similarly, in the case f is strictly decreasing and g is strictly increasing, we only need to consider the following identity,

$$\int_0^1 [u - C_{X,Y}(1-u,u)]du = \frac{1}{2} - \int_0^1 C_{X,Y}(1-u,u)\, du = \frac{1}{2} - \int_0^1 C_{X,Y}(t,1-t)\, dt,$$

where we have taken $t = 1 - u$.   □

Corollary 1. *The measure $\lambda(C)$ is invariant via f strictly increasing and g strictly decreasing, (or vice versa) if and only if copula C satisfies the Gini's index $\gamma(C) = 0$.*

Proof. In [7], the Gini's index $\gamma(C)$ might be written in another form,

$$\gamma(C) = 4 \int_0^1 [C(u, u) + C(u, 1 - u)] \, du - 2 \ .$$

Thus, $\gamma(C) = 0$ if and only if

$$\int_0^1 [C(u, u) + C(u, 1 - u)] \, du = \frac{1}{2} \ .$$

$\square$

4   Examples

Example 1. Let X and Y be random variables with copula C_θ,

$$C_\theta = \theta M + (1 - \theta)P, \text{ with } \theta \in [0; 1] \ . \tag{14}$$

This copula captures comonotonic dependence when the parameter θ tends to 1. Let's evaluate how to the measure $\lambda(C_\theta)$ is used effectively. In fact, the measure $\lambda(C_\theta)$ and $\sigma(C_\theta)$ (compare Table 1.) follow that

$$\lambda(C_\theta) = \frac{1}{3} \left(4\theta - \theta^2 \right), \text{ and } \sigma(C_\theta) = \theta \ . \tag{15}$$

Example 2. Let X and Y be random variables with copula C_θ,

$$C_\theta = \theta W + (1 - \theta)P, \text{ with } \theta \in [0; 1] \ . \tag{16}$$

Then, the measure $\lambda(C_\theta)$ and $\sigma(C_\theta)$ (compare Table 2.) are computed as follows

$$\lambda(C_\theta) = -\frac{1}{3} \left(2\theta + \theta^2 \right), \text{ and } \sigma(C_\theta) = \theta \ . \tag{17}$$

Clearly, the measure $\lambda(C_\theta)$ can distinguish countermonotonic and monotonicity. Indeed, when the parameter θ tends to 1, the copula models structure of countermonotonicity in which is measured by negative values.

Next examples are motivated to illustrate Theorem 6 and Theorem 7.

Table 1. Comparison between $\lambda(C_\theta)$ and $\sigma(C_\theta)$ with copula C_θ in (14)   .

θ	0.0	0.1	0.2	0.3	0.4	0.5	0.6	0.7	0.8	0.9	1.0
$\lambda(C_\theta)$	0.00	0.13	0.25	0.37	0.48	0.58	0.68	0.77	0.85	0.93	1.00
$\sigma(C_\theta)$	0.00	0.10	0.20	0.30	0.40	0.50	0.60	0.70	0.80	0.90	1.00

Table 2. Comparison between $\lambda(C_\theta)$ and $\sigma(C_\theta)$ with copula C_θ in (16).

θ	0.0	0.1	0.2	0.3	0.4	0.5	0.6	0.7	0.8	0.9	1.0
$\lambda(C_\theta)$	0.00	−0.07	−0.15	−0.23	−0.32	−0.42	−0.52	−0.63	−0.75	−0.87	−1.00
$\sigma(C_\theta)$	0.00	0.10	0.20	0.30	0.40	0.50	0.60	0.70	0.80	0.90	1.00

Example 3. Let $\theta \in [-1; 1]$, and we suppose X and Y are random variables with Mardia family of copulas C_θ,

$$C_\theta(u, v) = \frac{\theta^2(1+\theta)}{2} M(u, v) + (1 - \theta^2) P(u, v) + \frac{\theta^2(1-\theta)}{2} W(u, v) \ . \quad (18)$$

Then, a straightforward calculation shows that

$$\lambda(C_\theta) = \frac{1}{3}\theta^2 + \theta^3 - \frac{1}{12}\theta^4 - \frac{1}{4}\theta^6 \ . \quad (19)$$

So with $\theta \in [-1; 1]$, we have

$$\lambda(C_\theta) = 0 \Leftrightarrow \begin{bmatrix} \theta_1 = 0 \\ \theta_2 = -0.32249074 \end{bmatrix} \quad (20)$$

From this of view, we can see that if we take $\theta = \theta_2 \approx -0.3225$, then $\lambda(C_{\theta_2}) = 0$, but X and Y are actually not independent. However, if we define integral

$$\int_0^1 C_\theta(u, u) \, du = \frac{1}{3} + \frac{1}{24}\theta^2 + \frac{1}{8}\theta^3,$$

and set up $\frac{1}{3} + \frac{1}{24}\theta^2 + \frac{1}{8}\theta^3 = \frac{1}{3}$, then we get $\theta = 0$ or $\theta = -\frac{1}{3}$. Due to Theorem 6, there is only a value $\theta = 0$ which is implied independence. Indeed, when $\theta = 0$, we obtain copula $C_0(u, v) = P(u, v) = uv$.

Now, if we take a transformation $f(X)$ and $g(Y)$, where f is a strictly increasing and g is a strictly decreasing function (or vice versa), then copula of $f(X)$ and $g(X)$ is given by C_θ^* which is just made a permutation between M and W,

$$C_\theta^*(u, v) = \frac{\theta^2(1+\theta)}{2} W(u, v) + (1 - \theta^2) P(u, v) + \frac{\theta^2(1-\theta)}{2} M(u, v), \quad (21)$$

and the measure $\lambda(C_\theta^*)$ could be computed directly,

$$\lambda(C_\theta^*) = \frac{1}{3}\theta^2 - \theta^3 - \frac{1}{12}\theta^4 - \frac{1}{4}\theta^6 \ . \quad (22)$$

Clearly, we have $\lambda(C_{(-\theta)}) = \lambda(C_\theta^*)$. Hence, they are symmetric to the Y-axis. Similarly, the Gini's index $\gamma(C_\theta) = \theta^3$ and $\gamma(C_\theta^*) = -\theta^3$ (Compare Fig. 2). Note that, $\lambda(C_\theta) = \lambda(C_\theta^*)$, (i.e. invariant) iff $\theta = 0$ or the Gini's index $\gamma(C_\theta) = 0$.

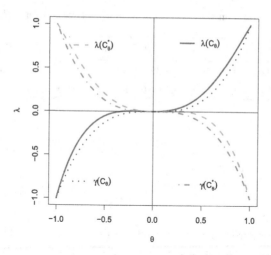

Fig. 2. Comparisons among the measures of Mardia's copulas.

Example 4. Let X and Y be random variables with copula C,

$$
C\left(u,v\right) = \begin{cases} \min\left(u,v\right), & |u-v| > \dfrac{1}{2}, \\[2mm] \max\left(u+v-1,0\right), & |u+v-1| > \dfrac{1}{2}, \\[2mm] \dfrac{u+v}{2} - \dfrac{1}{4}, & \text{elsewhere.} \end{cases} \tag{23}
$$

Then, the measure $\lambda(C)$ and $\sigma(C)$ are calculated as follows

$$
\lambda(C) = -\frac{1}{4}, \text{ and } \sigma(C) = \frac{1}{4} . \tag{24}
$$

However, the other measures of dependence are equal zero,

$$
\tau\left(C\right) = \rho\left(C\right) = \gamma\left(C\right) = 0 .
$$

From Corollary 1, the measure $\lambda(C)$ is actually invariant for all strictly transformations with respect to the copula in (23).

5   Conclusion

The measure $\lambda(C)$ has some more advantages such as $\lambda(C) \in [-1;1]$. In particular, if $\lambda(C) < 0$ then C tends to the countermonotonic copula, W. On the other hand, if $\lambda(C) > 0, C$ tends to the comonotonic copula, M. Moreover, by capturing the Sobolev metric, the measure $\lambda(C)$ can be used to overcome the inconsistency between MCD and the L^p-distance which is used in $\sigma(C)$. Also, for some families of copula C_θ, some popular measures can not detect structure of dependences, but the measure λ can be done (say, $\lambda(C_\theta) \neq 0$). Furthermore, this paper also established a necessary and sufficient condition for independence.

References

1. Cherubini, U., Luciano, E., Vecchiato, W.: Copula Methods in Finance. John Wiley & Sons Ltd., Chichester (2004)
2. Darsow, W.F., Nguyen, B., Olsen, E.T.: Copulas and Markov Processes. Illinois J. Math. **36**(4), 600–642 (1992)
3. Darsow, W.F., Olsen, E.T.: Norms for Copulas. Internat. J. Math. and Math. Sci. **18**(3), 417–436 (1995)
4. Joe, H.: Multivariate Models and Dependence Concepts. Chapman & Hall/CRC, London (1997)
5. Kimeldorf, G., Sampson, A.R.: Monotone Dependence. Ann. Statist. **6**(4), 895–903 (1978)
6. Mari, D.D., Kotz, S.: Correlation and Dependence. World Scientific Publishing Co. Pte. Ltd., Singapore (2004)
7. Nelsen, R.B.: An Introduction to Copulas, 2nd edn. Springer, New York (2006)
8. Nguyen, T.H.: A Copula Approach to Model Validation. IJITAS **4**(4), 531–547 (2011)
9. Scarsini, M.: On Measure of Concordance. Stochastica **8**(3), 201–218 (1984)
10. Schweizer, B., Wolff, E.F.: On Nonparametric Measures of Dependence for random Variables. Ann. Stat. **9**(4), 879–885 (1981)
11. Stoimenov, P.A.: A Measure of Mutual Complete Dependence. Ph.D. Thesis, TU Dormund (2008)

Why ARMAX-GARCH Linear Models Successfully Describe Complex Nonlinear Phenomena: A Possible Explanation

Hung T. Nguyen[1,2], Vladik Kreinovich[3(✉)], Olga Kosheleva[4],
and Songsak Sriboonchitta[2]

[1] Department of Mathematical Sciences, New Mexico State University, Las Cruces,
NM 88003, USA
hunguyen@nmsu.edu
[2] Department of Economics, Chiang Mai University, Chiang Mai, Thailand
songsakecon@gmail.com
[3] Department of Computer Science, University of Texas at El Paso, 500 W.
University, El Paso, TX 79968, USA
vladik@utep.edu
[4] University of Texas at El Paso, 500 W. University, El Paso, TX 79968, USA
olgak@utep.edu

Abstract. Economic and financial processes are complex and highly nonlinear. However, somewhat surprisingly, linear models like ARMAX-GARCH often describe these processes reasonably well. In this paper, we provide a possible explanation for the empirical success of these models.

1 Formulation of the Problem

Economic and Financial Processes are Very Complex. It is well known that economic and financial processes are very complex. The future values of the corresponding quantities are very difficult to predict, and many empirical dependencies are highly nonlinear.

Surprising Empirical Success of ARMAX-GARCH Models. In spite of the clear non-linearity of the economic and financial processes, linear models are surprisingly efficient in predicting the future values of the corresponding quantities. Specifically, if we are interested in the quantity X which is affected by the external quantity d, then good predictions can often be made based on the AutoRegressive-Moving-Average model with eXogenous inputs model (ARMAX) [3,4]:

$$X_t = \sum_{i=1}^{p} \varphi_i \cdot X_{t-i} + \sum_{i=1}^{b} \eta_i \cdot d_{t-i} + \varepsilon_t + \sum_{i=1}^{q} \theta_i \cdot \varepsilon_{t-i}, \qquad (1)$$

for appropriate parameters φ_i, η_i, and θ_i. Here, ε_t are random variables of the type $\varepsilon_t = \sigma_t \cdot z_t$, where z_t is white noise with 0 mean and standard deviation 1,

© Springer International Publishing Switzerland 2015
V.-N. Huynh et al. (Eds.): IUKM 2015, LNAI 9376, pp. 138–150, 2015.
DOI: 10.1007/978-3-319-25135-6_14

and the dynamics of standard deviations σ_t is described by the Generalized AutoRegressive Conditional Heteroscedasticity (GARCH) model [2–4]:

$$\sigma_t^2 = \alpha_0 + \sum_{i=1}^{\ell} \beta_i \cdot \sigma_{t-i}^2 + \sum_{i=1}^{k} \alpha_i \cdot \varepsilon_{t-i}^2. \tag{2}$$

What we do in this Paper. In this paper, we provide a possible explanation for the empirical success of the ARMAX-GARCH models.

Specifically, we start with simplest predictions models, in which many important aspects are ignored, and then show that by appropriately taking these aspects into account, we come up with the ARMAX-GARCH model.

2 First Approximation: Closed System

First Approximation: Description. Let us start with the simplest possible model, in which we ignore all outside effects on the system, be it deterministic or random. Such no-outside-influence systems are known as *closed systems*.

In such a closed system, the future state X_t is uniquely determined by its previous states:

$$X_t = f(X_{t-1}, X_{t-2}, \ldots, X_{t-p}). \tag{3}$$

So, to describe how to predict the state of a system, we need to describe the corresponding prediction function $f(x_1, \ldots, x_p)$.

In the remaining part of this section, we will describe the reasonable properties of this prediction function, and then we will show that these properties imply that the prediction function be linear.

First Reasonable Property of the Prediction Function $f(x_1, \ldots, x_p)$: Continuity. In many cases, the values X_t are only approximately known. For example, if we are interested in predicting Gross Domestic Product (GDP) or unemployment rate, we have to take into account that the existing methods of measuring these characteristics are approximate.

Thus, the actual values X_t^{act} of the quantity X may be, in general, slightly different from the observed values X_t. It is therefore reasonable to require that when we apply the prediction function to the observed (approximate) value, then the prediction $f(X_{t-1}, \ldots, X_{t-p})$ should be close to the prediction $f(X_{t-1}^{\mathrm{act}}, \ldots, X_{t-p}^{\mathrm{act}})$ based on the actual values X_t^{act}.

In other words, if the inputs to the function $f(x_1, \ldots, x_p)$ change slightly, the output should also change slightly. In precise terms, this means that the function $f(x_1, \ldots, x_p)$ should be *continuous*.

Second Reasonable Property of the Prediction Function $f(x_1, \ldots, x_p)$: Additivity. In many practical situations, we observe a joint effect of two (or more) different subsystems $X = X^{(1)} + X^{(2)}$. For example, the varying price of the financial portfolio can be represented as the sum of the prices corresponding to two different parts of this portfolio. In this case, the desired future value X_t also consists of two components: $X_t = X_t^{(1)} + X_t^{(2)}$.

In this case, we have two possible ways to predict the desired value X_t:

- first, we can come up with a prediction X_t by applying the prediction function $f(x_1, \ldots, x_p)$ to the joint values $X_{t-i} = X_{t-i}^{(1)} + X_{t-i}^{(2)}$;
- second, we can apply this prediction function to the first system, then apply it to the second subsystem, and then add the resulting predictions $X_t^{(1)}$ and $X_t^{(2)}$ to come up with the joint prediction $X_t = X_t^{(1)} + X_t^{(2)}$.

It makes sense to require that these two methods lead to the same prediction, i.e., that:

$$f\left(X_{t-1}^{(1)} + X_{t-1}^{(2)}, \ldots, X_{t-p}^{(1)} + X_{t-p}^{(2)}\right) =$$

$$f\left(X_{t-1}^{(1)}, \ldots, X_{t-p}^{(1)}\right) + f\left(X_{t-1}^{(?)}, \ldots, X_{t-p}^{(2)}\right). \tag{4}$$

In mathematical terms, this means that the predictor function should be *additive*, i.e., that

$$f\left(x_1^{(1)} + x_1^{(2)}, \ldots, x_p^{(1)} + x_p^{(2)}\right) = f\left(x_1^{(1)}, \ldots, x_p^{(1)}\right) + f\left(x_1^{(2)}, \ldots, x_p^{(2)}\right)$$

for all possible tuples $\left(x_1^{(1)}, \ldots, x_p^{(1)}\right)$ and $\left(x_1^{(2)}, \ldots, x_p^{(2)}\right)$.

Known Result. We have argued that the desired function $f(x_1, \ldots, x_p)$ should be continuous and additive. It is known (see, e.g., [1,5]) that every continuous additive function is a homogeneous linear function, i.e., it has the form

$$f(x_1, \ldots, x_p) = \sum_{i=1}^{p} \varphi_i \cdot x_i \tag{5}$$

for some values φ_i.

Indeed, for the tuples $(x_1, 0, \ldots, 0)$, $(0, x_2, 0, \ldots, 0)$, ..., $(0, \ldots, 0, x_n)$ that add up to $(x_1 \ldots, x_n)$, additivity implies that $f(x_1, \ldots, x_n) = \sum_{i=1}^{n} f_i(x_i)$, where we denoted $f_i(x_i) \stackrel{\text{def}}{=} f(0, \ldots, 0, x_i, 0, \ldots, 0)$.

For each function $f_i(x_i)$, additivity of the function $f(x_1, \ldots, x_n)$ implies that $f_i\left(x_i^{(1)} + x_i^{(2)}\right) = f_i\left(x_i^{(1)}\right) + f_i\left(x_i^{(2)}\right)$. In particular, for $x_i = 0$, we have $f_i(0) = f_i(0) + f_i(0)$ hence $f_i(0) = 0$.

For any integer $q > 0$, $1 = \dfrac{1}{q} + \ldots + \dfrac{1}{q}$ (q times), hence additivity implies that

$$f_i(1) = f_i\left(\frac{1}{q}\right) + \ldots + f_i\left(\frac{1}{q}\right) \quad (q \text{ times}),$$

so $f_i(1) = q \cdot f_i\left(\dfrac{1}{q}\right)$ and $f_i\left(\dfrac{1}{q}\right) = \dfrac{1}{q} \cdot f_i(1)$.

For any $p > 0$, we have $\dfrac{p}{q} = \dfrac{1}{q} + \ldots + \dfrac{1}{q}$ (p times), hence additivity implies that

$$f_i\left(\frac{p}{q}\right) = f_i\left(\frac{1}{q}\right) + \ldots + f_i\left(\frac{1}{q}\right) \quad (p \text{ times}),$$

i.e., that $f_i\left(\dfrac{p}{q}\right) = p \cdot f_i\left(\dfrac{1}{q}\right) = \dfrac{p}{q} \cdot f_i(1)$.

For negative integers p, for which $p = -|p|$, we have $\dfrac{p}{q} + \dfrac{|p|}{q} = 0$, hence additivity implies that $f_i\left(\dfrac{p}{q}\right) + f_i\left(\dfrac{|p|}{q}\right) = f_i(0) = 0$, so

$$f_i\left(\frac{p}{q}\right) = -f_i\left(\frac{|p|}{q}\right) = -\frac{|p|}{q} \cdot f_i(1) = \frac{p}{q} \cdot f_i(1).$$

Thus, for all rational values $x_i = \dfrac{p}{q}$, we get $f_i(x_i) = \varphi_i \cdot x_i$, where we denoted $\varphi_i \overset{\text{def}}{=} f_i(1)$. Every real number x_i can be represented as a limit of its rational approximations $x_i^{(k)} \to x_i$. For these rational approximations, we have $f_i\left(x_i^{(k)}\right) = \varphi_i \cdot x_i^{(k)}$.

Continuity of the prediction function $f(x_1, \ldots, x_n)$ implies that the function $f_i(x_i)$ is continuous as well. Thus, when $x_i^{(k)} \to x_i$, we get $f_i\left(x_i^{(k)}\right) \to f_i(x_i)$. So, in the limit $k \to \infty$, the formula $f_i\left(x_i^{(k)}\right) = \varphi_i \cdot x_i^{(k)}$ implies that the equality $f_i(x_i) = \varphi_i \cdot x_i$ holds for any real value x_i.

Thus, from $f(x_1, \ldots, x_n) = \sum\limits_{i=1}^{n} f_i(x_i)$, we conclude that indeed

$$f(x_1, \ldots, x_n) = \sum_{i=1}^{n} \varphi_i \cdot x_i.$$

Conclusion: We Must Consider Linear Predictors. Since the prediction function $f(x_1, \ldots, x_n)$ is continuous and additive, and all continuous additive functions have the form (5), the prediction formula (3) has the following form

$$X_t = \sum_{i=1}^{p} \varphi_i \cdot X_{t-i}. \tag{6}$$

Thus, for this case, we have indeed justified the use of *linear predictors*.

3   Second Approximation: Taking External Quantities Into Account

Second Approximation: Description. To get a more adequate description of the economic system, let us take into account that the desired quantity X

may also be affected by some external quantity d. For example, the stock price may be affected by the amount of money invested in stocks.

In this case, to determine the future state X_t, we need to know not only the previous states of the system $X_{t-1}, X_{t-2}, \ldots$, but also the corresponding values of the external quantity $d_t, d_{t-1}, \ldots$ Thus, the general prediction formula now takes the following form:

$$X_t = f(X_{t-1}, X_{t-2}, \ldots, X_{t-p}, d_t, d_{t-1}, \ldots, d_{t-b}). \tag{7}$$

So, to describe how to predict the state of a system, we need to describe the corresponding prediction function $f(x_1, \ldots, x_p, y_0, \ldots, y_b)$. Let us consider reasonable properties of this prediction function.

First Reasonable Property of the Prediction Function $f(x_1, \ldots, x_p, y_0, \ldots, y_b)$: **Continuity.** Similarly to the previous case, we can conclude that small changes in the inputs should lead to small changes in the prediction. Thus, the prediction function $f(x_1, \ldots, x_p, y_0, \ldots, y_b)$ should be continuous.

Second Reasonable Property of the Prediction Function $f(x_1, \ldots, x_p, y_0, \ldots, y_b)$: **Additivity.** As we have mentioned earlier, in many practical situations, we observe a joint effect of two (or more) different subsystems $X = X^{(1)} + X^{(2)}$. In this case, the overall external effect d can be only decomposed into two components $d = d^{(1)} + d^{(2)}$: e.g., investments into two sectors of the stock market.

In this case, just like in the first approximation, we have two possible ways to predict the desired value X_t:

- first, we can come up with a prediction X_t by applying the prediction function $f(x_1, \ldots, x_p, y_0, \ldots, y_b)$ to the joint values $X_{t-i} = X_{t-i}^{(1)} + X_{t-i}^{(2)}$ and $d_{t-i} = d_{t-i}^{(1)} + d_{t-i}^{(2)}$;
- second, we can apply this prediction function to the first system, then apply it to the second subsystem, and then add the resulting predictions $X_t^{(1)}$ and $X_t^{(2)}$ to come up with the joint prediction $X_t = X_t^{(1)} + X_t^{(2)}$.

It makes sense to require that these two methods lead to the same prediction, i.e., that:

$$f\left(X_{t-1}^{(1)} + X_{t-1}^{(2)}, \ldots, X_{t-p}^{(1)} + X_{t-p}^{(2)}, d_t^{(1)} + d_t^{(2)}, \ldots, d_{t-b}^{(1)} + d_{t-b}^{(2)}\right) =$$

$$f\left(X_{t-1}^{(1)}, \ldots, X_{t-p}^{(1)}, d_t^{(1)}, \ldots, d_{t-b}^{(1)}\right) + f\left(X_{t-1}^{(2)}, \ldots, X_{t-p}^{(2)}, d_t^{(2)}, \ldots, d_{t-b}^{(2)}\right). \tag{8}$$

Thus, the prediction function $f(x_1, \ldots, x_n, y_0, \ldots, y_b)$ should be additive.

Conclusion: We Must Consider Linear Predictors. We argued that the prediction function $f(x_1, \ldots, x_n, y_0, \ldots, y_b)$ should be continuous and additive. We have already proven that every continuous additive function is a homogeneous linear function, i.e., that each such function has the form

$$f(x_1, \ldots, x_p, y_0, \ldots, y_b) = \sum_{i=1}^{p} \varphi_i \cdot x_i + \sum_{i=0}^{b} \eta_i \cdot y_i \tag{9}$$

for some values φ_i and η_i. Thus, the prediction equation (7) takes the following form:

$$X_t = \sum_{i=1}^{p} \varphi_i \cdot X_{t-i} + \sum_{i=0}^{b} \eta_i \cdot d_{t-i}. \tag{10}$$

4 Third Approximation: Taking Random Effects into Account

Description of the Model. In addition to the external quantities d, the desired quantity X is also affected by many other phenomena. In contrast to the explicitly known quantity d, we do not know the values characterizing all these phenomena, so it is reasonable to consider them *random effects*. Let us denote the random effect generated at moment t by ε_t.

In this case, to determine the future state X_t, we need to know not only the previous states of the system X_{t-1}, X_{t-2}, ..., and the corresponding values of the external quantity d_t, d_{t-1}, ..., we also need to know the values of these random effects ε_t, ε_{t-1}, ... Thus, the general prediction formula now takes the form

$$X_t = f(X_{t-1}, X_{t-2}, \ldots, X_{t-p}, d_t, d_{t-1}, \ldots, d_{t-b}, \varepsilon_t, \ldots, \varepsilon_{t-q}). \tag{11}$$

So, to describe how to predict the state of a system, we need to describe the corresponding prediction function $f(x_1, \ldots, x_p, y_0, \ldots, y_b, z_0, \ldots, z_q)$. Let us consider reasonable properties of this prediction function.

First Reasonable Property of the Prediction Function $f(x_1, \ldots, x_p,$ $y_0, \ldots, y_b, z_0, \ldots, z_q)$: **Continuity.** Similarly to the previous cases, we can conclude that small changes in the inputs should lead to small changes in the prediction. Thus, the prediction function $f(x_1, \ldots, x_p, y_0, \ldots, y_b, z_0, \ldots, z_q)$ should be continuous.

Second Reasonable Property of the Prediction Function $f(x_1, \ldots, x_p,$ $y_0, \ldots, y_b, z_0, \ldots, z_q)$: **Additivity.** As we have mentioned earlier, in many practical situations, we observe a joint effect of two (or more) different subsystems $X = X^{(1)} + X^{(2)}$. In this case, the overall external effect d can be only decomposed into two components $d = d^{(1)} + d^{(2)}$, and the random effects can also be decomposed into effects affecting the two subsystems: $\varepsilon = \varepsilon^{(1)} + \varepsilon^{(2)}$.

In this case, just like in the first two approximations, we have two possible ways to predict the desired value X_t:

- first, we can come up with a prediction X_t by applying the prediction function $f(x_1, \ldots, x_p, y_0, \ldots, y_b)$ to the joint values $X_{t-i} = X_{t-i}^{(1)} + X_{t-i}^{(2)}$, $d_{t-i} = d_{t-i}^{(1)} + d_{t-i}^{(2)}$, and $\varepsilon_{t-i} = \varepsilon_{t-i}^{(1)} + \varepsilon_{t-i}^{(2)}$;
- second, we can apply this prediction function to the first system, then apply it to the second subsystem, and then add the resulting predictions $X_t^{(1)}$ and $X_t^{(2)}$ to come up with the joint prediction $X_t = X_t^{(1)} + X_t^{(2)}$.

It makes sense to require that these two methods lead to the same prediction, i.e., that:

$$f\left(X_{t-1}^{(1)} + X_{t-1}^{(2)}, \ldots, d_t^{(1)} + d_t^{(2)}, \ldots, \varepsilon_t^{(1)} + \varepsilon_t^{(2)}, \ldots\right) =$$

$$f\left(X_{t-1}^{(1)}, \ldots, d_t^{(1)}, \ldots, \varepsilon_t^{(1)}, \ldots\right) + f\left(X_{t-1}^{(2)}, \ldots, d_t^{(2)}, \ldots, \varepsilon_t^{(2)}, \ldots\right). \quad (12)$$

Thus, the prediction function $f(x_1, \ldots, y_0, \ldots, z_0, \ldots)$ should be additive.

Conclusion: We Must Consider Linear Predictors. We have argued that the prediction function $f(x_1, \ldots, y_0, \ldots, z_0, \ldots)$ should be continuous and additive. We have already proven that every continuous additive function is a homogeneous linear function. So, we have

$$f(x_1, \ldots, x_p, y_0, \ldots, y_b, z_0, \ldots, z_q) = \sum_{i=1}^{p} \varphi_i \cdot x_i + \sum_{i=0}^{b} \eta_i \cdot y_i + \sum_{i=0}^{q} \theta_i \cdot z_i \quad (13)$$

for some values φ_i, η_i, and θ_i; thus, the prediction formula (11) takes the following form:

$$X_t = \sum_{i=1}^{p} \varphi_i \cdot X_{t-i} + \sum_{i=0}^{b} \eta_i \cdot d_{t-i} + \sum_{i=0}^{q} \theta_i \cdot \varepsilon_{t-i}. \quad (14)$$

Deriving the Original ARMAX Formula (1). The formula (14) is almost identical to the ARMAX formula (1), the only difference is that in our formula (14), the value ε_t is multiplied by a coefficient θ_0, while in the ARMAX formula (1), this coefficient is equal to 1.

To derive the formula (1), let us first comment that it is highly improbable that the random quantity ε_t does not have any effect on the current value X_t of the desired quantity; thus, the parameter θ_0 describing this dependence should be non-zero.

Now, to describe the random effects, instead of the original values ε, we can consider the new values $\varepsilon' \stackrel{\text{def}}{=} \theta_0 \cdot \varepsilon$. In terms of thus re-scaled random effects, we have $\varepsilon = \dfrac{1}{\theta_0} \cdot \varepsilon'$. Thus, the corresponding linear combination of random terms takes the form

$$\sum_{i=0}^{q} \theta_i \cdot \varepsilon_{t-i} = \theta_0 \cdot \varepsilon_t + \sum_{i=1}^{q} \theta_i \cdot \varepsilon_{t-i} = \varepsilon_0' + \sum_{i=1}^{q} \theta_i \cdot \frac{1}{\theta_0} \cdot \varepsilon_{t-i}', \quad (15)$$

i.e., the form

$$\sum_{i=0}^{q} \theta_i \cdot \varepsilon_{t-i} = \varepsilon_0' + \sum_{i=1}^{q} \theta_i' \cdot \varepsilon_{t-i}', \quad (16)$$

where we denoted $\theta_i' \stackrel{\text{def}}{=} \theta_i \cdot \dfrac{1}{\theta_0}$.

Substituting the formula (16) into the expression (14), we get the desired ARMAX formula:

$$X_t = \sum_{i=1}^{p} \varphi_i \cdot X_{t-i} + \sum_{i=0}^{b} \eta_i \cdot d_{t-i} + \varepsilon_0' + \sum_{i=1}^{q} \theta_i' \cdot \varepsilon_{t-i}'. \tag{17}$$

Similar Arguments can be Used to Explain Formulas of Vector ARMAX (VARMAX). Similar arguments lead to a multi-D (version) of the formula (17), in which X, d, ε are vectors, and φ_i, η_i, and θ_i' are corresponding matrices.

5 Fourth Approximation: Taking Into Account that Standard Deviations Change with Time

Description of the Model. In the previous sections, we described how the desired quantity X changes with time. In the previous section, we showed how to take into account the random effects $\varepsilon_t = \sigma_t \cdot z_t$ that affect our system.

To complete the description of the system's dynamics, it is necessary to supplement this description with a description of how the corresponding standard deviation σ_t changes with time. So, now, instead of simply predicting the values X_t, we need to predict both the values X_t and the values σ_t.

To predict both values X_t and σ_t, we can use:

- the previous states of the system X_{t-1}, X_{t-2}, $\ldots$,
- the corresponding values of the external quantity d_t, d_{t-1}, $\ldots$,
- the values of these random effects ε_t, ε_{t-1}, $\ldots$, and
- the previous values of the standard deviation σ_{t-1}, σ_{t-2}, $\ldots$

Thus, the general prediction formulas now take the form

$$X_t = f(X_{t-1}, \ldots, d_t, \ldots, \varepsilon_t, \ldots, \sigma_{t-1}, \ldots); \tag{18}$$

$$\sigma_t = g(X_{t-1}, \ldots, d_t, \ldots, \varepsilon_t, \ldots, \sigma_{t-1}, \ldots). \tag{19}$$

So, to describe how to predict the state of a system, we need to describe the corresponding prediction functions $f(x_1, \ldots, y_0, \ldots, z_0, \ldots, t_1, \ldots)$ and $g(x_1, \ldots, y_0, \ldots, z_0, \ldots, t_1, \ldots)$. Let us consider reasonable properties of this prediction function.

First Reasonable Property of the Prediction Functions $f(x_1, \ldots, y_0, \ldots, z_0, \ldots, t_1, \ldots)$ **and** $g(x_1, \ldots, y_0, \ldots, z_0, \ldots, t_1, \ldots)$**: Continuity.** Similarly to the previous cases, we can conclude that small changes in the inputs should lead to small changes in the prediction. Thus, the prediction functions $f(x_1, \ldots, y_0, \ldots, z_0, \ldots, t_1, \ldots)$ and $g(x_1, \ldots, y_0, \ldots, z_0, \ldots, t_1, \ldots)$ should be continuous.

Second Reasonable Property of the Prediction Functions $f(x_1,\ldots,$ $y_0,\ldots,z_0,\ldots,t_1,\ldots)$ **and** $g(x_1,\ldots,y_0,\ldots,z_0,\ldots,t_1,\ldots)$: **Independence-Based Additivity.** As we have mentioned earlier, in many practical situations, we observe a joint effect of two (or more) different subsystems $X = X^{(1)} + X^{(2)}$. In this case, the overall external effect d can be only decomposed into two components $d = d^{(1)} + d^{(2)}$, and the random effects can also be decomposed into effects affecting the two subsystems: $\varepsilon = \varepsilon^{(1)} + \varepsilon^{(2)}$.

In our final model, we also need to take into the standard deviations σ; so, we need to know how to compute the standard deviation σ of the sum of two random variables based on their standard deviations $\sigma^{(1)}$ and $\sigma^{(2)}$ of the two components. In general, this is not possible: to know the standard deviation σ of the sum, we need to know not only the standard deviations $\sigma^{(1)}$ and $\sigma^{(2)}$, we also need to know the correlation between the random variables $\varepsilon^{(1)}$ and $\varepsilon^{(2)}$.

However, there are two reasonable cases when σ can be computed based on $\sigma^{(1)}$ and $\sigma^{(2)}$:

– the case when the random variables $\varepsilon^{(1)}$ and $\varepsilon^{(2)}$ are independent, and
– the case when the random variables $\varepsilon^{(1)}$ and $\varepsilon^{(2)}$ are strongly correlated.

In this section, we will consider both cases; in this subsection, we will consider the first case.

It is known that the variance $V = \sigma^2$ of the sum of two independent random variables is equal to the sum of the variances, so $V = V^{(1)} + V^{(2)}$. To utilize this property, it makes sense to use the variance V instead of standard deviation. In terms of variance, the predictions formulas take the form

$$X_t = f'(X_{t-1},\ldots,d_t,\ldots,\varepsilon_t,\ldots,V_{t-1},\ldots); \tag{20}$$

$$V_t = g'(X_{t-1},\ldots,d_t,\ldots,\varepsilon_t,\ldots,V_{t-1},\ldots), \tag{21}$$

for appropriate functions $f'(x_1,\ldots,y_0,\ldots,z_0,\ldots,t_1,\ldots)$ and $g'(x_1,\ldots,y_0,\ldots,z_0,\ldots,t_1,\ldots)$.

In this case, just like in the first three approximations, we have two possible ways to predict the desired values X_t and V_t:

– first, we can come up with predictions X_t and V_t by applying the prediction functions
  $f'(x_1,\ldots,y_0,\ldots,z_0,\ldots,t_1,\ldots)$ and $g'(x_1,\ldots,y_0,\ldots,z_0,\ldots,t_1,\ldots)$ to the joint values $X_{t-i} = X_{t-i}^{(1)} + X_{t-i}^{(2)}$, $d_{t-i} = d_{t-i}^{(1)} + d_{t-i}^{(2)}$, $\varepsilon_{t-i} = \varepsilon_{t-i}^{(1)} + \varepsilon_{t-i}^{(2)}$, and $V_{t-i} = V_{t-i}^{(1)} + V_{t-i}^{(2)}$;
– second, we can apply these prediction functions to the first system, then apply them to the second subsystem, and then add the resulting predictions $X_t^{(i)}$ and $V_t^{(i)}$ to come up with the joint predictions $X_t = X_t^{(1)} + X_t^{(2)}$ and $V_t = V_t^{(1)} + V_t^{(2)}$.

It makes sense to require that these two methods lead to the same prediction, i.e., that:

$$f'\left(X_{t-1}^{(1)} + X_{t-1}^{(2)},\ldots,d_t^{(1)} + d_t^{(2)},\ldots,\varepsilon_t^{(1)} + \varepsilon_t^{(2)},\ldots,V_{t-1}^{(1)} + V_{t-1}^{(2)},\ldots\right) =$$

$$f'\left(X_{t-1}^{(1)}, \ldots, d_t^{(1)}, \ldots, \varepsilon_t^{(1)}, \ldots, V_{t-1}^{(1)}, \ldots\right) + \qquad (22)$$

$$f'\left(X_{t-1}^{(2)}, \ldots, d_t^{(2)}, \ldots, \varepsilon_t^{(2)}, \ldots, V_{t-1}^{(2)}, \ldots\right);$$

$$g'\left(X_{t-1}^{(1)} + X_{t-1}^{(2)}, \ldots, d_t^{(1)} + d_t^{(2)}, \ldots, \varepsilon_t^{(1)} + \varepsilon_t^{(2)}, \ldots, V_{t-1}^{(1)} + V_{t-1}^{(2)}, \ldots\right) =$$

$$g'\left(X_{t-1}^{(1)}, \ldots, d_t^{(1)}, \ldots, \varepsilon_t^{(1)}, \ldots, V_{t-1}^{(1)}, \ldots\right) + \qquad (23)$$

$$g'\left(X_{t-1}^{(2)}, \ldots, d_t^{(2)}, \ldots, \varepsilon_t^{(2)}, \ldots, V_{t-1}^{(2)}, \ldots\right).$$

Thus, both prediction functions $f'(x_1, \ldots, y_0, \ldots, z_0, \ldots, t_1, \ldots)$ and $g'(x_1, \ldots, y_0, \ldots, z_0, \ldots, t_1, \ldots)$ should be additive.

Since every continuous additive function is a homogeneous linear function, we have

$$f'(x_1, \ldots, y_0, \ldots, z_0, \ldots, t_1, \ldots) = \sum_{i=1}^{p} \varphi_i \cdot x_i + \sum_{i=0}^{b} \eta_i \cdot y_i + \sum_{i=0}^{q} \theta_i \cdot z_i + \sum_{i=1}^{\ell} \beta_i' \cdot t_i \quad (24)$$

and

$$g'(x_1, \ldots, y_0, \ldots, z_0, \ldots, t_1, \ldots) = \sum_{i=1}^{p} \varphi_i' \cdot x_i + \sum_{i=0}^{b} \eta_i' \cdot y_i + \sum_{i=0}^{q} \theta_i' \cdot z_i + \sum_{i=1}^{\ell} \beta_i \cdot t_i. \quad (25)$$

for some values φ_i, φ_i', η_i, η_i', θ_i, θ_i', β_i, and β_i'.

Similarly to the previous case, without losing generality, we can take $\theta_0 = 1$. Thus, the prediction formulas (20) and (21) take the following form:

$$X_t = \sum_{i=1}^{p} \varphi_i \cdot X_{t-i} + \sum_{i=0}^{b} \eta_i \cdot d_{t-i} + \varepsilon_t + \sum_{i=1}^{q} \theta_i \cdot \varepsilon_{t-i} + \sum_{i=1}^{\ell} \beta_i' \cdot \sigma_{t-i}^2; \qquad (26)$$

$$\sigma_t^2 = \sum_{i=1}^{p} \varphi_i' \cdot X_{t-i} + \sum_{i=0}^{b} \eta_i' \cdot d_{t-i} + \sum_{i=0}^{q} \theta_i' \cdot \varepsilon_{t-i} + \sum_{i=1}^{\ell} \beta_i \cdot \sigma_{t-i}^2. \qquad (27)$$

Third Reasonable Property of the Prediction Functions: Dependence-Based Additivity. In the previous subsection, we considered the case when the random variables corresponding to two subsystems are independent. This makes sense, e.g., when we divide the stocks into groups by industry, so that different random factors affect the stocks from different groups. Alternatively, we can divide the stocks from the same industry by geographic location of the corresponding company, in which case the random factors affecting both types of stocks are strongly positively correlated.

For such random quantities, the standard deviation of the sum is equal to the sum of standard deviations $\sigma = \sigma^{(1)} + \sigma^{(2)}$. In this case, we can similarly use two different ways to predicting X_t and σ_t:

- first, we can come up with predictions X_t and V_t by applying the prediction formulas (26) and (27) to the joint values $X_{t-i} = X_{t-i}^{(1)} + X_{t-i}^{(2)}$, $d_{t-i} = d_{t-i}^{(1)} + d_{t-i}^{(2)}$, $\varepsilon_{t-i} = \varepsilon_{t-i}^{(1)} + \varepsilon_{t-i}^{(2)}$, and $\sigma_{t-i} = \sigma_{t-i}^{(1)} + \sigma_{t-i}^{(2)}$;
- second, we can apply these prediction formulas to the first system, then apply them to the second subsystem, and then add the resulting predictions $X_t^{(i)}$ and $V_t^{(i)}$ to come up with the joint predictions $X_t = X_t^{(1)} + X_t^{(2)}$ and $\sigma_t = \sigma_t^{(1)} + \sigma_t^{(2)}$.

It makes sense to require that these two methods lead to the same prediction.

Let us Use the Dependence-Based Additivity Property. Let us apply the dependence-based additivity property to the case when the two combined subsystems are identical, i.e., when $X_{t-i}^{(1)} = X_{t-i}^{(2)}$, $d_{t-i}^{(1)} = d_{t-i}^{(2)}$, $\varepsilon_{t-i}^{(1)} = \varepsilon_{t-i}^{(2)}$, and $\sigma_{t-i}^{(1)} = \sigma_{t-i}^{(2)}$. In this case, $X_{t-i}^{(1)} = X_{t-i}^{(2)} = 0.5 \cdot X_{t-i}$, $d_{t-i}^{(1)} = d_{t-i}^{(2)} = 0.5 \cdot d_{t-i}$, $\varepsilon_{t-i}^{(1)} = \varepsilon_{t-i}^{(2)} = 0.5 \cdot \varepsilon_{t-i}$, and $\sigma_{t-i}^{(1)} = \sigma_{t-i}^{(2)} = 0.5 \cdot \sigma_{t-i}$. Substituting these values $X_{t-i}^{(1)}$, $d_{t-i}^{(1)}$, $\varepsilon_{t-i}^{(1)}$, and $\sigma_{t-i}^{(1)}$ into the formula (26), we conclude that

$$X_t^{(1)} = \sum_{i=1}^{p} \varphi_i \cdot 0.5 \cdot X_{t-i} + \sum_{i=0}^{b} \eta_i \cdot 0.5 \cdot d_{t-i} + 0.5 \cdot \varepsilon_t +$$

$$\sum_{i=1}^{q} \theta_i \cdot 0.5 \cdot \varepsilon_{t-i} + \sum_{i=1}^{\ell} \beta_i' \cdot 0.25 \cdot \sigma_{t-i}^2. \tag{29}$$

Thus, for $X_t = X_t^{(1)} + X_t^{(2)} = 2X^{(1)}(t)$, we get

$$X_t = \sum_{i=1}^{p} \varphi_i \cdot X_{t-i} + \sum_{i=0}^{b} \eta_i \cdot d_{t-i} + \varepsilon_t + \sum_{i=1}^{q} \theta_i \cdot \varepsilon_{t-i} + 0.5 \cdot \sum_{i=1}^{\ell} \beta_i' \cdot \sigma_{t-i}^2. \tag{30}$$

We require that the prediction (26) based on the sums should be equal to the sum (30) of the predictions based on the individual subsystems. Thus, the right-hand sides of the expressions (26) and (30) should be equal for all possible values of the input quantities X_{t-i}, d_{t-i}, ε_{t-i}, and σ_{t-i}. By comparing these right-hand sides, we see that this is possible only if $\beta_i' = 0$.

Similarly, substituting the values $X_{t-i}^{(1)} = 0.5 \cdot X_{t-i}$, $d_{t-i}^{(1)} = 0.5 \cdot d_{t-i}$, $\varepsilon_{t-i}^{(1)} = 0.5 \cdot \varepsilon_{t-i}$, and $\sigma_{t-i}^{(1)} = 0.5 \cdot \sigma_{t-i}$ into the formula (27), we conclude that

$$\left(\sigma_t^{(1)}\right)^2 = \sum_{i=1}^{p} \varphi_i' \cdot 0.5 \cdot X_{t-i} + \sum_{i=0}^{b} \eta_i' \cdot 0.5 \cdot d_{t-i} +$$

$$\sum_{i=0}^{q} \theta_i' \cdot 0.5 \cdot \varepsilon_{t-i} + \sum_{i=1}^{\ell} \beta_i \cdot 0.25 \cdot \sigma_{t-i}^2. \tag{31}$$

Thus, for $\sigma_t^2 = \left(2\sigma_t^{(1)}\right)^2 = 4 \cdot \left(\sigma_t^{(1)}\right)^2$, we get

$$\sigma_t^2 = 2 \cdot \sum_{i=1}^{p} \varphi_i' \cdot X_{t-i} + 2 \cdot \sum_{i=0}^{b} \eta_i' \cdot d_{t-i} + 2 \cdot \sum_{i=0}^{q} \theta_i' \cdot \varepsilon_{t-i} + \sum_{i=1}^{\ell} \beta_i \cdot \sigma_{t-i}^2. \qquad (32)$$

We require that the prediction (27) based on the sums should be equal to the sum (32) of the predictions based on the individual subsystems. Thus, the right-hand sides of the expressions (27) and (32) should be equal for all possible values of the input quantities X_{t-i}, d_{t-i}, ε_{t-i}, and σ_{t-i}. By comparing these right-hand sides, we see that this is possible only if $\varphi_i' = 0$, $\eta_i' = 0$, and $\theta_i' = 0$.

Conclusion. Since $\varphi_i' = 0$, $\eta_i' = 0$, and $\theta_i' = 0$, the formulas (26) and (27) take the following form:

$$X_t = \sum_{i=1}^{p} \varphi_i \cdot X_{t-i} + \sum_{i=0}^{b} \eta_i \cdot d_{t-i} + \sum_{i=0}^{q} \theta_i \cdot \varepsilon_{t-i}; \qquad (33)$$

$$\sigma_t^2 = \sum_{i=1}^{\ell} \beta_i \cdot \sigma_{t-i}^2. \qquad (34)$$

Relation to the ARMAX-GRARCH Formula. We can see that the formula (33) is exactly the ARMAX formula, and that the formula (34) is a simplified version of the GARCH formula (our formula lack a constant term α_0 and the terms proportional to ε_{t-i}^2).

We have derived these empirically successful formulas from first principles. Thus, we indeed provide a reasonable explanation for the empirical success of these formulas.

6   Conclusions and Future Work

Conclusions. In this paper, we analyzed the following problem:

- on the one hand, economic and financial phenomena are very complex and highly nonlinear;
- on the other hand, in many cases, linear ARMAX-GARCH formulas provide a very good empirical description of these complex phenomena.

Specifically, we showed that reasonable first principles lead to the ARMAX formulas and to the (somewhat simplified version of) GARCH formulas. Thus, we have provided a reasonable explanation for the empirical success of these formulas.

Remaining Problem. While our approach explains the ARMAX formula, it provides only a partial explanation of the GARCH formula: namely, we only explain a simplified version of the GARCH formula (2). It is desirable to come up with a similar explanation of the full formula (2).

Intuitively, the presence of additional terms proportional to ε^2 in the formula (2) is understandable. Indeed, when the mean-0 random components $\varepsilon^{(1)}$ and $\varepsilon^{(2)}$ are independent, the average value of their product $\varepsilon^{(1)} \cdot \varepsilon^{(2)}$ is zero. Let us show that this makes the missing term $\sum_{i=1}^{k} \alpha_i \cdot \varepsilon_{t-i}^2$ additive – and thus, derivable from our requirements. Indeed, we have

$$\sum_{i=1}^{k} \alpha_i \cdot \left(\varepsilon_{t-i}^{(1)} + \varepsilon_{t-i}^{(2)} \right)^2 =$$

$$\sum_{i=1}^{k} \alpha_i \cdot \left(\varepsilon_{t-i}^{(1)} \right)^2 + \sum_{i=1}^{k} \alpha_i \cdot \left(\varepsilon_{t-i}^{(2)} \right)^2 + 2 \sum_{i=1}^{k} \alpha_i \cdot \left(\varepsilon_{t-i}^{(1)} \cdot \varepsilon_{t-i}^{(2)} \right).$$

Here, the last term – the average value of the product $\varepsilon^{(1)} \cdot \varepsilon^{(2)}$ – is practically 0:

$$\sum_{i=1}^{k} \alpha_i \cdot \left(\varepsilon_{t-i}^{(1)} \cdot \varepsilon_{t-i}^{(2)} \right) \approx 0,$$

so we indeed have independence-based additivity:

$$\sum_{i=1}^{k} \alpha_i \cdot \left(\varepsilon_{t-i}^{(1)} + \varepsilon_{t-i}^{(2)} \right)^2 \approx \sum_{i=1}^{k} \alpha_i \cdot \left(\varepsilon_{t-i}^{(1)} \right)^2 + \sum_{i=1}^{k} \alpha_i \cdot \left(\varepsilon_{t-i}^{(2)} \right)^2.$$

The term α_0 can also be intuitively explained: since there is usually an additional extra source of randomness which constantly adds randomness to the process.

It is desirable to transform these intuitive arguments into a precise derivation of the GARCH formula (2).

Acknowledgments. We acknowledge the partial support of the Center of Excellence in Econometrics, Faculty of Economics, Chiang Mai University, Thailand. This work was also supported in part by the National Science Foundation grants HRD-0734825 and HRD-1242122 (Cyber-ShARE Center of Excellence) and DUE-0926721. The authors are thankful to the anonymous referees for valuable suggestions.

References

1. Aczél, J., Dhombres, J.: Functional Equations in Several Variables. Cambridge University Press, Cambridge (2008)
2. Bollerslev, T.: Generalized autoregressive conditional heteroskedasticity. Journal of Econometrics **31**(3), 307–327 (1986)
3. Brockwell, P.J., Davis, R.A.: Time Series: Theories and Methods. Springer, New York (2009)
4. Enders, W.: Applied Econometric Time Series. Wiley, New York (2014)
5. Kreinovich, V., Nguyen, H.T., Sriboonchitta, S.: Prediction in econometrics: towards mathematical justification of simple (and successful) heuristics. International Journal of Intelligent Technologies and Applied Statistics (IJITAS) **5**(4), 443–460 (2012)

A Copula-Based Stochastic Frontier Model for Financial Pricing

Phachongchit Tibprasorn[1], Kittawit Autchariyapanitkul[2][(✉)],
Somsak Chaniam[3], and Songsak Sriboonchitta[1]

[1] Faculty of Economics, Chiang Mai University, Chiang Mai, Thailand
[2] Faculty of Economics, Maejo University, Chiang Mai, Thailand
kittawit_autchariya@cmu.ac.th
[3] Department of Mathematics, Faculty of Science, Chiang Mai University,
Chiang Mai, Thailand

Abstract. We use the concept of a stochastic frontier in production to analyses the problem of pricing in stock markets. By modifying the classical stochastic frontier model to accommodate for errors dependency, using copulas, we show that our extended stochastic frontier model is more suitable for financial analyses. The validation is achieved by using AIC in our model selection problem.

Keywords: Copula · Financial econometrics · Gaussian quadrature · Technical efficiency · Stochastic frontier

1 Introduction

Unless the economy is controlled by a monopolist, the pursuit of economic profits, through technology efficiency in the production industry, is a natural behavior of producers. A producer is like a machine: converting inputs to outputs. Given a technology (of production), the relationship between inputs and outputs is indicated as $x \to y$: the input x produces the output y.

Since various inputs x (vectors) can produce the same output y (scalar), the map $y \to \{x : x \Rightarrow y\}$ is set-valued. If we are interested in *cost functions*, then we will be interested in the boundary (frontier) of the set $\psi(y) = \{x : x \Rightarrow y\}$, whereas if we are interested in *production functions*, then we will look at $x \to \{y : x \Rightarrow y\}$ whose frontier $\varphi(x) = \max\{y : x \Rightarrow y\}$ (when appropriate, e.g. when $y \in \mathbb{R}$, and $\{y : x \Rightarrow y\}$ is compact). Multivariate extensions are possible. In this paper, we address the production problem, the cost problem is just its dual, and hence can be analyzed similarly.

The above function $\varphi : X \to Y$, $\varphi(x) = \max\{y : x \Rightarrow y\}$ is referred to as the (*production) frontier function* [1]. The set $\{y : x \Rightarrow y\}$ is the set of outputs that are feasible for an input x.

To specify further, the set $Q(y) = \{x : x \to y, \lambda x \nrightarrow y, \lambda < 1\}$, which is the set of inputs (for a given output y) capable of producing y, but when contracted, become incapable of doing so. This set $Q(y)$ is referred to as an (input) *isoquant*. The isoquant defines the boundary (frontier) of the input requirement set.

© Springer International Publishing Switzerland 2015
V.-N. Huynh et al. (Eds.): IUKM 2015, LNAI 9376, pp. 151–162, 2015.
DOI: 10.1007/978-3-319-25135-6_15

Among other things, the frontier function will be used to defined technology efficiency or inefficiency (in fact to quantify it), a characteristic of obvious importance. To arrive at the concept of "technical efficiency", we compare observed production and "theoretical" production. We can take the ratio of actual observed output to the optimal value as specified by the (theoretical) production function. Specifically, for $y \leq \varphi(x)$, then a measure of efficiency of a firm i can be taken as $TE_i(y, x) = \frac{y}{\varphi(x)}$ (which is ≤ 1).

The obvious problem is how to get the frontier function of technology? Well, as in many similar issues in science, in general, and in economics, in particular, only empirical data could shed light on them. Thus, we are heading to the problem of estimating frontier functions.

Let's begin with a model like $y_i = \varphi(x_i, \theta)TE_i$ where $0 < TE(y_i, x_i) \leq 1$, or equivalently

$$\log y_i = \log \varphi(x_i, \theta) + \log TE_i \tag{1}$$

If we let $u_i = -\log TE_i$ (or $TE_i = e^{-u_i}$), then $u_i \geq 0$, playing the role of a measure of "inefficiency" since $u_i \approx 1 - TE_i$.

Frontier functions as given above, in which deviations of observations from the theoretical maxima are only attributed to the inefficiency of firms, are called *deterministic frontier function.*

If the output that a producer can get is supposed to be restricted both by the production function and by random outside factors such as luck or unforeseen shocks. The model is called a *stochastic frontier (production) function.*

The probability of not being able to complete the production efficiency (maximum output) of a producer might be due to other random factors such as "bad weather" (say, in the context of agriculture). In other words, deviations from the deterministic frontier might not be entirely under a control of the producer. Note that traditionally, producers are capable of reaching the maximum outputs (the "efficient" frontier). Thus, a more realistic concept of production frontier should be formulated with such random factors in mind: the "frontier" is randomly placed by the whole collection of random variables that might enter the model outside the control of the producer.

Specifically, the relationship between input x and output y (in a given technology) should be of the form

$$y = \varphi(x) + v - u, \tag{2}$$

where $\varphi(x) = \max\{y : x \Rightarrow y\}$, v is a random error representing the stochastic characteristic of the frontier $\varphi(.)$ (e.g., measurement error with respect to x), and u is associated with inter-producer efficiency differences. As such, $u \geq 0$, and the **stochastic frontier** is $x \to \varphi(x) + v$ which is now (with the random disturbance v) stochastic (i.e., not always on target). It is still a "frontier" since $u \geq 0$ implies that, almost surely, $y \leq \varphi(x) + v$.

Given this formulation of a stochastic frontier, the problem (for applications) is the *specification of the distributions of the error terms* v and u. In fact, what we need to carry out a "stochastic frontier analysis" is the specification of the distribution of the error term $v - u$ in the model.

Technically speaking, we need the distribution of $v - u$ where v is a symmetric random variable and u is one-sided random variable ($u \geq 0$). Thus, theoretically, possible models (specifications) for v and u can be chosen among appropriate types of distributions (such as normals for v, half-normals/ exponentials/ gammas for u). One more important specification: with two specified *marginal distributions* $F_v(.)$ and $F_u(.)$, we still need another ingredient to obtain the *joint distribution of* (v, u), since the distribution of $v - u$ can only be derived if we know the joint distribution of (u, v). That extra ingredient is **copulas** [2]. The "standard" additional assumption that v and u are independent is in fact an assumption about a special copula in the model.

Previous studied in stochastic frontier models (SFM) of the Aigner et al. [3], Meeusen and Broeck [4], Olsen et al. [5], Stevenson [6] and Battese and Coelli [7] defined a model with a specific distributional form for the error term and used MLE to estimate the parameters. The gap between the frontier and the observed production is a measure of inefficiency. Such distance is modeled by a one-sided random variable and two-sided random variable capturing the measurement error of the frontier. (see, Sriboonchitta [8], Sanzidur et al. [9], and Aigner et al. [3]). In their studies, they imposed the strong assumption that the one-sided error term u and symmetric error term v were independent. A normal distribution is assumed for the symmetric error term v while a half-normal distribution is assigned to the one-sided random variable u. Thus, by the technical efficiency estimation, we mean "estimation of u". Note that u is a random variable with support on $\mathbb{R}_+$.

For applying the technical efficiency in financial institutions, from the studied of Hasan MZ et al. [10], they estimated the technical efficiency in Dhaka Stock Exchange (DSE) by assuming the Cobb-Douglas function. In their research, they assumed truncated normal and half-normal with time-variant and time-invariant inefficiency effects were calculated, and the component error terms were independent. The results are showed that the technical efficiency (TE) was big in the group of investment and quite small for the bank group. Liu and Chung [11] studied the Chinese A-share Seasoned Equity Offerings (SEOs) by comparing a novel under-pricing measure with an SFM. Moreover, again, the error terms were assumed independent. Now, we can relax this strong assumption by using copula to find the joint distribution of the two random variables (see, Wiboonpongse et al. [12], Amsler et al. [13], Burns [14]). There are many papers using copula such as Autchariyapanitkul et al. [15], Kreinovich et al. [16], Kreinovich et al. [17], Sirisrisakulchai and Sriboonchitta [18] and Tang et al. [19].

In this paper, we use the stochastic frontier model for financial pricing problem. Specifically, input x_i plays the role of historical price and volume of a stock, and the output y is the current price of a stock. The technical efficiency (TE) in this financial context measures the success level of the strategies to hold stocks for an investment. The goal of this study is to predict the frontier of prices as we may call "pricing frontier", which shows the highest possible price of a stock can be retrieved, it is quite interesting that the actual price beneath the frontier curve displays the inefficiency of the stock market. Also, we can use this method as a strategy to selects the stock in the portfolio. Besides, we use this to identify the determinants that influence the share prices of the selected stock and the level of influence in the stock market. Moreover, we need to find out if factors, such as historical price, volume, and market indices are significantly related to stock prices.

The maximum simulated likelihood method is used to estimate the unknown parameters (see, Wiboonpongse et al. [12], Green [20], Burns [14]). However, in here, we estimate integral by the *Gaussian quadrature* for the simplest way, and then, we use the optimization method to get these parameters.

This study focused on stocks in the Stock Exchange of Thailand (*SET*). As an emerging stock market, it is usually shown higher levels of asymmetric information due to less complicated regulation, the less care for investigation duty of institutional investors and large-block shareholders compared with developed markets. This study gives us an excellent opportunity to investigate price as it is more likely to be significant. SET has received considerable attention from many researchers and investors as the case studies and potential investment alternatives. Also, it is quite crucial to examine the valuation of stock prices.

The paper is arranged as follows. Section 2 provides the knowledge of stochastic frontier and copulas while Section 3 explains the empirical application to stock market. Section 4 employs the results, and final Section makes the conclusion and extension.

2   Copulas as Dependence Measures

Since the serious drawbacks of Pearson's linear correlation and cannot be used for "heavy-tailed" distributions, especially when applying to financial economics. First, dependence structures. Whatever they are, X and Y are related to each other in some fashion, linear or not. Can we classify all different types of dependence using copulas? Here are some examples of copulas.

(*i*) The independent copula

Recall that two real-valued random variables X and Y are said to be independent if, for any A, B in $\mathcal{B}(\overline{\mathbb{R}})$,

$$P(X \in A, Y \in B) = P(X \in A)P(Y \in B) \tag{3}$$

In terms of distribution functions $F_X, F_Y, F_{(X,Y)}$, the above is equivalent to, for any $x, y \in \mathbb{R}$

$$F_{(X,Y)}(x, y) = F_X(x)F_Y(y) \tag{4}$$

Thus, independence is described by the product copula $C(u, v) = uv$

(*ii*) The Gaussian copula

For C being the *Gaussian copula*, i.e.,

$$C_\theta(u,v) = \int_{-\infty}^{\Phi^{-1}(u)} \int_{-\infty}^{\Phi^{-1}(v)} \left(\frac{1}{2\pi\sqrt{1-\rho^2}}\right) \exp\left[-\frac{x^2 + y^2 - 2\theta xy}{2(1-\theta^2)}\right], \mathrm{d}x\mathrm{d}y \quad (5)$$

Since, $\theta \in (1,-1)$, we have $\tau(C_\theta, C_\theta) = \frac{2}{\pi}arcsin(\theta)$ and $\theta(X,Y) = \frac{6}{\pi}arcsin(\frac{\theta}{2})$. Φ represents the standard normal distribution function. Thus, the Gaussian copulas are strongly comprehensive.

(iii) t-copula
Recall that (X,Y) is said to follow a bivariate t distribution with ν degrees of freedom if its density function is of the form

$$h(x,y) = \frac{\Gamma\left(\frac{\nu+2}{2}\right)}{\Gamma\left(\frac{\nu}{2}\right)\sqrt{(\nu\pi)^2 \mid \Sigma \mid}} \left[1 + \frac{x^2 - 2\rho xy + y^2}{\nu(1-\rho^2)}\right]^{-\frac{\nu+2}{2}} \quad (6)$$

where we write x for (x,y) and Σ a positive definite matrix. The (bivariate) t-copula is

$$C_{\nu,\Sigma}(u,v) = \int_{-\infty}^{t_\nu^{-1}(u)} \int_{-\infty}^{t_\nu^{-1}(v)} h(x,y)\mathrm{d}x\mathrm{d}y \quad (7)$$

where t_ν^{-1} is the quantile function of the univariate t-distribution with ν degrees of freedom. It can be shown that t-copulas exhibit asymptotic dependence in the tail.

(iv) Clayton copula
Clayton copula can capture the lower tail dependence for $\theta > 0$ and given by

$$C_\theta^{Cl}(u,v) = (\max\{u^{-\theta} + v^{-\theta} - 1, 0\})^{-\frac{1}{\theta}}, \quad (8)$$

where $\theta \in [-1, \infty) - \{0\}$. For the limits $\theta \to 0$ we obtain the independence copula, while for $\theta \to \infty$ the Clayton copula arrives at the comonotonicity copula. For $\theta = -1$ we obtain the Fréchet-Hoeffding lower bound.

(v) Frank's copula
Note that *Frank's copulas* are *Archimedean copulas* with generator $\varphi_\theta(u) = -\log\frac{e^{-\theta u} - 1}{e^{-\theta} - 1}$ leads to the Frank's copula given by

$$C_\theta^{Fr}(u,v) = -\frac{1}{\theta} \ln\left(1 + \frac{(e^{-\theta u} - 1)*(e^{-\theta v} - 1)}{e^{-\theta} - 1}\right), \quad (9)$$

for $\theta \in \mathbb{R} - \{0\}$. Thus, this family is strongly comprehensive.

(*vi*) Gumbel copula

The *Gumbel copula* or *Gumbel-Hougaard copula* of (X, Y) is given in the following form:

$$C_\theta^{Gu}(u, v) = \exp(-[(\log u)^\theta + (-\log v)^\theta]^{\frac{1}{\theta}}), \tag{10}$$

where $\theta \in [1, \infty)$. It is and Archimedean copula with (additive) generator $\varphi_\theta(t) = (-\log t)^\theta$. For $\theta = 1$ that show the independence copula. And for $\theta \to \infty$ The Gumbel copula tends to the comonotonicity copula.

3   Copula-Based Stochastic Frontier Model (SFM)

Consider the stochastic frontier model given by

$$Y = f(X; \beta) + (V - U), \tag{11}$$

where X is a vector of pricing factors and β is the associated vector of parameters that we need to estimate. Thus, the error component V is the usual symmetric error term with *cdf* $F_V(v) = Pr(V \leq v)$ and U is a one-side error term which we define as pricing inefficiency with *cdf* $F_U(u) = Pr(U \leq u)$ and they are assumed to be independent. The pricing efficiency (TE) can be obtained by

$$TE = \exp(-U). \tag{12}$$

Since, $U \geqslant 0$ and follow a half-normal distribution. The density as following

$$f^{HN}(U) = \frac{2}{\sqrt{2\pi\sigma_U^2}} \exp\left\{-\frac{U^2}{2\sigma_U^2}\right\}. \tag{13}$$

And V is a normal distribution. The density is written as

$$f^N(V) = \frac{1}{\sqrt{2\pi\sigma_V^2}} \exp\left\{-\frac{V^2}{2\sigma_V^2}\right\}. \tag{14}$$

Assumed the independence assumption, the joint density function of U and V is the product of their individual density functions

$$f(U, V) = \frac{1}{\pi\sqrt{\sigma_U^2\sigma_V^2}} \exp\left\{-\frac{U^2}{2\sigma_U^2} - \frac{V^2}{2\sigma_V^2}\right\}. \tag{15}$$

Since, $\xi = V - U$, the joint density function of U and ξ is

$$f(U, \xi) = \frac{1}{\pi\sqrt{\sigma_U^2\sigma_V^2}} \exp\left\{-\frac{U^2}{2\sigma_U^2} - \frac{(\xi + U)^2}{2\sigma_V^2}\right\}. \tag{16}$$

The marginal density of ξ can be obtained by integrating U out of $f(U, \xi)$, which yields

$$f(\xi) = \int_0^\infty f(U, \xi)dU. \tag{17}$$

The strong assumption on U and V can be relaxed by using copula to get a better model. The copula-based stochastic frontier model sufficiently copes with the dependence structure between U and V. By Sklar's theorem, the joint *cdf* of U and V is represented by

$$H(u,v) = Pr(U \leq u, V \leq v) \tag{18a}$$

$$= C_\theta(F_U(u), F_V(v)), \tag{18b}$$

where $C_\theta(\cdot, \cdot)$ is the bivariate copula with unknown parameter θ. We can obtain the likelihood function from the composite error $\xi = V - U$, where $\xi = \varepsilon$, $(-\infty < \varepsilon < \infty)$. From Smith 2004, transforming $(U, V) \rightarrow (U, \xi)$, we obtain *pdf* of (U, ξ) as below

$$h(u, \varepsilon) = f_U(u)f_V(u + \varepsilon)c_\theta(F_U(u), F_V(u + \varepsilon)), \tag{19}$$

where $f_U(u)$ and $f_V(v)$ are the marginals of $H(u,v)$ and $v = u + \varepsilon$. Thus, the *pdf* of ξ given by

$$h_\theta(\varepsilon) = \int_{\mathbb{R}_+} h(u, \varepsilon)du \tag{20a}$$

$$= \int_0^\infty f_U(u)f_V(u + \varepsilon)c_\theta(F_U(u), F_V(u + \varepsilon))du. \tag{20b}$$

We can obtain the likelihood function by assuming that there are a cross-section of n observations, thus the likelihood given by

$$L(\beta, \sigma_u^2, \sigma_v^2, \theta) = \prod_{i=2}^n h_\theta(\log \frac{P_t}{P_{t-1}} - \beta_0 - x_t'\beta_i), \tag{21}$$

where P_t is the price at time t, x_t is the associated price factors at time t, σ_u^2, σ_v^2 are the scale parameter of marginal distributions of U and V, respectively. Smith [21] pointed that it is very hard to find the closed form solution for the composite error ξ. Such that, before proceeding to the optimization stage. We estimate equation (20) by transform semi-infinite integral to definite integral by letting $u = 0.5 + w/(1 - w), w \in (-1, 1)$. Now, we have

$$\int_{-1}^1 f_U(0.5 + \frac{w}{1 - w})f_V(0.5 + \frac{w}{1 - w} + \varepsilon) \times$$

$$c_\theta(F_U(0.5 + \frac{w}{1 - w}), F_V(0.5 + \frac{w}{1 - w} + \varepsilon))\frac{1}{(1 - w)^2}dw = \int_{-1}^1 g(w)dw. \tag{22}$$

By the *Gaussian quadrature*, we use 100 nodes to approximate the integral, we get

$$\int_{-1}^1 g(w)dw \approx \sum_{i=1}^{100} w_i(x_i)g(x_i) \tag{23}$$

We maximize the likelihood function in equation (21) by interior point algorithm using random points as starting values, we have

$$\text{Log}(L) = \sum \text{Log}(h_\theta(\varepsilon_t)). \tag{24}$$

The main point of SFM based copula is the technical efficiency TE_θ, that, following Battese and Coelli [7] can be specified as

$$TE_\theta = E(\exp(-U)|\xi = \varepsilon) \tag{25a}$$

$$= \frac{1}{h_\theta(\varepsilon)} \int_{\mathbb{R}_+} \exp(-u)h(u, \varepsilon)du \tag{25b}$$

$$= \frac{\int_0^\infty \exp(-u)f_U(u)f_V(u + \varepsilon)c_\theta(F_U(u), F_V(u + \varepsilon))du}{\int_0^\infty f_U(u)f_V(u + \varepsilon)c_\theta(F_U(u), F_V(u + \varepsilon))du}. \tag{25c}$$

We estimate above equation by using Monte Carlo simulation. Thus, we have

$$TE_\theta = \frac{\sum\limits_{i=1}^{N} \exp(-u_i)f_V(u_i + \varepsilon)c_\theta(f_U(u_i), F_V(u_i + \varepsilon))}{\sum\limits_{i=1}^{N} f_V(u_i + \varepsilon)c_\theta(f_U(u_i), F_V(u_i + \varepsilon))}, \tag{26}$$

where u_i is a random number from the half normal $(0, \sigma^2)$

4   Empirical Results for Model Selection

We are going to use several real data to show empirically, that our copula-based stochastic model is more suitable than the classical stochastic model by using Akaike Information Criterion (AIC).

In this paper, we used the stocks in SET50, which is a composite index. SET50 is the first large-cap index of Thailand to provide a benchmark of investment in Stock Exchange of Thailand. SET50 consist of seven sectors such as Financial, Resources, Services, Technology, Property & Construction, Agro & Food industry and Industrials. In this paper, we choose several companies from these sectors. There are PTT Public Company Limited (PTT) and Advanced Info Service Public Company Limited (ADVANC), due to the high volatility, significant market capitalization and high market value. All the weekly data are taken from March 2009 until Jan 2014 with a total of 260 observations for each series. Table 1 displays a summary of the variables.

Given the stochastic frontier in (11), the empirical version of SFM with the specification of the compound interest equation with decomposed errors can be expressed as:

$$P_{i,t} = f(P_{i,t-i}; V_{i,t-i}; P_{m,t-i}, \beta_{t-i}) + \varepsilon_t \tag{27a}$$

$$P_{i,t} = P_{i,t-1} \exp\left\{ \beta_0 + \beta_1 ln\frac{P_{i,t-1}}{P_{i,t-2}} + \beta_2 \ln\frac{V_{i,t-1}}{V_{i,t-2}} + \beta_3 \ln\frac{P_{m,t-1}}{P_{m,t-2}} \right\} \tag{27b}$$

$$\ln\left(\frac{P_{i,t}}{P_{i,t-1}}\right) = \beta_0 + \beta_1 \ln\frac{P_{i,t-1}}{P_{i,t-2}} + \beta_2 \ln\frac{V_{i,t-1}}{V_{i,t-2}} + \beta_3 \ln\frac{P_{m,t-1}}{P_{m,t-2}}, \tag{27c}$$

Table 1. Data descriptive and statistics

	Price PTT	Price ADVANC	Volume PTT	Volume VADVANC	SET50
Mean	0.0023	0.0037	−0.0034	−0.0048	0.0043
Median	0.0000	0.0053	−0.0203	−0.0390	0.0070
Maximum	0.1211	0.1285	1.3741	1.6360	0.0801
Minimum	−1.0920	−0.1957	−1.1999	−1.9024	−0.0813
Std. Dev.	0.0362	0.0402	0.4322	0.4885	0.0240
Skewness	0.2314	−0.5753	0.2831	0.1089	−0.3599
Kurtosis	4.1838	5.5080	3.3102	4.3679	3.7030
Obs.	260				

All values are the growth rate of price and volume.

Table 2. PTT's parameters estimation

Parameter	Case I	Case II	Case III	Case IV	Case V	Case VI	Case VII
β_0	0.0118	0.0167	0.0301	0.0444	0.0253	0.0058	0.0010
	(0.0446)	(0.0068)	(0.0161)	(0.3338)	(0.0075)	(0.1871)	(0.1366)
β_1	0.0202	0.0186	0.0173	0.0226	0.0706	0.0898	0.0185
	(0.1094)	(0.1037)	(0.1177)	(0.0445)	(0.1737)	(0.1416)	(0.1077)
β_2	−0.0117	−0.0122	−0.0121	−0.0122	−0.0151	−0.0111	−0.0116
	(0.0089)	(0.0042)	(0.0044)	(0.3535)	(0.0043)	(0.0547)	(0.5694)
β_3	0.1566	0.1636	0.1648	0.1558	0.0801	0.0955	0.1576
	(0.1144)	(0.1074)	(0.1461)	(0.1992)	(0.1495)	(0.1274)	(0.1489)
$\sigma(V)$	0.0292	0.0383	0.0342	0.0460	0.0421	0.0322	0.0304
	(0.0093)	(0.0029)	(0.0085)	(0.0147)	(0.0036)	(0.0015)	(0.0031)
$\sigma(U)$	0.0127	0.0001	0.0356	0.0536	0.0300	0.0037	0.0001
	(0.0554)	(0.0000)	(0.0203)	(0.0815)	(0.0069)	(0.0007)	(0.0000)
θ	−0.0009	3.6596	-	-	7.2063	1.7987	
	(1.6396)	(0.2168)	-	-	(1.2418)	0.9165	
ρ	-	−4.7062	3.6619	-	-	5.2630	0.0408
	-	(0.7738)	(2.6585)	-	-	(1.6044)	(0.3552)
ν	-	-	-	2	-	-	
	-	-	-	(0.7403)	-	-	
$LogL$	539.0571	596.2154	540.2937	539.2748	542.2003	539.5641	595.2099
AIC	−1090.1142	−1178.4309	−1064.5947	−1064.5496	−1070.4007	−1065.1282	−1176.4199

where $i = 1, 2, \cdots, n$, $P_{i,t}$ is the weekly close price at time t, $P_{m,t}$ is the weekly close SET50's index at time t and $V_{i,t}$ is the weekly trade volume of individual stock at time t.

The choice of families of copulas in the paper, including Gaussian copula, Gumbel copula, t-copula, Frank copula and Clayton copula. We used a likelihood function (21) to obtained all parameters using N=100 nodes and maximized using the Gaussian Quadrature algorithm. The empirical results show in the Table 2 and Table 3.

According to AIC criteria, the best model in this situation is the one based on the Frank copula with $\theta < 0$ for the case of PTT and t-copula for the

Table 3. ADVANC's parameters estimation

Parameter	Case I	Case II	Case III	Case IV	Case V	Case VI	Case VII
β_0	0.0360	0.0360	0.0553	0.0984	−0.0411	0.0195	0.0531
	(0.0218)	(0.0120)	(0.0108)	(0.0262)	(0.0348)	(0.0104)	(0.0086)
β_1	−0.1521	−0.1521	−0.1349	−0.1358	−0.1314	−0.1543	−0.1313
	(0.2422)	(0.0162)	(0.2106)	(0.0582)	(0.0386)	(0.0741)	(0.0561)
β_2	−0.0110	−0.0110	−0.0096	−0.0110	−0.0091	−0.0122	−0.0087
	(0.0052)	(0.0145)	(0.0043)	(0.0135)	(0.0043)	(0.0293)	(0.0043)
β_3	0.4203	0.4203	0.3914	0.3969	0.3893	0.4196	0.3966
	(0.0184)	(0.0730)	(0.2994)	(0.0497)	(0.1255)	(0.0667)	(0.0924)
$\sigma(V)$	0.0284	0.0284	0.0418	0.0857	0.1009	0.0504	0.0372
	(0.0041)	(0.0088)	(0.0071)	(0.0117)	(0.0650)	(0.0026)	(0.0078)
$\sigma(U)$	0.0424	0.0424	0.0664	0.1205	0.0001	0.0222	0.0637
	(0.0312)	(0.0049)	(0.0144)	(0.0268)	(0.0000)	(0.0020)	(0.0109)
θ	−0.0009	3.6596	-	-	7.2063	1.7987	
	(1.6396)	(0.2168)	-	-	(1.2418)	0.9165	
ρ	-	−0.0001	0.0664	-	-	3.9169	1.3047
	-	(0.0075)	(1.7915)	-	-	(0.1881)	(0.8388)
ν	-	-	-	2	-	-	
	-	-	-	(0.7403)	-	-	
$LogL$	480.1462	480.1460	483.7950	484.1054	540.5408	478.7202	484.1267
AIC	−972.2924	−972.2920	−951.5899	−954.2108	−1067.0815	−943.4404	−954.2534

Consistent standard errors () is in parenthesis.
Case I: N,NH and iid.
Case II: N,NH and Frank $[\theta < 0]$.
Case III: N,NH and Frank $[\theta > 0]$.
Case IV: N,NH and Gaussian $\{\theta \in [-1, 1]\}$
Case V: N,NH and t-copula $\{\theta \in [-1, 1], \nu > 2\}$
Case VI: N,NH and Gumbel $[\theta > 1]$.
Case VII: N,NH and Clayton $[\theta > 0]$.

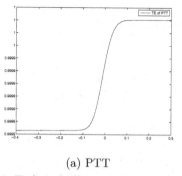

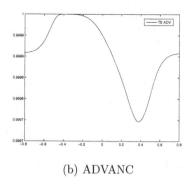

(a) PTT   (b) ADVANC

Fig. 1. Technical efficiencies for the Frank copula with $\theta < 0$ (a) and t-copula (b) based models

case of ADVANC. While, Gaussian copula and Gumbel copula are the worst case for PTT and ADVANC, respectively. Fig. 1 shows the technical efficiency calculated using the based model up on the Frank copula with $\theta < 0$ and t-copula based model. The efficiency values are close to 1 for both case of PTT and ADVANC for the best copula-based model. Clearly, this finding recommends that a considerable strategies to select stock quite efficiency. We can justify that large negative and positive dependence between the error term can not affect much on the technical efficiency in both cases.

5 Conclusions

We have applied and extended the stochastic frontier models that usually are used in the analysis of the production function in the agricultural science problem. However, in this paper we considered a production function in the term of the compound interest equation for the financial situation. The copula-based approach allows u and v to have a relationship showing the dependence structure between them. We used Gaussian quadrature to obtained all parameters. The information criterion, AIC is applied to exhibit the dependence between the random error term and technical inefficiency. Many families of copulas are used to combine with a stochastic frontier model, such as Gaussian, t, Frank, Claton, Gumbel, Joe copula.

Finally, we used the copula-based stochastic frontier model to analyze the prices of an interested stock. We justified copula families using the AIC criterion, and the results showed that Frank copula with $\theta < 0$ and t-copula are the best results for PTT and ADVANC, respectively. We can consider that the price of both stocks is not over or underestimated in terms of technical efficiency using the financial formula based on a stochastic frontier model. The intuition and economic meaning behind this study is of great interest for the future research.

Acknowledgments. we would like to thank Prof. Dr. Hung T. Nguyen for his comments and suggestions.

References

1. Subal, C.K., Lovell, C.A.K.: Stochastic Frontier Analysis. Cambridge University Press, Cambridge (2000)
2. Joe, H.: Dependence Modeling with Copulas. Chapman and Hall/CRC Press, London (2015)
3. Aigner, D., Lovell, K., Schmidt, P.: Formulation and estimation of stochastic frontier function models. Journal of Econometrics **6**, 21–37 (1977)
4. Meeusen, W., Van Den Broek, J.: Efficiency estimation from Cobb-Douglas production function with composed error. International Economic Review **8**, 435–444 (1977)
5. Olsen, J.A., Schmidt, P., Waldman, D.M.: Amonte Carlo study of estimators of stochastic frontier production functions. Journal of Econometrics **13**, 67–82 (1980)

6. Stevenson, R.E.: Likelihood functions for generalised stochastic frontier estimation. Journal of Econometrics **13**, 57–66 (1980)

7. Battese, G.E., Coelli, T.J.: Prediction of firm-level technical efficiencies with a generalized frontier production function and panel data. Journal of Econometrics **3**, 631–645 (1988)

8. Sriboonchitta, S., Nguyen, H.T., Wiboonpongse, A., Liu, J.: Modeling volatility and dependency of agricultural price and production indices of Thailand: Static versus time-varying copulas. Int. J. Approximate Reasoning **54**, 793–808 (2013)

9. Sanzidur, R., Wiboonpongse, A., Sriboonchitta, S., Chaovanapoonphol, Y.: Production efficiency of Jasmine rice producers in Northern and North-Eastern Thailand. J. Agric. Econ. **60**(2), 419–435 (2009)

10. Hansan, M.Z., Kamil, A.A., Mustafa, A., Baten, M.A.: Stochastic frontier model approach for measuring stock market efficiency with different distributions. PLos ONE **7**(5), e37047 (2012). doi:10.1371/journal.pone.0037047

11. Liu, C., Chung, C.Y.: SEO underpricing in China's stock market: a stochastic frontier approach. Applied Financial Economics **23**, 393–402 (2013)

12. Wiboonpongse, A., Liu, J., Sriboonchitta, S., Denoeux, T.: Modeling dependence between error components of the stochastic model using copula: Application to intercrop coffee production in Northern Thailand frontier. International Journal of Approximate Reasoning (2015). http://dx.doi.org/10.1016/j.ijar.2015.04.001

13. Amsler, C., Prokhotov, A., Schmidt, P.: Using copulas to model time dependence in stochastic frontier models. Econometric Reviews **33**(5–6), 497–522 (2014)

14. Burns, R.: The Simulated Maximum Likelihood Estimation of Stochastic Frontier Models with Correlated Error Components. The University of Sydney, Sydney (2004)

15. Autchariyapanitkul, K., Chainam, S., Sriboonchitta, S.: Portfolio optimization of stock returns in high-dimensions: A copula-based approach. Thai Journal Mathematics, 11–23 (2014)

16. Kreinovich, V., Nguyen, H.T., Sriboonchitta, S.: Why Clayton and Gumbel Copulas: a symmetry-based explanation. In: Huynh, V.N., Kreinovich, V., Sriboonchitta, S., Suriya, K. (eds.) TES 2013. AISC, vol. 200, pp. 79–90. Springer, Heidelberg (2013)

17. Kreinovich, V., Nguyen, H.T., Sriboonchitta, S.: How to detect linear dependence on the Copula level? In: Huynh, V.-N., Kreinovich, V., Sriboonchitta, S. (eds.) Modeling Dependence in Econometrics. AISC, vol. 251, pp. 87–101. Springer, Heidelberg (2014)

18. Sirisrisakulchai, J., Sriboonchitta, S.: Modeling dependence of accident-related outcomes using pair Copula constructions for discrete data. In: Huynh, V.-N., Kreinovich, V., Sriboonchitta, S. (eds.) Modeling Dependence in Econometrics. AISC, vol. 251, pp. 251–264. Springer, Heidelberg (2014)

19. Tang, J., Sriboonchitta, S., Ramos, V., Wong W.: Modelling dependence between tourism demand and exchange rate using the copula-based GARCH model. Current Issues in Tourism, 1–19 (2014)

20. Greene, W.: A stochastic frontier model with correction for sample selection. J. Prod. Anal. **34**(1), 15–24 (2010)

21. Smith, D.M.: Stochastic frontier model with dependent error components. Econometrics Journal **11**, 172–192 (2008)

Capital Asset Pricing Model
with Interval Data

Sutthiporn Piamsuwannakit[1,2], Kittawit Autchariyapanitkul[3]($\boxtimes$),
Songsak Sriboonchitta[1], and Rujira Ouncharoen[4]

[1] Faculty of Economics, Chiang Mai University, Chiang Mai, Thailand
[2] Faculty of Management Science, Chiang Rai Rajabhat University,
Chiang Rai, Thailand
[3] Faculty of Economics, Maejo University, Chiang Mai, Thailand
kittawit_autchariya@cmu.ac.th
[4] Department of Mathematics, Faculty of Science, Chiang Mai University,
Chiang Mai, Thailand

Abstract. We used interval-valued data to predict stock returns rather
than just point valued data. Specifically, we used these interval values in
the classical capital asset pricing model to estimate the beta coefficient
that represents the risk in the portfolios management analysis. We also
use the method to obtain a point valued of asset returns from the interval-
valued data to measure the sensitivity of the asset return and the market
return. Finally, AIC criterion indicated that this approach can provide
us better results than use the close price for prediction.

Keywords: CAPM · Interval-valued data · Least squares method · Linear regression

1 Introduction

Capital asset pricing model provides a piece of information of asset return related
to the market return via its systematic risk. In general, asset returns of any
interested asset and market returns are calculated from a single-valued data.
Most of the papers in financial econometrics use only closed price taking into
account for calculation but in the real world stock price is moving up and down
within the range of highest price and lowest price. So, in this paper we intend
to use all the points in the range of high and low to improve the results in
our calculations. We also put an assumption of a normal distribution on these
interval-valued data.

An enormous number of research on CAPM model with single-valued data
could be found in much financial research topic, the reader is referred to, e.g.,
William F. Sharpe [1] and John Lintner [2] only a single-valued of interest was
considered. Many various technics were applied to the original CAPM model that
we can found in the work from Autchariyapanitkul et al. [3], the authors used
quantile regression under asymmetric Laplace distribution (ALD) to quantify

© Springer International Publishing Switzerland 2015
V.-N. Huynh et al. (Eds.): IUKM 2015, LNAI 9376, pp. 163–170, 2015.
DOI: 10.1007/978-3-319-25135-6_16

the beta of the asset returns in CAPM model. The results showed that this method can capture the stylized facts in financial data to explain the return of stocks under quantile, especially under the middle quantile levels. In Barnes and Hughes [4], the beta risk is significant in both tails of the conditional distribution of returns. In Chen et al. [5], the authors used a couple of methods to obtain the time-varying market betas in CAPM to analyze stock in the Dow Jones Industrial for several quantiles. The results indicated that smooth transition quantile method performed better than others methods.

Interval-valued data has become popular in many research fields especially in the context of financial portfolio analysis. Most of the financial data are usually affected by imprecision, uncertainty, inaccuracy and incompleteness, etc. The uncertainty in the data may be captured with interval-valued data. There are several existing research in the literature for investigating this issue. see Billard [6], Carvalho [7], Cattaneo [8], Diamond [9], Gil [10], Körner [11], Manski [12], Neto [14]. However, In these research papers are lacking in a foundation and theoretical background to support this idea.

The connection between the classical linear regression and the interval-valued data that share the important properties could be found for the work by Sun and Li [15]. In their paper, they provided a theoretical support framework between the classical one and the interval-valued linear regression such as least squares estimation, asymptotic properties, variances estimation, etc. However, in their paper only one of an explanatory variable can use to described the responding variable. In this paper, we intend to apply the concept of the interval-valued data to the CAPM model. We replace a single value of market returns and asset returns with the range of high and low historical data into the model.

The rest of the paper is organized as follows. Section 2 gives a basics knowledge of a linear regression model for interval-valued data. In Section 3 discusses the empirical discovering and the solutions of the forecasting problem. The last section gives the conclusion and extension of the paper.

2   A Review of Real Interval-Valued Data

Now, take a close look at financial data (D_i). Suppose, we have a range of any numbers between a minimum and maximum prices given by $D_i = [min, \ldots, max] = [Low, \ldots, High]$, where the minimum price is the "lowest price", and the maximum price is the "highest price". Certainly, this range contains the point that we called "close price". In many research papers, they are usually using the close price for calculations. A close price is a number that takes any values in the range of D_i between the lowest and the highest prices, $D_i = [Low, \ldots Close, \ldots, High]$. The close price could be either the lowest price or the highest price.

In this paper, we try to find the better value for calculations rather than a close price that is the best-represented point in the range of D_i to improving our predictions. We considered a normal distribution on this interval-valued data.

3 An Interval-Valued Data in a Linear Regression Model

Suppose we can observe an i.i.d random paired intervals variables $x_i = [\underline{x_i}, \overline{x_i}]$ and $y_i = [\underline{y_i}, \overline{y_i}]$, $i = 1, 2, \ldots, n$ where $\overline{x_i}, \overline{y_i}$ are the maximum values of x_i and $\underline{x_i}, \underline{y_i}$ are the minimum values of y_i. Additionally, we can rewrite the value of x_i, y_i in the form of intervals as

$$x_i = [x_i^m - x_i^r, x_i^m + x_i^r], \tag{1a}$$

$$y_i = [y_i^m - y_i^r, y_i^m + y_i^r], \ i = 1, 2, \ldots, n, \tag{1b}$$

where x_i^m, y_i^m is the mid-points of x_i and y_i and x_i^r, y_i^r is the radii of x_i and y_i, satisfying $x_i^r, y_i^r \geq 0$. Suppose, we consider the following linear regression model given by

$$y_i = ax_i + b + \varepsilon_i, \ i = 1, 2, \cdots, n. \tag{2}$$

Analogously, it is easy to interpret the meaning of x_i, y_i by the distance of centers and radii as the following equations

$$x_i = x_i^m + \delta_{x_i}, \ \delta_{x_i} \in N(0, (k_0 \Delta x_i)^2) \tag{3a}$$

$$y_i = y_i^m + \delta_{y_i}, \ \delta_{y_i} \in N(0, (k_0 \Delta y_i)^2), \tag{3b}$$

where x_i^m, y_i^m are the centers of x_i and y_i, respectively. Then, $\Delta x_i = \frac{\overline{x_i} - \underline{x_i}}{2}, \Delta y_i = \frac{\overline{y_i} - \underline{y_i}}{2}$ are the radii of x_i and y_i, respectively and $x_i^m = \frac{\overline{x_i} + \underline{x_i}}{2}$, $y_i^m = \frac{\overline{y_i} + \underline{y_i}}{2}$ are the mid-point of x_i and y_i, respectively. Thus, given the linear regression for the interval valued data we have

$$y_i^m + \delta_{y_i} = ax_i^m + a\delta_{x_i} + b \tag{4a}$$

$$y_i^m = ax_i^m + b + (a\delta_{x_i} - \delta_{y_i}), \tag{4b}$$

where $(a\delta_{x_i} - \delta_{y_i}) \sim N(0, \sigma^2) \equiv N(0, k_0^2 a^2 \Delta x_i^2 + \Delta y_i^2)$. Assume that $a\delta_{x_i} - \delta_{y_i}$ is an independence. Thus, we can estimate parameters a, b, k_0 by the maximum likelihood function given by

$$\max_{a,b,k_0} L(a, b, k_0 | ([\underline{x_i}, \overline{x_i}], [\underline{y_i}, \overline{y_i}]), i = 1, \ldots, n)$$

$$= \max_{a,b,k_0} \prod_{i=1}^{n} \left(\frac{1}{\sqrt{2\pi k_0^2 (a^2 \Delta x_i^2 + \Delta y_i^2)}} \exp \left[-\frac{1}{2} \frac{(y_i^m - ax_i^m - b)^2}{k_0^2 (a^2 \Delta x_i^2 + \Delta y_i^2)} \right] \right) \tag{5}$$

This approach was already developed in Sun and Li [15]. And soften the criticisms of lack of theory, Manski has a whole book (see, Manski [12],[13]), this is finance not pure mathematics here. The proof of success is better fit not theorems.

3.1 Goodness of Fit in Linear Regression Model for an Interval-valued Data

In the deterministic linear regression model, we use variance to describe variation of the variable interested and so that as we knew the ratio $\frac{a^2 Var(X)}{Var(Y)} \in [0, 1]$ can be explained as an indication of goodness-of-fit. In this paper, we used the concept of the chi-squared test (χ^2) of the goodness of fit. Recall that $\sigma_{x_i} = k_0 \Delta x_i$ and $\sigma_{y_i} = k_0 \Delta y_i$ given the simple linear regression we have

$$y_i = ax_i + b \tag{6a}$$

$$y_i^m + \delta_{y_i} = ax_i^m + a\delta_{x_i} \tag{6b}$$

$$y_i^m - ax_i^m - b = a\delta_{x_i} - \delta_{y_i}, \tag{6c}$$

where $\delta_{x_i}, \delta_{y_i} \sim N(0, \sigma^2)$. Thus, we have $a^2\sigma_{x_i}^2 + \sigma_{y_i}^2$, by replacing $k_0^2(a^2 \Delta x_i^2 + \Delta y_i^2)$ to above equation 6. The empirical χ^2−test is obtained by estimated this following equation

$$\chi_{cal}^2 = \sum_{i=1}^{n} \frac{(y_i^m - ax_i^m - b)^2}{k_0^2(a^2 \Delta x_i^2 + \Delta y_i^2)}, \tag{7}$$

where the degree of freedom is $n - 2$.

4   An Application to the Stock Market

We consider the following financial model that is so called Capital Asset Pricing Model (CAPM). Only two sets of interval-valued data are used to explain the relationship of the asset. The fitted model is based on the least square estimation.

4.1   Capital Asset Pricing Model

The Capital Asset Pricing Model (CAPM) is a linear relationship that was created by William F. Sharpe [1] and John Lintner [2]. The CAPM use to calculate a sensitivity of the expected return on the asset to expected return on the market. The combination of a linear function of the security market line:

$$E(R_A) - R_F = \beta_0 + \beta_1 E(R_M - R_F), \tag{8}$$

where $E(R_A)$ explains the expected return of the asset, R_M represents the expected market portfolio return, β_0 is the intercept and R_F is the risk-free rate. $E(R_M - R_F)$ is the expected risk premium, and β_1 is the equity beta, denoting market risk. To measure the systematic risk of each stock via the beta takes form:

$$\beta_1 = \frac{cov(R_A, R_M)}{\sigma_M^2}, \tag{9}$$

where σ_M^2 represents the variance of the expected market return. Given that, the CAPM predicts portfolio's expected return should be about its risk and the market returns.

4.2   Beta Estimation with Interval Data

From the deterministic model in equation (8), we calculate the β coefficient through the likelihood by equation (5) instead. Suppose we have observed the realization interval stock return $[\underline{R_A i}, \overline{R_A i}]$ = $[(\overline{r_{a1}}, \underline{r_{a1}}), \ldots, (\overline{r_{an}}, \underline{r_{an}})], i = 1, 2, \ldots, n$ and return from market $[\underline{R_M i}, \overline{R_M i}]$ = $[(\overline{r_{m1}}, \underline{r_{m1}}), \ldots, (\overline{r_{mn}}, \underline{r_{mn}})], i = 1, 2, \ldots, n$ over the past N years. These observations will be assumed an independent random. From likelihood for an interval values we have

$$\max_{a,b,k_0} L(a, b, k_0 | ([\underline{R_{Mi}}, \overline{R_{Mi}}], [\underline{R_{Ai}}, \overline{R_{Ai}}]), i = 1, \ldots, n)$$

$$= \max_{a,b,k_0} \prod_{i=1}^{n} \left(\frac{1}{\sqrt{2\pi k_0^2 (a^2 \Delta Rm_i^2 + \Delta Ra_i^2)}} \exp\left[-\frac{1}{2} \frac{(Ra_i^m - aRm_i^m - b)^2}{k_0^2 (a^2 \Delta Rm_i^2 + \Delta Ra_i^2)} \right] \right)$$

(10)

4.3   Empirical Results

Our data contains 259 weekly interval-valued returns in total during 2010-2015 are obtained from Yahoo. We compute the log returns on the following stock, namely, Chesapeake Energy Corporation (CHK)and Microsoft Corporation (MSFT). Due to significant capitalization and high turnover volume.

In this paper, we use Treasury bills as a proxy. From Autchariyapanitkul et al. [3] and Mukherji [16] suggested that Treasury bills are better proxies for the risk-free rate, only related to the U.S. market.

Table 1 and Table 2 report the estimated results from equation (5). For example, the simple linear regression model for the asset returns (Y) and the market returns (X) for interval valued data for CHK is written to be

$$R_A = -0.0021 + 0.9873 R_M.$$

(11)

From the above linear equation, the return of a stock is likely to increase less than the return from the market. A non-parametric chi-square test is used

Table 1. Estimated parameter results for CHK

	Interval-Valued data		Point-Valued data	
parameters	values	std. Dev.	values	std. Dev.
β_0	−0.0021	0.0233	−0.0191	0.0055
β_1	0.9873	0.0914	0.7226	0.0713
k	0.4472	0.0845	-	-
MSE	-	-	0.036	
LL	525.7021	-	361.1400	-
χ^2	259.00	-	-	-
AIC	−1045.04	-	−716.28	-

Table 2. Estimated parameter results for MSFT

parameters	Interval-Valued data		Point-Valued data	
	values	std. Dev.	values	std. Dev.
β_0	−0.0004	0.0015	−0.0088	0.0035
β_1	1.0086	0.0220	0.8489	0.0005
k	0.4017	0.0170	-	-
MSE	-	-	0.0025	-
LL	692.3808	-	478.9365	-
χ^2	259.00	-	-	-
AIC	−1378.76	-	−951.87	-

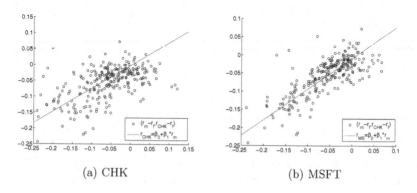

(a) CHK   (b) MSFT

Fig. 1. Securities characteristic line for point valued data

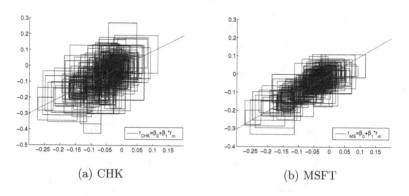

(a) CHK   (b) MSFT

Fig. 2. Securities characteristic line for interval valued data

to validate the method of interval-valued data. The theoretical χ^2_{n-2} gives the value of CHK, $\chi^2_{n-2} = 303.2984$ compare with the empirical value $\chi^2_{emp} = 259.00$ confirm that the market returns can be used to explain the asset returns. The model selection criteria Akaike information criterion (AIC) was employed to compare these two techniques. The AIC of interval-valued data gives a value of −1051.4402 is smaller than the AIC of pointed-valued data, which indicate that the results from the interval-valued method is more prefer than the deterministic one.

The relationship between market return and asset return are plotted in Figure 1 and Figure 2 for pointed-valued data and interval-valued data, respectively.

The rectangular are the high and low interval-valued data, and the straight line is the securities characteristic line, the slope of this straight line represent the systematic risk beta. All investments and portfolio of investments must lie along a straight line in the return beta space.

5   Conclusions and Extension

The systematic risk has played as the critical role of financial measurement in capital asset pricing model. Academic and practitioners attempt to estimate its underlying value accurately. Fortunately, there have been the novel approaches to evaluating the beta with interval-valued data. We used every price range of real world data to obtained the single value of the systematic risk same as the results from the conventional CAMP model.

In this paper, we use our approach to an interval-valued data in CAPM for only one stock in $S\&P500$ for a demonstration. With this, a method can be used to investigate the linear relationship between the expected asset returns and its asymmetric market risk by including all of the levels of prices in the range of an interval-valued data. The results clearly show that the beta can measure the responsiveness to the asset returns and market returns. However, only a systematic risk is calculated through the model, and we neglect the unsystematic risk under CAPM assumption. CAPM concludes that the expected return of a security or a portfolio equals the rate on a risk-free security plus a risk premium.

By AIC criterion, it should be noticed that the estimation by using interval-valued data more reasonable than just used the single valued in the calculations. Not only one explanatory variable can be used to explain the outcome variable but with this method also allowed us to use more than one covariate in the model.

For future research, we are interested to use this method to the time series models such as ARMA, GARCH model. Additionally, we can use this method to the model with more than one explanatory variables such as Fama and French (1993). A three-factor model can be extended the CAPM by putting size and value factors in the classical one.

Acknowledgments. The authors thank Prof. Dr. Vladik Kreinovich for his suggestions for an interval-valued data computation method. We would like to thank the referee for giving comments on manuscript.

References

1. William, F.: Capital Asset Prices A Theory of Market Equilibrium Under Conditions of Risk. Journal of Finance **19**(3), 425–442 (1964)
2. John, L.: The Valuation of Risk Assets and the Selection of Risky Investments in Stock. The Review of Economics and Statistics **47**(1), 13–37 (1965)
3. Autchariyapanitkul, K., Chainam, S., Sriboonchitta, S.: Quantile regression under asymmetric Laplace distribution in capital asset pricing model. In: Huynh, V.-N., Kreinovich, V., Sriboonchitta, S., Suriya, K. (eds.) Econometrics of Risk, vol. 583, pp. 219–231. Springer, Heidelberg (2015)
4. Barnes, L.M., Hughes, W.A.: A Quantile Regression Analysis of the Cross Section of Stock Market Returns. Federal Reserve Bank of Boston, working paper (2002)
5. Chen, W.S.C., Lin, S., Yu, L.H.P.: Smooth Transition Quantile Capital Asset Pricing Models with Heteroscedasticity. Computational Economics **40**, 19–48 (2012)
6. Billard, L.: Dependencies and variation components of symbolic interval-valued data. In: Brito, P., Cucumel, G., Bertrand, P., de Carvalho, F. (eds.) Selected Contributions in Data Analysis and Classification. Studies in Classification, Data Analysis, and Knowledge Organization, pp. 3–12. Springer, Heidelberg (2007)
7. de Carvalho, F.A.T., Lima Neto, E.A., Tenorio, C.P.: A new method to fit a linear regression model for interval-valued data. In: Biundo, S., Frühwirth, T., Palm, G. (eds.) KI 2004. LNCS (LNAI), vol. 3238, pp. 295–306. Springer, Heidelberg (2004)
8. Cattaneo, M.E.G.V., Wiencierz, A.: Likelihood-based imprecise regression. International Journal of Approximate Reasoning **53**, 1137–1154 (2012)
9. Diamond, P.: Least square fitting of compact set-valued data. J. Math. Anal. Appl. **147**, 531–544 (1990)
10. Gil, M.A., Lubiano, M.A., Montenegro, M., Lopez, M.T.: Least squares fitting of an affine function and strength of association for interval-valued data. Metrika **56**, 97–101 (2002)
11. Körner, R., Näther, W.: Linear regression with random fuzzy variables: extended classical estimates, best linear estimates, least squares estimates. Information Sciences **109**, 95–118 (1998)
12. Manski, C.F., Tamer, T.: Inference on regressions with interval data on a regressor or outcome. Econometrica **70**, 519–546 (2002)
13. Manski, C.F.: Partial Identification of Probability Distributions. Springer, New York (2003)
14. Neto, E.A.L., Carvalho, F.A.T.: Centre and range method for fitting a linear regression model to symbolic interval data. Computational Statistics and Data Analysis **52**, 1500–1515 (2008)
15. Sun, Y., Li, C.: Linear regression for interval-valued data: a new and comprehensive model, under review (2015). arXiv: 1401.1831
16. Mukherji, S.: The capital asset pricing model's risk-free rate. International Journal of Business and Finance Research **5**, 793–808 (2011)

Confidence Intervals for the Difference Between Normal Means with Known Coefficients

Suparat Niwitpong[⊠] and Sa-Aat Niwitpong

Department of Applied Statistics, Faculty of Applied Sciences, King Mongkut's University of Technology North Bangkok, Bangkok, Thailand
suparatn@kmutnb.ac.th

Abstract. Statistical estimation of the difference between normal means with known coefficients of variation has been investigated for the first time. This phenomenon occurs normally in environment and agriculture experiments when the scientist knows the coefficients of variation of their experiments. In this paper, we constructed new confidence intervals for the difference between normal means with known coefficients of variation. We also derived analytic expressions for the coverage probability and the expected length of each confidence interval. To confirm our theoretical results, Monte Carlo simulation will be used to assess the performance of these intervals based on their coverage probabilities and their expected lengths.

Keywords: Confidence interval · Coverage probability · Expected length · Known coefficient of variation

1 Introduction

Behrens-Fisher problem is the problem of interval estimation and hypothesis testing concerning the difference between the two independent normal means when the variances of the two populations are not assumed to be equal. In this paper, we are mainly concern about the confidence intervals for the difference between normal means with known coefficients of variation. In practice, there are situations in area of agricultural, biological, environmental and physical sciences that a coefficients of variation is known. For example, in environmental studies, Bhat and Rao [1] argued that there are some situations that show the standard deviation of a pollutant is directly related to the mean that means the coefficient of variation ($\tau = \sigma/\mu$) is known. Bhat and Rao [1] described that in clinical chemistry, "when the batches of some substance (chemicals) are to be analyzed, if sufficient batches of the substances are analyzed, their coefficients of variation will be known". In addition, Brazauskas and Ghorai [2] also gave some examples in medical, biological and chemical experiments shown that in practice there are problems concerning that coefficients of variation are known. Most of this statistical problem is due to the estimation of the mean of normal distribution with known coefficient of variation see e.g. Khan [4], Searls [8, 9] and

© Springer International Publishing Switzerland 2015
V.-N. Huynh et al. (Eds.): IUKM 2015, LNAI 9376, pp. 171–182, 2015.
DOI: 10.1007/978-3-319-25135-6_17

the references cited in the mentioned papers. Recent paper of Bhat and Rao [1] extended the mentioned papers to the test for the normal mean, when its coefficient of variation is known. Their simulation results shown that, for the two-sided alternatives the likelihood ratio test or the Wald test is the best test. This paper extends the recent work of Bhat and Rao [1] to the confidence interval for the normal population mean with known coefficient of variation. Niwitpong [7] proposed the new confidence interval of the normal mean with known coefficient of variation based on the best unbiased estimator proposed by Khan [4]. She compared her confidence interval with the confidence interval based on Searl [8] by using the expected length as a criterion to assess the confidence intervals. In this paper, we extend the paper work of Niwitpong and Niwitpong [6] to construct confidence intervals of normal mean with known coefficient of variation based on the maximum likelihood estimator and the new confidence interval based on the statistic t-test. Like Niwitpong [7], we also proved analytic expressions of the coverage probability and the expected length of each confidence interval. Monte Carlo simulation is used to assess these confidence intervals using the coverage probability and the expected lengths between confidence intervals. Typically, we prefer confidence interval with coverage probability at least the nominal value $(1 - \alpha)$ and its expected length is short.

2   Confidence Intervals for the Difference Between Two Normal Population Means

Let $X_1, \ldots X_n$ and $Y_1, \ldots Y_m$ be random samples from two independent normal distributions with means μ_x, μ_y and standard deviations σ_x and σ_y, respectively. The sample means and variances for X and Y are also denoted as $\bar{X}$, $\bar{Y}$, S_x^2 and S_y^2, respectively. We are interested in $100(1 - \alpha)\%$ confidence interval for $\theta = \mu_x - \mu_y$ when we know coefficients of variation.

2.1   The Confidence Interval for θ Based on Pooled Estimate of Variances and Welch-Satterthwaite Methods

When it is assumed that $\sigma_x^2 = \sigma_y^2$, it is well-known that, by using the pivotal quantity T_1 which is

$$T_1 = \frac{(\bar{X} - \bar{Y}) - (\mu_x - \mu_y)}{S_p\sqrt{\frac{1}{n} + \frac{1}{m}}}$$

the $100(1 - \alpha)\%$ confidence interval for θ is

$$CI_1 = \left[(\bar{X} - \bar{Y}) - t_{1-\alpha/2,n+m-2}S_p\sqrt{\tfrac{1}{n} + \tfrac{1}{m}}, (\bar{X} - \bar{Y}) + t_{1-\alpha/2,n+m-2}S_p\sqrt{\tfrac{1}{n} + \tfrac{1}{m}}\right]$$

where $S_p^2 = \frac{(n-1)S_x^2+(m-1)S_y^2}{n+m-2}$, $S_x^2 = (n - 1)^{-1}\sum_{i=1}^{n}(X_i - \bar{X})^2$, $S_y^2 = (m - 1)^{-1}$ $\sum_{j=1}^{m}(Y_j - \bar{Y})^2$ and $t_{1-\alpha/2}$ is the $(1 - \alpha/2)$th percentile of t-distribution with

$n+m-2$ degrees of freedom. Note that T_1 has an exact t-distribution with $n+m-2$ degrees of freedom.

In the case where the two variances differ, i.e. $\sigma_x^2 \neq \sigma_y^2$, the confidence interval for θ is constructed using the pivotal quantity T_2

$$T_2 = \frac{(\bar{X} - \bar{Y}) - (\mu_x - \mu_y)}{\sqrt{\frac{S_x^2}{n} + \frac{S_y^2}{m}}}$$

It is well-known that T_2 is *approximately* distributed as a t-distribution with degrees of freedom equal

$$\nu = \frac{(A+B)^2}{\frac{A^2}{n-1} + \frac{B^2}{m-1}}, A = \frac{S_x^2}{n}, B = \frac{S_y^2}{m}$$

An approximate $100(1-\alpha)\%$ confidence interval for θ is therefore

$$CI_{WS} = \left[(\bar{X} - \bar{Y}) - t_{1-\alpha/2,\nu} \sqrt{\frac{S_x^2}{n} + \frac{S_y^2}{m}}, (\bar{X} - \bar{Y}) + t_{1-\alpha/2,\nu} \sqrt{\frac{S_x^2}{n} + \frac{S_y^2}{m}} \right]$$

where $t_{1-\alpha/2,\nu}$ is the $(1 - \alpha/2)$ th percentile of T_2 distribution with degrees of freedom ν. The confidence interval CI_{WS} is known as the WS confidence interval.

2.2   Confidence Intervals for the Difference Between Normal Means with Known Coefficients of Variation

Niwitpong [7] proposed the confidence interval for the normal mean with known a coefficient of variation based on Searl [8] compared with Niwitpong [5] and Khan [4]. She found that for large sample size confidence interval the confidence interval based on Searl [8] outperforms other intervals. As a result, we now extend Niwitpong [7] to construct a confidence interval for θ with known coefficients of variation.

Searls [8] proposed the estimator $\bar{Y}^* = (m+\tau_y^2)^{-1} \sum_{j=1}^{m} Y_j$ where he showed that this estimator has lower mean squares error than that of the unbiased estimator $\bar{Y}$ and its variance is $\frac{\sqrt{m}S_y}{m+\tau_y^2}$.

Using the Central Limit Theorem (CLT), it is easy to see that the $100(1-\alpha)\%$ confidence interval for θ when τ_x, τ_y are known is given by

$$CI_S = \left[(\bar{X}^* - \bar{Y}^*) - d\sqrt{\frac{nS_x^2}{(n+\tau_x^2)^2} + \frac{mS_y^2}{(m+\tau_y^2)^2}}, (\bar{X}^* - \bar{Y}^*) + d\sqrt{\frac{nS_x^2}{(n+\tau_x^2)^2} + \frac{mS_y^2}{(m+\tau_y^2)^2}} \right]$$

where $\bar{X}^* = (n + \tau_x^2)^{-1} \sum_{i=1}^{n} X_i$ and $Var(\bar{X}^*) = \frac{\sqrt{n}S_x}{n+\tau_x^2}$, where d is $z_{1-\alpha/2}$, an upper $1 - \alpha/2$ percentiles of the standard normal-distribution.

Khan [4] also derived the maximum likelihood estimator of μ_y when $\tau_y = \sigma_y/\mu_y$ is known which is $\hat{\theta}_m = [\sqrt{4\tau_y^2 S_m^2 + (1 + 4\tau_y^2)\bar{Y}^2} - \bar{Y}]/2\tau_y^2$. The estimator

$\hat{\theta}_m$ is asymptotically normal $N(\mu_y, \tau_y^2 \mu_y^2/m(1 + 2\tau_y^2))$ where $\tau_y^2 \mu_y^2/m(1 + 2\tau_y^2)$ is the Cramer-Rao bound. It is easy to see that a copy of $\hat{\theta}_m$ and its variance, we have $\hat{\theta}_n$ and its variance. Using CLT, it is easy to show that the $100(1 - \alpha)\%$ confidence interval for θ when τ_x, τ_y are known is given by

$$CI_K = \left[(\hat{\theta}_n - \hat{\theta}_m) - d\sqrt{\frac{S_n^2}{n(1+2\tau_x^2)} + \frac{S_m^2}{m(1+2\tau_y^2)}}, (\hat{\theta}_n - \hat{\theta}_m) + d\sqrt{\frac{S_n^2}{n(1+2\tau_x^2)} + \frac{S_m^2}{m(1+2\tau_y^2)}}\right]$$

where $S_n^2 = n^{-1}\sum_{i=1}^n (X_i - \bar{X})^2$, $S_m^2 = m^{-1}\sum_{j=1}^m (Y_j - \bar{Y})^2$.

In the following section, we derived analytic expressions for the coverage probability and the expected length of each confidence interval.

3   Coverage Probabilities and Expected Lengths of Confidence Intervals for θ with Known Coefficients of Variation

Theorems 1-3, below, show explicit expressions for the coverage probabilities and the expected lengths of confidence intervals CI_{WS}, CI_S, CI_K respectively.

Theorem 1. *The coverage probability and the expected length of CI_{WS} with known coefficients of variation are respectively*

$$E[\Phi(W) - \Phi(-W)]$$

and

$$\begin{cases} 2d\sigma_x\sigma_y(nm)^{-1/2}\delta\sqrt{r_1}F\left[\frac{-1}{2}, \frac{m-1}{2}, \frac{m+n-2}{2}, \frac{r_1-r_2}{r_1}\right], if \quad r_2 < 2r_1 \\ 2d\sigma_x\sigma_y(nm)^{-1/2}\delta\sqrt{r_2}F\left[\frac{-1}{2}, \frac{n-1}{2}, \frac{m+n-2}{2}, \frac{r_2-r_1}{r_2}\right], if \quad 2r_1 \leq r_2 \end{cases}$$

where $W = \dfrac{d\sqrt{\frac{S_x^2}{n} + \frac{S_y^2}{m}}}{\sqrt{n^{-1}\sigma_x^2 + m^{-1}\sigma_y^2}}$, $d = t_{1-\alpha/2,\nu}$, $\delta = \dfrac{\sqrt{2}\Gamma(\frac{n+m-1}{2})}{\Gamma(\frac{m+n-2}{2})}$, $r_1 = \dfrac{m}{(n-1)\sigma_y^2}$,
$r_2 = \dfrac{n}{(m-1)\sigma_x^2}$, *$E(.)$ is an expectation operator, $F(a; b; c; k)$ is the hypergeometric function defined by $F(a; b; c; k) = 1 + \frac{ab}{c}\frac{k}{1!} + \frac{a(a+1)b(b+1)}{c(c+1)}\frac{k^2}{2!} + \dots$ where $|k| < 1$, see Press [10], $\Gamma[.]$ is the gamma function and $\Phi[.]$ is the cumulative distribution function of $N(0; 1)$.*

Proof. Since, for normal samples, $\bar{X}$, $\bar{Y}$, S_x^2 and S_y^2 are independent of one another. From CI_{WS}, we have

$$1 - \alpha = P\left[(\bar{X} - \bar{Y}) - d\sqrt{\frac{S_x^2}{n} + \frac{S_y^2}{m}} < \mu_x - \mu_y < (\bar{X} - \bar{Y}) + d\sqrt{\frac{S_x^2}{n} + \frac{S_y^2}{m}}\right]$$

$$= P\left[\frac{-d\sqrt{\frac{S_x^2}{n} + \frac{S_y^2}{m}}}{\sqrt{n^{-1}\sigma_x^2 + m^{-1}\sigma_y^2}} < \frac{(\mu_x - \mu_y) - (\bar{X} - \bar{Y})}{\sqrt{n^{-1}\sigma_x^2 + m^{-1}\sigma_y^2}} < \frac{d\sqrt{\frac{S_x^2}{n} + \frac{S_y^2}{m}}}{\sqrt{n^{-1}\sigma_x^2 + m^{-1}\sigma_y^2}}\right]$$

$$= P\left[\frac{-d\sqrt{\frac{S_x^2}{n} + \frac{S_y^2}{m}}}{\sqrt{n^{-1}\sigma_x^2 + m^{-1}\sigma_y^2}} < Z < \frac{d\sqrt{\frac{S_x^2}{n} + \frac{S_y^2}{m}}}{\sqrt{n^{-1}\sigma_x^2 + m^{-1}\sigma_y^2}}\right]$$

$$= E[I_{\{-W < Z < W\}}(\xi)], I_{\{-W < Z < W\}}(\xi) = \begin{cases} 1, if \xi \in \{-W < Z < W\} \\ 0, otherwise \end{cases}$$

$$= E[E[I_{\{-W < Z < W\}}(\xi)] | S], S = (S_x^2, S_y^2)'$$

$$= E[\Phi(W) - \Phi(-W)]$$

where $Z \sim N(0; 1)$.

The length of CI_{WS} is $2d\sqrt{\frac{S_x^2}{n} + \frac{S_y^2}{m}}$ and the expected length of CI_{WS} is

$$E\left[2d\sqrt{\frac{S_x^2}{n} + \frac{S_y^2}{m}}\right] = 2dE\left[\sqrt{\frac{mS_x^2 + nS_y^2}{nm}}\right]$$

$$2dE\left[\sqrt{\frac{mS_x^2 + nS_y^2}{nm}}\right] = 2d\sigma_x\sigma_y(nm)^{-1/2}E\left[\sqrt{\frac{mS_x^2 + nS_y^2}{\sigma_x^2\sigma_y^2}}\right]$$

$$= 2d\sigma_x\sigma_y(nm)^{-1/2}E\left[\sqrt{\frac{\left(\frac{m}{(n-1)\sigma_y^2}\right)(n-1)S_x^2}{\sigma_x^2} + \frac{\left(\frac{n}{(m-1)\sigma_x^2}\right)(m-1)S_y^2}{\sigma_y^2}}\right]$$

$$= 2d\sigma_x\sigma_y(nm)^{-1/2}E[\sqrt{r_1 Z_1 + r_2 Z_2}]$$

$$= \begin{cases} 2d\sigma_x\sigma_y(nm)^{-1/2}\delta\sqrt{r_1}F\left[\frac{-1}{2}, \frac{m-1}{2}, \frac{m+n-2}{2}, \frac{r_1-r_2}{r_1}\right], if \quad r_2 < 2r_1 \\ 2d\sigma_x\sigma_y(nm)^{-1/2}\delta\sqrt{r_2}F\left[\frac{-1}{2}, \frac{n-1}{2}, \frac{m+n-2}{2}, \frac{r_2-r_1}{r_2}\right], if \quad 2r_1 \leq r_2 \end{cases}$$

where $Z_1 = \frac{(n-1)S_x^2}{\sigma_x^2} \sim \chi_{n-1}^2$, $Z_2 = \frac{(m-1)S_y^2}{\sigma_y^2} \sim \chi_{m-1}^2$, $r_1 = \frac{m}{(n-1)\sigma_y^2}$, $r_2 = \frac{n}{(m-1)\sigma_x^2}$, and for more details of $E[\sqrt{r_1 Z_1 + r_2 Z_2}]$, see Press [10]. Thus we complete the proof.

Theorem 2. *The coverage probability and the expected length of CI_S with known coefficients of variation are respectively*

$$E[\Phi(W_1) - \Phi(-W_1)]$$

and

$$\begin{cases} 2d\sigma_x\sigma_y(nm)^{-1/2}\delta\sqrt{s_1}F\left[\frac{-1}{2}, \frac{m-1}{2}, \frac{m+n-2}{2}, \frac{s_1-s_2}{s_1}\right], if \quad s_2 < 2s_1 \\ 2d\sigma_x\sigma_y(nm)^{-1/2}\delta\sqrt{s_2}F\left[\frac{-1}{2}, \frac{n-1}{2}, \frac{m+n-2}{2}, \frac{s_2-s_1}{s_2}\right], if \quad 2s_1 \leq s_2 \end{cases}$$

where $W_1 = \dfrac{d\sqrt{\dfrac{nS_x^2}{(n+\tau_x^2)^2} + \dfrac{mS_y^2}{(m+\tau_y^2)^2}}}{\sqrt{\dfrac{n\sigma_x^2}{(n+\tau_x^2)^2} + \dfrac{m\sigma_y^2}{(m+\tau_y^2)^2}}}$, $\;d = z_{1-\alpha/2}$, $\;\delta = \dfrac{\sqrt{2}\Gamma(\frac{n+m-1}{2})}{\Gamma(\frac{m+n-2}{2})}$, $\;p_1 = $

$\dfrac{n}{(n-1)\sigma_x^2(n+\tau_x^2)^2}$, $\;p_2 = \dfrac{m}{(m-1)\sigma_x^2(m+\tau_y^2)^2}$, $\;E(.)$ is an expectation operator, $F(a;b;c;k)$ is the hypergeometric function defined by $F(a;b;c;k) = 1 + \frac{ab}{c}\frac{k}{1!} + \frac{a(a+1)b(b+1)}{c(c+1)}\frac{k^2}{2!} + \dots$ where $|k| < 1$, see Press [10].

Proof. Since, for normal samples, $\bar{X}^*$, $\bar{Y}^*$, S_x^2 and S_y^2 are independent of one another. From CI_S, we have

$$1 - \alpha = P\left[(\bar{X}^* - \bar{Y}^*) - d\sqrt{\frac{nS_x^2}{(n+\tau_x^2)^2} + \frac{mS_y^2}{(m+\tau_y^2)^2}} < \mu_x - \mu_y \right.$$

$$\left. < (\bar{X}^* - \bar{Y}^*) + d\sqrt{\frac{nS_x^2}{(n+\tau_x^2)^2} + \frac{mS_y^2}{(m+\tau_y^2)^2}} \right]$$

$$= P\left[\frac{-d\sqrt{\frac{nS_x^2}{(n+\tau_x^2)^2} + \frac{mS_y^2}{(m+\tau_y^2)^2}}}{\sqrt{\frac{n\sigma_x^2}{(n+\tau_x^2)^2} + \frac{m\sigma_y^2}{(m+\tau_y^2)^2}}} < \frac{(\mu_x - \mu_y) - (\bar{X}^* - \bar{Y}^*)}{\sqrt{\frac{n\sigma_x^2}{(n+\tau_x^2)^2} + \frac{m\sigma_y^2}{(m+\tau_y^2)^2}}} \right.$$

$$\left. < \frac{d\sqrt{\frac{nS_x^2}{(n+\tau_x^2)^2} + \frac{mS_y^2}{(m+\tau_y^2)^2}}}{\sqrt{\frac{n\sigma_x^2}{(n+\tau_x^2)^2} + \frac{m\sigma_y^2}{(m+\tau_y^2)^2}}} \right]$$

$$= P\left[\frac{-d\sqrt{\frac{nS_x^2}{(n+\tau_x^2)^2} + \frac{mS_y^2}{(m+\tau_y^2)^2}}}{\sqrt{\frac{n\sigma_x^2}{(n+\tau_x^2)^2} + \frac{m\sigma_y^2}{(m+\tau_y^2)^2}}} < Z < \frac{d\sqrt{\frac{nS_x^2}{(n+\tau_x^2)^2} + \frac{mS_y^2}{(m+\tau_y^2)^2}}}{\sqrt{\frac{n\sigma_x^2}{(n+\tau_x^2)^2} + \frac{m\sigma_y^2}{(m+\tau_y^2)^2}}} \right]$$

$$= E[I_{\{-W_1 < Z < W_1\}}(\xi)], \; I_{\{-W_1 < Z < W_1\}}(\xi) = \begin{cases} 1, & if\, \xi \in \{-W_1 < Z < W_1\} \\ 0, & otherwise \end{cases}$$

$$= E[E[I_{\{-W_1 < Z < W_1\}}(\xi)]|S], S = (S_x^2, S_y^2)'$$

$$= E[\Phi(W_1) - \Phi(-W_1)]$$

where $Z \sim N(0;1)$.

The length of CI_S is $2d\sqrt{\frac{nS_x^2}{(n+\tau_x^2)^2} + \frac{mS_y^2}{(m+\tau_y^2)^2}}$ and the expected length of CI_S is

$$E\left[2d\sqrt{\frac{nS_x^2}{(n+\tau_x^2)^2} + \frac{mS_y^2}{(m+\tau_y^2)^2}} \right] = 2dE\left[\sqrt{\frac{nS_x^2}{(n+\tau_x^2)^2} + \frac{mS_y^2}{(m+\tau_y^2)^2}} \right]$$

$$= 2d\sigma_x\sigma_y E\left[d\sqrt{\frac{nS_x^2}{(n+\tau_x^2)^2\sigma_x^2\sigma_y^2} + \frac{mS_y^2}{(m+\tau_y^2)^2\sigma_x^2\sigma_y^2}}\right]$$

$$= 2d\sigma_x\sigma_y E\left[d\sqrt{\frac{n}{(n-1)(n+\tau_x^2)^2\sigma_y^2}\frac{(n-1)S_x^2}{\sigma_x^2} + \frac{m}{(m-1)(m+\tau_y^2)^2\sigma_x^2}\frac{(m-1)S_y^2}{\sigma_y^2}}\right]$$

$$= 2d\sigma_x\sigma_y E[\sqrt{p_1 Z_1 + p_2 Z_2}]$$

$$= \begin{cases} 2d\sigma_x\sigma_y\delta\sqrt{s_1}F\left[\frac{-1}{2}, \frac{m-1}{2}, \frac{m+n-2}{2}, \frac{p_1-p_2}{p_1}\right], if \quad p_2 < 2p_1 \\ 2d\sigma_x\sigma_y\delta\sqrt{s_2}F\left[\frac{-1}{2}, \frac{n-1}{2}, \frac{m+n-2}{2}, \frac{p_2-p_1}{p_2}\right], if \quad 2p_1 \leq p_2 \end{cases}$$

where $Z_1 = \frac{(n-1)S_x^2}{\sigma_x^2} \sim \chi_{n-1}^2$, $Z_2 = \frac{(m-1)S_y^2}{\sigma_y^2} \sim \chi_{m-1}^2$, $p_1 = \frac{n}{(n-1)(n+\tau_x^2)^2\sigma_y^2}$, $p_2 = \frac{m}{(m-1)(m+\tau_y^2)^2\sigma_x^2}$. Thus we complete the proof.

Theorem 3. *The coverage probability and the expected length of CI_K with known coefficients of variation are respectively*

$$E[\Phi(W_2) - \Phi(-W_2)]$$

and

$$\begin{cases} 2d\sigma_x\sigma_y(nm)^{-1/2}\delta\sqrt{s_1}F\left[\frac{-1}{2}, \frac{m-1}{2}, \frac{m+n-2}{2}, \frac{s_1-s_2}{s_1}\right], if \quad s_2 < 2s_1 \\ 2d\sigma_x\sigma_y(nm)^{-1/2}\delta\sqrt{s_2}F\left[\frac{-1}{2}, \frac{n-1}{2}, \frac{m+n-2}{2}, \frac{s_2-s_1}{s_2}\right], if \quad 2s_1 \leq s_2 \end{cases}$$

where $W_2 = \frac{d\sqrt{\frac{S_n^2}{n(1+2\tau_x^2)} + \frac{S_m^2}{m(1+2\tau_y^2)}}}{\sqrt{\frac{\sigma_n^2}{n(1+2\tau_x^2)} + \frac{\sigma_m^2}{m(1+2\tau_y^2)}}}$, $d = z_{1-\alpha/2}$, $\delta = \frac{\sqrt{2}\Gamma(\frac{n+m-1}{2})}{\Gamma(\frac{m+n-2}{2})}$, $s_1 = \frac{m}{n(1+2\tau_x^2)\sigma_x^2}$, $s_2 = \frac{n}{m(1+2\tau_y^2)\sigma_x^2}$, $E(.)$ *is an expectation operator,* $F(a; b; c; k)$ *is the hypergeometric function defined by* $F(a; b; c; k) = 1 + \frac{ab}{c}\frac{k}{1!} + \frac{a(a+1)b(b+1)}{c(c+1)}\frac{k^2}{2!} + \ldots$ *where* $|k| < 1$, *see Press [10].*

Proof. Similarly to Theorem 3.1-3.2 for the proof of coverage probability of CI_K, The length of CI_K is $2d\sqrt{\frac{S_n^2}{n(1+2\tau_x^2)} + \frac{S_m^2}{m(1+2\tau_y^2)}}$ and the expected length of CI_K is

$$E\left[2d\sqrt{\frac{S_n^2}{n(1+2\tau_x^2)} + \frac{S_m^2}{m(1+2\tau_y^2)}}\right] = 2dE\left[\sqrt{\frac{S_n^2}{n(1+2\tau_x^2)} + \frac{S_m^2}{m(1+2\tau_y^2)}}\right]$$

$$= 2d\sigma_x\sigma_y(nm)^{-1/2}E\left[\sqrt{\frac{mS_n^2}{(1+2\tau_x^2)\sigma_x^2\sigma_y^2} + \frac{nS_m^2}{(1+2\tau_y^2)\sigma_x^2\sigma_y^2}}\right]$$

$$= 2d\sigma_x\sigma_y(nm)^{-1/2}E\left[\sqrt{\frac{m}{n(1+2\tau_x^2)\sigma_y^2}\frac{nS_n^2}{\sigma_x^2} + \frac{n}{m(1+2\tau_y^2)\sigma_x^2}\frac{mS_m^2}{\sigma_y^2}}\right]$$

$$= 2d\sigma_x\sigma_y(nm)^{-1/2}E[\sqrt{s_1Q_1 + s_2Q_2}]$$

$$= \begin{cases} 2d\sigma_x\sigma_y(nm)^{-1/2}\delta\sqrt{s_1}F\left[\frac{-1}{2}, \frac{m-1}{2}, \frac{m+n-2}{2}, \frac{s_1-s_2}{s_1}\right], if \quad s_2 < 2s_1 \\ 2d\sigma_x\sigma_y(nm)^{-1/2}\delta\sqrt{s_2}F\left[\frac{-1}{2}, \frac{n-1}{2}, \frac{m+n-2}{2}, \frac{s_2-s_1}{s_2}\right], if \quad 2s_1 \leq s_2 \end{cases}$$

where $Q_1 = \frac{nS_n^2}{\sigma_x^2} \sim \chi_n^2$, $Q_2 = \frac{mS_m^2}{\sigma_y^2} \sim \chi_m^2$, $s_1 = \frac{m}{n(1+2\tau_x^2)\sigma_y^2}$, $s_2 = \frac{n}{m(1+2\tau_y^2)\sigma_x^2}$. Thus we complete the proof.

4   Simulation Studies

In this section, we compute coverage probabilities and average length widths of three confidence intervals: CI_{WS}, CI_S, CI_K using Monte Carlo simulation. The R program is used with a number of simulation runs equal 10,000. We set sample sizes $n = 10$, 20 ,30, 50 and 100 and the pair of coefficients of variation (τ_x, τ_y) as shown in Table 1-4.

From Tables 1-4, it is shown that a confidence interval CI_S outperforms other confidence intervals for small and moderate sample sizes i.e. $n = 10$, 20, and 30 as its coverage probability is over to a nominal level 0.95. For moderate and large $n =$50 and 100, three confidence intervals were not different in terms of coverage probabilities but a confidence interval CI_K outperforms other confidence intervals based on its shortest average length width.

5   Application

We provide an application to illustrate the use of the proposed confidence intervals. Walpole et al. (2002, p. 255) give data concerning two kind of threads are being compared for strength. Fifty pieces of each type of thread are tested under similar conditions. Brand A had an average tensile strength $(\bar{x}) = 78.3$ kilograms with a standard deviation $(sd) = 5.6$ kilograms, while Brand B had an average tensile strength $\bar{y} = 87.2$ with a standard deviation $(sd) = 6.3$, we assume that $\tau_x = 0.0715198$ and $\tau_y = 0.07224771$, the lower limit, upper limit and average length widths of each interval, CI_{WS}, CI_S, CI_K, are as follow:

As can be seen from table 4, a confidence interval for CI_K has a shortest average length width for moderate sample size $n = 50$. This results are supported our theoretical results.

Confidence Intervals   179

Table 1. Coverage probabilities and average length widths of confidence intervals CI_{WS}, CI_S, CI_K at a 95% nominal level of a coefficient of confidence interval.

(n,m)	(τ_x, τ_y)	CI_{WS}	CI_S	CI_K
10	(0.05,0.05)	0.9360	0.9651	0.9224
		(0.0865)	(0.0998)	(0.0819)
	(0.05,0.15)	0.9248	0.9555	0.9094
		(0.1920)	(0.2212)	(0.1787)
	(0.10,0.10)	0.9294	0.9628	0.9182
		(0.1726)	(0.1990)	(0.1624)
	(0.10,0.20)	0.9247	0.9575	0.9135
		(0.2730)	(0.3139)	(0.2514)
	(0.20,0.20)	0.9362	0.9642	0.9242
		(0.3470)	(0.3988)	(0.3187)
	(0.20,0.30)	0.9309	0.9617	0.9204
		(0.4404)	(0.5042)	(0.3937)
	(0.30,0.30)	0.9392	0.9660	0.9284
		(0.5204)	(0.5950)	(0.4606)
	(0.30,0.40)	0.9332	0.9619	0.9258
		(0.6108)	(0.6950)	(0.5245)
	(0.40,0.40)	0.9338	0.9619	0.9251
		(0.9614)	(0.7849)	(0.5845)
	(0.40,0.50)	0.9343	0.9608	0.9285
		(0.7859)	(0.8870)	(0.6432)
	(0.50,0.50)	0.9351	0.9628	0.9300
		(0.8667)	(0.9748)	(0.6956)
	(0.50,0.60)	0.9347	0.9624	0.9308
		(0.9590)	(1.0715)	(0.7444)
	(0.60,0.60)	0.9367	0.9647	0.9263
		(1.0401)	(1.1572)	(0.7906)
20	(0.05,0.05)	0.9412	0.9562	0.9362
		(0.0615)	(0.0657)	(0.0598)
	(0.05,0.15)	0.9360	0.9501	0.9285
		(0.1370)	(0.1462)	(0.1310)
	(0.10,0.10)	0.9420	0.9551	0.9351
		(0.1231)	(0.1314)	(0.1190)
	(0.10,0.20)	0.9371	0.9517	0.9308
		(0.1942)	(0.2070)	(0.1837)
	(0.20,0.20)	0.9443	0.9581	0.9441
		(0.2462)	(0.2625)	(0.2323)
	(0.20,0.30)	0.9428	0.9565	0.9398
		(0.3131)	(0.3323)	(0.2875)
	(0.30,0.30)	0.9449	0.9580	0.9406
		(0.3694)	(0.3928)	(0.3357)
	(0.30,0.40)	0.9422	0.9556	0.9377
		(0.4348)	(0.4614)	(0.3832)
	(0.40,0.40)	0.9421	0.9545	0.9406
		(0.4926)	(0.5221)	(0.4274)
	(0.40,0.50)	0.9421	0.9564	0.9416
		(0.5586)	(0.5904)	(0.4693)
	(0.50,0.50)	0.9432	0.9567	0.9417
		(0.6181)	(0.6524)	(0.5088)
	(0.50,0.60)	0.9437	0.9570	0.9397
		(0.6805)	(0.7160)	(0.5434)
	(0.60,0.60)	0.9404	0.9546	0.9398
		(0.7401)	(0.7771)	(0.5785)

Table 2. Coverage probabilities and average length widths of confidence intervals CI_{WS}, CI_S, CI_K at a 95% nominal level of a coefficient of confidence interval.

(n,m)	(τ_x, τ_y)	CI_{WS}	CI_S	CI_K
30	(0.05,0.05)	0.9421	0.9518	0.9390
		(0.0504)	(0.0525)	(0.0494)
	(0.05,0.15)	0.9431	0.9530	0.9384
		(0.1123)	(0.1171)	(0.1083)
	(0.10,0.10)	0.9443	0.9522	0.9405
		(0.1009)	(0.1052)	(0.0983)
	(0.10,0.20)	0.9435	0.9521	0.9418
		(0.1589)	(0.1657)	(0.1517)
	(0.20,0.20)	0.9452	0.9543	0.9429
		(0.2016)	(0.2101)	(0.1918)
	(0.20,0.30)	0.9460	0.9552	0.9458
		(0.2563)	(0.2668)	(0.2373)
	(0.30,0.30)	0.9452	0.9538	0.9430
		(0.3023)	(0.3146)	(0.2771)
	(0.30,0.40)	0.9478	0.9573	0.9446
		(0.3565)	(0.3704)	(0.3169)
	(0.40,0.40)	0.9475	0.9555	0.9448
		(0.4036)	(0.4191)	(0.3530)
	(0.40,0.50)	0.9435	0.9527	0.9400
		(0.4555)	(0.4721)	(0.3860)
	(0.50,0.50)	0.9523	0.9586	0.9498
		(0.5037)	(0.5217)	(0.4186)
	(0.50,0.60)	0.9440	0.9529	0.9469
		(0.5564)	(0.5750)	(0.4480)
	(0.60,0.60)	0.9453	0.9539	0.9441
		(0.6051)	(0.6245)	(0.4761)
50	(0.05,0.05)	0.9498	0.9551	0.9465
		(0.0391)	(0.0400)	(0.0386)
	(0.05,0.15)	0.9404	0.9475	0.9406
		(0.0871)	(0.0893)	(0.0846)
	(0.10,0.10)	0.9490	0.9549	0.9466
		(0.0780)	(0.0800)	(0.0766)
	(0.10,0.20)	0.9469	0.9525	0.9437
		(0.1237)	(0.1267)	(0.1188)
	(0.20,0.20)	0.9466	0.9522	0.9461
		(0.1565)	(0.1603)	(0.1499)
	(0.20,0.30)	0.9470	0.9512	0.9489
		(0.1991)	(0.2039)	(0.1856)
	(0.30,0.30)	0.9456	0.9504	0.9464
		(0.2344)	(0.2399)	(0.2162)
	(0.30,0.40)	0.9438	0.9484	0.9446
		(0.2764)	(0.2827)	(0.2472)
	(0.40,0.40)	0.9501	0.9551	0.9463
		(0.3127)	(0.3197)	(0.2754)
	(0.40,0.50)	0.9460	0.9532	0.9443
		(0.3544)	(0.3619)	(0.3021)
	(0.50,0.50)	0.9458	0.9504	0.9432
		(0.3911)	(0.3992)	(0.3269)
	(0.50,0.60)	0.9483	0.9548	0.9473
		(0.4320)	(0.4405)	(0.3499)
	(0.60,0.60)	0.9480	0.9533	0.9500
		(0.4686)	(0.4774)	(0.3710)

Table 3. Coverage probabilities and average length widths of confidence intervals CI_{WS}, CI_S, CI_K at a 95% nominal level of a coefficient of confidence interval.

(n,m)	(τ_x, τ_y)	CI_{WS}	CI_S	CI_K
100	(0.05,0.05)	0.9502	0.9522	0.9494
		(0.0276)	(0.0280)	(0.0274)
	(0.05,0.15)	0.9465	0.9498	0.9449
		(0.0619)	(0.0626)	(0.0604)
	(0.10,0.10)	0.9481	0.9505	0.9464
		(0.0553)	(0.0560)	(0.0546)
	(0.10,0.20)	0.9497	0.9529	0.9503
		(0.0876)	(0.0886)	(0.0845)
	(0.20,0.20)	0.9452	0.9482	0.9433
		(0.1107)	(0.1120)	(0.1065)
	(0.20,0.30)	0.9483	0.9511	0.9479
		(0.1410)	(0.1427)	(0.1321)
	(0.30,0.30)	0.9458	0.9506	0.9483
		(0.1660)	(0.1679)	(0.1539)
	(0.30,0.40)	0.9464	0.9497	0.9470
		(0.1958)	(0.1979)	(0.1759)
	(0.40,0.40)	0.9422	0.9460	0.9467
		(0.2216)	(0.2240)	(0.1959)
	(0.40,0.50)	0.9487	0.9515	0.9530
		(0.2506)	(0.2532)	(0.2147)
	(0.50,0.50)	0.9450	0.9473	0.9502
		(0.2769)	(0.2797)	(0.2323)
	(0.50,0.60)	0.9489	0.9511	0.9491
		(0.3057)	(0.3086)	(0.2488)
	(0.60,0.60)	0.9515	0.9535	0.9510
		(0.3320)	(0.3351)	(0.2644)

Table 4. Lower limit, upper limit and average length widths of confidence intervals CI_{WS}, CI_S, CI_K at a 95% nominal level of a coefficient of confidence interval.

CIs	$lowerlimit$	$upperlimit$	$length$
CI_{WS}	-11.27869	-6.521306	4.757387
CI_S	-11.29383	-6.503207	4.790621
CI_K	-9.605	-8.227906	1.377094

6   Conclusions

In this paper we proposed new confidence intervals for the difference between normal means with known coefficients of variation. We derived, mathematically, coverage probabilities and expected lengths of these intervals: CI_{WS}, CI_S, CI_K. It is shown in sections 4 that the coverage probabilities of CI_{WS} and CI_K cannot maintain the nominal value 0.95 when sample size $n < 50$. So we choose the confidence interval CI_S as its coverage probability is over the nominal level at 0.95. A confidence interval CI_K is preferable for moderate and large n as its has coverage probability which approaches the nominal level at 0.95 and has a shortest length for sample sizes $n = 50, 100$ especially for a large pairs of coefficients of variation. This is confirmed from our application.

Acknowledgments. The first author is grateful to grant number KMUTNB-GOV-57-21 from King Mongkut's University of Technology North Bangkok.

References

1. Bhat, K.K., Rao, K.A.: On Tests for a Normal Mean with Known Coefficient of Variation. International Statistical Review **75**, 170–182 (2007)
2. Brazauskas, V., Ghorai, J.: Estimating the Common Parameter of Normal Models with Known Coefficients of Variation: A Sensitivity Study of Asymptotically Efficient Estimators. Journal of Statistical Computation and Simulation **77**, 663–681 (2007)
3. Casella, G., Berger, R.: Statistical Inferences, 2nd edn. Duxbury, California, USA (2002)
4. Khan, R.A.: A Note on Estimating the Mean of a Normal Distribution with Known Coefficient of Variation. Journal of the American Statistical Association **63**, 1039–1041 (1968)
5. Niwitpong, S.: Confidence Interval for the Normal Means with Known a Coefficient of Variation. World Academic of Science, Engineering and Technology **69**, 677–680 (2012)
6. Niwitpong, S., Niwitpong, S.: On Simple Confidence Interval for the Normal Means withKnown a Coefficient of Variation. World Academy of Science, Engineering and Technology, International Journal of Mathematical, Computational Science and Engineering **7**, 723–726 (2013)
7. Niwitpong, S.: Confidence Interval for the Difference Normal Means with Known Coefficient of Variation. (to be appear in Far East Journal of Mathematical Sciences)
8. Searls, D.T.: The Utilization of a Known Coefficient of Variation in the Estimation Procedure. Journal of the American Statistical Association **59**, 1225–1226 (1964)
9. Searls, D.T.: A Note on the Use of an Approximately Known Coefficient of Variation. The American Statistician **21**, 20–21 (1967)
10. Press, S.J.: A Confidence Interval Comparison of Two Test Procedures for the Behrens-Fisher Problems. Journal of the American Statistical Association. **61**, 454–466 (1966)

Approximate Confidence Interval for the Ratio of Normal Means with a Known Coefficient of Variation

Wararit Panichkitkosolkul[✉]

Department of Mathematics and Statistics, Faculty of Science and Technology,
Thammasat University, Pathumthani 12121, Thailand
wararit@mathstat.sci.tu.ac.th

Abstract. An approximate confidence interval for the ratio of normal population means with a known coefficient of variation is proposed. This has applications in the area of bioassay and bioequivalence when the scientist knows the coefficient of variation of the control group. The proposed confidence interval is based on the approximate expectation and variance of the estimator by Taylor series expansion. A Monte Carlo simulation study was conducted to compare the performance of the proposed confidence interval with the existing confidence interval. Simulation results show that the proposed confidence interval performs as well as the existing one in terms of coverage probability and expected length. However, the approximate confidence interval is very easy to calculate compared with the exact confidence interval.

Keywords: Interval estimation · Central tendency · Standardized measure of dispersion · Coverage probability · Expected length

1 Introduction

Statistical analysis for the ratio of normal means is applied in the area of bioassay and bioequivalence (see, for example, [1-12]). The ratio of normal means is defined by

$$\theta = \frac{\mu_x}{\mu_y},$$

where μ_x and μ_y are the population means of X and Y, respectively. The confidence interval for the ratio of normal means has also been studied by many researchers. For example, Fieller's theorem [4-5] constructs the confidence interval for the ratio means. Koschat [13] has also demonstrated that the coverage probability of a confidence interval constructed using Fieller's theorem is exact for all parameters when a common variance assumption is assumed.

Niwitpong et al. [14] proposed two confidence intervals for the ratio of normal means with a known coefficient of variation. Their confidence intervals can be applied in some situations, for instance when the coefficient of variation of a control group is known. One of their confidence intervals is developed based on an exact method in which this confidence interval is constructed from the pivotal statistics Z,

© Springer International Publishing Switzerland 2015
V.-N. Huynh et al. (Eds.): IUKM 2015, LNAI 9376, pp. 183–192, 2015.
DOI: 10.1007/978-3-319-25135-6_18

where Z follows the standard normal distribution. The other confidence interval is constructed based on the generalized confidence interval [15]. Simulation results show that the coverage probabilities of the two confidence intervals are not significantly different. However, the confidence interval based on the exact method is shorter than the generalized confidence interval. The exact method uses Taylor series expansion to find the expectation and variance of the estimator of θ and uses these results for constructing the confidence interval for θ. The lower and upper limits of the confidence interval based on the exact method are difficult to compute since they depend on an infinite summation. Therefore, our main aim in this paper is to propose an approximate confidence interval for the ratio of normal means with a known coefficient of variation. The computation of the new proposed confidence interval is easier than the exact confidence interval proposed by Niwitpong et al. [14]. In addition, we also compare the estimated coverage probabilities and average lengths of the new proposed confidence interval and existing confidence interval using a Monte Carlo simulation.

The paper is organized as follows. In Section 2, the theoretical background of the existing confidence interval for θ is discussed. We provide the theorem for constructing the approximate confidence interval for θ in Section 3. In Section 4, the performance of the confidence intervals for θ is investigated through a Monte Carlo simulation study. Conclusions are provided in the final section.

2   Existing Confidence Interval

In this section, we review the theorem and corollary proposed by Niwitpong et al. [14] and use these to construct the exact confidence interval for θ.

Theorem 1. *(Niwitpong et al. [14]) Let* $X_1,...,X_n$ *be a random sample of size* n *from a normal distribution with mean* μ_x *and variance* σ_x^2 *and* $Y_1,...,Y_m$ *be a random sample of size* m *from a normal distribution with mean* μ_y *and variance* σ_y^2. *The estimator of* θ *is* $\hat{\theta} = \dfrac{\overline{X}}{\overline{Y}}$ *where* $\overline{X} = n^{-1}\sum_{i=1}^{n}X_i$ *and* $\overline{Y} = m^{-1}\sum_{j=1}^{m}Y_j$. *The expectation of* $\hat{\theta}$ *and* $\hat{\theta}^2$ *when a coefficient of variation,* $\tau_y = \dfrac{\sigma_y}{\mu_y}$ *is known, are respectively*

$$E(\hat{\theta}) = \theta\left[1+\sum_{k=1}^{\infty}\frac{(2k)!}{2^k k!}\left(\frac{\tau_y^2}{n}\right)^k\right] \tag{1}$$

and

$$E(\hat{\theta}^2) = \theta^2\sum_{k=1}^{\infty}\frac{(2k+1)!}{2^k k!}\left(\frac{\tau_y^2}{n}\right)^k.$$

From (1), $\lim\limits_{n\to\infty} E(\hat{\theta}) = \theta$ and $E\left(\dfrac{\hat{\theta}}{w}\right) = \theta$, where $w = 1 + \sum\limits_{k=1}^{\infty} \dfrac{(2k)!}{2^k k!}\left(\dfrac{\tau_y^2}{n}\right)^k$. Thus,

the unbiased estimator of θ is $\dfrac{\hat{\theta}}{w} = \dfrac{\bar{X}}{w\bar{Y}}$.

Corollary 1. *From Theorem 1,* $\operatorname{var}(\hat{\theta}) \approx \dfrac{\sigma_x^2}{n\mu_y^2} + \left(\dfrac{\theta^2}{m} + \dfrac{\sigma_x^2}{nm\mu_y^2}\right)\tau_y^2.$

Now we will use the fact that, from the central limit theorem,

$$Z = \frac{\hat{\theta} - \theta}{\sqrt{\operatorname{var}(\hat{\theta})}} \sim N(0,1).$$

Based on Theorem 1 and Corollary 1, we get

$$Z = \frac{\dfrac{\hat{\theta}}{w} - \theta}{\sqrt{\dfrac{\sigma_x^2}{n\mu_y^2} + \left(\dfrac{\theta^2}{m} + \dfrac{\sigma_x^2}{nm\mu_y^2}\right)\tau_y^2}} \sim N(0,1). \tag{2}$$

Therefore, the $100(1-\alpha)\%$ exact confidence interval for θ based on Equation (2) is

$$CI_{exact} = \frac{\hat{\theta}}{w} \pm z_{1-\alpha/2}\sqrt{\frac{S_x^2}{n\bar{Y}^2} + \left(\frac{\hat{\theta}^2}{m} + \frac{S_x^2}{nm\bar{Y}^2}\right)\tau_y^2},$$

where $S_x^2 = (n-1)^{-1}\sum\limits_{i=1}^{n}(X_i - \bar{X})^2$, $\quad w = 1 + \sum\limits_{k=1}^{\infty}\dfrac{(2k)!}{2^k k!}\left(\dfrac{\tau_y^2}{n}\right)^k$ and $z_{1-\alpha/2}$ is the

$100(1-\alpha/2)$ percentile of the standard normal distribution.

3   Proposed Confidence Interval

To find a simple approximate expression for the expectation of $\hat{\theta}$, we use a Taylor

series expansion of $\dfrac{x}{y}$ around μ_x, μ_y:

$$\frac{x}{y} \approx \frac{x}{y}\bigg|_{\mu_x,\mu_y} + (x-\mu_x)\frac{\partial}{\partial x}\left(\frac{x}{y}\right)\bigg|_{\mu_x,\mu_y} + (y-\mu_y)\frac{\partial}{\partial y}\left(\frac{x}{y}\right)\bigg|_{\mu_x,\mu_y} + \frac{1}{2}(x-\mu_x)^2\frac{\partial^2}{\partial x^2}\left(\frac{x}{y}\right)\bigg|_{\mu_x,\mu_y}$$

$$+\frac{1}{2}(y-\mu_y)^2\frac{\partial^2}{\partial y^2}\left(\frac{x}{y}\right)\bigg|_{\mu_x,\mu_y} + (x-\mu_x)(y-\mu_y)\frac{\partial^2}{\partial x\partial y}\left(\frac{x}{y}\right)\bigg|_{\mu_x,\mu_y}$$

$$+O\left(\left((x-\mu_x)\frac{\partial}{\partial x}+(y-\mu_y)\frac{\partial}{\partial y}\right)^3\left(\frac{x}{y}\right)\right). \tag{3}$$

Theorem 2. *Let* $X_1,...,X_n$ *be a random sample of size* n *from a normal distribution with mean* μ_x *and variance* σ_x^2 *and* $Y_1,...,Y_m$ *be a random sample of size* m *from a normal distribution with mean* μ_y *and variance* σ_y^2. *The estimator of* θ *is*

$\hat{\theta}=\dfrac{\overline{X}}{\overline{Y}}$ *where* $\overline{X}=n^{-1}\sum\limits_{i=1}^{n}X_i$ *and* $\overline{Y}=m^{-1}\sum\limits_{j=1}^{m}Y_j$. *The approximate expectation and*

variance of $\hat{\theta}$ *when a coefficient of variation,* $\tau_y=\dfrac{\sigma_y}{\mu_y}$ *is known, are respectively*

$$E(\hat{\theta}) \approx \theta\left(1+\frac{\tau_y^2}{m}\right)$$

and

$$\mathrm{var}(\hat{\theta}) \approx \frac{\sigma_x^2}{n\mu_y^2}+\frac{\theta^2}{m}\tau_y^2.$$

Proof of Theorem 2

Consider random variables $\overline{X}$ and $\overline{Y}$ where $\overline{Y}$ has support $(0,\infty)$. Let $\hat{\theta}=\dfrac{\overline{X}}{\overline{Y}}$. Find approximations for $E(\hat{\theta})$ and $\mathrm{var}(\hat{\theta})$ using Taylor series expansion of $\hat{\theta}$ around μ_x,μ_y as in Equation (3). The mean of $\hat{\theta}$ can be found by applying the expectation operator to the individual terms (ignoring all terms higher than two),

$$E(\hat{\theta}) \quad = \quad E\left(\frac{\bar{X}}{\bar{Y}}\right)$$

$$= \quad E\left(\frac{\bar{X}}{\bar{Y}}\right)\bigg|_{\mu_x,\mu_y} + E\left[\frac{\partial}{\partial \bar{X}}\left(\frac{\bar{X}}{\bar{Y}}\right)(\bar{X}-E(\bar{X}))\right]\bigg|_{\mu_x,\mu_y} + E\left[\frac{\partial}{\partial \bar{Y}}\left(\frac{\bar{X}}{\bar{Y}}\right)(\bar{Y}-E(\bar{Y}))\right]\bigg|_{\mu_x,\mu_y}$$

$$+\frac{1}{2}E\left[\frac{\partial^2}{\partial \bar{X}^2}\left(\frac{\bar{X}}{\bar{Y}}\right)(\bar{X}-E(\bar{X}))^2\right]\bigg|_{\mu_x,\mu_y} + \frac{1}{2}E\left[\frac{\partial^2}{\partial \bar{Y}^2}\left(\frac{\bar{X}}{\bar{Y}}\right)(\bar{Y}-E(\bar{Y}))^2\right]\bigg|_{\mu_x,\mu_y}$$

$$+E\left[\frac{\partial^2}{\partial \bar{X}\partial \bar{Y}}\left(\frac{\bar{X}}{\bar{Y}}\right)(\bar{X}-E(\bar{X}))(\bar{Y}-E(\bar{Y}))\right]\bigg|_{\mu_x,\mu_y} + O(n^{-1})$$

$$\approx \quad \frac{\mu_x}{\mu_y}+0+0+0+\frac{1}{2}\left(\frac{2E(\bar{X})}{(E(\bar{Y}))^3}Var(\bar{Y})\right)-\frac{\text{cov}(\bar{X},\bar{Y})}{(E(\bar{Y}))^2}$$

$$\approx \quad \frac{\mu_x}{\mu_y}+\text{var}(\bar{Y})\frac{\mu_x}{\mu_y^3}-\frac{\text{cov}(\bar{X},\bar{Y})}{\mu_y^2}$$

$$= \quad \frac{\mu_x}{\mu_y}+\text{var}(\bar{Y})\frac{\mu_x}{\mu_y^3}, \quad \text{cov}(\bar{X},\bar{Y})=0$$

$$= \quad \frac{\mu_x}{\mu_y}\left(1+\frac{\sigma_y^2}{m\mu_y^2}\right)$$

$$= \quad \theta\left(1+\frac{\tau_y^2}{m}\right). \tag{4}$$

An approximation of the variance of $\hat{\theta}$ is obtained by using the first-order terms of the Taylor series expansion:

$$\text{var}(\hat{\theta}) \quad = \quad \text{var}\left(\frac{\bar{X}}{\bar{Y}}\right)$$

$$= \quad E\left[\left(\frac{\bar{X}}{\bar{Y}}-E\left(\frac{\bar{X}}{\bar{Y}}\right)\right)^2\right]$$

$$\approx \quad E\left[\left(\frac{\bar{X}}{\bar{Y}}-\frac{\mu_x}{\mu_y}\right)^2\right]$$

$$\approx \quad E\left[\left(\frac{\mu_x}{\mu_y}+\frac{\partial}{\partial \bar{X}}\left(\frac{\bar{X}}{\bar{Y}}\right)(\bar{X}-E(\bar{X}))+\frac{\partial}{\partial \bar{Y}}\left(\frac{\bar{X}}{\bar{Y}}\right)(\bar{Y}-E(\bar{Y}))-\frac{\mu_x}{\mu_y}\right)^2\right]\bigg|_{\mu_x,\mu_y}$$

$$= \quad \left(\frac{\partial}{\partial \bar{X}}\left(\frac{\bar{X}}{\bar{Y}}\right)\right)^2\text{var}(\bar{X})+\left(\frac{\partial}{\partial \bar{Y}}\left(\frac{\bar{X}}{\bar{Y}}\right)\right)^2\text{var}(\bar{Y})+2\frac{\partial}{\partial \bar{X}}\left(\frac{\bar{X}}{\bar{Y}}\right)\frac{\partial}{\partial \bar{Y}}\left(\frac{\bar{X}}{\bar{Y}}\right)\text{cov}(\bar{X},\bar{Y})\bigg|_{\mu_x,\mu_y}$$

$$\approx \quad \frac{\mathrm{var}(\overline{X})}{\mu_y^2} + \frac{\mu_x^2\,\mathrm{var}(\overline{Y})}{\mu_y^4} - \frac{2\mu_x\,\mathrm{cov}(\overline{X},\overline{Y})}{\mu_y^3}$$

$$= \quad \frac{\mathrm{var}(\overline{X})}{\mu_y^2} + \frac{\mu_x^2\,\mathrm{var}(\overline{Y})}{\mu_y^4}, \quad \mathrm{cov}(\overline{X},\overline{Y}) = 0$$

$$= \quad \frac{\sigma_x^2}{n\mu_y^2} + \frac{\mu_x^2\sigma_y^2}{m\mu_y^4}$$

$$= \quad \frac{\sigma_x^2}{n\mu_y^2} + \frac{\theta^2}{m}\cdot\tau_y^2. \tag{5}$$

∎

It is clear from Equation (4) that $\hat{\theta}$ is asymptotically unbiased $\left(\lim\limits_{m\to\infty} E(\hat{\theta}) = \theta\right)$

and $E\left(\dfrac{\hat{\theta}}{v}\right) = \theta$, where $v = 1 + \dfrac{\tau_y^2}{m}$. Therefore, the unbiased estimator of θ is

$\dfrac{\hat{\theta}}{v} = \dfrac{\overline{X}}{v\overline{Y}}$. From Equation (5), $\hat{\theta}$ is consistent $\left(\lim\limits_{n,m\to\infty} \mathrm{var}(\hat{\theta}) = 0\right)$.

We then apply the central limit theorem and Theorem 2,

$$Z \quad = \quad \frac{\dfrac{\hat{\theta}}{v} - \theta}{\sqrt{\dfrac{\sigma_x^2}{n\mu_y^2} + \dfrac{\theta^2}{m}\tau_y^2}} \sim N(0,1).$$

Therefore, it is easily seen that the $(1-\alpha)100\%$ approximate confidence interval for θ is

$$CI_{approx} \quad = \quad \frac{\hat{\theta}}{v} \pm z_{1-\alpha/2}\sqrt{\frac{S_x^2}{n\overline{Y}^2} + \frac{\hat{\theta}^2}{m}\tau_y^2},$$

where $S_x^2 = (n-1)^{-1}\sum\limits_{i=1}^{n}(X_i - \overline{X})^2$, $v = 1 + \dfrac{\tau_y^2}{m}$ and $z_{1-\alpha/2}$ is the $100(1-\alpha/2)$ percentile of the standard normal distribution.

4   Simulation Study

A Monte Carlo simulation was conducted using the R statistical software [16] version 3.1.3 to compare the estimated coverage probabilities and average lengths of the new proposed confidence interval and the exact confidence interval. Three sets of normal data were generated with $\theta = 0.5, 1$ and 2, and the ratio of variances $\sigma_x^2/\sigma_y^2 = 0.25$, 0.5, 0.8, 1, 2, 3, 4, 5 and 10. The sample sizes were set at $(n,m) = (10,10), (10,20),$ $(20,10)$ and $(20,20)$. The number of simulation runs was 10,000 and the nominal confidence level $1-\alpha$ was fixed at 0.95. The results are demonstrated in Tables 1-3. Both confidence intervals have estimated coverage probabilities close to the nominal confidence level in the case of small ratio of variances. Additionally, the estimated coverage probabilities of all confidence intervals decrease as the values of σ_x^2/σ_y^2 get larger (i.e., for exact confidence interval, $\theta = 0.5$, $(n,m) = (10,10)$, 0.94258 for $\sigma_x^2/\sigma_y^2 = 0.25$; 0.93096 for $\sigma_x^2/\sigma_y^2 = 5$; 0.92200 for $\sigma_x^2/\sigma_y^2 = 5$).

Table 1. Estimated coverage probabilities and expected lengths of confidence intervals for the ratio of normal means with a known coefficient of variation when $\theta = 0.5$.

n	m	σ_x^2/σ_y^2	Coverage Probabilities		Expected Lengths	
			Exact	Approx.	Exact	Approx.
10	10	0.25	0.94258	0.94253	0.08720	0.08718
		0.5	0.93837	0.93835	0.07501	0.07500
		0.8	0.93259	0.93257	0.06995	0.06995
		1	0.93096	0.93090	0.06821	0.06820
		2	0.92797	0.92797	0.06442	0.06442
		3	0.92449	0.92449	0.06305	0.06305
		4	0.92167	0.92166	0.06236	0.06236
		5	0.92200	0.92199	0.06202	0.06202
		10	0.92221	0.92220	0.06104	0.06104
10	20	0.25	0.93690	0.93678	0.07509	0.07508
		0.5	0.93153	0.93143	0.06809	0.06808
		0.8	0.92769	0.92763	0.06529	0.06529
		1	0.92560	0.92567	0.06443	0.06443
		2	0.92246	0.92243	0.06232	0.06232
		3	0.92088	0.92090	0.06169	0.06169
		4	0.92063	0.92056	0.06133	0.06133
		5	0.91990	0.91986	0.06109	0.06109
		10	0.91845	0.91844	0.06069	0.06069
20	10	0.25	0.94708	0.94724	0.07589	0.07588
		0.5	0.94615	0.94632	0.06181	0.06181
		0.8	0.94370	0.94373	0.05564	0.05563
		1	0.94384	0.94382	0.05337	0.05336
		2	0.94044	0.94058	0.04863	0.04863
		3	0.93937	0.93944	0.04688	0.04688
		4	0.93811	0.93808	0.04601	0.04601
		5	0.93778	0.93780	0.04551	0.04551
		10	0.93589	0.93592	0.04439	0.04439
20	20	0.25	0.94717	0.94714	0.06183	0.06182
		0.5	0.94357	0.94357	0.05341	0.05341
		0.8	0.94213	0.94213	0.04985	0.04985
		1	0.94174	0.94173	0.04861	0.04860
		2	0.93825	0.93825	0.04600	0.04600
		3	0.93819	0.93817	0.04509	0.04509
		4	0.93719	0.93719	0.04464	0.04464
		5	0.93639	0.93639	0.04435	0.04435
		10	0.93569	0.93568	0.04385	0.04385

The estimated coverage probabilities and the expected lengths of the proposed confidence interval are not significantly different from these of the existing confidence interval in any situation. However, the approximate confidence interval is very easy to calculate compared with the exact confidence interval because the exact confidence interval is based on an infinite summation.

Table 2. Estimated coverage probabilities and expected lengths of confidence intervals for the ratio of normal means with a known coefficient of variation when $\theta = 1$.

n	m	σ_x^2 / σ_y^2	Coverage Probabilities		Expected Lengths	
			Exact	Approx.	Exact	Approx.
10	10	0.25	0.94720	0.94717	0.27803	0.27792
		0.5	0.94485	0.94479	0.21450	0.21443
		0.8	0.94394	0.94388	0.18529	0.18524
		1	0.94304	0.94300	0.17447	0.17442
		2	0.93781	0.93774	0.15024	0.15022
		3	0.93385	0.93384	0.14108	0.14106
		4	0.93099	0.93099	0.13646	0.13645
		5	0.93047	0.93047	0.13322	0.13321
		10	0.92530	0.92530	0.12720	0.12719
10	20	0.25	0.94488	0.94547	0.21468	0.21461
		0.5	0.94335	0.94339	0.17454	0.17450
		0.8	0.93924	0.93950	0.15662	0.15659
		1	0.93648	0.93660	0.15023	0.15020
		2	0.92933	0.92931	0.13611	0.13610
		3	0.92785	0.92783	0.13146	0.13145
		4	0.92580	0.92571	0.12857	0.12856
		5	0.92614	0.92625	0.12723	0.12723
		10	0.92309	0.92312	0.12379	0.12379
20	10	0.25	0.94951	0.94935	0.26411	0.26405
		0.5	0.94931	0.94934	0.19635	0.19631
		0.8	0.94907	0.94900	0.16408	0.16405
		1	0.94831	0.94818	0.15174	0.15172
		2	0.94603	0.94611	0.12370	0.12369
		3	0.94457	0.94435	0.11268	0.11267
		4	0.94452	0.94452	0.10681	0.10680
		5	0.94233	0.94225	0.10300	0.10299
		10	0.94130	0.94141	0.09523	0.09523
20	20	0.25	0.94795	0.94792	0.19637	0.19633
		0.5	0.94803	0.94800	0.15181	0.15178
		0.8	0.94576	0.94573	0.13127	0.13125
		1	0.94615	0.94612	0.12369	0.12367
		2	0.94428	0.94422	0.10684	0.10683
		3	0.94233	0.94232	0.10053	0.10052
		4	0.94080	0.94079	0.09715	0.09715
		5	0.93958	0.93958	0.09517	0.09517
		10	0.93833	0.93833	0.09103	0.09103

Table 3. Estimated coverage probabilities and expected lengths of confidence intervals for the ratio of normal means with a known coefficient of variation when

n	m	σ_x^2/σ_y^2	Coverage Probabilities		Expected Lengths	
			Exact	Approx.	Exact	Approx.
10	10	0.25	0.94976	0.94973	0.51319	0.51313
		0.5	0.94889	0.94888	0.37262	0.37258
		0.8	0.94849	0.94847	0.30378	0.30375
		1	0.94850	0.94848	0.27719	0.27716
		2	0.94623	0.94619	0.21421	0.21419
		3	0.94575	0.94574	0.18862	0.18860
		4	0.94200	0.94197	0.17431	0.17430
		5	0.94134	0.94134	0.16503	0.16502
		10	0.93721	0.93719	0.14481	0.14481
10	20	0.25	0.94873	0.94917	0.37253	0.37249
		0.5	0.94783	0.94856	0.27720	0.27717
		0.8	0.94763	0.94799	0.23145	0.23143
		1	0.94758	0.94785	0.21414	0.21412
		2	0.94242	0.94246	0.17430	0.17429
		3	0.93990	0.93994	0.15856	0.15856
		4	0.93740	0.93731	0.15001	0.15000
		5	0.93508	0.93509	0.14463	0.14462
		10	0.92890	0.92882	0.13338	0.13338
20	10	0.25	0.95042	0.94996	0.50551	0.50548
		0.5	0.94969	0.94930	0.36209	0.36207
		0.8	0.95026	0.95007	0.29102	0.29100
		1	0.94965	0.94966	0.26318	0.26317
		2	0.95004	0.94987	0.19603	0.19602
		3	0.94756	0.94742	0.16773	0.16772
		4	0.94769	0.94764	0.15164	0.15163
		5	0.94830	0.94824	0.14109	0.14108
		10	0.94581	0.94590	0.11714	0.11714
20	20	0.25	0.94871	0.94872	0.36212	0.36209
		0.5	0.94861	0.94861	0.26320	0.26319
		0.8	0.95021	0.95019	0.21475	0.21474
		1	0.94861	0.94860	0.19602	0.19601
		2	0.94816	0.94815	0.15169	0.15169
		3	0.94843	0.94840	0.13360	0.13360
		4	0.94751	0.94751	0.12359	0.12358
		5	0.94667	0.94666	0.11716	0.11715
		10	0.94156	0.94155	0.10303	0.10302

5   Conclusions

In this paper, we proposed an approximate confidence interval for the ratio of normal population means with a known coefficient of variation. Normally, this arises when the scientist knows the coefficient of variation of the control group. The approximate confidence interval proposed uses the approximation of the expectation and variance of the estimator. The proposed new confidence interval is compared with the exact confidence interval constructed by Niwitpong et al. [14] through a Monte Carlo simulation study. The approximate confidence interval performs as efficiently as the exact confidence interval in terms of coverage probability and expected length. Moreover, approximate confidence interval also is easy to compute compared with the exact confidence interval.

Acknowledgements. The author is grateful to anonymous referees for their valuable comments, which have significantly enhanced the quality and presentation of this paper.

References

1. Bliss, C.I.: The Calculation of the Dose-Mortality Curve. Ann. Appl. Biol. **22**, 134–167 (1935)
2. Bliss, C.I.: The Comparison of Dose-Mortality Data. Ann. Appl. Biol. **22**, 309–333 (1935)
3. Irwin, J.O.: Statistical Method Applied to Biological Assay. J. R. Stat. Soc. Ser B. **4**, 1–60 (1937)
4. Fieller, E.C.: A Fundamental Formula in the Statistics of Biological Assay and Some Applications. Q. J. Pharm. Pharmacol. **17**, 117–123 (1944)
5. Fieller, E.C.: Some Problems in Interval Estimation. J. R. Stat. Soc. Ser B. **16**, 175–185 (1954)
6. Finney, D.J.: The Principles of the Biological Assay (with discussion). J. R. Stat. Soc. Ser B. **1**, 46–91 (1947)
7. Finney, D.J.: The Meaning of Bioassay. Biometrics **21**, 785–798 (1965)
8. Cox, C.P.: Interval Estimates for the Ratio of the Means of Two Normal Populations with Variances Related to the Means. Biometrics **41**, 261–265 (1985)
9. Srivastava, M.S.: Multivariate Bioassay, Combination of Bioassays and Fieller's Theorem. Biometrics **42**, 131–141 (1986)
10. Kelly, G.E.: The Median Lethal Dose-Design and Estimation. The Statistician **50**, 41–50 (2000)
11. Lee, J.C., Lin, S.H.: Generalized Confidence Intervals for the Ratio of Means of Two Normal Populations. J. Stat. Plan. Infer. **123**, 49–60 (2004)
12. Vuorinen, J., Tuominen, J.: Fieller's Confidence Intervals for the Ratio of Two Means in the Assessment of Average Bioequivalence from Crossover Data. Stat. Med. **13**, 2531–2545 (1994)
13. Koschat, M.A.: A Characterization of the Fieller Solution. Ann. Statist. **15**, 462–468 (1987)
14. Niwitpong, S., Koonprasert, S., Niwitpong, S.: Confidence Intervals for the Ratio of Normal Means with a Known Coefficient of Variation. Adv. Appl. Stat. **25**, 47–61 (2011)
15. Weerahandi, S.: Generalized Confidence Intervals. J. Amer. Statist. Assoc. **88**, 899–905 (1993)
16. Ihaka, R., Gentleman, R.: R: A Language for Data Analysis and Graphics. J. Comp. Graph. Stat. **5**, 299–314 (1996)

Confidence Intervals for the Ratio of Coefficients of Variation of the Gamma Distributions

Patarawan Sangnawakij, Sa-Aat Niwitpong$^{(\boxtimes)}$, and Suparat Niwitpong

Department of Applied Statistics, Faculty of Applied Science,
King Mongkut's University of Technology North Bangkok, Bangkok, Thailand
patarawan@mathstat.sci.tu.ac.th, {snw,suparatn}@kmutnb.ac.th

Abstract. One of the most useful statistical measures is the coefficient of variation which is widely used in many fields of applications. Not only in a single population, the coefficients of variation are applied in two populations. In this paper, we proposed two new confidence intervals for the ratio of coefficients of variation in the gamma distributions based on the method of variance of estimates recovery with the methods of Score and Wald intervals. Moreover, the coverage probability and expected length of the proposed confidence intervals are evaluated via a Monte Carlo simulation.

Keywords: Confidence interval · Coefficient of variation · Gamma distribution · Simulation

1 Introduction

In statistical analysis, the population coefficient of variation is defined as a ratio of the population standard deviation to the mean. Since the coefficient of variation is a unit-free measure, it is often used to compare the distributions of variables even though the variables have different units [1]. In application, the sample coefficient of variation is used in many areas of science, engineering and economics. For example, the coefficient of variation is used to summarize the variability in the enzyme-linked immunosorbent assay (ELISA) and other methods of chemical analysis in the laboratory [2]. Moreover, the coefficient of variation is used as a measure of risk sensitivity for the human and animal decision making under risk in a meta-analysis [3]. In economics, the ratio of coefficients of variation is reported the variations about the staple food price of the people in Indonesia [4].

Although the point estimator is a useful summary statistics, however the confidence interval is more meaningful and provides the information respecting to the parameter of interest than point estimator [5]. Therefore, many studies have constructed the confidence intervals for the coefficient of variation, particularly in the special problem of the ratio of coefficients of variation. For example, we refer to the paper of Verrill and Johnson [6]. They introduced the asymptotic

© Springer International Publishing Switzerland 2015
V.-N. Huynh et al. (Eds.): IUKM 2015, LNAI 9376, pp. 193–203, 2015.
DOI: 10.1007/978-3-319-25135-6_19

confidence interval for the ratio of coefficients of variation in the normal distributions. However, by simulation, it was found that their confidence interval has the coverage probability less than the nominal level in some cases. Furthermore, Donner and Zou [7] proposed the confidence intervals for the functions of parameters by using the method of variance of estimates recovery (MOVER). One of their proposed the confidence intervals is the confidence interval for the coefficient of variation in the normal distribution. Then, Buntao and Niwitpong [8] introduced the confidence intervals for the ratio of coefficients of variation in the delta-lognormal distribution based on the methods of the MOVER and the generalized confidence interval (GCI). By simulation, it was found that the their confidence intervals provide the coverage probabilities close to the nominal level. From the reviewed literatures, we can see that many studies have been interested in constructing the confidence intervals for the ratio of coefficients of variation in the normal distributions. However, inference problems for the lifetime distributions are also important in the practical applications.

The gamma distribution is a probability model which is frequently used in the actuarial science as the waiting time until death [9,10]. The probability density function of the two-parameter gamma distribution, $Gamma\ (\alpha, \beta)$, is defined as

$$f_X(x; \alpha, \beta) = \frac{1}{\Gamma(\alpha)\beta^\alpha} \exp^{-\frac{x}{\beta}} x^{\alpha-1}, x > 0, \alpha > 0, \beta > 0,$$

where α and β are the shape and scale parameters, respectively. The mean of X is $E(X) = \alpha\beta$ and the variance of X is $Var(X) = \alpha\beta^2$. Thus, the population coefficient of variation is given by $\tau = \frac{1}{\sqrt{\alpha}}$. To estimate the population coefficient of variation, the sample coefficient of variation is defined as a point estimator $\hat{\tau} = \frac{1}{\sqrt{\hat{\alpha}}}$, where $\hat{\alpha}$ is the estimator of α.

Recently, Sangnawakij et al. [11] proposed the confidence intervals for the coefficient of variation using two well-known methods, the Score interval and the Wald interval. By simulation, it was found that their confidence intervals perform well in terms of coverage probability in almost all cases. That research is the study in the single coefficient of variation in gamma distribution. However, the confidence interval for the ratio of coefficients of variation is interested and there have not been research studies done in the case of two gamma populations. Therefore, the objective of this paper is to construct the new confidence intervals for the ratio of coefficients of variation in the gamma distributions based on the method of variance of estimates recovery described by Donner and Zou [7] with the confidence intervals of Sangnawakij et al. [11].

This paper is organized as a follows. In Section 2, we consider the confidence intervals for the single coefficient of variation of Sangnawakij et al. [11]. The methods to establish the proposed confidence intervals for the ratio of coefficients of variation in the gamma distributions based on the MOVER approach by Donner and Zou [7] are indicated in Section 3. The simulation results and a numerical illustration are shown in Section 4 and Section 5, respectively. Finally, some concluding remarks are given in Section 6.

2 The Confidence Intervals for the Coefficient of Variation

Let $X = (X_1, X_2, \ldots X_n)$ and $Y = (Y_1, Y_2, \ldots Y_m)$ be random samples from two gamma distributions, i.e., $X \sim Gamma(\alpha_1, \beta_1)$ and $Y \sim Gamma(\alpha_2, \beta_2)$. The coefficients of variation for the variable X and Y are denoted as τ_1 and τ_2, respectively, where $\tau_1 = \frac{1}{\sqrt{\alpha_1}}$ and $\tau_2 = \frac{1}{\sqrt{\alpha_2}}$.

In the recent, Sangnawakij et al. [11] introduced the $(1 - \gamma)100\%$ confidence intervals for the single coefficient of variation in the gamma distribution by the methods of the Score interval and the Wald interval. Their confidence intervals are as follows:

1. The confidence interval for the coefficient of variation, τ_1, from the method of Score confidence interval

$$CI_{s1} = [l_{s1}, u_{s1}],$$

where

$\bar{X}$ is the sample mean of X, $l_{s1} = \sqrt{\frac{2}{n}(n \ln \bar{X} - \sum_{i=1}^{n} \ln X_i - Z_{\frac{\gamma}{2}} \sqrt{\frac{n}{2\hat{\alpha}_1^2}})}$,

$u_{s1} = \sqrt{\frac{2}{n}(n \ln \bar{X} - \sum_{i=1}^{n} \ln X_i + Z_{\frac{\gamma}{2}} \sqrt{\frac{n}{2\hat{\alpha}_1^2}})}$

2. The confidence interval for the coefficient of variation, τ_1, from the method of Wald confidence interval

$$CI_{w1} = [l_{w1}, u_{w1}],$$

where $l_{w1} = (\sqrt{\hat{\alpha}_1 + Z_{\frac{\gamma}{2}} \sqrt{\frac{2\hat{\alpha}_1^2}{n}}})^{-1}$, $u_{w1} = (\sqrt{\hat{\alpha}_1 - Z_{\frac{\gamma}{2}} \sqrt{\frac{2\hat{\alpha}_1^2}{n}}})^{-1}$, $Z_{\frac{\gamma}{2}}$ is the $\frac{\gamma}{2}$ quantile of the standard normal distribution and $\hat{\alpha}_1 = \frac{1}{2[\ln \bar{X} - \frac{1}{n} \sum_{i=1}^{n} \ln X_i]}$ is the maximum likelihood estimator of α_1

Similarly, the confidence intervals for τ_2 based on the methods of the Score interval and the Wald interval are

$$CI_{s2} = [l_{s2}, u_{s2}]$$

and

$$CI_{w2} = [l_{w2}, u_{w2}],$$

respectively, where $\bar{Y}$ is the sample mean of Y, $\hat{\alpha}_2 = \frac{1}{2[\ln \bar{Y} - \frac{1}{m} \sum_{i=1}^{m} \ln Y_i]}$ is the maximum likelihood estimator of α_2, $l_{s2} = \sqrt{\frac{2}{m}(m \ln \bar{Y} - \sum_{i=1}^{m} \ln Y_i - Z_{\frac{\gamma}{2}} \sqrt{\frac{m}{2\hat{\alpha}_2^2}})}$,

$u_{s2} = \sqrt{\frac{2}{m}(m \ln \bar{Y} - \sum_{i=1}^{m} \ln Y_i + Z_{\frac{\gamma}{2}} \sqrt{\frac{m}{2\hat{\alpha}_2^2}})}$, $l_{w2} = (\sqrt{\hat{\alpha}_2 + Z_{\frac{\gamma}{2}} \sqrt{\frac{2\hat{\alpha}_2^2}{m}}})^{-1}$ and

$u_{w2} = (\sqrt{\hat{\alpha}_2 - Z_{\frac{\gamma}{2}} \sqrt{\frac{2\hat{\alpha}_2^2}{m}}})^{-1}$.

Then, we use the confidence intervals for the single coefficient of variation in this section to establish the confidence intervals for the ratio of coefficients of variation in the gamma distributions. The important methods are presented in the next section.

3   The Confidence Intervals for the Ratio of Coefficients of Variation

From two gamma distributions, the ratio of coefficients of variation is defined as

$$\eta = \frac{\tau_1}{\tau_2} = \frac{\frac{1}{\sqrt{\alpha_1}}}{\frac{1}{\sqrt{\alpha_2}}}.$$

The confidence interval for the parameter η can be constructed by the following method.

The closed form method of variance estimation or the method of variance of estimates recovery (MOVER) is the method for constructing the confidence interval for the functions of parameters, i.e., in the form $\theta_1 + \theta_2$ and $\frac{\theta_1}{\theta_2}$. It was introduced by Zou and his colleague in many papers [7], [13], [14]. The concept of this method is to find the separate confidence intervals for two single parameters and then recover the variance estimates from the confidence intervals, after that to form the confidence interval for the function of parameters.

For more information, we explain the MOVER method based on the central limit theorem (CLT) to construct the confidence interval for $\theta_1 + \theta_2$. Thus, the general form of two-sided confidence interval, under the assumption of independent between the estimators $\hat{\theta}_1$ and $\hat{\theta}_2$, is given by

$$[L, U] = [(\hat{\theta}_1 + \hat{\theta}_2) \mp Z_{\frac{\gamma}{2}} \sqrt{Var(\hat{\theta}_1) + Var(\hat{\theta}_2)}],$$

where $Var(\hat{\theta}_1)$ and $Var(\hat{\theta}_2)$ are the unknown variances of $\hat{\theta}_1$ and $\hat{\theta}_2$, respectively. Zou et al. [13] assumed that $[l_i', u_i']$ are the $(1 - \gamma)100\%$ confidence intervals for θ_i, for $i = 1, 2$. Moreover, they described that the value of $l_1' + l_2'$ is similar to L and $u_1' + u_2'$ is similar to U. To estimate $Var(\hat{\theta}_i)$, using the CLT and under the conditions $\theta_1 = l_1'$ and $\theta_2 = l_2'$, the estimated variances recovered from l_i' to obtain L can be derived as $\widehat{Var}(\hat{\theta}_i) \approx \frac{(\hat{\theta}_i - l_i')^2}{Z_{\frac{\gamma}{2}}^2}$. On the other hand, under the conditionals $\theta_1 = u_1'$ and $\theta_2 = u_2'$, the estimated variances recovered from u_i' to obtain U can be derived as $\widehat{Var}(\hat{\theta}_i) \approx \frac{(u_i' - \hat{\theta}_i)^2}{Z_{\frac{\gamma}{2}}^2}$. Replacing the corresponding estimated variances into the interval $[L, U]$, then we will obtain the $(1 - \gamma)100\%$ two-sided confidence interval for $\theta_1 + \theta_2$.

Similarly, the confidence interval for the difference of parameters can be constructed by changing $\theta_1 - \theta_2$ into the form $\theta_1 + (-\theta_2)$ and then recovering the variance estimates by following the above method. In this study, we can see that the ratio of parameters, $R = \frac{\theta_1}{\theta_2}$ is equivalent to the form $\theta_1 - R\theta_2 = 0$. Hence, the confidence interval for $\theta_1 - R\theta_2$ is obtained. Setting the lower and upper limits for $\theta_1 - R\theta_2$ equal to zero, we can solve the confidence interval for R.

Following Donner and Zou [7], the $(1-\gamma)100\%$ two-sided confidence interval for $\frac{\theta_1}{\theta_2}$ is given by $[L_r', U_r']$

$$L_r' = \frac{\hat{\theta}_1\hat{\theta}_2 - \sqrt{(\hat{\theta}_1\hat{\theta}_2)^2 - l_1'u_2'(2\hat{\theta}_1 - l_1')(2\hat{\theta}_2 - u_2')}}{u_2'(2\hat{\theta}_2 - u_2')} \tag{1}$$

and

$$U_r' = \frac{\hat{\theta}_1\hat{\theta}_2 + \sqrt{(\hat{\theta}_1\hat{\theta}_2)^2 - u_1'l_2'(2\hat{\theta}_1 - u_1')(2\hat{\theta}_2 - l_2')}}{l_2'(2\hat{\theta}_2 - l_2')}, \tag{2}$$

where $[l_i', u_i']$ are the $(1-\gamma)100\%$ confidence intervals and $\hat{\theta}_i$ are the point estimators of θ_i, for $i = 1, 2$.

In order to construct the new confidence intervals for $\eta = \frac{\tau_1}{\tau_2}$, we apply the confidence interval of Donner and Zou [7] to achieve the objective of this paper. Now, we set $\hat{\theta}_1 = \hat{\tau}_1 = \frac{1}{\sqrt{\hat{\alpha}_1}}$ is the estimator of τ_1, $\hat{\theta}_2 = \hat{\tau}_2 = \frac{1}{\sqrt{\hat{\alpha}_2}}$ is the estimator of τ_2, $l_1' = l_1$, $u_1' = u_1$, $l_2' = l_2$ and $u_2' = u_2$. Then, we replace them into equations (1) and (2). It easy to see that the general form of the $(1-\gamma)100\%$ confidence interval for η is $[L_r, U_r]$,

$$
\begin{aligned}
L_r &= \frac{\hat{\tau}_1\hat{\tau}_2 - \sqrt{(\hat{\tau}_1\hat{\tau}_2)^2 - l_1 u_2(2\hat{\tau}_2 - u_2)(2\hat{\tau}_1 - l_1)}}{u_2(2\hat{\tau}_2 - u_2)} \\
&= \frac{\frac{1}{\sqrt{\hat{\alpha}_1\hat{\alpha}_2}} - \sqrt{\frac{1}{\hat{\alpha}_1\hat{\alpha}_2} - l_1 u_2(\frac{2}{\sqrt{\hat{\alpha}_2}} - u_2)(\frac{2}{\sqrt{\hat{\alpha}_1}} - l_1)}}{u_2(\frac{2}{\sqrt{\hat{\alpha}_2}} - u_2)}
\end{aligned} \tag{3}
$$

and

$$
\begin{aligned}
U_r &= \frac{\hat{\tau}_1\hat{\tau}_2 + \sqrt{(\hat{\tau}_1\hat{\tau}_2)^2 - u_1 l_2(2\hat{\tau}_2 - l_2)(2\hat{\tau}_1 - u_1)}}{l_2(2\hat{\tau}_2 - l_2)} \\
&= \frac{\frac{1}{\sqrt{\hat{\alpha}_1\hat{\alpha}_2}} + \sqrt{\frac{1}{\hat{\alpha}_1\hat{\alpha}_2} - u_1 l_2(\frac{2}{\sqrt{\hat{\alpha}_2}} - l_2)(\frac{2}{\sqrt{\hat{\alpha}_1}} - u_1)}}{l_2(\frac{2}{\sqrt{\hat{\alpha}_2}} - l_2)},
\end{aligned} \tag{4}
$$

where $[l_i, u_i]$ are the $(1-\gamma)100\%$ confidence intervals for τ_i, and $\hat{\alpha}_i$ are the MLEs of α_i, for $i = 1, 2$.

From equations (3) and (4), we can see that the lower and upper limits of the confidence interval for η depend on the interval $[l_i, u_i]$, for $i = 1, 2$. This point is important. Therefore, in the next subsection, the confidence intervals for the ratio of coefficients of variation based on the MOVER with the Score interval and the Wald interval are considered.

3.1 The Confidence Interval for the Ratio of Coefficients of Variation Using the Score Method

In section 2, we know that $[l_{s1}, u_{s1}]$ and $[l_{s2}, u_{s2}]$ are the $(1-\gamma)100\%$ confidence intervals based on the method of Score interval for τ_1 and τ_2, respectively. Using the concept of the MOVER approach, we replace $l_1 = l_{s1}$, $l_2 = l_{s2}$, $u_1 = u_{s1}$ and $u_2 = u_{s2}$ into equations (3) and (4). Therefore, the $(1-\gamma)100\%$ confidence interval for η is given by $CI_{rs} = [l_{rs}, u_{rs}]$, where

$$l_{rs} = \frac{\frac{1}{\sqrt{\hat{\alpha}_1\hat{\alpha}_2}} - \sqrt{\frac{1}{\hat{\alpha}_1\hat{\alpha}_2} - l_{s1}u_{s2}(\frac{2}{\sqrt{\hat{\alpha}_2}} - u_{s2})(\frac{2}{\sqrt{\hat{\alpha}_1}} - l_{s1})}}{u_{s2}(\frac{2}{\sqrt{\hat{\alpha}_2}} - u_{s2})} \tag{5}$$

and

$$u_{rs} = \frac{\frac{1}{\sqrt{\hat{\alpha}_1\hat{\alpha}_2}} + \sqrt{\frac{1}{\hat{\alpha}_1\hat{\alpha}_2} - l_{s2}u_{s1}(\frac{2}{\sqrt{\hat{\alpha}_2}} - l_{s2})(\frac{2}{\sqrt{\hat{\alpha}_1}} - u_{s1})}}{l_{s2}(\frac{2}{\sqrt{\hat{\alpha}_2}} - l_{s2})}. \tag{6}$$

3.2 The Confidence Interval for the Ratio of Coefficients of Variation Using the Wald Method

Like the previous subsection, since we know that $[l_{w1}, u_{w1}]$ and $[l_{w2}, u_{w2}]$ are the $(1-\gamma)100\%$ confidence intervals based on the method of Wald interval for τ_1 and τ_2, respectively. Then, we set $l_1 = l_{w1}$, $l_2 = l_{w2}$, $u_1 = u_{w1}$ and $u_2 = u_{w2}$ into equations (3) and (4). Therefore, the $(1-\gamma)100\%$ confidence interval for η is derived as $CI_{rw} = [l_{rw}, u_{rw}]$, where

$$l_{rw} = \frac{\frac{1}{\sqrt{\hat{\alpha}_1\hat{\alpha}_2}} - \sqrt{\frac{1}{\hat{\alpha}_1\hat{\alpha}_2} - l_{w1}u_{w2}(\frac{2}{\sqrt{\hat{\alpha}_2}} - u_{w2})(\frac{2}{\sqrt{\hat{\alpha}_1}} - l_{w1})}}{u_{w2}(\frac{2}{\sqrt{\hat{\alpha}_2}} - u_{w2})} \tag{7}$$

and

$$u_{rw} = \frac{\frac{1}{\sqrt{\hat{\alpha}_1\hat{\alpha}_2}} + \sqrt{\frac{1}{\hat{\alpha}_1\hat{\alpha}_2} - l_{w2}u_{w1}(\frac{2}{\sqrt{\hat{\alpha}_2}} - l_{w2})(\frac{2}{\sqrt{\hat{\alpha}_1}} - u_{w1})}}{l_{w2}(\frac{2}{\sqrt{\hat{\alpha}_2}} - l_{w2})}. \tag{8}$$

4 Performance of the Confidence Intervals

To examine the confidence interval, we prefer the confidence interval which has the values of coverage probability at least or close to the nominal coverage level $(1-\gamma)$, and has the shortest expected length. In this section, the simulations are carried out using the R statistical program [12] to evaluate the coverage probabilities and the expected lengths of the proposed confidence intervals with 10,000 simulations at significance level $\gamma = 0.05$. In this simulation study, the first step, the data were generated from two independent gamma distributions with (α_1, β_1) and (α_2, β_2), respectively, where $\beta_1 = \beta_2 = 2$ are fixed and α_i are adjusted to get the required coefficients of variation, i.e., $\alpha_i = \frac{1}{\tau_i^2}$, for $i = 1, 2$. A set values of the coefficients of variation, $(\tau_1, \tau_2) = (0.05, 0.05)$, $(0.05, 0.10)$, $(0.05, 0.30)$, $(0.10, 0.10)$, $(0.10, 0.20)$, $(0.10, 0.40)$, $(0.20, 0.20)$, $(0.20, 0.30)$, $(0.20, 0.50)$ and sample sizes $(n, m) = (20, 20)$, $(40, 40)$, $(100, 100)$, $(200, 200)$. Then, the 95% of proposed confidence intervals for the ratio of coefficients of variation, CI_{ds} and CI_{dw} are calculated. Finally, the estimated coverage probability and the expected length are given by

Table 1. The coverage probabilities and expected lengths of 95% confidence intervals for the ratio of coefficients of variation of gamma distributions.

n	m	(τ_1, τ_2)	Method	CP^a	EL^b	n	m	(τ_1, τ_2)	Method	CP^a	EL^b
20	20	(0.05,0.05)	Score	0.9760	1.1508	100	100	(0.05,0.05)	Score	0.9530	0.4106
			Wald	0.9779	1.1687				Wald	0.9543	0.4119
		(0.05,0.10)	Score	0.9771	0.5781			(0.05,0.10)	Score	0.9571	0.2052
			Wald	0.9786	0.5871				Wald	0.9583	0.2058
		(0.05,0.30)	Score	0.9762	0.1905			(0.05,0.30)	Score	0.9521	0.0678
			Wald	0.9787	0.1935				Wald	0.9529	0.0681
		(0.10,0.10)	Score	0.9742	1.1577			(0.10,0.10)	Score	0.9554	0.4113
			Wald	0.9756	1.1756				Wald	0.9559	0.4126
		(0.10,0.20)	Score	0.9758	0.5788			(0.10,0.20)	score	0.9532	0.2046
			Wald	0.9783	0.5878				Wald	0.9538	0.2053
		(0.10,0.40)	Score	0.9727	0.2849			(0.10,0.40)	Score	0.9538	0.1013
			Wald	0.9751	0.2893				Wald	0.9544	0.1017
		(0.20,0.20)	Score	0.9741	1.1553			(0.20,0.20)	Score	0.9526	0.4112
			Wald	0.9759	1.1732				Wald	0.9535	0.4126
		(0.20,0.30)	Score	0.9760	0.7677			(0.20,0.30)	Score	0.9529	0.2726
			Wald	0.9776	0.7796				Wald	0.9538	0.2734
		(0.20,0.50)	Score	0.9777	0.4530			(0.20,0.50)	Score	0.9512	0.1612
			Wald	0.9792	0.4600				Wald	0.9517	0.1617
40	40	(0.05,0.05)	Score	0.9664	0.7003	200	200	(0.05,0.05)	Score	0.9497	0.2834
			Wald	0.9679	0.7059				Wald	0.9502	0.2839
		(0.05,0.10)	Score	0.9618	0.3506			(0.05,0.10)	Score	0.9558	0.1417
			Wald	0.9639	0.3534				Wald	0.9564	0.1420
		(0.05,0.30)	Score	0.9633	0.1160			(0.05,0.30)	Score	0.9511	0.0469
			Wald	0.9643	0.1169				Wald	0.9513	0.0470
		(0.10,0.10)	Score	0.9604	0.7011			(0.10,0.10)	Score	0.9511	0.2839
			Wald	0.9615	0.7067				Wald	0.9513	0.2843
		(0.10,0.20)	Score	0.9636	0.3501			(0.10,0.20)	Score	0.9542	0.1413
			Wald	0.9656	0.3529				Wald	0.9546	0.1415
		(0.10,0.40)	Score	0.9599	0.1733			(0.10,0.04)	Score	0.9464	0.0700
			Wald	0.9613	0.1747				Wald	0.9467	0.0701
		(0.20,0.20)	Score	0.9623	0.7015			(0.20,0.20)	Score	0.9533	0.2834
			Wald	0.9634	0.7071				Wald	0.9536	0.2839
		(0.20,0.30)	Score	0.9624	0.4654			(0.20,0.30)	Score	0.9526	0.1883
			Wald	0.9643	0.4691				Wald	0.9530	0.1886
		(0.20,0.50)	Score	0.9620	0.2755			(0.20,0.50)	Score	0.9444	0.1114
			Wald	0.9638	0.2777				Wald	0.9449	0.1116

[a] is the coverage probability for the confidence interval.
[b] is the expected length for the confidence interval.

$$1 - \gamma \cong \frac{c(L \leq \eta \leq U)}{10,000}$$

and

$$Length = \frac{\sum_{j=1}^{1000}(U_j - Lj)}{10,000},$$

where $c(L \leq \eta \leq U)$ is the number of simulation runs for which the the ratio of coefficients of variation, η, lie within the confidence interval. The simulation results for all combination parameters are shown in Table 1.

As a results, the proposed confidence intervals based on the MOVER method with the Score interval, CI_{rs}, and the Wald interval, CI_{rw}, provide the coverage probabilities more than or close to the nominal value at 0.95. Moreover, the expected lengths of either proposed confidence interval are not different in all cases. Furthermore, the expected lengths of CI_{rs} and CI_{rw} tend to decrease when sample size increases. From the simulation result, it means that 95% of the proposed confidence intervals accurately cover the true parameter.

5   Data Example

In this section, we illustrate the real data provided by Proschan [15] on the time of successive failures of the air conditioning system of Boeing 720 airplanes. The

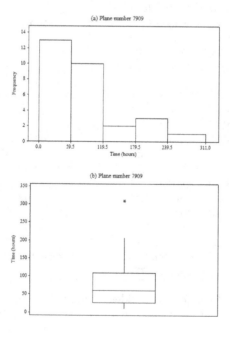

Fig. 1. (a) a histogram and (b) a Box-and-Whisker plot of the air conditioning system of plane number 7909

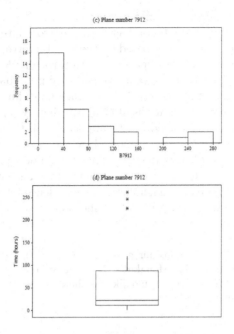

Fig. 2. (a) a histogram and (b) a Box-and-Whisker plot of the air conditioning system of plane number 7912

data in operating hours for jet plane number 7909, 29 observations, and number 7912, 30 observation, are shown by the histograms and the Box-and-Whisker plots in Fig. 1 and Fig. 2, respectively.

In this samples, we observe the statistic as a follow. For plane number 7909, the sample mean $\bar{x} = 83.5172$, $\sum_{i=1}^{29} \ln x_i = 118.8105$, the MLE $\hat{\alpha}_1 = 1.5238$. For plane number 7912, the sample mean $\bar{y} = 59.6000$, $\sum_{i=1}^{30} \ln y_i = 100.7427$, the MLE $\hat{\alpha}_2 = 0.6853$. Hence, the 95% confidence intervals by the Score method for τ_1 and τ_2 are $[l_{s1}, u_{s1}] = [0.5643, 0.9970]$ and $[l_{s2}, u_{s2}] = [0.8489, 1.4824]$, respectively. Moreover, we have the 95% confidence intervals by the Wald method for τ_1 and τ_2 are $[l_{w1}, u_{w1}] = [0.6582, 1.1629]$ and $[l_{w2}, u_{w2}] = [0.9843, 1.7188]$, respectively.

Therefore, it easy to see that the 95% confidence interval CI_{rs} for η is $[0.4437, 1.0078]$ with the length of interval 0.5641. On the other hand, the 95% confidence interval CI_{rw} for η is $[0.4446, 1.0184]$ with the length of interval 0.5783. Clearly, the expected lengths of CI_{rs} and CI_{rw} are not different as the simulation result in the previous section.

6   Conclusions

The new confidence intervals for the ratio of coefficients of variation which are proposed in this paper extend the recent paper of Sangnawakij et al. [11] based on the method of variance of estimates recovery (MOVER) of Donner and Zou [7]. The proposed confidence interval based on the Score interval, CI_{rs}, is compared with the confidence interval based on the Wald interval, CI_{rw}, in terms of the coverage probability and the expected length through a Monte Carlo simulation. The results indicate that the values of coverage probability of all proposed confidence intervals maintain the nominal coverage level. Furthermore, for overall comparisons, the expected lengths of CI_{rs} are slightly different from that of CI_{rw}. These results are confirmed by the application in the real data example. We also note that the results are supported by the asymptotic equivalent theory of the Score and Wald statistics [16]. That is, the new proposed confidence intervals are acceptable. Moreover, our confidence intervals have clearly formulas and it can be calculated easily. Therefore, the confidence intervals in this paper can be chosen to estimate the ratio of coefficients of variation in the gamma distributions.

Acknowledgments. The first author gratefully acknowledge the financial support from the Graduate Studies, King Mongkut's University of Technology North Bangkok, Thailand. Moreover, the authors would like to thank to the referees for providing valuable comments and suggestions.

References

1. Albatineh, A.N., Kibria, B.M.G., Zogheib, B.: Asymptotic Sampling Distribution of Inverse Its Applications: Revisited. International Journal of Advanced Statistics and Probability Statistics and Probability **2**, 15–20 (2014)
2. Reed, G.F., Lynn, F., Meade, B.D.: Use of Coefficient of Variation in Assessing Variability of Quantitative Assays. Clina. Diagn. Lab. Immunol. **9**, 1235–1239 (2002)
3. Weber, E.U., Shafir, S., Blais, A.: Predicting Risk Sensitivity in Humans and Lower Animals: Risk as Variance or Coefficient of Variation. Psychological Review **111**, 430–445 (2004)
4. Jati, K.: Staple Food Balance Sheet, Coefficient of Variation, and Price Disparity in Indonesia. Journal of Advanced Management Science **2**, 65–71 (2014)
5. Casella, G., Berger, R.L.: Statistical Inference, 2nd edn. Duxbury Press, California (2002)
6. Verrill, S., Johnson, R.A.: Confidence Bounds and Hypothesis Tests for Normal Distribution Coefficients of Variation. Communications in Statistics-Theory and Method **36**, 2187–2206 (2007)
7. Donner, A., Zou, G.Y.: Closed-form Confidence Intervals for Functions of the Normal Mean and Standard Deviation. Statistical Methods in Medical Research 1–13 (2010)
8. Buntao, N., Niwitpong, S.: Confidence Intervals for the Ratio of Coefficients of Variation of Delta-Lognormal Distribution. Applied Mathematical Sciences **7**, 3811–3818 (2013)
9. Gupta, B.C., Guttman, I.: Statistics and Probability with Applications for Engineers and Scientists. John Wiley and Sons, New Jersey (2013)
10. Nabendu, P., Chun, J., Wooi, K.L.: Handbook of Exponential and Related Distribution for Engineers and Scientists. Chapman & Hall/CRC, Baco Raton (2006)

11. Sangnawakij, P., Niwitpong, S., Niwitpong, S.: Confidence Intervals for the Coefficient of Variation in a Gamma Distribution (submitted for publication)
12. An Introduction to R. Notes on R: A Programming Environment for Data Analysis and Graphics. http://cran.r-project.org/
13. Zou, G.Y., Huang, W., Zhang, X.: A Note on Confidence Interval Estimation for a Linear Function of Binomial Proportions. Computational Statistics and Data Analysis **53**, 1080–1085 (2009)
14. Zou, G.Y., Taleban, J., Huo, C.Y.: Confidence Interval Estimation for Lognormal Data with Application to Health Economics. Computational Statistics and Data Analysis **53**, 3755–3764 (2009)
15. Proschan, F.: Theoretical Explanation of Observed Decreasing Failure Rate. Technometrics **5**, 375–383 (1963)
16. Shao, J.: Mathematical Statistics. Springer-Verlag, New York (1999)

A Deterministic Clustering Framework in MMMs-Induced Fuzzy Co-clustering

Shunnya Oshio, Katsuhiro Honda$^{(\boxtimes)}$, Seiki Ubukata, and Akira Notsu

Osaka Prefecture University, Sakai, Osaka 599-8531, Japan
{honda,subukata,notsu}@cs.osakafu-u.ac.jp

Abstract. Although various FCM-type clustering models are utilized in many unsupervised classification tasks, they often suffer from bad initialization. The deterministic clustering approach is a practical procedure for utilizing a robust feature of very fuzzy partitions and tries to converge the iterative FCM process to a plausible solution by gradually decreasing the fuzziness degree. In this paper, a novel framework for implementing the deterministic annealing mechanism to fuzzy co-clustering is proposed. The advantages of the proposed framework against the conventional statistical co-clustering model are demonstrated through some numerical experiments.

Keywords: Fuzzy co-clustering · Deterministic annealing · Initialization problem

1 Introduction

While Fuzzy c-Means (FCM)-type clustering models [1,2] can effectively achieve unsupervised classification in simple iterative processes, they often suffer from the initialization problem, in which the algorithms may converge to inappropriate local solutions because of bad initialization. Deterministic annealing (DA) [3] is a possible way for avoiding local solutions in fuzzy clustering, in which a deterministic procedure of gradually degrading fuzziness degrees is implemented for utilizing the robust feature of very fuzzy partitions. In [3], the entropy-based fuzzification term [4] first plays a role for regularizing the k-Means objective function with a very large fuzification weight so that it brings a unique solution regardless random initialization, and then, the fuzzification weight is gradually degraded in order to find a plausible solution with the intended fuzzy degree.

In this paper, the DA scheme is implemented to a fuzzy co-clustering model. Fuzzy Clustering for Categorical Multivariate data (FCCM) [5] is an FCM variant, in which co-clusters of object-item pairs are extracted from cooccurrence information among objects and items. Co-cluster structures are represented by two different kinds of fuzzy memberships: *object memberships* and *item memberships*. In order to extract familiar object-item pairs, the FCM-type objective function is defined by the aggregation degree of objects and items instead of the within-cluster-error measure of FCM. So, the fuzzy co-clustering model is

V.-N. Huynh et al. (Eds.): IUKM 2015, LNAI 9376, pp. 204–213, 2015.
DOI: 10.1007/978-3-319-25135-6_20

a prototype-less clustering model. Considering the maximization principle of the aggregation measure, in [5], membership fuzzification was achieved by the entropy-based regularization approach [4]. Although the iterative algorithm has similar form to the conventional FCM, the dual fuzzification model has no comparative statistical models and often needs very careful tuning of two penalty weights.

Recently, a statistical model-induced fuzzy co-clustering model was proposed, which is induced from the statistical Multinomial Mixture Models (MMMs) [6]. Fuzzy Co-Clustering induced by MMMs (FCCMM) [7,8] modified the FCCM objective function by replacing the linear aggregation measure with a non-linear one induced by the pseudo-log-likelihood function of MMMs. In FCCMM, the object and item memberships are identified with the probability of each generative class for an object and the probability of appearance of an item in a class, respectively, and they are fuzzified based on different fuzzification principles.

In this paper, the DA scheme is utilized in both object and item membership fuzzification processes in the fuzzy co-clustering context. The fuzziness degree of object memberships are tuned by the K-L information-based regularization approach [9], and DA is implemented by decreasing the penalty weight for the K-L information term. On the other hand, the fuzziness degree of item memberships are tuned by a weighting exponent approach [10], and DA is implemented by decreasing the weight.

The remaining parts of this paper are organized as follows: Section 2 presents a brief review on the DA scheme in FCM clustering. In Section 3, the DA scheme is implemented to FCCMM. The characteristic features are demonstrated through a numerical experiment in Section 4 and a summary conclusion is given in Section 5.

2 FCM Clustering and Deterministic Annealing

Assume that we have n objects with their m-dimensional vector observations $\boldsymbol{x}_i$, $i = 1, \ldots, n$. In FCM clustering [1,2], the objects are partitioned into C fuzzy clusters, which are represented by prototypical centroids $\boldsymbol{b}_c$, $c = 1, \ldots, C$, in such a way that the within-cluster-errors are minimized. Cluster assignment of each object is represented by their fuzzy memberships u_{ci}, which indicate the degree of belongingness of object i to cluster c and is constrained such that $\sum_{c=1}^{C} u_{ci} = 1$. Besides the linear objective function of crisp k-Means clustering [11], membership fuzzification is performed by introducing non-linear nature with respect to u_{ci} in the FCM objective function.

Instead of the weighting exponent of the standard FCM [1], Miyamoto and Mukaidono [4] introduced the entropy-based regularization concept to fuzzification of k-Means objective function as follows:

$$L_{efcm} = \sum_{c=1}^{C} \sum_{i=1}^{n} u_{ci} \|\boldsymbol{x}_i - \boldsymbol{b}_c\|^2 + \lambda_u \sum_{c=1}^{C} \sum_{i=1}^{n} u_{ci} \log u_{ci}, \tag{1}$$

where λ_u tunes the degree of fuzziness of object memberships. The larger the λ_u, the fuzzier the object memberships.

Rose et al. [3] introduced a deterministic annealing (DA) concept to data clustering. Although the DA clustering model was purely constructed on a probabilistic framework, the updating formula is equivalent to that of the entropy-based FCM. In the DA clustering model, the fuzzification penalty λ_u is regarded as the temperature parameter and the FCM cost function is deterministically optimized at each temperature sequentially, starting at high temperature and going down. In the FCM context, it has been shown that the fuzzification penalty term can play a role for regularizing the k-Means objective function and a very fuzzy model with a huge λ_u often brings a unique solution with a smoothened cost function. For example, with a huge λ_u, all u_{ci} tend to have homogeneous values of $u_{ci} = 1/C$ [12], which is a unique solution regardless of random initializations.

3   DA-Based Fuzzy Co-clustering

3.1   FCCM and Its Connection with Statistical Co-clustering Model

Assume that we have cooccurence information among n objects and m items with their frequencies of co-appearance r_{ij} such that the information is summarized in an $n \times m$ cooccurrence matrix $R = \{r_{ij}\}$. FCCM [5] is an FCM-type co-clustering model, in which fuzzy co-clusters of mutually familiar object-item pairs are extracted by estimating two different fuzzy memberships of u_{ci} for object i and w_{cj} for item j to cluster c. The aggregation quality of C co-clusters are measured by the aggregation degree $\sum_{c=1}^{C} \sum_{i=1}^{n} \sum_{j=1}^{m} u_{ci} w_{cj} r_{ij}$ to be maximized instead of the within-cluster-errors in FCM, so that familiar object-item pairs having a large r_{ij} take large memberships in a same cluster. Then, adopting the entropy-based fuzzification penalties, the FCCM objective function to be maximized was defined as:

$$L_{fccm} = \sum_{c=1}^{C} \sum_{i=1}^{n} \sum_{j=1}^{m} u_{ci} w_{cj} r_{ij} - \lambda_u \sum_{c=1}^{C} \sum_{i=1}^{n} u_{ci} \log u_{ci} - \lambda_w \sum_{c=1}^{C} \sum_{j=1}^{m} w_{cj} \log w_{cj}, \quad (2)$$

where λ_u and λ_w are the independent fuzzification weight for object and item memberships, respectively.

Avoiding a trivial solution, u_{ci} and w_{cj} are estimated under different constraints: $\sum_{c=1}^{C} u_{ci} = 1$ for u_{ci} and $\sum_{j=1}^{m} w_{cj} = 1$ for w_{cj}. These constraints brings a statistical interpretation of co-cluster estimation such that u_{ci} represents the probability of the c-th generative distribution for object i while w_{cj} is identified with the occurrence probability of item j in the c-th generative distribution.

The statistical interpretation of FCCM implies a clustering interpretation on the basic statistical co-clustering model of MMMs, whose pseudo-log-likelihood function is given as:

$$L_{mmms} = \sum_{c=1}^{C} \sum_{i=1}^{n} \sum_{j=1}^{m} u_{ci} r_{ij} \log w_{cj} + \sum_{c=1}^{C} \sum_{i=1}^{n} u_{ci} \log \frac{\alpha_c}{u_{ci}}, \quad (3)$$

where α_c is the a priori probability of the c-th generative distribution. From the object clustering view point, the pseudo-log-likelihood function can be divided into two components. The first term $\sum_{c=1}^{C} \sum_{i=1}^{n} \sum_{j=1}^{m} u_{ci} r_{ij} \log w_{cj}$ is the aggregation measure among objects and items in each cluster, which is a linear function with respect to u_{ci}. The second term of K-L information $\sum_{c=1}^{C} \sum_{i=1}^{n} u_{ci} \log \frac{\alpha_c}{u_{ci}}$ is identified with the regularization term for achieving soft partition [9]. On the other hand, from the item clustering view point, the pseudo-log-likelihood function has no fuzzification penalty term and the non-linearity with respect to w_{cj} is achieved by log function in the first aggregation term.

3.2 Fuzzy Co-clustering Induced by MMMs and Deterministic Annealing

Supported by the statistical interpretation of FCCM, a fuzzy counterpart of MMMs was proposed, in which the FCCM objective function was modified by extending the pseudo-log-likelihood function of MMMs. Based on the K-L information regularization concept [9], it is expected that the degree of fuzziness of object partition can be arbitrary adjusted by tuning the effect of the fuzzification penalty, i.e., we can adopt an adjustable weight on the K-L information term. On the other hand, the degree of fuzziness of item partition is expected to be adjustable by tuning the non-linear degree of the log-based term. Following these considerations, Fuzzy Co-Clustering induced by MMMs (FCCMM) [8] introduced the objective function as:

$$
L_{fccmm} = \sum_{c=1}^{C} \sum_{i=1}^{n} \sum_{j=1}^{m} \frac{1}{\lambda_w} u_{ci} r_{ij} \left((w_{cj})^{\lambda_w} - 1 \right)
$$

$$
+ \lambda_u \sum_{c=1}^{C} \sum_{i=1}^{n} u_{ci} \log \frac{\alpha_c}{u_{ci}}, \tag{4}
$$

where λ_u and λ_w are the fuzzification penalty weight for object and item memberships, respectively. The K-L information-based penalty term is a direct extension of the MMMs-based soft object partition. The larger the λ_u, the fuzzier the object partition.

For item membership fuzzification, $\frac{1}{\lambda_w} \left((w_{cj})^{\lambda_w} - 1 \right)$ was adopted instead of $\log w_{cj}$, where

$$
\log w_{cj} \sim \lim_{\lambda_w \to 0} \frac{1}{\lambda_w} \left((w_{cj})^t - 1 \right). \tag{5}
$$

When $\lambda_w = 1$, $\frac{1}{\lambda_w} \left((w_{cj})^{\lambda_w} - 1 \right)$ is reduced to a linear function with respect to w_{cj} and we have crisp memberships only. As λ_w is smaller, $\frac{1}{\lambda_w} \left((w_{cj})^{\lambda_w} - 1 \right)$ becomes much more non-linear and we have fuzzier item memberships [10]. When $0 < \lambda_w < 1$, the fuzziness degree is weaker then MMMs. On the other hand, when $\lambda_w < 0$, we have much fuzzier item partitions than MMMs.

In [8], it was demonstrated that the frequency of the best object partition becomes larger as the fuzziness degrees of object memberships are larger while too much larger or smaller penalty weights cause computational instabilities. On the other hand, tuning of item fuzziness degree λ_w brought inconsistent effects on object and item partitions. As is expected, a smaller λ_w gave a fuzzier item memberships but caused a crisper object partition. It may be because a fuzzier item memberships can effectively conceal noise and can contribute to fine interpretation of object memberships. Then, in this paper, two different DA strategies are adopted to object and item membership fuzzification mechanisms with the goal of reducing the inappropriate influences of initialization problems.

First, for object membership fuzzification, a DA process starts from slightly fuzzier situation than the intended fuzziness degrees and is degraded until the model is reduced to the intended one. In general simulated annealing approaches, a practical way for decreasing the temperature parameter T_k with iteration index k is

$$T_{k+1} = \gamma T_k \ (0.8 \le \gamma < 1), \tag{6}$$

where γ is the depletion rate. Based on the same concept, the fuzzification parameters are adjusted. Because the object membership fuzzifier λ_u is directly identifiable with the temperature parameter of the conventional DA clustering model, it can be degraded as:

$$\lambda_{u,k+1} = \min\{\gamma_u \lambda_{u,k}, \lambda_u^{min}\}, \tag{7}$$

where $0 < \gamma_u < 1$, and $\lambda_u = 1$ corresponds to MMMs.

Second, tuning of item membership fuzziness is performed based on a different concept. This paper focusses on the stability of object partitions and adopts a DA framework for reducing effects of bad initializations from the object clustering view point. As demonstrated in the previous work [8], a crisper item membership can contribute to reducing bad effects of inappropriate initializations. Because item partition becomes fuzzier as λ_w is smaller, λ_w should be decreased from a slightly larger value to the intended one.

Besides the case of λ_u, it should be noted that λ_w can take both positive and negative values while the above annealing schedules are designed only for positive ones.

A possible way for tuning λ_w among $[\lambda_w^{min}, \lambda_w^{max}]$ is:

$$\lambda_{w,k+1} = \min\{\gamma_w(\lambda_{w,k} - 2 \times \lambda_w^{min} + \lambda_w^{max}) + 2 \times \lambda_w^{min} - \lambda_w^{max}, \lambda_w^{min}\}, \tag{8}$$

so that $T_{k+1} = \gamma^k \times T^{max}$ in the interval $[0, T^{max}]$ is virtually realized in $[\lambda_w^{min} - (\lambda_w^{max} - \lambda_w^{min}), \lambda_w^{max}]$ with the center λ_w^{min}.

Then, a sample procedure of the proposed algorithm is written as follows:

Algorithm: Fuzzy Co-Clustering induced by Multinomial Mixture models with Deterministic Annealing (FCCMM-DA)

Step 1. Initialize fuzzy memberships u_{ci} and w_{cj} such that they satisfy $\sum_{c=1}^{C} u_{ci} = 1, \forall i$ and $\sum_{j=1}^{m} w_{cj} = 1, \forall c$. Choose the possible interval of fuzziness penalty weights $\lambda_u \in [\lambda_u^{min}, \lambda_u^{max}]$ and $\lambda_w \in [\lambda_w^{min}, \lambda_w^{max}]$, and termination criterion ε. Let the initial penalties be $\lambda_u = \lambda_u^{min}$ and $\lambda_w = \lambda_w^{max}$.

Step 2. Update cluster volumes α_c, $c = 1, \ldots, C$ by

$$\alpha_c = \frac{1}{n} \sum_{i=1}^{n} u_{ci}. \tag{9}$$

Step 3. Update w_{cj}, $c = 1, \ldots, C$, $j = 1, \ldots, m$ by

$$w_{cj} = \left(\sum_{\ell=1}^{m} \left(\frac{\sum_{i=1}^{n} r_{ij} u_{ci}}{\sum_{i=1}^{n} r_{i\ell} u_{ci}} \right)^{\frac{1}{\lambda_w - 1}} \right)^{-1}. \tag{10}$$

Step 4. Update u_{ci}, $c = 1, \ldots, C$, $i = 1, \ldots, n$ by followings:
For $\lambda_w \neq 0$,

$$u_{ci} = \frac{\alpha_c \exp\left(\frac{1}{\lambda_u \lambda_w} \sum_{j=1}^{m} r_{ij} (w_{cj})^{\lambda_w} \right)}{\sum_{\ell=1}^{C} \alpha_\ell \exp\left(\frac{1}{\lambda_u \lambda_w} \sum_{j=1}^{m} r_{ij} (w_{\ell j})^{\lambda_w} \right)}. \tag{11}$$

For $\lambda_w = 0$,

$$u_{ci} = \frac{\alpha_c \prod_{j=1}^{m} (w_{cj})^{r_{ij}/\lambda_u}}{\sum_{\ell=1}^{C} \alpha_\ell \prod_{j=1}^{m} (w_{\ell j})^{r_{ij}/\lambda_u}}. \tag{12}$$

Step 5. Update λ_u and λ_w by Eqs.(7) and (8)
Step 6. If $\max_{c,i} | u_{ci}^{NEW} - u_{ci}^{OLD} | < \varepsilon$,
   then stop. Otherwise, return to Step 2.

4   Numerical Experiments

In this section, the characteristic features of the proposed method are demonstrated in a numerical experiment. A noisy 100×60 artificial cooccurrence matrix R shown in Fig. 1-(b) was used in [7,8], which was generated from a noise-less R_0 with 100 objects ($n = 100$) and 60 items ($m = 60$) shown in Fig. 1-(a). $R_0 = \{r_{ij}^0\}$ and $R = \{r_{ij}\}$ are the base matrix without noise and its noisy variant, whose elements are depicted by black and white cells as $r_{ij} = 1$ and $r_{ij} = 0$, respectively. The noisy matrix R, which includes roughly 4 co-clusters ($C = 4$) in diagonal blocks while some items are shared by multiple clusters, was generated from R_0 by replacing $r_{ij}^0 = 1$ with $r_{ij} = 0$ at a rate of 50% and $r_{ij}^0 = 0$ elements with $r_{ij} = 1$ at a rate of 10%.

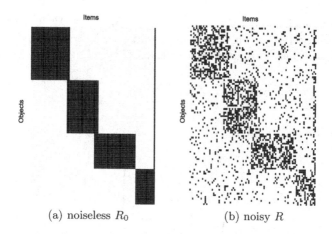

(a) noiseless R_0 (b) noisy R

Fig. 1. Artificial coocurrance matrices [7,8]

The initialization sensitivity of the FCCMM algorithm is compared with and without annealing mechanisms, where the initial item membership vectors $w_c = (w_{c1}, \ldots, w_{cm})^\top$ of $C = 4$ clusters were constructed from normalized cooccurrence information vectors $r_i = (r_{i1}, \ldots, r_{im})^\top$ of 4 objects such that $w_c = (r_{i1}^*, \ldots, r_{im}^*)^\top$ and $\sum_j r_{ij}^* = 1$. The FCCMM algorithm with various penalty weight values was applied to the noisy cooccurrence matrix R with initial partitions given by 210 different 4-objects combinations constructed from 10 pre-selected objects, i.e., all trials started from common 210 initialization candidates for fair comparisons. The partition quality is compared with Rand Index (RI) of maximum membership partitions, where RI_u implies the ratio of matching with the ideal *object* partition of Fig. 1-(a) after maximum membership *object* partition.

In the previous research [8], without the DA scheme, the frequencies of the clustering results with $RI_u > 0.9$ *object* c-partition was reported as Table 1, where '—' means that the algorithm cannot work because of overflow with too fuzzy or too crisp penalty settings. The initialization sensitivity was reduced with larger λ_u and λ_w, i.e., a fuzzier object partition and a crisper item partition can contribute to stable co-clustering.

Table 1. Comparison of initialization sensitivity without DA in artificial data: the frequencies of $RI_u > 0.9$ in 210 different trials [8]

		object penalty λ_u		
		0.5	1.0	2.0
item	0.3	185 (88%)	202 (96%)	—
penalty	0.0	40 (19%)	108 (51%)	171 (81%)
λ_w	-0.3	—	0 (0%)	21 (10%)

Table 2. Comparison of initialization sensitivity with object fuzziness annealing in artificial data: the frequencies of $RI_u > 0.9$ in 210 different trials

		object penalty λ_u		
	λ_u^{max}	1.0	2.0	3.0
	$\downarrow$	$\downarrow$	$\downarrow$	$\downarrow$
	λ_u^{min}	0.5	1.0	2.0
item	0.3	200 (95%)	207 (99%)	—
penalty	0.0	86 (41%)	161 (77%)	198 (94%)
λ_w	-0.3	—	21 (10%)	29 (14%)

Table 3. Comparison of initialization sensitivity with item fuzziness annealing in artificial data: the frequencies of $RI_u > 0.9$ in 210 different trials

		object penalty λ_u		
	$\lambda_w^{max} \rightarrow \lambda_w^{min}$	0.5	1.0	2.0
item	0.5 → 0.3	196 (93%)	208 (99%)	—
penalty	0.3 → 0.0	182 (87%)	196 (93%)	207 (99%)
λ_w	0.0 → -0.3	—	51 (24%)	90 (43%)

In the following, the proposed annealing scheme is introduced with the goal of achieving the stable co-clustering features by gradual tuning of fuzziness degrees. First, annealing of object membership fuzziness is considered with fixed item fuzziness, where the fuzziness penalty of object partition is reduced as:

$$\lambda_u = \max\{\lambda_u^{min}, 0.99^k \lambda_u^{max}\}, \tag{13}$$

where k is the iteration index, and the final value was always be guaranteed as $\lambda_u = \lambda_u^{min}$ in this experiment for comparison purposes.

Table 2 shows that the initialization sensitivity of the FCCMM algorithm was efficiently reduced by introducing annealing of object fuzziness degrees and higher quality was achieved with smaller fuzziness degrees compared with Table 1.

Second, annealing of item membership fuzziness is considered with fixed object fuzziness, where the fuzziness penalty of item partition is tuned as:

$$\lambda_w = \max\{\lambda_w^{min}, 0.99^t \times 2(\lambda_w^{max} - \lambda_w^{min}) + \lambda_w^{min} - (\lambda_w^{max} - \lambda_w^{min})\}. \tag{14}$$

Here, λ_w was replaced with $\lambda_w \rightarrow 0$ in case of $|\lambda_w| < 0.05$ for avoiding computational overflow. The final value was always be guaranteed as $\lambda_w = \lambda_w^{min}$ in this experiment for comparison purposes.

Table 3 shows again that the initialization sensitivity of the FCCMM algorithm was efficiently reduced by introducing annealing of item fuzziness degrees and higher quality was achieved with higher fuzziness degrees compared with Table 1.

5 Conclusion

In this paper, a novel framework for introducing DA schemes to the MMMs-induced fuzzy co-clustering algorithm was proposed by tuning the degree of fuzziness of object and item memberships. With the goal of reducing the effects of bad initialization in object partitions, the degree of object partition fuzziness was decreased with a deterministic process, which is based on the similar concept to the conventional DA-based FCM model. On the other hand, tuning of item membership fuzziness was implemented in a different strategy, which is designed for utilizing a side effect of item membership fuzzification, i.e., crisper item memberships can contribute to stable object partition. The experimental results demonstrated that the both proposed DA schemes can work for deriving appropriate solutions more often than the conventional model without DA schemes.

A possible future work includes the development of a better design of annealing schedules for achieving more effective operation of the DA framework. Another direction of future study is to investigate the influences of the DA schemes on the interpretability of co-cluster solutions especially from the item fuzziness tuning view point.

Acknowledgments. This work was supported by the Ministry of Education, Culture, Sports, Science and Technology, Japan, under Grant-in-Aid for Scientific Research (#26330281).

References

1. Bezdek, J.C.: Pattern Recognition with Fuzzy Objective Function Algorithms. Plenum Press (1981)
2. Miyamoto, S., Ichihashi, H., Honda, K.: Algorithms for Fuzzy Clustering. Springer (2008)
3. Rose, K., Gurewitz, E., Fox, G.: A deterministic annealing approach to clustering. Pattern Recognition Letters **11**, 589–594 (1990)
4. Miyamoto, S., Mukaidono, M.: Fuzzy c-means as a regularization and maximum entropy approach. In: Proc. of the 7th International Fuzzy Systems Association World Congress, vol. 2, pp. 86–92 (1997)
5. Oh, C.-H., Honda, K., Ichihashi, H.: Fuzzy clustering for categorical multivariate data. In: Proc. of Joint 9th IFSA World Congress and 20th NAFIPS International Conference, pp. 2154–2159 (2001)
6. Rigouste, L., Cappé, O., Yvon, F.: Inference and evaluation of the multinomial mixture model for text clustering. Information Processing and Management **43**(5), 1260–1280 (2007)
7. Honda, K., Oshio, S., Notsu, A.: FCM-type fuzzy co-clustering by K-L information regularization. In: Proc. of 2014 IEEE International Conference on Fuzzy Systems, pp. 2505–2510 (2014)
8. Honda, K., Oshio, S., Notsu, A.: Fuzzy co-clustering induced by multinomial mixture models. Journal of Advanced Computational Intelligence and Intelligent Informatics **19** (to appear, 2015)

9. Honda, K., Ichihashi, H.: Regularized linear fuzzy clustering and probabilistic PCA mixture models. IEEE Transactions on Fuzzy Systems **13**(4), 508–516 (2005)

10. Honda, K., Oshio, S., Notsu, A.: Item membership fuzzification in fuzzy co-clustering based on multinomial mixture concept. In: Proc. of 2014 IEEE International Conference on Granular Computing, pp. 94–99 (2014)

11. MacQueen, J. B.: Some methods of classification and analysis of multivariate observations. In: Proc. of 5th Berkeley Symposium on Math. Stat. and Prob., pp. 281–297 (1967)

12. Liu, Z.-Q., Miyamoto, S. (eds.): Soft Computing and Human-Centered Machines. Springer-Verlag (2000)

FCM-Type Co-clustering Transfer Reinforcement Learning for Non-Markov Processes

Akira Notsu$^{(\boxtimes)}$, Takanori Ueno, Yuichi Hattori, Seiki Ubukata, and Katsuhiro Honda

Graduate School of Engineering, Osaka Prefecture University,
Gakuen 1-1, Naka, Sakai, Osaka 599-8531, Japan
notsu@cs.osakafu-u.ac.jp
http://www.cs.osakafu-u.ac.jp/hi/

Abstract. In applying reinforcement learning to continuous space problems, discretization or redefinition of the learning space can be a promising approach. Several methods and algorithms have been introduced to learning agents to respond to this problem. In our previous study, we introduced an FCCM clustering technique into Q-learning (called QL-FCCM) and its transfer learning in the Markov process. Since we could not respond to complicated environments like a non-Markov process, in this study, we propose a method in which an agent updates his Q-table by changing the trade-off ratio, Q-learning and QL-FCCM, based on the damping ratio. We conducted numerical experiments of the single pendulum standing problem and our model resulted in a smooth learning process.

Keywords: FCM-type Co-clustering · Reinforcement learning · Transfer learning

1   Introduction

Reinforcement learning (RL) is learning by interacting with an environment. An RL agent learns from the consequences of his actions, rather than from being explicitly taught, and selects his actions based on his past experiences and by new choices, which is essentially trial and error learning. RL presupposes that agents can visit a finite number of states, and when visiting a state, a numerical reward will be collected [1]. Each state has a changeable value attached to it. From every state, subsequent states can be reached by actions. The value of a given state is defined by the averaged future reward that can be accumulated by selecting actions from this particular state.

When considering the problems in the Markov decision process with a limited state and a limited action, RL can be achieved by identifying the best actions that lead to the greatest reward. Actions selected by the agent affect not only direct reward but also subsequent states. The following are the two main RL

© Springer International Publishing Switzerland 2015
V.-N. Huynh et al. (Eds.): IUKM 2015, LNAI 9376, pp. 214–225, 2015.
DOI: 10.1007/978-3-319-25135-6_21

problems: many learnings are required and performing enough reliable learnings is very time-consuming [2][3]. Various RL methods have been proposed to solve these problems. For example, neural networks [4] and fuzzy rules extracted from data [5] are employed by RLs to reduce the hours spent in learning. We also proposed an automatic adaptive space segmentation algorithm [6] and simple reinforcement learning [7] for small-memory agents.

In our previous study [8], we improved learning speed by applying clustering to the Q-table of the learning agent and classified his states and actions into several classes. We proposed a learning method (QL-FCCM) that takes membership values into account to simplify learning and obtained agent's knowledge representation to reuse and analyze the results. In this method, even though we improved the learning speed in a simple problem, we could not respond to complicated environments. Therefore, in this study, we propose a method in which an agent updates his Q-table by changing the use ratio of two methods, Q-learning and QL-FCCM, and address the learning speed in a non-Markov process.

2   Fundamental Theory

2.1   Co-clustering

FCM-Type Co-clustering. When considering whether a certain individual belongs to a cluster, the individual does not completely belong to just one cluster in the real world; sometimes one individual belongs to two or more clusters. Fuzzy c-Means (FCM)[9], which expresses the degree of affiliation to a cluster, simultaneously allows data to belong to two or more clusters. The division situation is denoted by the degree of affiliation.

Fuzzy clustering for categorical multivariate data (FCCM) [10] is an FCM-type co-clustering model whose goal is to extract co-clusters of objects and items from co-occurrence matrices. In FCCM, two different types of fuzzy memberships are used for objects and items. The membership of objects is derived in a similar concept to the conventional FCM in which each object is forced to belong to a solo cluster based on an exclusive condition called the row-sum condition: the sum of memberships with respect to clusters is 1. On the other hand, the membership of the items is given with the column-sum condition, where the sum of memberships with respect to the items in each cluster must be 1. The membership of items only evaluates their mutual responsibility in each cluster and is not necessarily applicable for revealing to which cluster the item belongs because items can be shared by multiple clusters or ejected from all clusters.

We consider an approach that applies exclusive features to the memberships of both items and objects in FCM-type co-clustering. To force all objects to belong to a solo cluster, we introduce a penalty for avoiding items not to be shared, where the aggregation degree of each cluster is maximized with items that are exclusively assigned. The idea of adding a penalty to cluster sharing is based on sequential fuzzy cluster extraction from object-type data in which compact clusters are sequentially extracted one by one, and a penalty is considered so that objects cannot be assigned to multiple clusters. Although the penalty was used in a sequential procedure [11], the model applies it to a batch iteration process.

FCM-Type Co-clustering with Exclusive Partition. By extending FCMfs fuzzy clustering model, Honda et al. proposed fuzzy clustering for FCCM with exclusive partition in which the objective function is defined by considering the aggregation degree of each cluster [12].

A co-clustering model with exclusive partition was developed by introducing to FCCM the penalization concept of sequential cluster extraction. To avoid multiple cluster assignments, the sequential co-clustering model estimated fuzzy membership w_{cj} $(c > 1)$ by minimizing penalty weight $w_{cj} \cdot w_{tj}$ $(t = 1, \ldots, c-1)$. $w_c = (w_{c1}, \ldots, w_{cm})^\top$ is obtained by solving an eigenvalue problem of a full-adjacency matrix in which $\sum_{l=1}^{c-1} \beta_t w_{tj}^2$ is subtracted from the diagonal elements [13]. In FCCM with an exclusive partition model, the following similar penalization scheme is introduced into its objective function:

$$
\begin{aligned}
L_{fccm} = & \sum_{c=1}^{C} \sum_{i=1}^{n} \sum_{j=1}^{m} u_{ci} w_{cj} r_{ij} \\
& - \beta \sum_{c=1}^{C} \sum_{j=1}^{m} \sum_{t \neq c} w_{cj} \cdot w_{tj}^* \\
& - \lambda_u \sum_{c=1}^{C} \sum_{i=1}^{n} u_{ci} \log u_{ci} \\
& - \lambda_w \sum_{c=1}^{C} \sum_{j=1}^{m} w_{cj} \log w_{cj}.
\end{aligned}
\tag{1}
$$

The second term is the penalty for avoiding multiple cluster assignments and weight β tunes the degree of exclusive features. If item j already belongs to another cluster and w_{tj}^* $(t \neq c)$ is large, w_{cj} becomes small. w_{tj}^*, which is the current value of w_{tj}, is temporally fixed in the FCM-type iterative algorithm. In a similar manner to relational fuzzy c-means (RFCM) [14], where the objective function includes the dot products of memberships and part of the memberships are temporally fixed, the updating rule for w_{cj} is given by temporally fixed w_{tj}^*.

The updating rules for the two types of memberships are based on the partial optimality of the above objective function. u_{ci} is updated in the same formula as the conventional FCCM model:

$$
u_{ci} = \frac{\exp \left(\lambda_u^{-1} \sum_{j=1}^{m} w_{cj} r_{ij} \right)}{\sum_{\ell=1}^{C} \exp \left(\lambda_u^{-1} \sum_{j=1}^{m} w_{\ell j} r_{ij} \right)}.
\tag{2}
$$

On the other hand, w_{cj} is updated by considering the penalty:

$$w_{cj} = \frac{\exp\left(\lambda_w^{-1}\left(\sum_{i=1}^n u_{ci}r_{ij} - \beta \sum_{t \neq c} w_{tj}^*\right)\right)}{\sum_{\ell=1}^m \exp\left(\lambda_w^{-1}\left(\sum_{i=1}^n u_{ci}r_{i\ell} - \beta \sum_{t \neq c} w_{t\ell}^*\right)\right)}. \tag{3}$$

An iterative algorithm is repeated until it becomes convergent. Weight β is first set to '0' and is slowly increased so that the cluster partition gradually transits from a conventional to an exclusive one. This soft transition approach clarifies the co-cluster structure of the conventional FCCM by gradually forcing item memberships to be exclusive [14] [9].

2.2 Q-Learning

Q-learning is a reinforcement learning technique that works by learning an action-value function that gives the expected utility of action a in given state s and following a fixed policy [15][16]. One Q-learning strength is that it can compare the expected utility Q-value of the available actions without requiring a model of the environment.

In the learning, the agent selects action a, and next state s_{t+1} of the agent followed after taking action a is observed, and reward r is received by the agent after the action. The expected utility Q-value is iteratively updated by them. For each state s from state set S and for each action a from action set A, we update the Q-value that depends on current state s of the agent. Q-learning requires exploratory actions to adequately sample all of the available state transitions. We selected action a and applied it to state s by Boltzmann distribution [17]:

$$p(a_t|s) = \frac{e^{Q(s,a_t)/T}}{\sum_{a \in A} e^{Q(s,a)/T}}, \tag{4}$$

where T is the computational temperature.

An update is calculated with the following expression [17]:

$$\begin{aligned} Q(s_t, a) \leftarrow \quad & Q(s_t, a) \\ +\alpha\Big[r_{t+1} + \gamma \max_{a' \in A(s_{t+1})} & Q(s_{t+1}, a') - Q(s_t, a)\Big], \end{aligned} \tag{5}$$

where α is the learning rate and γ is the discount factor. Once Q-learning is finished, the optimal policy and the optimal value function were found without continuously updating the policy during learning.

2.3 QL-FCCM

Objects and items in FCCM are regarded as states and actions in Q-learning. FCCM with exclusive partition is applied to the Q-table of the learning agent who learns based on the FCCM result [8]. We proposed an updated equation using the memberships of states and actions. We verified the learning performance in our proposed method by comparing it to Q-learning.

Q-Value Updates of Learning Process. In Q-learning, the agent selects the action, gets a reward, and updates the Q-table. Therefore, only one Q-value of a state at which the agent arrived is updated. In QL-FCCM, the Q-values are updated in all states (except for the next state) with the equation based on the membership value each time the agent acts. The updated equation in QL-FCCM is shown:

Table 1. Algorithm

Algorithm: QL-FCCM
1: Input discount rate γ ,learning rate α
2: Initialize Q(s,a) arbitrarily, for all s,a
3: Loop (for each episode)
4:   Initialize s
5:   Repeat (for each episode)
6:     Choose a from using policy derived from Q
7:     Take action a, observe return r, and next state s_{t+1}
8:     For all s $(s \neq s_{t+1})$
9:       If $s = s_t$
10:        $Q\left(s_t, a\right) \leftarrow Q\left(s_t, a\right) + \frac{\alpha}{n_s}\left[r + \gamma \max_{a'} Q\left(s_{t+1}, a'\right) - Q\left(s_t, a\right)\right]$
11:       Else
12:        $Q\left(s, a\right) \leftarrow Q\left(s, a\right) + \frac{\alpha}{n_s}\left[\max_{c' \in C} w\left(a, c'\right) \sum_{c=1}^{C} u_{cs} u_{cs_t}\right]$ $\times \left[r + \gamma \max_{a'} Q\left(s_{t+1}, a'\right) - Q\left(s, a\right)\right]$
13:     End if
14:   End for
15:   Until s is terminal
16: End loop

Update Equation Based on Membership

$$Q\left(s, a\right) \leftarrow Q\left(s, a\right) + \frac{\alpha}{n_s}\left[\max_{c' \in C} w\left(a, c'\right) \sum_{c=1}^{C} u_{cs} u_{cs_t}\right]$$
$$\times \left[r + \gamma \max_{a' \in A(s_{t+1})} Q\left(s_{t+1}, a'\right) - Q\left(s, a\right)\right], \tag{6}$$

$$\left(s \in S - s_{t+1}\right)$$

Where C is the number of clusters and u and w are the membership values of the states and actions. s are all the states (except for the subsequent state). In the following experiment, compared with Q-learning, we divided the update amount by the total number of states n_s to set the following: the amount of updates in Q-learning $\geq$ amount of updates in QL-FCCM. The learning rate is changed by the membership value.

3   Proposed Method

In our proposed method, an agent updates his Q-table by changing the ratio of two methods: Q-learning and QL-FCCM. The updated equation in this method is shown below:

$$Q\left(s,a\right) \leftarrow \quad Q\left(s,a\right) + \alpha\Big[\textstyle\sum_{c=1}^{C} u_{cs}u_{cs_t}\Big]$$
$$\times \Big[r_{t+1} + \gamma \max_{a' \in A(s_{t+1})} Q\left(s_{t+1}, a'\right) - Q\left(s,a\right)\Big], \tag{7}$$

where $s \in S$, $s \neq s_t$.

C indicates the number of clusters, u indicates the value of the membership of the states and α indicates the learning rate. Though we set α's initial value to 1, we decrease its value in the exponential function by increasing the number of episodes. From this, although the agent can update his Q-table in QL-FCCM in an early stage, in the closing stages he can update it in Q-learning. When α is 0, the agent only updates his Q-table in the Q-learning. In this study, α is multiplied by 0.99 by increasing the number of episodes, and α is set to 0 since the number of episodes reaches 1000. In Fig. 1, we show the changes of α's value.

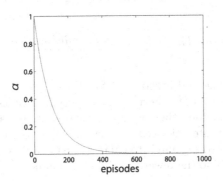

Fig. 1. Changes of α's value

4   Transfer Learning Simulation

4.1   Single Pendulum Standing Problem

We simulated the single pendulum standing problem (Fig. 2) twice. Each simulation has slightly different parameters. At the second trial, agents can use the clustering data about the environment extracted from the first trial data.

The pendulum is connected to a truck by a link, and force can only be added to the truck from the left and right. Our goal is to balance the pendulum on the top by adding force to the truck.

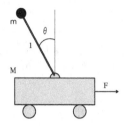

Fig. 2. Single pendulum simulation

The mass of the truck is $M[kg]$, the length of the link is $l[m]$, the weightfs mass is $m[kg]$, and the mass of the link is $0[kg]$. The angle of the pendulum is $\theta[rad]$, the angular velocity is $\dot\theta[rad/s]$, and the angular acceleration is $\ddot\theta[rad/s^2]$. In later experiments, we set $M = 1[kg]$, $m = 0.2[kg]$, $l = 1[m]$, and gravitational acceleration $g = 9.8[m/s^2]$. The space of the state is defined as $\theta(-\pi < \theta \leq \pi)$ and $\dot\theta(-3\pi \leq \dot\theta \leq 3\pi)$. The action is defined as the force on the truck: $F(-20 \leq F \leq 20)$.

To decide the state and action, the angle, the angular velocity, and the force on the truck are divided into 20 pieces. This physical calculation is performed by the Runge-Kutta 4th order method. The following motion equation is simulated in increments of 0.05 seconds:

$$\ddot\theta = \frac{(m + M)g \sin\theta - F \cos\theta - ml\dot\theta^2 \sin\theta \cos\theta}{l(M + m \sin^2\theta)}. \tag{8}$$

The Runge-Kutta method is shown as follows.

Agents obtain rewards of the current state every four times the motion equation is calculated. Therefore, they obtain rewards every 0.2 seconds. Moreover, in the next 0.2-second interval, agents select the force to add to the truck and update Q-table. The learning aim is to balance the pendulum on the truckfs top. To achieve this, we define the reward in time t as follows:

$$r_t = \pi - |\theta|. \tag{9}$$

This equation means that the nearer the pendulum is to the top ($\theta = 0$), the more reward the agents obtain. The initial state is $\theta = 0$, $\dot\theta = 0$ (the pendulum is on the top), and one episode is 60 seconds. One simulation consists of 5000 episodes, and we calculated the average of ten simulations. In this experiment, $\alpha = 0.1$, discount factor $\gamma = 0.9$, and agents' action choice method was ε-greedy ($\varepsilon = 0.2$).

Procedure of Runge-Kutta method
Begin:
Initialize $\frac{d^2x}{dt^2} = F(x,v,t)$, $v = \frac{dx}{dt}$;
Set initial state x_0, v_0, t_0;
For cycle := 1 to MAXCYCLE do
$k_1 = \Delta t \times v_0$;
$m_1 = \Delta t \times F(x_0, v_0, t_0)$;
$k_2 = \Delta t \times (v_0 + \frac{m_1}{2})$;
$m_2 = \Delta t \times F(x_0 + \frac{k_1}{2}, v_0 + \frac{m_1}{2}, t_0 + \frac{\Delta t}{2})$;
$k_3 = \Delta t \times (v_0 + \frac{m_2}{2})$;
$m_3 = \Delta t \times F(x_0 + \frac{k_2}{2}, v_0 + \frac{m_2}{2}, t_0 + \frac{\Delta t}{2})$;
$k_4 = \Delta t \times (v_0 + m_3)$;
$m_4 = \Delta t \times F(x_0 + k_3, v_0 + m_3, t_0 + \Delta t)$;
$k = \frac{k_1 + 2k_2 + 2k_3 + k_4}{6}$;
$m = \frac{m_1 + 2m_2 + 2m_3 + m_4}{6}$;
$x_0 = x_0 + k$;
$v_0 = v_0 + m$;
$t_0 = t_0 + dt$;
End
End

4.2 Evaluation of Partition Quality of Clustering

To evaluate the partition quality of the clustering, PC is applied to FCCM. PC, which can be used for measuring the degree of the crispness of memberships, increases as a good split is conducted [9][18]. PC is shown below:

$$PC_u = \frac{1}{n} \sum_{c=1}^{C} \sum_{i=1}^{n} u_{ci}^2, \tag{10}$$

$$PC_w = \frac{1}{C} \sum_{c=1}^{C} \sum_{j=1}^{m} w_{cj}^2, \tag{11}$$

$$PC_{CO} = PC_u \times PC_w. \tag{12}$$

In this study, we calculated PC when we set the learning time to 60 seconds and the number of episodes to 1000. The PC value is shown below.

The larger the value of PC_{CO} is, the higher we evaluate the partition quality of clustering. Therefore we set the number of clusters C to 6.

5 Simulation Result

First, we simulated the Q-learning agent during a single pendulum problem with 5000 episodes, where $M = 1[kg]$, $m = 0.2[kg]$, and $l = 1[m]$. We applied FCCM to the Q-table obtained by agents in that simulation and simulated QL-FCCM

Table 2. Evaluating partition quality of clustering

Number of clusterings	PC_u	PC_w	PC_{co}
2	0.96	0.49	0.47
3	0.94	0.42	0.40
4	0.92	0.50	0.46
5	0.91	0.50	0.45
6	0.92	0.54	**0.49**
7	0.92	0.43	0.39
8	0.90	0.42	0.38
9	0.90	0.27	0.24
10	0.91	0.26	0.24

and Q-learning agents at similar situations, where $M = 1[kg]$, $m = 0.5[kg]$, and $l = 1[m]$. We show the simulation results in Fig. 3.

The horizontal axis shows the number of episodes, and the vertical axis shows the average rewards in ten simulations.

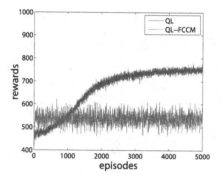

Fig. 3. QL-FCCM

Figure 3 shows that the QL-FCCM agents were not able to obtain higher rewards than the QL agents. Updating by membership adversely affected the learning. Since the QL-FCCM agents learned roughly based on the clustering data, they sometimes failed to anticipate well the appropriate values of the actions in such different situations.

Referring to this result, for our next simulation, we applied QL-FCCM in the early learning stage and subsequently applied Q-learning to anticipate appropriate values after rough learning. We show the simulation result in Fig. 4. We simulated two types. The timing of the learning changes was 500 episodes (QL-FCCM (500)) and 1000 episodes (QL-FCCM (1000)) with clustering data extracted by different QL agents at the 1000th episode.

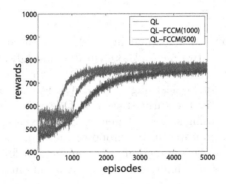

Fig. 4. Change from QL-FCCM to QL

Figure 4 shows that the average rewards sharply increased after changing the learning method. This result means that agents' learning was promoted by the clustering data, although the rewards were low in the early stage. This figure also shows that the clustering data, which were extracted from slightly different learning processes, have quite different results. How to extract the stable clustering data for the learning and ways of coping are future tasks.

In our next experiment, we applied our proposed method that gradually changes QL-FCCM to Q-learning.

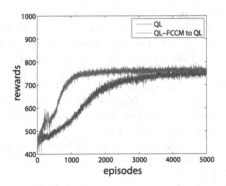

Fig. 5. Gradual change from QL-FCCM to QL

We applied our proposed method to the same situation and show the simulation result in Fig. 5. Agents obtained better learning than Q-learning in the early learning stage. Although the clustering data are inaccurate because they were extracted from a different environment, they promoted learning faster than the others.

6 Conclusion

In this study, we proposed a method in which an agent updates his Q-table by changing the ratio of two methods, Q-learning and QL-FCCM, to improve the learning speed in complicated problems. Our agent changed the ratio of his learning methods. In his learning process, QL-FCCM encouraged him to study comprehensively in the early stage and Q-learning encouraged him to study expertly in the closing stages. Therefore, in our experiment, we found the agent's learning speed improved in the proposed method.

Future work will develop an action selection method for clustering that is applied to the results from reinforcement learning agents in progress.

Acknowledgement. This work was supported by JSPS KAKENHI Grant Number 15K00344.

References

1. Sutton, R.S., Bart, A.G.: Generalization in Reinforcement Learning-An Introduction. The MIT Press (1998)
2. Notsu, A., Honda, H., Ichihashi, H., Wada, H.: Contraction algorithm in state and action space for Q-learning. In: 10th International Symposium on Advanced Intelligent Systems, pp. 93–96 (2009)
3. Komori, Y., Notsu, A., Honda, K., Ichihashi, H.: Determination of the change timing of space segmentation using PCA for reinforcement learning. In: The 6th International Conference on Soft Computing and Intelligent Systems The 13th International Symposium on Advanced Intelligent Systems, pp. 2287–2290 (2012)
4. Kosko, B.: Neural Networks and Fuzzy Systems: A Dynamical Systems Approach to Machine Intelligence. Prentice Hall, Englewood Cliffs (1992)
5. Hammell, R.J., Sudkamp, T.: Learning Fuzzy Rules from Data. http://ftp.rta.nato.int/public/pubfulltext/rto/mp/rto-mp-003/mp-003-08.pdf
6. Komori, Y., Notsu, A., Honda, K., Ichihashi, H.: Automatic Adaptive Space Segmentation for Reinforcement Learning. International Journal of Fuzzy Logic and Intelligent Systems **12**(1), 36–41 (2012)
7. Notsu, A., Honda, K., Ichihashi, H., Komori, Y.: Simple reinforcement learning for small-memory agent. In: 10th International Conference on Machine Learning and Applications, vol. 1, pp. 458–461 (2011)
8. Ueno, T., Notsu, A., Honda, K.: Application of FCM-type co-clustering to an agent in reinforcement learning. In: 1st IIAI International Conference on Advanced Information Technologies, vol. 12, pp. 1–5 (2013)
9. Bezdek, J.C.: Pattern Recognition with Fuzzy Objective Function Algorithms. Plenum Press (1981)
10. Oh, C.H., Honda, K., Ichihashi, H.: Fuzzy clustering for categorical multivariate data. In: Joint 9th IFSA World Congress and 20th NAFIPS International Conference, pp. 2154–2159 (2001)
11. Tsuda, K., Minoh, M., Ikeda, K.: Extracting straight lines by sequential fuzzy clustering. Pattern Recognition Letters. **17**, 643–649 (1996)

12. Matsumoto, Y., Honda, K., Notsu, A., Ichihashi, H.: Exclusive Partition in FCM-type Co-clustering and Its Application to Collaborative Filtering. International Journal of Computer Science and Network Security **12**(12), 52–58 (2012)
13. Honda, K., Notsu, A., Ichihashi, H.: Collaborative Filtering by Sequential User-Item Co-cluster Extraction from Rectangular Relational Data. International Journal of Knowledge Engineering and Soft Data Paradigms(IJKESDP) **2**(4), 312–327 (2010)
14. Hathaway, R.J., Davenport, J.W., Bezdek, J.C.: Relational duals of the c-means clustering algorithms. Pattern Recognition **22**(2), 205–212 (1989)
15. Watkins, C., Dayan, P.: Technical note: Q-learning. Machine Learning **3**(8), 279–292 (1992)
16. Rummery, G.A., Niranjan, M: On-line Q-learning using connectionist systems, Technical Report CUED/F-INFENG/TR 166. Engineering Department, Cambridge University (1994)
17. Jaakkola, T., Shingh, S.P., Jordan, M.: I: Reinforcement Learning Algorithm for Partially Observable Markov Decision Process. Advances in Neural Information Processing System **7**, 345–352 (1994)
18. Miyamoto, S., Ichihashi, H., Honda, K.: Algorithms for fuzzy clustering. Springer (2008)

MMMs-Induced Fuzzy Co-clustering with Exclusive Partition Penalty on Selected Items

Takaya Nakano, Katsuhiro Honda[✉], Seiki Ubukata, and Akira Notsu

Osaka Prefecture University, Sakai, Osaka 599-8531, Japan
{honda,subukata,notsu}@cs.osakafu-u.ac.jp

Abstract. Fuzzy co-clustering is a powerful tool for summarizing co-occurrence information while some intrinsic knowledge on meaningful items may be concealed by the dominant items shared by multiple clusters. In this paper, the conventional fully exclusive item partition model is modified such that exclusive penalties are forced only on some selected items. Its advantages are demonstrated through two numerical experiments. In a document clustering task, the proposed model is utilized for emphasizing cluster-wise meaningful keywords, which are useful for effectively summarizing document clusters. In an unsupervised classification task, the classification quality is improved by efficiently selecting promising items based on the item-wise single penalization test.

Keywords: Fuzzy co-clustering · Exclusive partition · Co-occurrence information

1 Introduction

Fuzzy c-Means (FCM) and its variants [1,2] are the basic techniques for unsupervised soft classification, in which objects with multivariate vector-form observations are partitioned into clusters with their fuzzy membership degrees. The FCM-type clustering models have also been extended to relational data analysis such as Relational Fuzzy c-Means (RFCM) [3] and Non-Euclidean Relational Fuzzy c-Means (NERF) [4], in which data objects are characterized by mutual similarity/dissimilarity measures.

Besides the object partitioning models, co-clustering is another direction of data clustering, in which the goal is to estimate dual-partitions of objects and items considering mutual co-occurrence degrees. Co-clustering is a popular technique in such tasks as document-keyword analysis and customer-products purchase history data analysis. For example, in document clustering tasks, a document set is summarized into several document clusters formed by familiar documents and their keywords by analyzing a co-occurrence matrix composed of frequencies of keywords in each document.

Fuzzy Clustering for Categorical Multivariate data (FCCM) [5] is an FCM-type co-clustering model, in which memberships of both objects and items are estimated by maximizing the clustering criterion of the aggregation degree of

© Springer International Publishing Switzerland 2015
V.-N. Huynh et al. (Eds.): IUKM 2015, LNAI 9376, pp. 226–235, 2015.
DOI: 10.1007/978-3-319-25135-6_22

objects and items in clusters. Soft memberships are given by introducing entropy-based fuzzification penalties [6] for both object and item memberships while different probabilistic constraints are forced to them such that row-sum-one condition for object memberships and column-sum-one condition for item memberships. Then, object memberships are mainly contribute to exclusive object partitioning while item memberships are used for characterizing each cluster by emphasizing the cluster-wise meaningful items. Although the partition concept has some similarity with statistical co-clustering models, there is no comparative statistical models and it is often difficult to carefully tune the fuzziness degrees of dual partitions without comparison with statistical counterpart.

A fuzzy co-clustering model was induced from a statistical co-clustering concept, which is a fuzzy variant of Multinomial Mixture Models (MMMs) [7]. Fuzzy Co-Clustering Model induced by MMMs concept (FCCMM) [8,9] defined a fuzzy clustering objective function from the pseudo-log-likelihood function of MMMs by introducing the K-L information-based regularization concept [10], and can tune the degree of partition fuzziness. The degree of fuzziness can be carefully tuned by comparing with MMMs and a fuzzier or crisper models were shown to have possibility of outperforming MMMs from the partition quality view point [8,9].

In FCCMM, the same row-sum and column-sum conditions with FCCM were adopted. So, item partition does not have explicit exclusive nature and item memberships does not necessarily suit item classification. In the previous work [11], the interpretability of fuzzy co-cluster structures were shown to be improved by introducing some explicit exclusive nature to item memberships. Additional penalties for avoiding all items not to be shared were introduced and it was demonstrated that we can find meaningful keywords in document clustering tasks, where memberships of some dominant keywords shared by multiple clusters were degraded and the efficient summary of each document cluster can be constructed by emphasizing cluster-wise typical keywords.

In this paper, the FCCMM model is further improved by introducing exclusive partition penalties to some selected items only. If we have some a priori knowledge on item characteristics, exclusive penalties should be forced by reflecting them, i.e., some popular items should be shared without exclusive nature while other distinctive items can be exclusive by emphasizing their belongingness.

The remaining parts of this paper are organized as follows: Section 2 presents a brief review on FCM-type co-clustering and MMMs-induced fuzzy co-clustering. In Section 3, the exclusive item partition model is proposed by introducing the item-wise penalization scheme to the MMMs-induced model. The characteristic features are demonstrated through two numerical experiments in Section 4 and a summary conclusion is given in Section 5.

2   MMMs-Induced Fuzzy Co-clustering

Supported by the close connection with FCM-type clustering [1,2] and statistical Gaussian Mixture Models (GMMs), a fuzzy counter part of GMMs-based clustering model was proposed, where membership fuzzification was achieved by K-L information-based regularization [10], and the flexible fuzzification model was shown to have higher classification qualities than GMMs with careful tuning of fuzzification degrees of object partition. Based on a similar fuzzification concept, FCCMM [8,9] was constructed induced from a statistical co-clustering model of MMMs, in which soft partition of objects are achieved by introducing K-L information-based fuzzification mechanism.

Assume that we have an $n \times m$ co-occurrence information matrix $R = \{r_{ij}\}$ composed of the frequency of item j, $j = 1, \ldots, m$ in object i, $i = 1, \ldots, n$, and the goal is to extract C co-clusters of familiar objects and items. In the fuzzy co-clustering context, the dual partition is represented by object memberships $u_{ci} \in [0, 1]$ and item memberships $w_{cj} \in [0, 1]$, which are the degree of belongingness to cluster c and are constrained such that $\sum_{c=1}^{C} u_{ci} = 1$ and $\sum_{j=1}^{m} w_{ij} = 1$, respectively. From the statistical view point, u_{ci} and w_{cj} can be regarded as the probabilities of class c given object i and item j given the c-th generative model, respectively. The objective function of FCCMM was constructed by modifying the pseudo-log-likelihood function in MMMs as:

$$L_{fccmm} = \sum_{c=1}^{C} \sum_{i=1}^{n} \sum_{j=1}^{m} u_{ci} r_{ij} \log w_{cj} + \lambda_u \sum_{c=1}^{C} \sum_{i=1}^{n} u_{ci} \log \frac{\alpha_c}{u_{ci}}, \qquad (1)$$

where α_c represents the volume of cluster c, which can be identified with the a priori probability of class c in the MMMs context. The first term measures the degree of aggregation of objects and items in cluster c such that it becomes larger as familiar objects and items having larger frequency r_{ij} take large u_{ci} and w_{cj} in the same cluster c. The K-L information term [10] is maximized by assigning similar values to both α_c and u_{ci} rather than hard memberships of $u_{ci} \in \{0, 1\}$.

When fuzzification weight λ_u is $\lambda_u = 1$, the objective function is equivalent to the pseudo-log-likelihood function to be maximized in MMMs. If $\lambda_u > 1$, FCCMM gives a much fuzzier object partition than MMMs while it becomes more crisp with $\lambda_u < 1$.

Here, it should be noted that, in the clustering context, u_{ci} is mainly utilized for object partition under the explicit exclusive constraint of $\sum_{c=1}^{C} u_{ci} = 1$ while w_{cj} is designed only for characterizing the significance degree of each item in the independent clusters and can have large values in multiple clusters or small values in all clusters.

3   Exclusive Partition of Items in Fuzzy Co-clustering and Sharing Penalties on Selected Items

From the item partition view point, characteristic items of clusters should be exclusively assigned for emphasizing the differences among clusters. In the previous work [11], the interpretability of fuzzy co-cluster partitions were shown to be improved by introducing exclusive penalties on both object and item memberships. In this paper, the exclusive partition model is further modified by forcing exclusive nature only on some selected items.

The FCCMM objective function of Eq.(1) implies that w_{cj} becomes larger when $\sum_{i=1}^{n} u_{ci}r_{ij}$ is large in cluster c. So, item j can take large w_{cj} in two or more clusters in the case where item j is popular in multiple clusters having large r_{ij}. This kind of item sharing may make it difficult to find the cluster-wise key items, which can be used in cluster interpretation. In order to avoid w_{cj} having large memberships in multiple clusters, $\sum_{i=1}^{n} u_{ci}r_{ij}$ should be degraded when $\sum_{t \neq c} w_{tj}$ is large. In [11], the trade-off between $\sum_{i=1}^{n} u_{ci}r_{ij}$ and $\sum_{t \neq c} w_{tj}$ is implemented by multiplying a sharing penalty s_{cj} on item j in cluster c:

$$s_{cj} = \exp\left(-\beta \sum_{t \neq c} w_{tj}^*\right), \tag{2}$$

where the temporal value w_{tj}^* of the current iteration is utilized for computational simplicity. s_{cj} takes 1 in no-sharing case while it becomes small with large $\sum_{t \neq c} w_{tj}^*$. Then, w_{cj} can have large values in at most one cluster. β tunes the sensitivity of the sharing penalty weight, which drastically decreases with a larger β while $\beta = 0$ reduces to the conventional non-penalized model. In order to follow and modify the conventional fuzzy co-cluster structures, β is first initialized as $\beta = 0$ and is increased such that $\beta = \min\{0.1 \times (t-1), \beta_{max}\}$ with iteration index t, where β_{max} is the final exclusive degree of item partition.

Although the sharing penalty weight s_{cj} was multiplied to all items for estimating fully exclusive item partitions in [11], it is considered only with some selected items in this paper. Assume that EI is a set of items to be exclusively assigned to co-clusters and other items can be shared by multiple clusters. Item sharing penalty s_{cj} is modified as:

$$s_{cj} = \begin{cases} \exp\left(-\beta \sum_{t \neq c} w_{tj}^*\right) & ; j \in EI \\ 1 & ; \text{otherwise} \end{cases} \tag{3}$$

and the objective function is revised as:

$$L_{fccmm'} = \sum_{c=1}^{C} \sum_{j=1}^{m} \left(\sum_{i=1}^{n} u_{ci}r_{ij}s_{cj}\right) \log w_{cj} + \lambda_u \sum_{c=1}^{C} \sum_{i=1}^{n} u_{ci} \log \frac{\alpha_c}{u_{ci}}. \tag{4}$$

Following the necessary condition for the optimality of the objective function under the probabilistic constraints, the updating rules for the modified FCCMM are derived as follows:

$$\alpha_c = \frac{1}{n} \sum_{i=1}^{n} u_{ci}, \tag{5}$$

$$u_{ci} = \frac{\alpha_c \prod_{j=1}^{m} (w_{cj})^{(r_{ij}s_{cj})/\lambda_u}}{\sum_{\ell=1}^{C} \alpha_\ell \prod_{j=1}^{m} (w_{\ell j})^{(r_{ij}s_{\ell j})/\lambda_u}}, \tag{6}$$

$$w_{cj} = \frac{\sum_{i=1}^{n} r_{ij} s_{cj} u_{ci}}{\sum_{\ell=1}^{m} \left(\sum_{i=1}^{n} r_{i\ell} s_{c\ell} u_{ci} \right)}. \tag{7}$$

In the iterative optimization algorithm, after random initialization, α_c, u_{ci} and w_{cj} are updated repeatedly until convergence.

4   Numerical Experiments

In this section, two experimental results are shown to demonstrate the characteristic features of the proposed penalization model.

4.1   Document Analysis

First, the proposed model is applied to a document analysis task. The document-keyword cooccurrence matrix used in [12] was constructed from a Japanese novel "Kokoro" written by Soseki Natsume, which can be downloaded from Aozora Bunko (http://www.aozora.gr.jp). The novel is composed of 3 chapters consisting of 36, 18 and 56 sections each. The cooccurrence matrix R includes the frequencies of 83 most frequently used nouns and verbs (items, $m = 83$) in the 110 sections (objects, $n = 110$), in which each element r_{ij} corresponds to the tf-idf weight [13] of each section-document pair. Withholding chapter information of each section, unsupervised fuzzy co-clustering was performed with the goal of extracting section-keyword (object-item) co-cluster structures. These chapter-wise meaningful keywords are expected to be useful in producing the abstracts of each document cluster, i.e., intrinsic chapters. Before application, cooccurrence information r_{ij} was normalized so that it has zero-minimum and unit-variance for each item.

In the previous work [11], the fuzzy co-cluster structures given by the conventional non-exclusive and the fully exclusive FCCMM were compared from the view point of interpretability of intrinsic chapter information, where the number of clusters were set as $C = 4$ considering the two sub-stories in the third chapter, i.e., clusters 1 and 2 correspond to Chapters 1 and 2 while Chapter 3 was divided into clusters 3 and 4. The fuzziness degree was $\lambda_u = 1.5$, which is slightly fuzzier than MMMs and can contribute to handling weak overlapping of chapter components.

Tables 1 and 2 compare the top 5 keywords (items) having 5 largest item memberships with their English translation in ( ), i.e., they were selected as meaningful keywords. The conventional non-exclusive FCCMM mainly listed some common keywords and failed to extract cluster-wise characteristic keywords. On the other hand, in the fully exclusive model, some common keywords were disappeared from the list and cluster-wise meaningful keywords were emphasized.

By the way, from the document summarization view point, nouns may be more informative than general verbs and can be shared by multiple clusters for fairly characterizing each cluster without illegal distortion. In the following,

Table 1. Top 5 Keywords in Non-exclusive Clusters [11]

c	1st	2nd	3rd	4th	5th
1	(it)	(I)	(become)	(do)	(be)
2	(do)	(father)	v (think)	(it)	(become)
3	(I)	(do)	(become)	(myself)	v (think)
4	(I)	(do)	(become)	(it)	v (think)

Table 2. Top 5 Keywords in Fully Exclusive Clusters [11]

c	1st	2nd	3rd	4th	5th
1	(master)	(become)	(it)	(I)	(be)
2	(father)	(mother)	(letter)	o (send)	aC (illness)
3	f (uncle)	(wife)	(do)	(I)	(myself)
4	K (a name)	(lady)	(I)	(he)	(face)

Table 3. Top 5 Keywords in Partially Exclusive Clusters ($\beta_{max} = 1$)

c	1st	2nd	3rd	4th	5th
1	(it)	(I)	(become)	(do)	(teacher)
2	(do)	(father)	(it)	(mother)	v (think)
3	(I)	(do)	(myself)	(become)	(think)
4	(I)	(do)	(become)	(it)	v (think)

the advantage of the exclusive item partition is further investigated by forcing exclusive penalties only on some selected items. Here, the exclusive penalties are forced only on 33 verbs, and the remaining nouns are allowed to be shared by multiple clusters. The fuzzification degree was again set as $\lambda_u = 1.5$. In order to test the role of the exclusive penalty weight, the list of selected keywords are compared with $\beta_{max} \in \{1, 5, 10\}$.

Tables 3, 4 and 5 list top 5 keywords when $\beta_{max} = 1, 5$ and 10, respectively. As exclusive penalty weight β_{max} was larger, the priority degrees of verbs were gradually degraded in the list, and the list of $\beta_{max} = 10$ shown in Table 5

Table 4. Top 5 Keywords in Partially Exclusive Clusters ($\beta_{max} = 5$)

c	1st	2nd	3rd	4th	5th
1	(it)	(I)	(teacher)	(become)	(do)
2	(father)	(it)	(mother)	(do)	(I)
3	(I)	(do)	(myself)	(it)	(wife)
4	(I)	(it)	K (a name)	(do)	(become)

Table 5. Top 5 Keywords in Partially Exclusive Clusters ($\beta_{max} = 10$)

c	1st	2nd	3rd	4th	5th
1	(it)	(I)	(teacher)	(person)	(this)
2	(father)	(it)	(mother)	(I)	(letter)
3	(I)	(myself)	(it)	(wife)	(uncle)
4	(I)	(it)	K (a name)	(myself)	(lady)

Table 6. Chapter-Cluster Cross Tabulation in Partially Exclusive Partition ($\beta_{max} = 10$)

Chap.	$c = 1$	$c = 2$	$c = 3$	$c = 4$
1	31	5	0	0
2	0	18	0	0
3	1	2	14	39

includes meaningful nouns only. These results fairly support the intended effect of the sharing penalty weight. Table 6 shows the cross tabulation of the proposed partially exclusive partition model with the exclusive penalty weight $\beta_{max} = 10$, which indicates that the chapter information could still be successfully revealed.

The plausibility of the selected keywords can be investigated as follows: Chapter 1 and Chapter 2 are titled as '(Master and I)' and 'e (My parents and I)', respectively, in the original novel. The selected keywords are available for easily estimating the contents of the two chapters. Then, the third chapter is titled as '(Master and His Testament)' and consists of the two sub-stories of the hero's monolog and the master's testament. In the same manner with the previous chapters, the selected keywords of clusters 3 and 4 are also available for revealing these sub-contents.

Table 7. Comparison of Unsupervised Classification Quality

model	$\lambda_u = 0.5$		$\lambda_u = 1.0$	
	$\beta_{max} = 1$	$\beta_{max} = 10$	$\beta_{max} = 1$	$\beta_{max} = 10$
Non-exclusive	0.791		0.794	
Fully exclusive	0.798	0.734	0.795	0.714
Partially exclusive	0.801	0.848	0.802	0.855
List of exclusive items	80, 97	11,12,31,36,40,41 53,64,80,81,82,86 97,100	81, 97	2,6,11,17,31,35 36,41,55,64,80,81 82,86,97
model	$\lambda_u = 1.5$		$\lambda_u = 2.0$	
	$\beta_{max} = 1$	$\beta_{max} = 10$	$\beta_{max} = 1$	$\beta_{max} = 10$
Non-exclusive	0.800		0.801	
Fully exclusive	0.797	0.702	0.793	0.438
Partially exclusive	0.804	0.841	0.804	0.760
List of exclusive items	17,41,82	6,11,12,17,31,35 36,41,53,55,61 64,80,82,86,97	31,35,41 46,70,86	3,4,6,11,12,15,31,35 36,41,43,46,49,53,54,61 70,80,81,82,86,88,99,100

In this way, the partially exclusive partition model can contribute to detailed co-cluster structure analysis by utilizing some a prior knowledge such as characteristics of nouns and verbs.

4.2   Unsupervised Classification

Next, the proposed model is applied to a social network dataset. Terrorist Attacks data set was downloaded from LINQS webpage of Statistical Relational Learning Group @ UMD (http://linqs.cs.umd.edu/projects//index.shtml), and consists of 1293 terrorist attacks each assigned to one of 6 labels indicating the type of attacks. Each attack is characterized by 106 distinct features with a 0/1-valued vector of attributes, whose entries indicate the absence/presence of a feature. In this experiment, withholding the actual class information, unsupervised co-clustering is performed with the goal of revealing the intrinsic classes. Here, major three labeled classes of *bombing*, *kidnapping* and *weapon-attack* are considered with the cluster number of $C = 3$.

In order to fairly compare the best classification qualities, the initial object cluster memberships were given following the correct class labels. The ratios of correct classification by maximum membership classification are compared in Table 7.

First, the conventional non-exclusive and the fully exclusive FCCMM were applied to the data set using four different fuzzification degrees $\lambda_u \in \{0.5, 1.0, 1.5, 2.0\}$. The penalty weights for exclusive item assignment were chosen from $\beta_{max} \in \{1, 10\}$, which realize the weak and hard exclusive situations, respectively. The classification ratios are shown in the top and second rows of Table 7. In non-exclusive FCCMM, the fuzzier model could achieve a slightly better classification quality. On the other hand, in the fully exclusive case, the classification quality was degraded because the fully exclusive penalty may bring a distorted co-cluster structure. Then, exclusive penalties should be forced only on some selected items.

Second, the proposed selective penalization model is applied with two phases. In the first phase, the applicability of exclusive partition for each item was evaluated by forcing an item-wise exclusive penalty on only one of 106 items in each trial. As listed in the last row of Table 7, the classification quality was improved 0.001 or more from the non-exclusive model by forcing the item-wise exclusive penalty on some of 106 items. Because a larger exclusive penalty weight $\beta_{max} = 10$ could emphasize the items to be exclusive more clearly, the improvement of the classification quality was found more frequently than a smaller weight $\beta_{max} = 1$. For example, in the $\lambda_u = 0.5$ case, the classification quality was improved only with two items (90 and 97) in the weak exclusive situation but with 14 items in the hard situation. Additionally, the number of items to be exclusive was increased as the fuzzification degree λ_u became larger. It is implied that the item partition should be crisper as object partition becomes fuzzier.

In the second phase, the proposed model is applied again by forcing exclusive penalties to the all items, which were selected by the item-wise single penalization test. The third row of Table 7 shows that the better classification quality was achieved with this two-phase selection scheme except for the case of $(\lambda_u, \beta_{max}) = (2.0, 10)$. Especially, the larger weight $\beta_{max} = 10$ significantly improved the classification quality and the best performance was provided with $(\lambda_u, \beta_{max}) = (2.0, 10)$.

These results imply the advantage of the proposed model and a promising two stage implementation procedure composed of the item selection based on item-wise penalization test and the successive partial exclusive penalization on the selected items.

5   Conclusion

In this paper, the conventional fully exclusive approach on item partition was modified such that the exclusive penalties are forced only on some selected items. Some meaningful items can be selected based on a priori knowledge on item characteristics and fuzzy co-cluster structures are estimated utilizing such a priori knowledge. Two experimental results demonstrated the effective document summarization ability and the improvement of classification ability of the proposed framework.

A possible future work includes the consideration of how to automatically tune the sensitivity weight β_{max}.

Acknowledgement. This work was supported by the Ministry of Education, Culture, Sports, Science and Technology, Japan, under Grant-in-Aid for Scientific Research (#26330281).

References

1. Bezdek, J.C.: Pattern Recognition with Fuzzy Objective Function Algorithms. Plenum Press (1981)
2. Miyamoto, S., Ichihashi, H., Honda, K.: Algorithms for Fuzzy Clustering. Springer (2008)
3. Hathaway, R.J., Davenport, J.W., Bezdek, J.C.: Relational duals of the c-means clustering algorithms. Pattern Recognition $22(2)$, 205–212 (1989)
4. Hathaway, R.J., Bezdek, J.C.: Nerf c-means: non-Euclidean relational fuzzy clustering. Pattern Recognition $27(3)$, 429–437 (1994)
5. Oh, C.-H., Honda, K., Ichihashi, H.: Fuzzy clustering for categorical multivariate data. In: Proc. of Joint 9th IFSA World Congress and 20th NAFIPS International Conference, pp. 2154–2159 (2001)
6. Miyamoto, S., Mukaidono, M.: Fuzzy c-means as a regularization and maximum entropy approach. In: Proc. of the 7th International Fuzzy Systems Association World Congress, vol. 2, pp. 86–92 (1997)
7. Rigouste, L., Cappé, O., Yvon, F.: Inference and evaluation of the multinomial mixture model for text clustering. Information Processing and Management $43(5)$, 1260–1280 (2007)
8. Honda, K., Oshio, S., Notsu, A.: FCM-type fuzzy co-clustering by K-L information regularization. In: Proc. of 2014 IEEE International Conference on Fuzzy Systems, pp. 2505–2510 (2014)
9. Honda, K., Oshio, S., Notsu, A.: Fuzzy co-clustering induced by multinomial mixture models. Journal of Advanced Computational Intelligence and Intelligent Informatics 19 (to appear, 2015)
10. Honda, K., Ichihashi, H.: Regularized linear fuzzy clustering and probabilistic PCA mixture models. IEEE Transactions on Fuzzy Systems $13(4)$, 508–516 (2005)
11. Honda, K., Oh, C.-H., Notsu, A.: Exclusive condition on item partition in fuzzy co-clustering based on K-L information regularization. In: Proc. Joint 7th International Conference on Soft Computing and Intelligent Systems and 15th International Symposium on Advanced Intelligent Systems, pp. 1413–1417 (2014)
12. Honda, K., Notsu, A., Ichihashi, H.: Fuzzy PCA-guided robust k-means clustering. IEEE Transactions on Fuzzy Systems $18(1)$, 67–79 (2010)
13. Salton, G., Buckley, C.: Term-weighting approaches in automatic text retrieval. Information Processing and Management $24(5)$, 513–523 (1988)

Clustering Data and Vague Concepts Using Prototype Theory Interpreted Label Semantics

Hanqing Zhao[1,2] and Zengchang Qin[1]([✉])

[1] Intelligent Computing and Machine Learning Lab, School of ASEE,
Beihang University, Beijing, P.R. China
{hzhao,zcqin}@buaa.edu.cn
[2] École d'Ingénieur Généraliste, École Centrale de Pékin, Beihang University,
Beijing, P.R. China

Abstract. Clustering analysis is well-used in data mining to group a set of observations into clusters according to their similarity, thus, the (dis)similarity measure between observations becomes a key feature for clustering analysis. However, classical clustering analysis algorithms cannot deal with observation contains both data and vague concepts by using traditional distance measures. In this paper, we proposed a novel (dis)similarity measure based on a prototype theory interpreted knowledge representation framework named label semantics. The new proposed measure is used to extend classical K-means algorithm for clustering data instances and the vague concepts represented by logical expressions of linguistic labels. The effectiveness of proposed measure is verified by experimental results on an image clustering problem, this measure can also be extended to cluster data and vague concepts represented by other granularities.

Keywords: Clustering · Label semantics · Prototype theory · K-means · Distance measure

1 Introduction

Clustering analysis (or clustering) is a main task of exploratory data mining, and a common technique for statistical data analysis [11]. Cluster analysis groups a set of observations (data) into several "clusters", and observations in the same "cluster" are considered as "similar" observations and they are "dissimilar" to those belong to other clusters. Besides, conceptual clustering is another type of clustering analysis for unsupervised classification, in which the observations are grouped according to their fitness to descriptive concepts, but not simple similarity measures. To our knowledge, these two types of clustering are rarely studied together though there are actual needs for grouping data and concepts [4].

Clustering algorithms are widely used in several fields, including machine learning, pattern recognition, image analysis, bioinformatics and so on. There are many successful classical cluster algorithms, such as K-means, fuzzy C-means

© Springer International Publishing Switzerland 2015
V.-N. Huynh et al. (Eds.): IUKM 2015, LNAI 9376, pp. 236–246, 2015.
DOI: 10.1007/978-3-319-25135-6_23

and rough C-means [1] for similarity based clustering, and hierarchical clustering algorithms for connectivity based clustering. Yet these classical clustering
algorithms using classical distance measures (e.g. Euclidian and Mahalanobis
distance) cannot cluster *vague concepts*, and their clustering results are heavily
depend on the distance measure between observations. Thus, in past decade,
many clustering algorithms using customized distance measure are proposed
in literature, for example, belief K-modes method (BKM) proposed by Hariz
et al. [2] and possibilistic K-modes method (PKM) proposed by Ammar and
Elouediare [3] are effective methods for clustering numerical data described by
categorical attributes (labels), where the distance measure between objects is
defined by the total mismatches of the corresponding attribute.

However, these clustering algorithms are restricted to numerical or discrete
data. However, in order to simulate the knowledge generation process, we hope
to deal with clustering some high-level knowledge, vague concepts or linguistic
expressions. For example, we have two sets of observations, including a set of
data of human heights in meters

$$hight = \{1.0, 1.3, 1.4, 1.6, 1.7, 1.9, 2.0\}$$

and a set of descriptive vague concepts $concepts = \{short, medium, tall\}$ in
which elements are defined by a set of prototypical elements. Given the numbers
of cluster centers $k = 3$, these observations can be clustered into three following clusters: $\{short, 1.0, 1.3, 1.4\}, \{medium, 1.6, 1.7\}, \{tall, 1.9, 2.0\}$. In order to
accomplish the above purpose, we need a suitable distance measure for measuring
the dissimilarity between numerical data and descriptive vague concepts. Label
semantics [5] can be used to construct distance measure between numerical data
and descriptive concepts, where descriptive concepts are represented by a set of
linguistic labels, Zhang and Qin [4] proposed such a distance measure based on
fuzzy set interpreted label semantics, where linguistic labels are represented by
fuzzy membership functions defined on a universe of discourse containing data
to be described. The prototype theory based interpretation of label semantics is
proposed by Lawry and Tang [6], where linguistic labels are represented by a set
of prototypical data. Based on this interpretation, in this paper, we proposed a
novel distance measure which makes it possible to cluster a set of observations
including numerical data, descriptive concepts and linguistic expressions, and
it's effectiveness is verified by applying it to the classical K-means clustering
algorithm.

This paper is structured as the following. Section 2 gives a general introduction of label semantics. In Section 3, we propose the new distance measure based
on prototype theory. Section 4 gives the extended K-means based on the new
measure and Section 5 gives the experimental results and compared to previous
research. Section 6 gives the final conclusion and discussions.

2   Label Semantics Framework

Label semantics [5] is a random set framework for modeling with words, which
encodes the semantic meaning of linguistic labels according to how they are

used by a population of individuals to convey information. Otherwise, it can also be regarded as a simulation of knowledge generation process, in order to acquire knowledge, an intelligent has to identify which label or logical expression is appropriate to describe a value or an observation, thus, the appropriateness of using a subset of labels to describe a certain object is named *appropriateness degrees* in label semantics framework.

Definition 1. *(Label expression). Given a finite set of labels $LA = \{L_1, ..., L_n\}$, the set of label expression LE is generated by logical expression of labels in LA as below:*

- *For $\forall L \in LA$, we have $L \in LE$*

- *For $\forall(\theta, \varphi) \in LA^2$, we have $(\theta \vee \varphi, \theta \wedge \varphi, \neg\theta) \in LE$*

For set of labels $S \subseteq LA$, an observation in the universe of discourse $x \in \Omega$ when an individual in a population $I \in V$ makes an assertion of the form "x is θ", which provides information about "what label is appropriate for describing observation x", this information is named label description of x, it is a random set from a population V to the power set of LA, denoted by D_x, the associated distribution of D_x is referred to mass assignment, denoted by m_x as follow:

Definition 2. *(Mass Assignment). Mass assignment is agent's subjective belief in a population V that the subset S contains all and only appropriate label(s) for describing object x:*

$$\forall S \subseteq LA, \ m_x(S) = P(I \in V: \ D_x^I = S) \tag{1}$$

Thus, the mass assignment m_x can be also regarded as a mass function defined as $m_x : P(LA) \to [0,1]$ where $P(LA)$ is the power set of LA and

$$\sum_{S \subseteq LA} m_x(S) = 1$$

Furthermore, to evaluate the how appropriate a single label $L \in LA$ is for describing a certain observation $x \in \Omega$, the appropriateness degree is defined as follows:

Definition 3. *(Appropriateness Degree). Appropriateness degree is a function defined as $\mu : LA \times \Omega \to [0, 1]$ satisfying:*

$$\forall x \in \Omega, \ \forall L \in LA, \ \mu_L(x) = \sum_{S \subseteq LA: L \in S} m_x(S) \tag{2}$$

Example 1. Given a finite set of labels for human age description: $LA_{Age} = \{young, middle\text{-}aged, old\}$ and a population of 10 individuals, Suppose 4 of 10 individuals consider that "young" is appropriated label for describing age 42, and other 6 support that both "young" and "middle-aged" are appropriate labels, according to Definition 2, the mass assignment for age 42 is:

$$m_{42} = \{middle - aged\} : 0.4, \{young, middle - aged\} : 0.6$$

Based on Definition 3, appropriateness degrees of each label for describing age 42 are:

$$\mu_{young}(42) = 0.6 \quad \mu_{middle-aged}(42) = 0.4 + 0.6 = 1$$

After defining the appropriateness degree evaluation method of single label, we may also interest in evaluating the appropriateness degree of a logical expression $\theta \in LE$, for this propose, it is necessary to identify what information is provided by a logical expression θ regarding the appropriateness of labels, thus, the λ function is defined to transform the information provided by a logical expression as below:

Definition 4. *(λ-Function). λ-function is a mapping from linguistic expression to the power set of labels: $\lambda : LE \rightarrow P(LA)$, which is defined as follow, for $\forall(\theta, \varphi) \in LE^2$:*

- $\forall L_i \in LA, \quad \lambda(L_i) = \{F \subseteq LA : L_i \in F\}$
- $\lambda(\theta \wedge \varphi) = \lambda(\theta) \cap \lambda(\theta)$
- $\lambda(\theta \vee \varphi) = \lambda(\theta) \cup \lambda(\theta)$
- $\lambda(\neg\theta) = \overline{\lambda(\theta)}$

Label semantics theory is a powerful tool for modeling with words, which has been well applied in machine learning and data mining, further details on using label semantics for data mining are available in [6].

3   Distance Measure Based on Logical Expressions

3.1   Prototype Theory Interpretation of Label Semantics

The proposed distance measure deals with labels and linguistic expressions interpreted by the prototype theory interpretation of label semantics framework. The label semantics framework is a random set framework for modeling with vagueness, where a set of labels is used by individuals vary across a population, such a theory cannot result in a truth-functional calculation [6]. In order to generate a functional calculus for appropriateness degrees, Lawry and Tang [6] have proposed an interpretation based on prototype theory. In this interpretation, each label $L_i \in LA$ is represented by a set of prototypical elements $P_i \in \Omega$, given a classical distance function $d(\cdot)$ define on the universe of discourse: $d : \Omega^2 \rightarrow [0, \infty)$, and δ is a probability density function which is defined on $[0, \infty)$, in our experiment, we consider d as the Euclidean distance. The appropriateness degree $\mu_{L_i}(x)$ of describing a data $x \in \Omega$ by using a certain label $L_i \in LA$ can be calculated as below:

$$\forall L_i \in LA, \ \forall x \in \Omega, \ \mu_{L_i}(x) = \int_{d(x,P_i)}^{\infty} \delta(t)dt \tag{3}$$

where $d(x, P_i) = min\{d(x, y) : \forall y \in P_i\}$. More details on the prototype theory interpretation of label semantics can be found in [6,9].

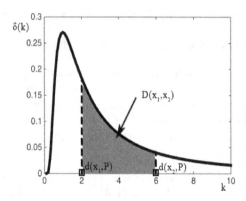

Fig. 1. Illustration of distance between two data points.

3.2   Distance between Vague Concepts

During the decision making process, an individual has to identify which label or logical expression can actually be used to describe an observation. The prototype theory interpretation of label semantics generates a functional calculus of appropriateness degree, thus, we can propose a measure based on appropriateness degrees to calculate dissimilarities between two observations of labels and logical expressions.

Definition 5. *(Distance Between Data Points). Given two observations in a universe Ω, and N labels $L_i \in LA$, $i \in [1, |LA|]$ and $i \in \mathbb{Z}$, for each label $L_i \in LA$, let there is a set $P_i \subseteq \Omega$ corresponding to prototypical elements for which L_i is certainly an appropriate description [6]. The distance between two observations (data points) is defined as a function define as $D(x_1, x_2) : \Omega^2 \to [0, \infty)$:*

$$D(x_1, x_2) = \sum_{i=1}^{N} |\mu_{L_i}(x_1) - \mu_{L_i}(x_2)| \tag{4}$$

where $\mu_{L_i}(x_j)$ is the appropriateness degree of describing data point x_j using label L_i, as defined in Definition 3. Given a single label L which is represented by a set of prototypical elements $P \in \Omega$, an illustration of distance between two data points is shown in Fig. 1, where the distance is defined as the integral of the density function δ from $d(x_1, L)$ to $d(x_2, L)$. Further more, the above distance measure has these following properties:

Theorem 1. *(Symmetric). Given $(x_1, x_2) \in \Omega^2$, the distance between two data points is symmetric*

$$D(x_1, x_2) = D(x_2, x_1)$$

Proof. According to Definition 1, for each $(x_1, x_2) \in \Omega^2$ we have

$$D(x_1, x_2) = \sum_{i=1}^{N} |\int_{d(x_1, P_i)}^{d(x_2, P_i)} \delta(t)dt| = \sum_{i=1}^{N} |\int_{d(x_2, P_i)}^{d(x_1, P_i)} \delta(t)dt| = D(x_2, x_1)$$

The proof is completed. The proof of this theorem is very intuitive as the distance is defined by the area between a range, it is symmetric as the area keeps the same from either the left to the right or from the right to the left.

Theorem 2. *(Triangular inequality). Given* $(x_1, x_2, x_3) \in \Omega^3$, *we have*

$$D(x_1, x_3) \leq D(x_1, x_2) + D(x_2, x_3)$$

Proof. According to Definition 1, for each $(x_1, x_2, x_3) \in \Omega^3$ we have

$$D(x_1, x_3) = \sum_{i=1}^{N} |\int_{d(x_1, P_i)}^{d(x_3, P_i)} \delta(t)dt| = \sum_{i=1}^{N} |\int_{d(x_1, P_i)}^{d(x_2, P_i)} \delta(t)dt + \int_{d(x_2, P_i)}^{d(x_3, P_i)} \delta(t)dt|$$

$$D(x_1, x_2) + D(x_1, x_2) = \sum_{i=1}^{N} |\int_{d(x_1, P_i)}^{d(x_2, P_i)} \delta(t)dt| + |\int_{d(x_2, P_i)}^{d(x_3, P_i)} \delta(t)dt|$$

According to the triangular inequality in the real number space where $\forall (a, b) \in \mathbb{R}^2$, $|a + b| \leq |a| + |b|$, as a result, for $\forall (x_1, x_2, x_3) \in \Omega^3$ and $\forall P_i \subseteq \Omega$ we have:

$$|\int_{d(x_1, P_i)}^{d(x_2, P_i)} \delta(t)dt + \int_{d(x_2, P_i)}^{d(x_3, P_i)} \delta(t)dt| \leq |\int_{d(x_1, P_i)}^{d(x_2, P_i)} \delta(t)dt| + |\int_{d(x_2, P_i)}^{d(x_3, P_i)} \delta(t)dt|$$

As a result, for $\forall (x_1, x_2, x_3) \in \Omega^3$:

$$D(x_1, x_3) \leq D(x_1, x_2) + D(x_2, x_3)$$

In conclusion, the distance between data points follows the triangular inequality.

Above definitions construct a functional calculus for measuring dissimilarity between two data points referring to labels which are represented by sets of prototypes defining on the universe of discourse. One step further, we consider how can we measure the dissimilarity between a certain label and a data point in the same universe.

Definition 6. *(Distance between point and label). Given a data point $x \in \Omega$ and a certain label $L_i \in LA$ represented by a set of prototypical elements $P_i \subseteq \Omega$, the distance between point and label is defined as below:*

$$D(x, L_i) = min\{D(x, y), \forall y \in P_i\} \tag{5}$$

where $D(x, y)$ is the distance between points as defined in Definition 1.

Specifically, when there is only one label $L \in LA$, $|LA| = 1$ which can be used to describe elements in Ω,we have:

$$D(x, L) = 1 - \mu_L(x) \tag{6}$$

Where $\mu_L(x)$ is the appropriateness degree of describing data point x using label L, thus, the distance $D(x, L)$ can be interpreted as the probability of "label L can not be used to describe data x". Furthermore, the distance between two sets of labels can be defined by:

Definition 7. *(Distance between set of labels). Given two sets of labels $(S_1, S_2) \in LA^2$, each label $L_i \in LA$ can be represented by a set of prototypical elements $P_i \in \Omega$, we have:*

$$D(S_1, S_2) = \frac{\sum_{L_i \in S_1} \sum_{L_j \in S_2} min\{D(x, y), \ \forall (x, y) \in P_i \times P_j\}}{|S_1| \cdot |S_2|} \tag{7}$$

where $|S_1|$ and $|S_2|$ are cardinalities of sets S_1 and S_2. $D(x, y)$ is the distance between points as defined by Definition 1. Based on the properties of distance between data points, it is obviously that the distance between set of labels is also symmetric and satisfies the triangular inequality.

The above distance measure is one dimensional, for an object with more than one feature to be described by labels. The distance measure between set of labels can be extended into multi-dimensional as shown in Definition 8.

Definition 8. *(Distance between multi-dimensional set of labels). The set of n-dimensional labels $MLA^{(n)}$ is a combination of descriptive labels of n different features $MLA(n) = LA_1 \times LA_2 \times ... \times LA_n$, where LA_i is the set of descriptive labels for describing the ith feature. For two multi-dimensional labels $(ML_1, ML_2) \in MLA(n)^2$ where:*

- *$ML_1 = (L_{11}, L_{12}, ..., L_{1n})$, $L_{1i} \in LA_i$*
- *$ML_2 = (L_{21}, L_{22}, ..., L_{2n})$, $L_{2i} \in LA_i$*

we have:

$$D(ML_1, ML_2) = \sqrt{\sum_{i=1}^{n} D(L_{1i}, L_{2i})^2} \tag{8}$$

In Definition 4, the λ-function provides an application from logical expressions to set of labels, utilizing this function and distance measure between set of labels, we can define the distance between logical expressions intuitively:

Definition 9. *(Distance between logical expressions). Given two logical expressions $(\theta, \varphi) \in LE$, the distance between θ and φ is:*

$$D(\theta, \varphi) = D(\mathbb{S}^{\theta \wedge \neg \varphi}, \mathbb{S}^{\varphi}) + D(\mathbb{S}^{\varphi \wedge \neg \theta}, \mathbb{S}^{\theta}) \tag{9}$$

$D(\theta, \lambda)$ *is the distance between label sets as defined in Definition 7, where label sets $\mathbb{S}^{\theta \wedge \neg \varphi}, \mathbb{S}^{\varphi}, \mathbb{S}^{\varphi \wedge \neg \theta}, \mathbb{S}^{\theta}$ are defined as follow:*

- *$\mathbb{S}^{\theta} = \{S | S \in \lambda(\theta)\}$*
- *$\mathbb{S}^{\varphi} = \{S | S \in \lambda(\varphi)\}$*
- *$\mathbb{S}^{\theta \wedge \neg \varphi} = \{S | S \in \lambda(\theta) \cap \overline{\lambda(\varphi)})\}$*
- *$\mathbb{S}^{\varphi \wedge \neg \theta} = \{S | S \in \lambda(\varphi) \cap \overline{\lambda(\theta)})\}$*

specifically, when $|(\mathbb{S}^{\theta \wedge \neg \varphi})| = 0$,

$$D(\theta, \varphi) = D(\mathbb{S}^{\varphi \wedge \neg \theta}, \mathbb{S}^{\theta}) \tag{10}$$

and when $|(\mathbb{S}^{\varphi \wedge \neg \theta})| = 0$,

$$D(\theta, \varphi) = D(\mathbb{S}^{\theta \wedge \neg \varphi}, \mathbb{S}^{\varphi}) \tag{11}$$

4   Clustering Mixed Objects

First proposed by MacQueen [7] in 1967, the K-means is regarded as the simplest yet effective technique for clustering analysis. The classical K-means algorithm using Euclidean distance cannot cluster vague concepts (e.g. linguistic descriptions). Based on the above distance measure, classical K-means algorithm can be extended for clustering mixed objects, including data points, labels which are represented by sets of prototypical elements, as defined in Section 3.1 and linguistic expressions.

The main objective of K-means clustering is to minimize the sum of squared distance between objects in each cluster and their mean, given objects

$$(x_1, x_2 ... x_N) \in \Omega^N$$

and k clusters, and let m_j as the mean of objects in cluster j, we define $x \in j$ if $m_j = \{m|\ min||x, m_j||, \ \forall j \in [1, k], \ j \in \mathbb{N}\}$, which is also equivalent to minimizing the following objective function:

$$S = \sum_{j=1}^{k} \sum_{x \in k} ||x, m_j||^2 \tag{12}$$

With the same objective, given an unlabeled data set of mixed objects $(obj_1, ..., obj_N) \in (\Omega \bigcup LE)^N$ the extended K-means algorithm for clustering mixed objects can be described as pseudo-codes in Table 1.

Table 1. Pseudo-code of extended K-means algorithm for clustering mixed objects.

Given a finite set of mixed objects $\mathbb{S} = \{obj_1 ... obj_N\}$ and a number of cluster k,
A set of randomly initialized centers $\mathbb{K} = \{c_1, ..., c_k\}$, a threshold $\varepsilon > 0$, and counter p

While $||c^{(p)} - c^{(p-1)}|| > \varepsilon$
p++

Step1. For each object $obj_i \in \mathbb{S}$, determine the cluster $obj_i \leftarrow c_i^{(p-1)}$ of each object, if:
$D(obj_i, c_i^{(p-1)}) = min\{D(obj_i, c_t^{(p-1)}) : t = 1, ..., k\}$

Step2. Calculate new clusters $c_t^{(p)}$, $x = 1, ..., k$, which satisfy:
$\sum_{obj \in cluster\ t} D(c_x^{(p)}, obj) = min\{\sum_{obj \in cluster\ t} D(x, obj)\}$

The new proposed distance measure is used both in Step 1 and 2 to calculate distances between each two objects, thus, this algorithm can be used to cluster mixed objects including numerical data and linguistic labels, necessarily, the cluster centers of each cluster should be numerical data in implement.

5 Experimental Studies

5.1 Distance Variation

Given a continue universe of discourse $[1, 20]$ of numerical data points defined on $\mathbb{R}$, and three labels L_1, L_2, L_3 which can be represented respectively by three sets of prototypical elements, $P_1 = \{1\}$, $P_2 = \{5, 5.5, 7\}$ and $P_3 = \{8, 8.5\}$. Given fixed $x_1 = 7.5$, when x_2 varies from 0 to 20, the variation of distance $D(x_1, x_2)$ is shown illustratively in Fig. 2.

This illustration indicates that the distance $D(x_1, x_2)$ varies rapidly when the data point x_2 is close to prototypes, in contrast, the variation becomes more and more slowly when the data point x_2 moves away from the prototypes. This phoneme can be interpreted as when data points are close to a linguistic concept, we can determine their dissimilarity according to the appropriateness of describing these objects using this concept more precisely than these data points are far away from this concept.

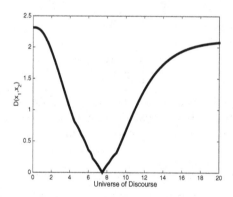

Fig. 2. Illustration of variation of distance between two data points: x_1 is fixed to 7.5 and x_2 varies from 0 to 20.

5.2 Clustering Images and Labels

In order to validate the performance of the novel distance measure for clustering images and vague concepts. We apply this measure in an extended K-means algorithm as introduced in Section 4 for clustering images and linguistic labels. We select 100 images from the Corel image data set[8] in 4 categories and 25 images in each category, each image is resized into 192×128, we chose 4 descriptive linguistic labels to describe images, including "sunset", "beach", "garden" and "polar bear". In our experiment, each image is represented as a 3-dimensional numerical data point according to its average HSV (Hue, Saturation, Value) [10] feature, besides, each label is represented as a set of 5 images which are randomly selected from the same category.

After designing labels and extracting image features, the data set of mixed objects including images represented as numerical data points and labels (as defined in Definition 8) represented as sets of 5 prototypical elements (images) for which the label is certainly an appropriate description. In our experiment, each label is represented by 5 images in same category, thus, the set which we have to cluster is constructed by 80 images (20 images in each category) and 4 labels, we cluster this set of mixed objects into 4 clusters. In our experiment, the 4 cluster centers are 4 data points which are randomly selected from the components of the 4 mutually different labels. Furthermore, a 20% cross-validation is used to evaluate the performance of proposed algorithm, each time we randomly change the 20 images (5 images of each label) which are considered as prototypical elements for representing labels, during the cross-validation process, each image is used as a component of label only once. The average accuracy of five times of experiment is regarded as the final experiment result, the comparison of accuracy between this method and the existing method proposed by Zhang and Qin [4], and their execution time to build the 4 clusters under the same hardware condition are shown in Table 2.

Further more, the illustration of above result and its variation is shown in Fig. 3.

Table 2. Performance of clustering mixed objects in terms of classification accuracy.

	Sunset	Beach	Garden	Polar-bear	Execution time
Our Model	81%	94%	67%	87%	18.2s
Zhang and Qin [4]	72%	60%	64%	96%	6318.1s

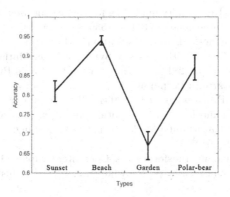

Fig. 3. Illustration of experimental result and its variation.

246 H. Zhao and Z. Qin

6 Conclusion

In this paper we proposed a novel distance measure based on prototype theory interpreted label semantics framework. This distance measure differs from the other distance measure by focusing on the difference of logical meanings which conveyed by the object. The new proposed distance measure is applicated to extend classical K-means algorithm for clustering numerical data and vague concepts which are in the form of linguistic labels. Experimental studies on a image clustering problem validated the effectiveness of our new proposed measure.

With a similar idea of measuring the dissimilarities according to the logical meaning which the object conveys, the proposed measure is extendable to measuring distances between any granularities, In future work, This measure can be applied to other applications and clustering vague concepts represented by other granularities.

Acknowledgments. This work is supported by the National Science Foundation of China No. 61305047.

References

1. Momin, B., Yelmar, P.: Modifications in K-Means Clustering Algorithm. International Journal of Soft Computing and Engineering (IJSCE) **2**(3) (2012)
2. Ben Hariz, S., Elouedi, Z., Mellouli, K.: Clustering Approach using belief function theory. In: Euzenat, J., Domingue, J. (eds.) AIMSA 2006. LNCS (LNAI), vol. 4183, pp. 162–171. Springer, Heidelberg (2006)
3. Ammar, A., Elouedi, Z.: A new possibilistic clustering method: the possibilistic K-modes. In: Pirrone, R., Sorbello, F. (eds.) AI*IA 2011. LNCS, vol. 6934, pp. 413–419. Springer, Heidelberg (2011)
4. Zhang, W., Qin, Z.: Clustering data and imprecise concepts. In: 2011 IEEE International Conference on Fuzzy Systems (FUZZ), pp. 603–608 (2011)
5. Lawry, J.: A framework for linguistic modelling. Artificial Intelligence **155**(1), 1–39 (2004)
6. Lawry, J., Tang, Y.: Uncertainty modelling for vague concepts: A prototype theory approach. Artificial Intelligence **173**(18), 1539–1558 (2009)
7. MacQueen, J.: Some methods for classification and analysis of multivariate observations. In: Proceedings of 5th Berkeley Symposium on Mathematical Statistics and Probability, vol. 1, pp. 281–297. University of California Press (1967)
8. He, X., Zemel, R., Carreira-Perpindn, M.: Multiscale conditional random fields for image labeling. In: Proceedings of the 2004 IEEE Computer Society Conference on Computer Vision and Pattern Recognition, vol. 2, p. 695 (2004)
9. Qin, Z., Tang, Y.: Uncertainty Modeling for Data Mining: A Label Semantics Approach. Springer (2014)
10. Smith, A.R.: Color gamut transform pairs. Computer Graphics **12**(3), 12–19 (1978)
11. Wikipedia. Cluster analysis. http://en.wikipedia.org/wiki/Cluster_analysis

An Ensemble Learning Approach Based on Rough Set Preserving the Qualities of Approximations

Seiki Ubukata[1](✉), Taro Miyazaki[2], Akira Notsu[1], Katsuhiro Honda[1], and Masahiro Inuiguchi[2]

[1] Graduate School of Engineering, Osaka Prefecture University,
Gakuencho 1-1, Sakai, Osaka 599-8531, Japan
{subukata,notsu,honda}@cs.osaka-fu.ac.jp
http://www.cs.osakafu-u.ac.jp/hi/
[2] Graduate School of Engineering Science, Osaka University,
Toyonaka, Osaka 560-8531, Japan
miyazaki@inulab.sys.es.osaka-u.ac.jp, inuiguti@sys.es.osaka-u.ac.jp

Abstract. In this paper, we confirm the effects of ensemble learning approaches in classification problems based on rough set. Furthermore, we propose an ensemble learning approach based on rough set preserving the qualities of approximations. The proposed method stands on a policy that subsets of attributes whose quality of lower approximation is less than the threshold value is not tolerate. We carried out numerical experiments in order to confirm the classification performance of the proposed method and confirmed its effectiveness.

Keywords: Rough sets · Ensemble learning · The qualities of approximations · MLEM2

1   Introduction

Rough set theory have been applied to various issues such as decision rule induction form decision tables and classification problems based on decision rules [3], [4], [5]. In classification problems, ensemble learning approaches such as bagging and random forests have attracted as techniques in handling the over-fitting problem.

In this paper, we confirm the effects of ensemble approaches in classification problems based on rough set. We apply bagging which include object sampling (bootstrap sampling), RF (random forests) which includes object sampling and attribute sampling, and ASE (attribute sampling ensemble) which includes only attribute sampling. We use decision rules extracted by MLEM2 rule induction algorithm as a base classifier.

Furthermore, we propose an ensemble learning approach based on rough set preserving the qualities of approximations. The reason that we introduce the preserving procedure is that decreases of qualities of approximation of decision

© Springer International Publishing Switzerland 2015
V.-N. Huynh et al. (Eds.): IUKM 2015, LNAI 9376, pp. 247–253, 2015.
DOI: 10.1007/978-3-319-25135-6_24

tables which are generated by the attribute sampling in an ensemble learning procedure may cause decreases in the classification performance. In ensemble approaches which include attribute sampling such as RF and ASE, how much condition attributes should be extracted becomes an important issue. We introduce the quality of approximation as an aid for automatic determination of the number of attribute sampling in sampling processes of ensemble learning approaches.

We carried out numerical experiments in order to confirm the classification performance of the proposed method and confirmed its effectiveness.

2   Preliminaries

In this section, we provide a basic explanation of classifications based on rough sets and ensemble learning approaches.

2.1   Classifications Based on Rough Sets

In rough set analyses, a data set is often expressed as a decision table. A decision table is defined by $\langle U, C \cup \{d\}, V, \rho \rangle$, where U is a non-empty finite set of objects, C is a finite set of condition attributes, $\{d\}$ is a singleton of a decision attribute, $V = \bigcup \{V_a\}_{a \in C \cup \{d\}}$ is a set of attribute values, and $\rho : U \times C \cup \{d\} \rightarrow V$ is a information function which assigns an attribute value of an object. In rough sets, binary relations are used for granulation of a universe and approximations of subsets of objects. A indiscernibility relation is defined by

$$IND(B) = \{(x, y) \in U \times U \mid \forall a \in B, \rho(x, a) = \rho(y, a)\},$$

where $B \subseteq C \cup \{d\}$. Using this relation, the lower and upper approximations of X are obtained as follows:

$$IND(B)[X] = \{x \in U \mid [x]_{IND(B)} \subseteq X\},$$

$$IND(B)\langle X \rangle = \{x \in U \mid [x]_{IND(B)} \cap X \neq \emptyset\},$$

where $[x]_{IND(B)}$ is the equivalence class of x with respect to $IND(B)$. The indiscernibility relation means that two objects are indistinguishable each other by use of focused attributes. The quality of approximation of a partition P by a subset B of attributes is defined as

$$\gamma_B(P) = \sum_{X \in P} \frac{card(IND(B)[X])}{card(U)}.$$

J. W. Grzymala-Busse has been developed the data mining system LERS and proposed a framework of decision rule induction based on rough sets and classification by means of decision rules [3], [4]. MLEM2 is a subsystem of LERS and an algorithm which induces the minimal set of decision rules which have minimal condition parts from the given decision table. Decision rules which are extracted from lower approximations are certain rules. On the other hand, decision rules which are extracted from upper approximations are possible rules.

2.2 Ensemble Learning Approaches

In classification problems, ensemble learning approaches are attracting attention as useful ways to address over-fitting problems and improve the classification performance. Let us explain typical ensemble learning approaches, that is, bagging and random forests.

Bagging. Bagging (bootstrap aggregating) was proposed by L. Breiman in 1996 [1]. In bagging approach, multiple new tables are generated by bootstrap sampling from the given data. Bootstrap sampling is random sampling of objects with replacement. Using a certain method, weak classifiers are obtained from each generated decision table. Finally, one prediction is determined by aggregating predictions of these weak classifiers. A majority vote is a common aggregating method. Aggregations of predictions of weak classifiers (cursory classifiers) realize low-variance and better classification performance for unknown objects.

Random Forests (RF). Random forests was also proposed by L. Breiman in 2001 [2]. In random forests, attribute sampling is carried out in addition to bootstrap sampling in bagging. Predictions are executed in the same way as bagging. Random forests is expected to cause lower variance and better classification performance than bagging.

3 An Ensemble Learning Approach Based on Rough Set Preserving the Qualities of Approximations

In this paper, we propose an ensemble learning approach based on rough set preserving the qualities of approximations. The proposed method is applied to ensemble learning approaches which include an attribute sampling process. In such approaches, how much attributes should be extracted becomes an important issue.

In order to cope with the issue, we focus on the quality of approximations which is a basic concept of rough sets. Decision tables which have few attributes may cause decreases in the qualities of lower approximations of the partition of the decision classes. Hence, the covering range of decision rules which are induced form the lower approximations become reduced in size. Moreover, it worsens explanation capability of known objects and unknown objects, and the performance of classification. The proposed method stands on a policy that subsets of attributes whose quality of lower approximation is less than the threshold level is not tolerate.

In the previous section, we introduced random forest as an ensemble approach which include attribute sampling process. In addition, we use following ensemble approach as an ensemble approach which include attribute sampling process.

Table 1. MLEM2 and Ensemble Learning Combinations

	No bootstrap sampling	Bootstrap sampling
No attribute sampling	MLEM2	MB
Attribute sampling	MA	MRF
Attribute sampling preserving the qualities of approximations	MAQ	MRFQ

Attribute Sampling Ensemble (ASE). In RF, both bootstrap sampling of objects and attribute sampling are executed in order to generate new decision tables for weak classifiers. In this paper, we call the method in which only attribute sampling is executed as attribute sampling ensemble (ASE).

3.1 The Procedure of the Proposed Method

The proposed method is applied to ensemble learning approaches which include attribute sampling, that is, to RF and ASE. In the proposed method, sets of attributes whose qualities of approximation are more than a certain threshold γ_{Th} are extracted in the attribute sampling process. We describe the procedure of the proposed method as follows.

Step 1. Let s be an empty attribute sequence, γ_{Th} be the threshold value of the quality of lower approximation, and n_e be the number of ensemble.
Step 2. $C_{temp} := C$.
Step 3. Randomly choose an attribute $a \in C_{temp}$ and add it to the end of sequence s, and update $C_{temp} := C_{temp} \setminus \{a\}$.
Step 4. Let B_s be the set of attributes which are components of the sequence s.
Step 5. Calculate $\gamma_{temp} = \gamma_{B_s}(U/IND(\{d\}))$.
Step 6. If $\gamma_{temp} < \gamma_{Th}$, repeat Step 3 to Step 5. Otherwise, go to Step 7.
Step 7. Add further $m - card(B_s)$ attributes randomly in the range of $card(B_s) \leq m \leq card(C)$.
Step 8. Generate the decision tables based on the attribute sequence s.
Step 9. Execute Step 2 to Step 8 in n_e times and generate n_e different decision tables, and create weak classifiers based on the tables.
Step 10. Aggregate the predictions of the weak classifiers just like ordinary ensemble methods.

In this research, γ_{Th} was set to $\gamma_{MAX} = \gamma_C(U/IND(\{d\}))$.

Note that MLEM2 algorithm depends on the alignment sequence of attributes in case of the coincidence of priority of conditional propositions. Thus we generate new decision tables based on attribute sequences in order to grow in diversity of decision rules.

Table 1 shows the summery of characteristics of introduced classification approaches and their abbreviated expressions.

· **Table 2.** Data set summary.

Data set	#Objects	#Attributes	#Classes	Attribute Type
Ecoli	336	7	8	numerical
Glass	214	9	6	numerical
Iris	150	4	3	numerical
Soybean_small	47	35	4	nominal
Wine	178	13	3	numerical
Zoo	101	16	7	nominal

4 Numerical Experiments

We carried out numerical experiments in order to confirm the classification performance of the proposed method. In this section, we present the experimental results and some considerations of it.

We used six data sets shown in Table 2. The data sets were retrieved from the UCI machine learning repository [6]. The application was implemented using Java and a desktop computer equipped with CPU Intel(R) Core(TM) i7-4770 @3.40GHz.

In order to measure the classification performance, we determined the average of error rates in 10 times 10-fold cross-validation. The following it the list of classifiers and their abbreviation which are used in the experiments.

MLEM2. The classifier based on MLEM2 decision rules and the LERS classification regulation.

MB. The bagging classifier based on MLEM2.

MRF. The RF classifier based on MLEM2.

MA. The ASE classifier based on MLEM2.

MRFQ. The MRF classifier preserving the qualities of approximations.

MAQ. The MA classifier preserving the qualities of approximations.

In all ensemble methods, we set up the ensemble number $n_e = 10$. In MB and MRF, the number of bootstrap samples is $n_b = card(U)$.

4.1 Results

Fig. 1 shows the classification error rates of each method for each data set. The vertical axis indicates classification error rates. The horizontal axis indicates the sampling number of attributes. Note that the classification performances of MRF and MA depend on a sampling number of attributes.

Let us consider the results.

- In all data sets except for 'Soybean_small', MB provides better classification accuracy than MLEM2.
- In ensemble approaches which include attribute sampling such as MRF and MA, the classification performances are depend on the number of sampling

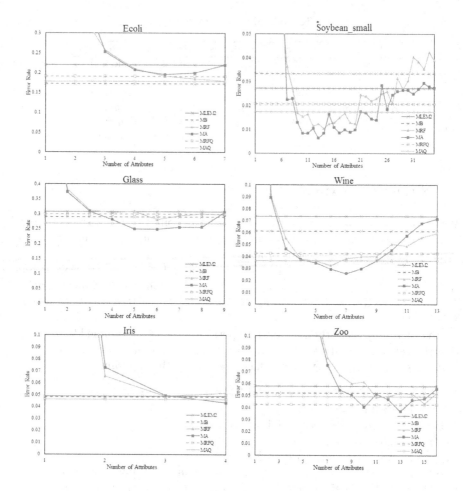

Fig. 1. Classification error rates of each method for each data set.

attributes. If almost every attributes are extracted, MRF and MA provide almost the same performance as MB and MLEM2, respectively. Gradually decreasing the number of sampling attributes, the classification performances get better gradually. This phenomenon appears prominently in 'Soybean_small' and 'Wine'.

- However, extreme declines of the number of sampling attributes cause extremely poor performance.
- There are data sets in which bootstrap sampling or attribute sampling have poor efficacy in improving the performance, e. g. 'Ecoli' and 'Iris'.
- In data sets in which attribute sampling works, MRFQ and MAQ provide better results than MB and MLEM2 and can close their performance to the best results of MRF and MA in all trial of numbers of attributes sampling.

The proposed method do not need to try all the number of sampling attributes and thus is much more efficient than MRFQ and MAQ in terms of time. Although the performances of MRFQ and MAQ fall just one step short of the best results of MRF and MA, their performances are well on to the best results. Thus, the proposed method is considered a valuable method which provide reasonably good performance effectively.

5   Concluding Remarks

In this paper, we confirmed the effects of ensemble approaches in classification problems based on rough set. Furthermore, we proposed an ensemble learning approach preserving the qualities of approximations and confirmed its effectiveness through numerical experiments. In this method, we introduced the quality of approximation as an aid for semi-automatic determination of the number of attribute sampling in ensemble learnings. The method stands on a policy that subsets of attributes whose quality of lower approximation is less than a certain threshold value is not tolerate. We carried out numerical experiments in order to confirm the classification performance of the proposed method and confirmed its effectiveness. As a result, the proposed method is considered a valuable method which provide a reasonably good performance effectively.

At present stage, the proposed method excludes sets of attributes whose cardinality is not enough. However, excessive set of attributes may also provide mediocre performances. In terms of future issues, we plan to consider a method which can also excludes excessive set of attributes.

References

1. Breiman, L.: Bagging predictors. Machine Learning **24**, 123–140 (1996)
2. Breiman, L.: Random Forests. Machine Learning **45**, 5–32 (2001)
3. Grzymala-Busse, J.W., Slowinski, R. (ed.): LERS-A System for Learning from Examples Based on Rough Sets. Handbook of Application and Advances of the Rough Set Theory, vol. 11, pp. 3–18. Kluwer Cademic Publishers (1992)
4. Grzymala-Busse, J.W.: MLEM2-discretization during rule induction. In: Proceeding of International Conference on Intelligent Information Processing and WEB Mining Systems, vol. 22, pp. 499–508 (2003)
5. Pawlak, Z.: Rough Sets. International Journal of Computer and Information Sciences **11**, 5 (1982)
6. UCI machine learning repository. http://archive.ics.uci.edu/ml/

Minimum Description Length Principle
for Compositional Model Learning

Radim Jiroušek[1,2]([✉]) and Iva Krejčová[1]

[1] Faculty of Management, University of Economics,
Jindřichův Hradec, Czech Republic
radim@utia.cas.cz, iva.krejcova@gmail.com
[2] Institute of Information Theory and Automation,
Prague Czech Academy of Sciences, Prague, Czech Republic

Abstract. Information-theoretic viewpoint at the data-based model construction is anchored on the assumption that both source data and a constructed model comprises certain information. Not having another source of information than source data, the process of model construction can be viewed at as the transformation of information representation. The combination of this basic idea with the Minimum Description Length principle brings a new restriction on the process of model learning: avoid models containing more information than source data, because these models must comprise an additional undesirable information. In the paper, the idea is explained and illustrated on the data-based construction of multidimensional probabilistic compositional models.

Keywords: Machine learning · Multidimensional models · Probability distributions · Composition · Information theory · Lossless encoding

1 Introduction

Minimum Description Length (MDL) principle has been used for model learning by a whole range of authors. In connection with Bayesian network learning let us mention for example Lam and Bacchus [10], (for general sources see also [2], and [3]). These authors regarded MDL as an application of a Occam's razor philosophical principle, which says that the best solution is more likely the simplest. In this paper we will study this approach also from another point of view, from the point of view of information theory.

Data-based model learning is usually based on the following simple idea: the data in question were generated by a generator whose probabilistic characteristics are unknown, but, in a way, stabile. If we do not have another source of information (such as, for example, some theoretical knowledge about the field of interest) all we know about this generator is encoded in the data file. So, when reconstructing the generator we should exploit as much of information contained in the data file as possible, but we should avoid adding any other undesirable information. In this sense, the process of model construction can be

© Springer International Publishing Switzerland 2015
V.-N. Huynh et al. (Eds.): IUKM 2015, LNAI 9376, pp. 254–266, 2015.
DOI: 10.1007/978-3-319-25135-6_25

viewed as a transformation of the data file into the constructed model. Since it is well-known that during any transformation process the amount of information cannot increase, we should check what is the amount of information before the transformation (i.e., the information contained in the input data file) and after the transformation (i.e., the information contained in the constructed model). Using this idea, models containing more information than the input data file will be considered unacceptable because, obviously, some undesirable information was added.

Accepting the above mentioned principle, a new problem arises: how to measure the above mentioned information amounts. Our proposal is to measure this information in bits necessary for the optimum lossless encoding of the data/model. In this context we take advantage of the old ideas of von Mises [13] and Kolmogorov [8] who both explored relations interconnecting randomness, complexity and information. So, we accept the principle: the more complex model, the more information it comprises. Nevertheless, realize that looking for the optimum lossless encoding would be in practical situations intractable. Therefore, we use in this paper some heuristics and also a famous Huffman's encoding [4], which is known to be optimal under some conditions. Thus, though the encoding used in this paper is only suboptimal, it will serve well to the purpose of this paper: to show that application of MDL principle is not as straightforward as it can seem at the first glance. We will show that the users should find a reasonable equilibrium balancing the complexity of the model structure and the preciseness of specified parameters.

The proposed approach is fully sensible also from the statistical point of view. The less data we have, the less amount of bits we may use to encode the model. It means, among others, that for small data files we cannot consider probability values specified with a high precision. This fully corresponds with the fact that having a small number of data, the confidence intervals for the estimates of probability parameters are rather wide. Therefore it does not have a sense to specify these estimates with a high precision, with a great number of digits.

Thus, the goal of this paper is not to introduce a new algorithm for data-based model learning. The paper presents two simple ideas that should be incorporated into any data-based learning algorithm and that we have not found in the literature. First, the amount of input data determines the upper limit to the complexity of the constructed model. It is against a common sense (and also against the information-theoretic principles) to construct a model whose encoding requires more bits than the input data file. The other idea is that the users should decide whether it is more advantageous to consider either simpler structure models with more precise parameters, or models with more complex structures, i.e., more parameters specified with lower precision.

The application of the above mentioned ideas will be illustrated on learning compositional models [6] that will be briefly introduced in the next section. Sections 3 and 4 will be devoted to the discussion of possibilities how to encode data and models, respectively, and Section 5 briefly describes two ways how to simplify constructed models to meet the upper limit given by the size of the input data.

2   Compositional Models

As said above, in this section, we will briefly introduce the models to be constructed from data; for more details and the properties of these models the reader is refereed to [6].

In the whole paper we consider a finite set of finite-valued random variables $N = \{X_1, X_2, \ldots, X_n\}$. Probability distributions (measures) will be denoted by the characters of Greek alphabet, as e.g., $\pi(N)$. Its marginal distribution for variables $M \subseteq N$ will be denoted either $\pi(M)$, or $\pi^{\downarrow M}$. Let $\mathbb{X}_i$ denote the set of values of variable X_i. It means that a probability distribution $\pi(N) : \mathbb{X}_N \to [0,1]$ is defined with the help $|\mathbb{X}_N|$ numbers (probabilities), where $\mathbb{X}_N$ denotes the Cartesian product $\mathbb{X}_N = \mathbb{X}_1 \times \mathbb{X}_2 \times \ldots \times \mathbb{X}_n$, e.g., the space of all states of variables N. Analogously, for a subset of variables $K \subset N$, $\mathbb{X}_K = \times_{u \in K} \mathbb{X}_u$.

For two distributions $\pi(N)$ and $\kappa(N)$, we say that κ dominates π (in symbol $\pi \ll \kappa$) if for all $x \in \mathbb{X}_N$, for which $\kappa(x) = 0$ also $\pi(x) = 0$. As a measure of similarity of two distributions we will consider their *Kullback-Leibler divergence* [9] (or crossentropy) defined

$$
Div(\pi; \kappa) = \begin{cases} \sum_{x \in \mathbb{X}_N : \pi(x) > 0} \pi(x) \log \frac{\pi(x)}{\kappa(x)} & \text{if } \pi \ll \kappa, \\ +\infty & \text{otherwise,} \end{cases}
$$

which is known to be zero if and only if $\pi = \kappa$.

Compositional models considered in this paper are multidimensional probability distributions that are assembled (*composed*) from its low-dimensional marginals with the help of a so called *operator of composition*. This operator realizes an operation in a way inverse to marginalization. For a probability distribution $\mu(N)$ and $J, K \subset N$, such that $J \cup K = N$, the respective marginal distribution $\mu^{\downarrow J}$ and $\mu^{\downarrow K}$ are unique. On the other side, if $J \neq N$ and $K \neq N$ then there are (infinitely) many distributions $\nu(N)$ such that $\nu^{\downarrow J} = \mu^{\downarrow J}$ and $\nu^{\downarrow K} = \mu^{\downarrow K}$. All these distributions ν are called *join extensions of* $\mu^{\downarrow J}$ *and* $\mu^{\downarrow K}$. One of them can be got by the application of the following operator of composition.

Definition 1. *For two arbitrary distributions $\pi(M)$ and $\lambda(L)$, for which $\pi^{\downarrow M \cap L} \ll \lambda^{\downarrow M \cap L}$, their composition is, for each $x \in \mathbb{X}_{L \cup M}$, given by the following formula[1]*

$$
(\pi \triangleright \lambda)(x) = \frac{\pi(x^{\downarrow M}) \lambda(x^{\downarrow L})}{\lambda^{\downarrow M \cap L}(x^{\downarrow M \cap L})}.
$$

In case $\pi^{\downarrow M \cap L} \not\ll \lambda^{\downarrow M \cap L}$, the composition remains undefined.

---

[1] In this paper we take $\frac{0.0}{0} = 0$ by definition.

Notice that the composition $\mu^{\downarrow J} \triangleright \mu^{\downarrow K}$ is always defined (because the marginals $\mu^{\downarrow J}$ and $\mu^{\downarrow K}$ are consistent), and that the distribution $\mu^{\downarrow J} \triangleright \mu^{\downarrow K}$ need not coincide with $\mu(N)$. It is easy to show (see [6]) that for the composed distribution $\mu^{\downarrow J} \triangleright \mu^{\downarrow K}$ variables $J \setminus K$ and $K \setminus J$ are conditionally independent[2] given variables $K \cap J$.

If we compose two general distributions $\pi(M)$ and $\lambda(L)$, and the composition $\pi \triangleright \lambda$ is defined, then the result is a distribution of variables $M \cup L$, and it is an extension of distribution π (see [6]), which is as similar as possible to a given distribution λ in the following sense (see Theorem 6.2 in[6])

$$\pi \triangleright \lambda = \arg \min_{\kappa(L \cup M): \kappa^{\downarrow M} = \pi} Div(\kappa^{\downarrow L}; \lambda).$$

Notice that if $\pi \triangleright \lambda$ is defined, then this minimum is unique. This is also the reason why we can say that $\pi \triangleright \lambda$ is a *projection* of λ into the set (space) of all the extensions of π for variables $L \cup M$ [1].

In this paper we are not interested in computational properties of distributions represented in a form of (iterative) compositions, so we need not present the algebraic properties of the operator of composition; for them, the reader is referred to [6]. Instead, let us present the definition of a compositional model.

Definition 2. *Distribution $\kappa(N)$ is a* compositional model *if there exists a cover $K_1, K_2, \ldots, K_m$ (i.e., $K_1 \cup \ldots \cup K_m = N$), such that[3]*

$$\kappa(N) = \kappa^{\downarrow K_1} \triangleright \kappa^{\downarrow K_2} \triangleright \ldots \triangleright \kappa^{\downarrow K_m}. \qquad (1)$$

Let us conclude this section by stating that the class of compositional models is exactly the same as the class of Bayesian networks [5].

3 Coding Data

The goal of this and the next section is not to find algorithms encoding compositional models and/or data files but just to estimate how many bits are necessary for such encodings. These numbers will be used to measure complexity of the respective models (data). More precisely, these numbers will be used when we will compare the complexity of two models, or the complexity of a model and

---

[2] Recall that for distribution $\kappa(N)$ variables K and L are conditionally independent given variables M $(K, L, M \subseteq N$ are assumed to be disjoint) if

$$\kappa(K \cup L \cup M) \cdot \kappa(M) = \kappa(K \cup M) \cdot \kappa(L \cup M).$$

[3] Since the operator of composition is not associative, we have to say how to understand the expression (1): If not specified otherwise by parentheses, the operator is always performed from left to right, i.e.,

$$\kappa^{\downarrow K_1} \triangleright \kappa^{\downarrow K_2} \triangleright \ldots \triangleright \kappa^{\downarrow K_m} = \left(\ldots \left((\kappa^{\downarrow K_1} \triangleright \kappa^{\downarrow K_2}) \triangleright \kappa^{\downarrow K_3} \right) \triangleright \ldots \triangleright \kappa^{\downarrow K_{m-1}} \right) \triangleright \kappa^{\downarrow K_m}.$$

the complexity of data. This is why we will not consider coding the number of variables, variable names and the cardinality of their value sets. Coding this information would just increase all the derived complexity measures by a constant. Therefore, without loss of generality we can assume in this paper that variable X_i is identified by its index i, and their values are $\mathbb{X}_i = \{0, \ldots, h_{i-1}\}$.

Under the above assumption when encoding a data file $\mathbf{D}$ we have to encode a matrix of nonnegative integers with d rows (records of the data file) and n columns (variables). For this we will consider several simple procedures. Let us repeat once more that we are aware of the fact that using more sophisticated types of codes, such as e.g. arithmetic codes [14], we could achieve even more economic encoding. The following codes are selected as a trade-off between precision and simplicity of the following exposition.

Direct Encoding. For a binary variable we need just one bit for each entry of the matrix. If the respective $h_i > 2$ then we use[4] $\ell_i = \lceil \log_2 h_i \rceil$ bits to encode the value of variable X_i. Therefore, for the direct encoding of the data file we need

$$c_d(\mathbf{D}) = d \times (\ell_1 + \ell_2 + \ldots + \ell_n) + c$$

bits, where c denotes the number of bits necessary to encode the number of records d (the number of rows in the matrix).

Frequency Encoding. For this coding we will take advantage of the fact that we need not consider the ordering of records in the data file. We increase the data matrix by one column into which we insert the number of repetition of each state (by *state* we understand the combination of values of all variables) in the data file. It enables us to keep in the matrix each state only once. Thus, denoting d_{red} the number of different states appearing in the original data file, and denoting f_{max} the maximal number of occurrences of the same state in the data file, then for this type of encoding we need

$$c_f(\mathbf{D}) = d_{red} \times (\ell_1 + \ell_2 + \ldots + \ell_n + \lceil \log_2(f_{max} - 1) \rceil) + 2 \times c$$

bits. $\lceil f_{max} - 1 \rceil$ appears in the formula, because all the numbers of repetition in the $(n+1)$th column are numbers from $1, \ldots, f_{max}$, and thus we can encode them as numbers from $0, \ldots, f_{max} - 1$, and $2 \times c$ bits are necessary to encode d_{red} a f_{max}.

Huffman Frequency Encoding. By this term we understand coding of the same table like in the previous case but for coding the numbers of occurrence we use the famous Huffman code [4]. The number of necessary bits for this code will be denoted by $c_{fH}(\mathbf{D})$ (see an example below).

---

[4] $\lceil r \rceil$ denotes the smallest integer, which is not less than r.

Lexicographic Encoding. Analogously to preceding type of encoding, consider an extended data matrix in which each state appears maximally once, and the $(n+1)$th column contains the number expressing how many times the state appears in the original data file. If the number of variables is rather small, it may happen that the following encoding of the considered matrix is more economic than that by frequency encoding: add to the matrix all states that do not appear in data (with number of repetition equaling 0), sort all the states in the lexicographic order, and then we can encode only the numbers from the $(n+1)$th column. This coding requires

$$c_l(\mathbf{D}) = |\mathbb{X}_N| \times \lceil \log_2 f_{max} \rceil + c$$

bits (realize the last c bits are used to encode f_{max}).

Huffman Lexicographic Encoding. As in the previous case we code only frequencies for all $|\mathbb{X}_N|$ combinations, for which we use the Huffman encoding. For real data files, Huffman process usually yields a code with the average length less that two bits per number (this is because in practical situations numbers of repetition greater than 1 are rare).

Naturally, the readers can extend the list of the considered data encoding possibilities by as many other approaches as they want (e.g. see [12]). In this paper we consider the complexity measure for the data file just

$$c(\mathbf{D}) = \min\{c_d(\mathbf{D}), c_f(\mathbf{D}), c_{fH}(\mathbf{D}, c_l(\mathbf{D}), c_{lH}(\mathbf{D})\}.$$

Example. The ideas presented in this paper will be illustrated by an example with artificially generated data. For the sake of simplicity we consider in this example just eight binary variables (with values $0, 1$), and a data files with 100 records (binary vectors). In spite of this we fix the number of necessary bits to encode the length of the data file to $c = 32$, because we made experiments with much bigger data files (up to 100 000 records). Recall that we neglect coding the information about the model.

To apply the direct encoding approach, when taking into account the considered small data file we need to encode the following table

$$d = 100 \begin{cases} 0 & 0 & 1 & 1 & 0 & 1 & 0 & 1 \\ 1 & 0 & 1 & 1 & 0 & 0 & 0 & 1 \\ & & & \vdots & & & \\ 1 & 1 & 0 & 0 & 0 & 0 & 1 & 0 \end{cases}$$

which means that we need $c_d(\mathbf{D}) = 100 \times 8 + 32 = 832$ bits.

To encode the same data file with the frequency encoding, transform first the data file into the form, in which all the rows (states) are unique and the last column contains the number of occurrences of the respective state in the original data file. For the considered data file we get the following table

$$d_{red} = 38 \begin{cases} \begin{array}{cccccccc|c} 0 & 0 & 1 & 1 & 0 & 1 & 0 & 1 & 27 \\ 1 & 0 & 1 & 1 & 0 & 0 & 0 & 1 & 10 \\ & & & \vdots & & & & & \\ 1 & 1 & 0 & 0 & 0 & 0 & 1 & 0 & 1 \end{array} \end{cases}$$

Thus we get $c_f(\mathbf{D}) = 38 \times (8+5) + 2 \times 32 = 558$ bits.

To get what we call Huffman version of frequency encoding we need to find Huffman code for the numbers of occurrences. In our case such a code is the following (the numbers in parentheses - the last column - read how many times the respective frequency number appears in the above table)

27	11111	(1×)
10	11110	(1×)
5	1110	(2×)
3	110	(5×)
2	10	(9×)
1	0	(20×)

Thus, using Huffman version of frequency encoding we have to encode the above coding table (which can easily be done with $6 \times (5+5) = 60$ bits, and for coding the numbers of occurrences we need only $2 \times 5 + 2 \times 4 + 5 \times 3 + 9 \times 2 + 20 \times 1 = 71$ bits (instead of $38 \times 5 = 190$, which is needed for the frequencies encoding in the previous case). So, we get $c_{fH}(\mathbf{D}) = 38 \times (8) + 60 + 71 + 2 \times 32 = 499$ bits.

To get the lexicographic encoding we have to consider all 2^8 states lexicographically ordered

$$256 \begin{cases} \begin{array}{cccccccc|c} 0 & 0 & 0 & 0 & 0 & 0 & 0 & 0 & 1 \\ 0 & 0 & 0 & 0 & 0 & 0 & 0 & 1 & 0 \\ 0 & 0 & 0 & 0 & 0 & 0 & 1 & 0 & 0 \\ & & & \vdots & & & & & \\ 0 & 0 & 1 & 1 & 0 & 1 & 0 & 1 & 27 \\ & & & \vdots & & & & & \\ 1 & 1 & 1 & 1 & 1 & 1 & 1 & 1 & 0 \end{array} \end{cases}$$

So, lexicographic encoding of the framed frequencies requires $c_l(\mathbf{D}) = 256 \times 5 + 32 = 1312$ bits. However, if we use Huffman approach to encode all the numbers appearing in the frame, i.e., if we use the following code

27	111111	(1×)
10	111110	(1×)
5	11110	(2×)
3	1110	(5×)
2	110	(9×)
1	10	(20×)
0	0	(218×)

Table 1. Requirements for coding the data files

	c_d	c_f	c_{fH}	c_l	c_{lH}
D_{100}	832	558	499	1,312	**404**
D_{1000}	8,032	1,680	**976**	2.080	992
D_{10000}	80,032	4,084	2,713	3,104	**2,362**
D_{100000}	800,032	5,676	5,800	**3,872**	4,775

we need only $7 \times (5 + 6) = 77$ bits to encode this coding table, and $c_{lH}(\mathbf{D}) = 77 + 2 \times 6 + 2 \times 5 + 5 \times 4 + 9 \times 3 + 20 \times 2 + 218 \times 1 = 404$ bits.

To illustrate the way how these complexity measures increase with the amount of the considered data we generated (using the same generator) another three data files with 1 000, 10 000 and 100 000 records. A summary of the bit requirements to encode all these data files is in Table 1.

4   Coding Models

To encode a compositional model given by Formula (1) we have to encode marginal distributions $\kappa^{\downarrow K_1}, \kappa^{\downarrow K_2}, \ldots, \kappa^{\downarrow K_m}$ in a proper order. Each of these distributions $\kappa^{\downarrow K_i}$ is described by the list of variables, i.e.,

number of variables $|K_i|$ $\lceil \log_2 n \rceil$ bits
list of variables $|K_i| \times \lceil \log_2 n \rceil$ bits

and the respective probabilities, whose total number is $\prod_{u \in K_i} h_u$. Obviously, encoding the probabilities is, as a rule, much more space demanding that encoding the variables, for which the respective marginal is defined. The latter encoding requires, as presented above, only $(|K_i| + 1) \times \lceil \log_2 n \rceil$ bits.

Naturally, the space requirements for the probability encoding is closely connected with the precision with which the respective probabilities should be specified. A simple way, which is used in this paper, is the following.

Select a positive integer, denote it *base*, and express all the considered probabilities as a ratio of two nonnegative integers

$$\frac{a}{base}.$$

This means that the respective probability will be encoded by integer a. From the obvious reasons it does not have a sense to choose $base > d$ (recall that d is the number of records in the input data file). However, $base$ may be much smaller than d and can be defined with respect to the size of confidence intervals computed for the probability estimates, or it can be reduced when we want to reduce the complexity of the constructed compositional model (as shown in the next section).

By employing the idea of representing probabilities by integers we get, in fact, exactly the same situation as that in the previous section: marginal distribution

$\kappa^{\downarrow K_i}$ is fully described by those states $x \in \mathbb{X}_{K_i}$, for which the probability $\kappa^{\downarrow K_i}(x)$ is positive and by the respective integer representing value $\kappa^{\downarrow K_i}(x)$. It means that for encoding the marginal distributions $\kappa^{\downarrow K_i}$ we can employ any of the techniques described in the previous section (perhaps, application of the *direct encoding* comes into consideration in very specific and unusual situations, though). As a rule, the most economic encoding is yielded by *Huffman lexicographic encoding*. *Frequency encoding* (both plain and Huffman's) may be applicable only for more-dimensional distributions, which are positive on a small part of the respective space $\mathbb{X}_{K_i}$.

Thus, when encoding compositional models we will face the only problem: whether it is more economic to construct a Huffman code specially for each marginal distribution (and thus also code the respective coding table), or construct one code for coding all the marginals from which the model is composed.

An analogous problem is connected with the selection of the number *base*. In this paper we consider only simple models and therefore we use one number *base* for the whole model. However, the reader certainly realizes that in some situations a greater chances to decrease the complexity of the model can be achieved when defining different $base_i$ for different marginals. Namely, the necessity of coding one number $base_i$ for each marginal distribution can be payed back by the savings achieved for coding the respective probabilities.

Example Continued. Let us illustrate the principles described above by coding a model

$$\mathbf{M_1} : \ \mu_1 = \kappa^{\downarrow\{1,2\}} \triangleright \kappa^{\downarrow\{3,4\}} \triangleright \kappa^{\downarrow\{3,5\}} \triangleright \kappa^{\downarrow\{1,4,5,6\}} \triangleright \kappa^{\downarrow\{5,6,8\}} \triangleright \kappa^{\downarrow\{2,5,6,7,8\}} \quad (2)$$

constructed from the considered data file $\mathbf{D}$ with 100 records. To describe a structure of the model we need to specify the number of marginal distributions $m = 6$, number $base = 100$.

Thus, the structure of the model (2) can be described with the help of $\lceil \log_2 n \rceil + c = 3 + 32 = 35$ bits, and to encode k-dimensional distribution by lexicographic encoding we need either:

number of variables k	$\lceil \log_2 n \rceil$ bits,
list of variables	$k \times \lceil \log_2 n \rceil$ bits,
frequencies (probabilities)	$2^k \times \lceil \log_2 base \rceil$ bits,

or, in the case that specification of the maximal frequencies for each marginal $f_{max,i}$ pays back by savings gained for more economic specification of all frequencies,

number of variables k	$\lceil \log_2 n \rceil$ bits,
list of variables	$k \times \lceil \log_2 n \rceil$ bits,
maximal frequency $f_{max,i}$	$\lceil \log_2 base \rceil$ bits,
frequencies (probabilities)	$2^k \times \lceil \log_2 f_{max,i} \rceil$ bits.

In our case the two approaches differ just by 18 bits, so let us consider the simpler (the former) approach. Thus we need for $\kappa^{\downarrow\{1,2\}}$, $\kappa^{\downarrow\{3,4\}}$, $\kappa^{\downarrow\{3,5\}}$: $3 + 6 + 28 = 37$ bits,

for $\kappa^{\downarrow\{1,4,5,6\}}$: $3 + 12 + 112 = 127$ bits,

for $\kappa^{\downarrow\{5,6,8\}}$: $3 + 9 + 56 = 68$ bits,

for $\kappa^{\downarrow\{2,7,5,6,8\}}$: $3 + 15 + 224 = 242$ bits,

which means that $c_l(\mathbf{M_1}) = 583$.

Taking into account the fact that among the 68 frequencies (probabilities) needed to represent the respective six marginals there appears twenty times "0" and sixteen times "1", it is not surprising that a more economic encoding is achieved by Huffman's version of lexicographic encoding, which yields for this model $c_{lH}(\mathbf{M_1}) = 423$. In any case, whatever type of encoding we may take into consideration we cannot reach the coding requirements sufficient to encode data $c_{lH}(\mathbf{D}) = 404$. This means that the model $\mathbf{M_1}$ described by formula (2) with probabilities specified with the help of $base = 100$ is unacceptable, and therefore, to meet the information-theoretic viewpoint at MDL principle described in Introduction, we have to simplify the considered model by any of the possibilities described in the next section.

5 Model Simplification

Perhaps the easiest way how to simplify the constructed model is to roughen the probability estimates by decreasing the constant $base$. Considering model $\mathbf{M_1}$ with $base = 100$ means that we take all the probability estimates with two digits of precision. Rounding these estimates to one decimal digit means to consider $base = 10$. Nevertheless, it is important to realize that we can consider finer roughening choosing any $10 < base < 100$. Denote $c_{lH}(\mathbf{M_{1:50}})$, $c_{lH}(\mathbf{M_{1:40}})$ and $c_{lH}(\mathbf{M_{1:32}})$ complexity of Huffman lexicographic encoding of model $\mathbf{M_1}$ with $base$ equaling 50, 40 and 32, respectively. Then for the probability estimates got from data file $\mathbf{D}$ we have $c_{lH}(\mathbf{M_{1:50}}) = 408$, $c_{lH}(\mathbf{M_{1:40}}) = 397$, and $c_{lH}(\mathbf{M_{1:32}}) = 284$. Thus, both the latter two models are acceptable from the information-theoretic viewpoint at MDL principle. Let us also note that a greater simplification achieved when changing $base$ from 40 to 32 than when changing $base$ from 50 to 40 is due to the fact that $\lceil\log_2 40\rceil > \lceil\log_2 32\rceil$ and $\lceil\log_2 50\rceil = \lceil\log_2 40\rceil$.

Another way how to simplify the considered model is to simplify its structure. Obviously, in the sense of space requirements the most costly is the five-dimensional marginal $\kappa^{\downarrow\{2,5,6,7,8\}}$. Let us consider two simplifications of $\mathbf{M_1}$ consisting only of two- and three-dimensional marginals:

$$\mathbf{M_2}: \ \mu_2 = \kappa^{\downarrow\{1,2\}} \triangleright \kappa^{\downarrow\{3,4\}} \triangleright \kappa^{\downarrow\{3,8\}} \triangleright \kappa^{\downarrow\{5,8\}} \triangleright \kappa^{\downarrow\{2,7,8\}} \triangleright \kappa^{\downarrow\{1,5,6\}}, \quad (3)$$

and

$$\mathbf{M_3}: \ \mu_3 = \kappa^{\downarrow\{3,4\}} \triangleright \kappa^{\downarrow\{3,5\}} \triangleright \kappa^{\downarrow\{1,5,6\}} \triangleright \kappa^{\downarrow\{5,6,8\}} \triangleright \kappa^{\downarrow\{6,7,8\}} \triangleright \kappa^{\downarrow\{1,2,7\}}. \quad (4)$$

Repeating computations described in the preceding section we get $c_l(\mathbf{M_2}) = 306$ and $c_l(\mathbf{M_3}) = 356$ bits, and $c_{lH}(\mathbf{M_2}) = 267$ and $c_{lH}(\mathbf{M_3}) = 304$ bits. Let us

Table 2. Kullback-Leibler divergences

	M_1	$M_{1:50}$	$M_{1:40}$	$M_{1:32}$	M_2	M_3
complexity	423	408	397	284	267	304
K-L divergence	0.2736	0.2795	0.2846	0.2881	0.2964	0.3036

stress that these complexities are computed for models with with $base = 100$. So, comparing these values with $c_{lH}(\mathbf{D}) = 404$ we see that both these models are from our point of view acceptable.

Nevertheless, it is clear that we cannot evaluate models just on the basis of MDL principle, just according to the number of bits necessary for their encoding. We also need a criterion evaluating to what extent each model carries the information contained in the considered data. For this, we use the Kullback-Leibler divergence between the sample probability distribution defined by data and the probability distribution defined by the model. So, for each considered model we can compute the Kullback-Leibler divergence between the eight-dimensional sample distribution κ defined by the considered data file with 100 records, and the distribution defined by the respective model. For example, for model $\mathbf{M_1}$ it is $Div(\kappa; \mu_1)$, where κ is the sample distribution, and μ_1 is the distribution defined from κ by Formula (2). The values of these divergences for all the considered models are in Table 2.

From Table 2 we can see that the simplification of a model by decreasing the value of constant $base$, i.e., by roughening the estimates of probabilities, leads to the decrease of complexity of the model and simultaneous increase of the Kullback-Leibler divergence. The greater this type of simplification, the greater the respective Kullback-Leibler divergence. A precise version of this statement can be expressed in a form of mathematical theorems whose presentation is beyond the scope of this paper. On the other hand, from the last two columns of Table 2 the reader can see that a similar relation valid for the simplification of a model by decreasing the complexity of a model would be much more complex. This is based on the fact that though both models $\mathbf{M_2}$ and $\mathbf{M_3}$ are the simplification of $\mathbf{M_1}$, no one is a simplification of the other. This means that for structure simplification the strength of simplification cannot be measured just by one parameter, by the amounts of bits necessary for the model encoding but we have to introduce also some partial order on the set of all potential simplifications, which is a topic for future research.

6   Conclusions

The novelty of this paper lies in the detailed analysis of the complexity of probabilistic models. We do not take into account only the structure of a model but also the precision of probabilities describing the model in question. It means that the final selection of the model is based on a trade-off between the complexity of model structure and the precision of probability estimates; the simplification of a model structure makes it possible to consider more precise probability estimates and vice versa. On the other side it also means that employing these ideas into the process of model construction substantially increases the space of possible solutions in comparison with the approaches when only the structure is optimized. Fortunately, a rather great part of the models are "forbidden" because their complexity is greater than the upper limit determined by the input data. It is a topic for the future research to design tractable algorithms taking advantage of this property.

In this paper, the new ideas are illustrated on the data based construction of probabilistic multidimensional compositional models. Naturally, it can be applied also to the construction of other probabilistic multidimensional models (like e.g., Bayesian networks), and also to construction of models in other uncertainty theories (see e.g., Shenoy's valuation based systems [7]).

Acknowledgments. This research was partially supported by GAČR under Grant No. 15-00215S.

References

1. Csiszár, I.: I-divergence Geometry of Probability Distributions and Minimization problems. Ann. probab. **3**, 146–158 (1975)
2. Grünwald, P.: A Tutorial Introduction to the Minimum Description Length Principle, p. 80 (2004). [cit. 2014–07–15] http://eprints.pascal-network.org/archive/00000164/01/mdlintro.pdf
3. Hansen, M.H., Bin, Y.U.: Minimum Description Length Model Selection Criteria for Generalized Linear Models, p. 20. http://www.stat.ucla.edu/cocteau/papers/pdf/glmdl.pdf
4. Huffman, D.A.: A Method for the Construction of Minimum-Redundancy Codes. Proceedings of the I.R.E., 1098–1102 (1952)
5. Jensen, F.V.: Bayesian Networks and Decision Graphs. IEEE Computer Society Press, New York (2001)
6. Jiroušek, R.: Foundations of compositional model theory. Int. J. General Systems **40**(6), 623–678 (2011)
7. Jiroušek, R., Shenoy, P.P.: Compositional models in valuation-based systems. Int. J. Approx. Reasoning **53**(8), 1155–1167 (2012)
8. Kolmogorov, A.N.: Tri podchoda k opredeleniju ponjatija 'kolichestvo informacii'. Problemy Peredachi Informacii **1**, 3–11 (1965)
9. Kullback, S., Leibler, R.A.: On information and sufficiency. Annals of Mathematical Statistics **22**, 76–86 (1951)
10. Lam, W., Bacchus, F.: Learning Bayesian Belief Networks: An approach based on the MDL Principle. Computational Intelligence **10**, 269–293 (1994). http://citeseerx.ist.psu.edu/viewdoc/download?doi=10.1.1.127.5504&rep=rep1&type=pdf

11. Lauritzen, S.L.: Graphical models. Oxford University Press (1996)
12. Mahdi, O.A., Mohammed, M.A., Mohamed, A.J.: Implementing a Novel Approach an Convert Audio Compression to Text Coding via Hybrid Technique. Int. J. Computer Science **3**(6), 53–9 (2012)
13. Von Mises, R.: Probability, statistics, and truth. Courier Corporation, Mineola (1957). [Originaly published in German by Springer, 1928]
14. Witten, I.H., Neal, R.M., Cleary, J.G.: Arithmetic Coding for Data Compression. Communications of the ACM **30**(6), 520–540 (1987)

On the Property of SIC Fuzzy Inference Model with Compatibility Functions

Hirosato Seki[(⊠)]

Graduate School of Engineering Science, Osaka University, Toyonaka, Osaka, Japan
seki@sys.es.osaka-u.ac.jp

Abstract. The single input connected fuzzy inference model (SIC model) can decrease the number of fuzzy rules drastically in comparison with the conventional fuzzy inference models. However, the inference results obtained by the SIC model were generally simple comapred with the conventional fuzzy inference models. In this paper, we propose a *SIC model with compatibility functions*, which weights the rules of the SIC model. Moreover, this paper shows that the inference results of the proposed model can be easily obtained even as the proposed model uses involved compatibility functions.

Keywords: Approximate reasoning · SIC fuzzy inference model · Compatibility function · Fuzzy function

1 Introduction

Since Mamdani [1] applied the concept of fuzzy inference to steam engine experimental device, relevant research and applications have been executed in various fields. Especially, researches on the T–S inference model [2], which is widely used as fuzzy control method and so on, are reported in many papers.

Fuzzy inference plays a significant role in fuzzy applications. However, as for the fuzzy rules in the traditional fuzzy inference models, all the input items of the system are set to the antecedent part, and all output items are set to the consequent part. Therefore, the problem is that the number of fuzzy rules becomes very huge; hence, the setup and adjustment of fuzzy rules become difficult. On the other hand, the *Single Input Connected fuzzy inference model* (SIC model) by Hayashi et al. [3,4] can reduce the number of fuzzy rules drastically compared with conventional fuzzy inference models. However, since the number of rules of the SIC model is limited compared to the conventional inference models, inference results gained by the SIC model are simple in general.

From the above reason, this paper proposes a *fuzzy functional SIC model* in which the real value of the consequent parts are extended to fuzzy function, and a *SIC model with compatibility functions* in which the compatibility functions are weighted to consequent parts of the fuzzy functional SIC model. Moreover, it shows that the SIC model with compatibility functions can be transformed into the weighted SIC fuzzy inference model.

© Springer International Publishing Switzerland 2015
V.-N. Huynh et al. (Eds.): IUKM 2015, LNAI 9376, pp. 267–278, 2015.
DOI: 10.1007/978-3-319-25135-6_26

This paper is organized as follows. In Section 2, the conventional fuzzy inference models are briefly reviewd. The fuzzy functional SIC model is proposed in Section 3. The properties of the fuzzy functional SIC model are clarified, and the SIC model with compatibility functions are porposed in Section 4. Finally, concluding remarks are given in Section 5.

2   Fuzzy Inference Models

In this section we review the min–max–gravity model, product–sum–gravity model and fuzzy functional inference model.

2.1   Min–Max–Gravity Model

We firstly explain the *min–max–gravity model* as *Mamdani's fuzzy inference model* [1] for the fuzzy inference form (see Fig. 1). The rules of the min–max–gravity model are given as follows:

$$\text{Rule } R_i = \begin{cases} x_1 = A_i^1, x_2 = A_i^2, \ldots, x_n = A_i^n \\ \longrightarrow y = B_i \end{cases} \tag{1}$$

where $x_1, x_2, \ldots, x_n$ are variables of the antecedent part, $A_i^1, A_i^2, \ldots, A_i^n$ fuzzy sets, B_i fuzzy sets of the consequent part, $i = 1, 2, \ldots, M$ and M is the total number of rules.

Each inference result B_i' which is infered from the fact "$x_1^0, x_2^0, \ldots, x_n^0$" and the fuzzy rule "$A_i^1, A_i^2, \ldots, A_i^n \longrightarrow B_i$" is given in the following.

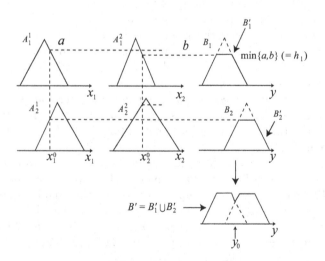

Fig. 1. Min–max–gravity model.

The degree of fitness, h_i, of the fact "$x_1^0, x_2^0, \ldots, x_n^0$" to the antecedent parts "$A_i^1, A_i^2, \ldots, A_i^n$" is given as

$$h_i = \min\{A_i^1(x_1^0), A_i^2(x_2^0), \ldots, A_i^n(x_n^0)\} \tag{2}$$

Thus, the inference result B_i' is given as

$$B_i'(y) = \min\{h_i, B_i(y)\} \tag{3}$$

The final consequence B' of (1) is aggregated from B_1', B_2', $\ldots$, B_M' by using the max. Namely,

$$B'(y) = \max\{B_1'(y), B_2'(y), \ldots, B_M'(y)\} \tag{4}$$

The representative point y_0 for the resulting fuzzy set B' is obtained as the center of gravity of B':

$$y_0 = \frac{\displaystyle\int y \cdot B'(y)dy}{\displaystyle\int B'(y)dy} \tag{5}$$

2.2   Product–Sum–Gravity Model

We secondly explain a fuzzy inference model called *product–sum–gravity model* [5–7] for the fuzzy inference form (see Fig. 2). The rules of the product–sum–gravity model are also given as (1).

Each inference result B_i' which is infered from the fact "$x_1^0, x_2^0, \ldots, x_n^0$" and the fuzzy rule "$A_i^1, A_i^2, \ldots, A_i^n \longrightarrow B_i$" is given in the following.

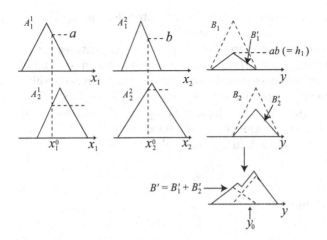

Fig. 2. Product–sum–gravity model.

The degree of fitness, h_i, of the fact "$x_1^0, x_2^0, \ldots, x_n^0$" to the antecedent parts "$A_i^1, A_i^2, \ldots, A_i^n$" is given as

$$h_i = A_i^1(x_1^0) \cdot A_i^2(x_2^0) \cdots A_i^n(x_n^0) \tag{6}$$

where $\cdot$ stands for the algebraic product. Thus, the inference result B_i' is given as

$$\begin{aligned} B_i'(y) &= A_i^1(x_1^0) \cdots A_i^n(x_n^0) \cdot B_i(y) \\ &= h_i \cdot B_i(y) \end{aligned} \tag{7}$$

The final consequence B' of (1) is aggregated from B_1', B_2', ..., B_M' by using the algebraic sum ($+$). Namely,

$$B'(y) = B_1'(y) + B_2'(y) + \cdots + B_M'(y) \tag{8}$$

The representative point y_0 for the resulting fuzzy set B' is obtained as the center of gravity of B':

$$y_0 = \frac{\displaystyle\int y \cdot B'(y)dy}{\displaystyle\int B'(y)dy} \tag{9}$$

2.3   Fuzzy Functional Inference Model

The rules in the fuzzy functional inference model [8] are constituted as

$$\text{Rule } R_i = \begin{cases} x_1 \text{ is } A_i^1, \ x_2 \text{ is } A_i^2, \ \ldots, \ x_n \text{ is } A_i^n \\ \longrightarrow y = F_i(x_1, x_2, \ldots, x_n) \end{cases} \tag{10}$$

where $x_1, x_2, \ldots, x_n$ are variables of the antecedent part, $A_i^1, A_i^2, \ldots, A_i^n$ fuzzy sets, and $F_i(x_1, x_2, \ldots, x_n)$ are fuzzy functions.

Since the consequent parts $F_i(x_1, x_2, \ldots, x_n)$ constitutes fuzzy sets, the final output y_0 of the fuzzy functional inference model should be obtained from Mamdani type fuzzy inference model. For example, the inference result y_0 is obtained as follows when the product–sum–gravity model is used (See Fig. 2).

Given the rules of (10), inputs $x_1^0, x_2^0, \ldots, x_n^0$ and each antecedent part $A_1^i, A_2^i, \ldots, A_n^i$, the degree h_i is obtained as (6), and the inference result $F_i(x_1^0, x_2^0, \ldots, x_n^0)'$ of each fuzzy rule is obtained as

$$F_i(x_1^0, x_2^0, \ldots, x_n^0)'(y) = h_i \cdot F_i(x_1^0, x_2^0, \ldots, x_n^0)(y) \tag{11}$$

where $\cdot$ is the algebraic product.

The final output F' of (10) is obtained by the summation of each inference result $F_i(x_1^0, x_2^0, \ldots, x_n^0)'$ as in (8). Namely,

$$F'(y) = F_1(x_1^0, x_2^0, \ldots, x_n^0)'(y) + \cdots$$
$$+ F_M(x_1^0, x_2^0, \ldots, x_n^0)'(y) \tag{12}$$

The representative point y_0 of conclusion F' is obtained as follows:

$$y_0 = \frac{\displaystyle\int y \cdot F'(y) dy}{\displaystyle\int F'(y) dy} \tag{13}$$

Note that it is possible to apply the *min–max–gravity model* by replacing algebraic product with min in (11) and algebraic sum with max in (12).

2.4   Single Input Connected (SIC) Fuzzy Inference Model

In this subsection we review the *Single Input Connected fuzzy inference model* (SIC model) for the single input type fuzzy inference model proposed by Hayashi et al. [3,4,7].

The SIC model has n rule modules. Rule modules of the SIC model are given as

$$\text{Rules-1} : \{x_1 = A_j^1 \longrightarrow y_1 = y_j^1\}_{j=1}^{m_1}$$

$$\vdots$$

$$\text{Rules-}i : \{x_i = A_j^i \longrightarrow y_i = y_j^i\}_{j=1}^{m_i} \tag{14}$$

$$\vdots$$

$$\text{Rules-}n : \{x_n = A_j^n \longrightarrow y_n = y_j^n\}_{j=1}^{m_n}$$

where Rules-i stands for the ith single input rule module, the ith input item x_i is the sole variable of the antecedent part of the Rules-i, and y_i stands for the variable of its consequent part. A_j^i means the fuzzy set of the jth rule of the Rules-i, y_j^i stands for a real value of consequent part, $i = 1, 2, \ldots, n$, $j = 1, 2, \ldots, m_i$, and m_i is the number of rules in Rules-i.

The SIC model sets up rule modules to each input item. The final inference result of SIC model is obtained by the weighted average of the degrees of the antecedent part and consequent part of each rule module. The degree h_j^i of the ith rule of the SIC model is given as

$$h_j^i = A_j^i(x_i^0) \tag{15}$$

The final inference result y_0 is given as follows by using degrees of antecedent part and consequent part from each rule module.

$$y^0 = \frac{\sum_{j=1}^{m_1} h_j^1 y_j^1 + \cdots + \sum_{j=1}^{m_n} h_j^n y_j^n}{\sum_{j=1}^{m_1} h_j^1 + \cdots + \sum_{j=1}^{m_n} h_j^n}$$

$$= \frac{\sum_{i=1}^{n} \sum_{j=1}^{m_i} h_j^i y_j^i}{\sum_{i=1}^{n} \sum_{j=1}^{m_i} h_j^i} \tag{16}$$

3   Fuzzy Functional SIC Inference Model

Seki [9] has proposed the *functional-type SIC model* in which the consequent part of the SIC model is a function, where the system has n inputs and 1 output, and each rule module corresponds to one of the n input items and has only the input item in its antecedent. However, although the antecedent parts of the functional-type SIC model are fuzzy sets, the consequent parts are function. Thus, this structure is hard to understand from the difference of antecedent parts and consequent parts. Rather, it may be completely natural that the consequent parts are also fuzzy.

In this section, we propose a *fuzzy functional SIC model* in which the consequent parts of the functional-type SIC model are generalized to fuzzy function. This model can make it easier to understand the structure of rules compared with the ordinary fuzzy inference models.

The rules of the fuzzy functional SIC model are given as follows.

$$\text{Rules-1} : \{x_1 = A_j^1 \longrightarrow y_1 = F_j^1(x_1)\}_{j=1}^{m_1}$$

$$\vdots$$

$$\text{Rules-}i : \{x_i = A_j^i \longrightarrow y_i = F_j^i(x_i)\}_{j=1}^{m_i} \tag{17}$$

$$\vdots$$

$$\text{Rules-}n : \{x_n = A_j^n \longrightarrow y_n = F_j^n(x_n)\}_{j=1}^{m_n}$$

where the Rules-i stands for the "ith single input rule module," x_i corresponding to the ith input item is the sole variable of the antecedent part of the Rules-i, and y_i is the variable of its consequent part. A_j^i and $F_j^i(x_i)$ are, respectively, fuzzy set and fuzzy function of the jth rule of the Rules-i, where $i = 1, 2, \ldots, n$; $j = 1, 2, \ldots, m_i$, and m_i stands for the number of rules in the Rules-i. An example as one-dimensional case of fuzzy function is shown in Fig. 3.

Given an input x_i^0 to the Rules-i, the degree of the antecedent part in the jth rule in the Rules-i is given by (18).

$$h_j^i = A_j^i(x_i^0) \tag{18}$$

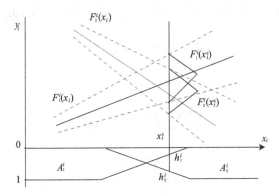

Fig. 3. Fuzzy function.

The consequent parts $F_j^i(x_i)$ constitutes fuzzy sets as well as the fuzzy functional inference model. Therefore, the inference results y_i^0 from rule modules of the fuzzy functional SIC model should be also obtained from the product–sum–gravity model, min-max–gravity model or fuzzy functional inference model. Namely, the inference result y_i^0 from rule modules Rules-i is given as

$$y^0 = \frac{\int y \cdot F_j^i(x)'(y)dy}{\int F_j^i(x)'(y)dy} \tag{19}$$

Final inference result y_0 of the fuzzy functional SIC model is given by

$$y_0 = \sum_{i=1}^{n} w_i y_i^0 \tag{20}$$

where w_i stands for the importance degree for each input item x_i ($i = 1, 2, \ldots, n$).

4   Additive SIC Inference Models

In this section, we show the theoretical properties of the fuzzy functional SIC model from point of view of equivalence.

We use the following symbol for short:

$$F_i' = F_j^i(x_i^0)' \tag{21}$$

$$f_i' = f_j^i(x_i^0)' \tag{22}$$

$$S_i' = S_j^i(x_i^0)' \tag{23}$$

where f_i' is center of gravity of F_i' and S_i' represents the area of inference result F_i' from each rule.

The inference result y_0 of F' in (12) is given as follows [7,10]:

$$y^0 = \frac{\displaystyle\int y \cdot F_i'(y)dy}{\displaystyle\int F_i'(y)dy}$$

$$= \frac{\displaystyle\sum_{i=1}^{n}\sum_{j=1}^{m_i}\{h_j^i * S_j^i(x_i)\}f_j^i(x_i)}{\displaystyle\sum_{i=1}^{n}\sum_{j=1}^{m_i}h_j^i * S_j^i(x_i)}$$

$$= \frac{\displaystyle\sum_{i=1}^{n}\sum_{j=1}^{m_i}S_j^i(x_i)'f_j^i(x_i)}{\displaystyle\sum_{i=1}^{n}\sum_{j=1}^{m_i}S_j^i(x_i)'} \tag{24}$$

It is found that the inference reult y_i^0 is obtained from the area S_i' and the center of gravity f_i' of each inference result F_i' and its center of gravity.

In the following discussion, we state the fuzzy functional SIC inference model in case of an *additive fuzzy inference model*, which uses the addition as in (12) in the aggregation of each inference result.

In (11), the relation of (24) holds, as long as the addition is used in the aggregation, even if instead of using a algebraic product, min is used to the operation. Because the area, form and center of gravity of fuzzy sets in the consequent parts are only changed, as shown in Fig. 4.

In order to obtain the inference result F_i' as in (11), the operetion $*$ can be generally used as follows [8]:

$$F_i'(y) = h_i * F_i(x_i^0) \tag{25}$$

The operation $*$ can be applied to (25) if it is under geometric mean. Therefore, the following operations as the t-norm and average operations can be considered:

$$\text{bounded} - \text{product} : 0 \vee (h + B(y) - 1) \tag{26}$$

$$\text{algebraicproduct} : \quad h \cdot B(y) \tag{27}$$

$$\text{min} : \quad h \wedge B(y) \tag{28}$$

$$\text{harmonicmean} : \quad \frac{2}{\frac{1}{h} + \frac{1}{B(y)}} \tag{29}$$

$$\text{geometricmean} : \quad \sqrt{h \cdot B(y)} \tag{30}$$

When the above operations are applied to (25), the inference result F_i' is obtained as shown in Fig. 4, and the areas of F_i' are given as follows:

$$
S' = \begin{cases} h^2 \cdot S & \cdots \text{ bounded} - \text{product} \\ h \cdot S & \cdots \text{ algebraicproduct} \\ h(2-h) \cdot S & \cdots \text{ min} \\ 4h(h \log \frac{h}{h+1} + 1) \cdot S & \cdots \text{ harmonicmean} \\ 4/3\sqrt{h} \cdot S & \cdots \text{ geometricmean} \end{cases} \tag{31}
$$

where $h = h_i$, $S' = S_i'$, and S represents the area $S_i(x_i^0)$ of the consequent parts $F_i(x_i^0)$.

The area S_i' of the inference result F_i' is obtained by (31) as

$$
S_i' = g_i(h_i) \cdot S_i(x_i^0). \tag{32}
$$

Namely, the area S_i' can be expressed as product of the functional degree named as *compatibility function* g_i [11], and the area $S_i(x_i^0)$ of the consequent part.

From the above-mentioned results, as long as the addition is used in aggregation, the inference result y_0 calculated from (10) and (24) is given as follows:

$$
y^0 = \frac{\sum_{i=1}^n g_1^i(h_1^i)S_1^i \cdot f_1'^i + \cdots + g_{M_i}^i(h_{M_i}^i)S_{M_i}^i \cdot f_M'^i}{\sum_{i=1}^n g_1^i(h_1^i)S_1^i + \cdots + g_{M_i}^i(h_{M_i}^i)S_{M_i}^i} \tag{33}
$$

where $S_i(x_i^0)$ is represented as S_i for short. Note that $f_j'^i (= f_j^i(x_i^0)')$ is the center of gravity of the inference result F_i' in (25).

Example 1. In the fuzzy functional SIC inference model, we consider the case of using the algebraic product in (25). Namely,

$$
S_j'^i = h_j^i \cdot S_j^i(x_i^0) \tag{34}
$$

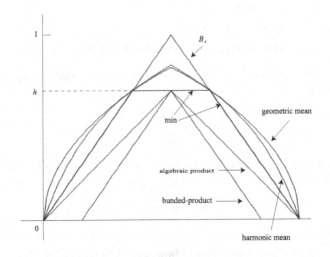

Fig. 4. Inference results F_i' under the operations (26)–(30)[8].

Moreover, since the center of gravity $f_j^i(x_i^0)'$ of F_i' is equal to the center of gravity $f_j^i(x_i^0)$ of consequent parts $F_j^i(x_i^0)$ in the case using the algebraic product, the ith inference result y_i^0 is obtained from (33) as

$$y^0 = \frac{\sum_{i=1}^n h_1^i S_1^i f_1^i + h_2^i S_2^i f_2^i + \cdots + h_M^i S_M^i f_M^i}{\sum_{i=1}^n h_1^i S_1^i + h_2^i S_2^i + \cdots + h_M^i S_M^i} \tag{35}$$

5 Fuzzy Functional SIC Inference Model with Compatibility Function

In this section, we formulate a *fuzzy functional SIC inference method with compatibility function* which represents the characteristics as the compatibility function [11] and fuzzy rule.

For the compute of (33), although $S_j^i(x_i^0)$ have shown the area of consequent parts, we can also use the characteristics of fuzzy rule, e.g., base length of the consequent part, importantness of the fuzzy rule and so on.

Moreover, we can use any compatibility function if $g(h)$ is nondecreasing function and $g(0) = 0$.

From the above-mentioned discussion, we introduce a *generalized compatibility function* which adds the characteristics as the compatibility function and fuzzy rule.

By (32), we consider the following function which unifies compatibility function $g(h)$ and area $S(x_i)$.

$$g(h, x_i) = g(h) \cdot S(x_i) \tag{36}$$

We have the following properties of the generalized compatibility function $g(h, x_i)$.

(1) $g(h, x_i) \geq 0$ (37)

(2) $g(h, x_i) = 0$ at $h = 0$ (38)

(3) $g(h, x_i)$ is nondecreasing function for h (39)

For example, we can introduce the base length $l(x_i)$ or weight $w(x_i)$, instead of the area. Namely,

$$g(h, x_i) = g(h) \cdot l(x_i),$$
$$g(h) \cdot w(x_i) \tag{40}$$

Moreover, when the operation is used as a t-norm, instead of algebraic product, the generalized compatibility function $g(h, x_i)$ is obtained as follows:

$$g(h, x_i) = 0 \vee (g(h) + S(x_i) - 1),$$
$$g(h) \wedge l(x_i) \tag{41}$$

In the above generalized compatibility functions, the properties (37)–(39) are all satisfied.

Here, we call the function $g(h, x_i)$ satisfing (37)–(39) as *generalized compatibility function* tentatively.

By introducing a generalized compatibility function $g(h, x_i)$, we can propose a *fuzzy functional SIC inference model with compatibility function*. Rules in the fuzzy functional SIC inference method with compatibility function are constituted as

$$\text{Rules-1} : \{x_1 = A_j^1 \longrightarrow y_1 = f_j^1(x_1) \text{ with } g_j^i(x_1)\}_{j=1}^{m_1}$$

$$\vdots$$

$$\text{Rules-}i : \{x_i = A_j^i \longrightarrow y_i = f_j^i(x_i) \text{ with } g_j^i(x_i)\}_{j=1}^{m_i} \tag{42}$$

$$\vdots$$

$$\text{Rules-}n : \{x_n = A_j^n \longrightarrow y_n = f_j^n(x_n) \text{ with } g_j^i(x_n)\}_{j=1}^{m_n}$$

Given inputs x_i^0, the ith inference result y_i^0 is given as

$$y^0 = \frac{\sum_{i=1}^n g_1^i(h_1^i, x_i^0) f_1^i(x_i^0) + \cdots + g_M^i(h_M^i, x_i^0) f_M^i(x_i^0)}{\sum_{i=1}^n g_1^i(h_1^i, x_i^0) + \cdots + g_M^i(h_M^i, x_i^0)} \tag{43}$$

The fuzzy functional SIC inference model with compatibility function can be reduced to the conventional SIC inference model and weighted SIC inference model when $g_1^i(h_1^i, x_i^0) = h_j^i$ and $h_j^i \cdot w_j^i(x_i^0)$, respectively.

6 Conclusion

This paper has proposed the fuzzy functional SIC model and SIC model with compatibility functions. It has shown the inference results by the fuzzy functional SIC inference model is easily calculated by introducing the compatibility function, if the aggregation from each inference result and center of gravity is used as addition.

Moreover, the SIC inference model with a compatibility function, which represents the characteristics such as area, base length and importantness of the fuzzy functinoal rule, has been proposed in this paper. This model includes the SIC inference model and weighted SIC model as a special case.

Further studies are required to apply the functional SIC inference model with a compatibility function to real systems.

Acknowledgement. This work was partially supported by a Grant-in-Aid for Scientific Research (Project No. 15K16065) from the Japan Society for the Promotion Science (JSPS).

References

1. Mamdani, E.H.: Application of fuzzy algorithms for control of simple dynamic plant. In: Proc. IEE, vol. 121, no. 12, pp. 1585–1588 (1974)
2. Takagi, T., Sugeno, M.: Fuzzy identification of systems and its applications to modeling and control. IEEE Trans. Syst., Man, Cybern. **SMC–15**(1), 116–132 (1985)

3. Hayashi, K., Otsubo, A., Murakami, S., Maeda, M.: Realization of nonlinear and linear PID controls using simplified indirect fuzzy inference method. Fuzzy Sets Syst. **105**, 409–414 (1999)

4. Hayashi, K., Otsubo, A., Shiranita, K.: Improvement of conventional method of PI fuzzy control. IEICE Trans. Fundamentals **E84–A**(6), 1588–1592 (2001)

5. Mizumoto, M.: Fuzzy controls under various fuzzy reasoning methods. Inf. Sci. **45**, 129–151 (1988)

6. Hu, B.-G., Mann, G.K.I., Gosine, R.G.: A systematic study of fuzzy PID controllers–function-based evaluation approach. IEEE Trans. Fuzzy Syst. **9**(5), 699–712 (2001)

7. Seki, H., Mizumoto, M.: On the equivalence conditions of fuzzy inference methods-part 1: basic concept and definition. IEEE Trans. on Fuzzy Sust. **19**(6), 1097–1106 (2011)

8. Seki, H., Mizumoto, M.: Fuzzy functional inference method. In: Proc. 2010 IEEE World Congress on Computational Intelligence, FUZZ-IEEE 2010, Barcelona, Spain, pp. 1643–1648, July 2010

9. Seki, H.: A learning method of SIC inference model and its application. In: Proc. the 15th Czech-Japan Seminar on Data Analysis and Decision Making under Uncertainty, Osaka, Japan, pp. 187–191, September 2012

10. Seki, H.: On the single input connected fuzzy inference model with consequent fuzzy sets. In: Proc. the 7th International Conference on Soft Computing and Pattern Recognition, Fukuoka, Japan, November 2015 (submitted)

11. Seki, H., Mizumoto, M.: Additive fuzzy functional inference methods. In: Proc. 2010 IEEE International Conference on Systems, Man, and Cybernetics, Istanbul, Turkey, pp. 4304–4309, October 2010

Applying Covering-Based Rough Set Theory to User-Based Collaborative Filtering to Enhance the Quality of Recommendations

Zhipeng Zhang[1], Yasuo Kudo[2]([✉]), and Tetsuya Murai[3]

[1] Course of Advanced Information and Electronic Engineering,
Graduate School of Engineering, Muroran Institute of Technology, Mizumoto 27-1,
Muroran 050-8585, Japan
14096507@mmm.muroran-it.ac.jp
[2] College of Information and Systems, Muroran Institute of Technology,
Mizumoto 27-1, Muroran 050-8585, Japan
kudo@csse.muroran-it.ac.jp
[3] Graduate School of Information Science and Technology, Hokkaido University,
Kita 14, Nishi 9, Kita-ku, Sapporo, Hokkaido 060-0814, Japan
murahiko@ist.hokudai.ac.jp

Abstract. Recommender systems provide personalized information by learning user preferences. Collaborative filtering (CF) is a common technique widely used in recommendation systems. User-based CF utilizes neighbors of an active user to make recommendations; however, such techniques cannot simultaneously achieve good values for accuracy and coverage. In this study, we present a new model using covering-based rough set theory to improve CF. In this model, relevant items of every neighbor are regarded as comprising a common covering. All common coverings comprise a covering for an active user in a domain, and covering reduction is used to remove redundant common coverings. Our experimental results suggest that this new model could simultaneously present improvements in accuracy and coverage. Furthermore, comparing our model with the unreducted model using all neighbors, our model utilizes fewer neighbors to generate almost the same results.

Keywords: Covering-based rough set · Recommender systems · Collaborative filtering · Covering reduction

1 Introduction

With the development of the internet and artificial intelligence (AI), the recommender system (RS) has become very popular recently. The RS can advise users when making decisions on the basis of personal preferences and help users discover items they might not find by themselves [1,2]. Collaborative filtering (CF) is a significant component of the recommendation process [4], that is based on the ways in which humans have made decisions throughout history [7,8]. User-based CF uses information of the active user's neighbors to make predictions

© Springer International Publishing Switzerland 2015
V.-N. Huynh et al. (Eds.): IUKM 2015, LNAI 9376, pp. 279–289, 2015.
DOI: 10.1007/978-3-319-25135-6_27

and recommendations [3]. Accuracy and coverage are two crucial metrics for evaluating the RS, but it is difficult to simultaneously achieve good values for these metrics when using user-based CF to make recommendations.

Rough set theory was first presented by Pawlak in the early 1980s [5]. Covering-based rough set has been regarded as a meaningful extension of the classical rough set to handle vague and imperfect knowledge better, which extends the partition of rough set to a covering [13,14]. The notion of reduction for covering is one of the most important results in covering-based rough set [17]. Currently, much of the literature has been focused on providing the theory behind covering-based rough set [13–16], but there is little regrading applications, especially for RSs.

In this study, covering-based rough set theory is first applied to the RS. We present a new model called covering-based collaborative filtering (CBCF) that uses the covering-based rough set to improve the user-based CF approach. The covering reduction notion is included to extract the relative effective neighbors from all neighbors, thus defining the active user's reduct-neighbors. According to these reduct-neighbors, we obtain predictions and recommendations more efficiently.

The remainder of our paper is organized as follows. In Section 2, we review the basic notions and knowledge of covering-based rough sets and covering reductions. In Section 3, we propose the covering-based rough set model for user-based CF. Next, in Section 4, we present our experiments and compare our results with a model that does not employ the covering reduction as well as a model using CF with the same number of neighbors. Finally, in Section 5, we note our conclusions and define areas for future work.

2 Background

2.1 Covering-Based Rough Set

In this subsection, we present the basic knowledge of covering and the covering approximation space. More details can be found in [13,14].

Definition 1. *Let U be the domain of discourse and C be a family of subsets of U. If none of the subsets in C is empty, and $\cup C = U$, C is called a covering of U.*

Definition 2. *Let U be a non-empty set and C be a covering of U. We call the ordered pair $\langle U, C \rangle$ a covering approximation space.*

We note here the definition of covering is an extension of the definition of partitions. Different lower and upper approximation operations would generate different types of covering-based rough set. The covering-based rough set was first presented by Zakowski [12], who extended Pawlak's rough set theory from a partition to a covering. Pomykala gave the notion of the second type of covering-based rough set [6], while Tsang presented the third type [9], Zhu defined the fourth [16] and fifth [15] types of covering-based rough set models, and Wang studied the sixth type of covering-based approximations [10].

2.2 Reduction Theory of Covering-Based Rough Set

The reduction of covering concept was presented by Zhu in [17]; Zhu also presented other covering reduction approaches in [15,18]. In [11], Yang initially constructed a new reduction theory by redefining the approximation spaces and the reductions of covering generalized rough set, which was applicable to all types of covering generalized rough set. In our present study, we focus only on one type of covering reduction algorithm, i.e., the one presented by Zhu in [18], because this algorithm could remove redundant neighbors more efficiently. Definition 3 defines this type of covering reduction algorithm, which corresponds to the definition of *exclusion(C)* in [18].

Definition 3. *Let C be a covering of a domain U and $\mathrm{K} \in C$. If there exists another element K' of C such that $\mathrm{K} \subset \mathrm{K}'$, we say that K is reducible in C; Otherwise, K is irreducible. When we remove all reducible elements from C, the new irreducible covering is called reduct of C and denoted by reduct(C).*

3   Covering-Based Rough Set Model for User-Based Collaborative Filtering

3.1   Purpose

For user-based CF, if fewer users in the top of a similarity list are selected as neighbors of an active user, high-accuracy items could be recommended for the active user; however, the types of recommendations will be decreased, even in just making the most popular items as recommendations. If more types of items are to be recommended, more users should be selected as neighbors of the active user, but the accuracy will decrease as the number of neighbors grows. Therefore, it is difficult for CF to simultaneously obtain good values for metrics of accuracy and coverage. To solve this problem, the relative effective neighbors should be selected from all neighbors such that the recommendations not only maintain good values of accuracy but also obtain satisfactory values of coverage.

3.2   Model Constructions

In this subsection, we present detailed information and the steps comprising CBCF, which does not use any user demographic data. In short, CBCF needs the following information:

The users set U: $U = \{u_1, u_2 ... u_E\}$, where E is the number of users.

The items set S: $S = \{s_1, s_2 ... s_I\}$, where I is the number of items.

The rating function f: $U \times S \rightarrow R$, $r_{x,i} = f(x, i)$ represents the rating score of user x for item i. Here,

$$r_{x,i} = \begin{cases} \gamma, & \text{the rating score,} \\ \star, & \text{no rating score.} \end{cases}$$

Furthermore, θ is set as the threshold for rating score, and items with $r_{x,i} \geq \theta$ are defined as items relevant to user x.

The item's attributes set A: $A = \{\alpha_1, \alpha_2...\alpha_P\}$, where α_n is an attribute of the item and P is the number of attributes.

Input: — The query of the active user in the following form:

$$[\alpha_1 = v_1] \wedge [\alpha_2 = v_2] \wedge ... \wedge [\alpha_m = v_m],$$

where v_m is a value of α_m.

Output: — A set of recommended items $Rec \subset S$

Step 1: Set $Rec = \emptyset$.

Step 2: Use the rating information and cosine-based similarity approach to compute the similarity between the active user and remaining users. The top K % of users in the similarity list, which are defined as L, are selected as neighbors of the active user. Here, we have

$$sim(x, y) = \frac{\sum_{i \in S_{xy}} r_{x,i} r_{y,i}}{\sqrt{\sum_{i \in S_{xy}} r_{x,i}^2} \sqrt{\sum_{i \in S_{xy}} r_{y,i}^2}}, \tag{1}$$

where $sim(x, y)$ indicates the similarity between users x and y and S_{xy} is the set of all items rated by both users x and y.

Step 3: The decision class X consists of all items that fit the active user's query options. Otherwise, some options will be rejected as minimally as possible until an adequate decision class X is obtained.

Step 4: Setting decision class X as the domain, and for each neighbor $j \in L$, relevant items of the neighbor j in domain X are a common covering C_j, where

$$C_j = \{i \in X | r_{j,i} \geq \theta\}. \tag{2}$$

Here, we define $C* = X - \cup C_j$; then, $C = \{C_1, C_2...C_n, C*\}$ is a covering for the active user in domain X. If the set $\cup C_j$ is empty, the domain is enlarged from X to S.

Step 5: On the basis of covering reduction in the covering-based rough set, redundant neighbors are removed from covering C of the active user to obtain reduct(C). The active user's reduct-neighbors, which are defined as L^*, consist of all users in reduct(C), and the relevant items of L^* comprise the candidates D for the active user.

Step 6: Depending on the ratings of L^*, scores are predicted for D using the adjusted weighted sum approach, i.e.,

$$P_{x,i} = \bar{r}_x + k \sum_{y \in U^*} sim(x,y) * (r_{y,i} - \bar{r}_y), \qquad (3)$$

where U^* represents the neighbors in L^* who have rated item i. Then, the average rating $\bar{r}_x$ of user x is defined as

$$\bar{r}_x = \frac{\sum_{i \in S_x} r_{x,i}}{card(S_x)}, where \quad S_x = \{i \in S | r_{x,i} \neq \star\}.$$

In the above, $P_{x,i}$ is the prediction of item i for user x, and multiplier k serves as a normalizing factor and is selected as

$$k = \frac{1}{\sum_{y \in U^*} sim(x,y)}. \qquad (4)$$

When all predictions are completed, the top N items in the prediction list defined as D_N are selected as the recommended items.

Step 7: Set $Rec = D_N$; output Rec.

3.3 Model Discussion

Reduction of covering is a core component of CBCF, which applies the notion of covering reduction to select relative effective neighbors. To remove redundant neighbors to the extent possible, decision class X is defined as the domain of the active user; therefore, the covering of the active user can be small enough. Based on the definition of reduction in covering-based rough set theory, for common covering C_i, if there exists another common covering C_j for which $C_i \subset C_j$, C_i will be considered as reducible and therefore removable. In this model, C_i denotes the relevant items of neighbor i, and $C_i \subset C_j$ indicates that neighbor j has more relevant items than neighbor i. In other words, neighbor j will be more efficient than neighbor i in domain X for making recommendations; therefore, neighbor i can be removed. Removing all reducible common coverings means that all relative effective neighbors are selected from all neighbors such that this model could just use the relative effective neighbors to make recommendations.

4 Experiments and Evaluation

4.1 Experimental Setup and Evaluation Metrics

For our experiments, we used the MovieLens [4] popular dataset, as it has often been utilized to evaluate RSs. The ratings dataset consists of 1682 movies, 943 users and a total of 100,000 ratings on a scale of 1 to 5. Each user has rated at least 20 movies, and for our study, movies rated above 3 were treated as the user's relevant movies. Furthermore, the covering reduction algorithm defined by Definition 3 was used for our experiments.

We also used the conventional leave-one-out procedure to evaluate the performance of our model. Items that the active user has rated were treated as unrated items, and our model predicted a rating score for every unrated item using information obtained from the remaining users. We summed every attribute's value in the relevant movies dataset, and two attributes with the largest sums are selected as query options of the active user.

To measure the performance of our new model, we use mean absolute error (MAE), root mean square error (RMSE), coverage, precision, recall, and F1 as evaluation metrics, all of which are popular metrics for evaluating RSs. Moreover, the reduction rate is defined as an evaluation metric, which measures the capability of removing redundant neighbors from all neighbors and is given as

$$ReductionRate = \frac{1}{card(U)} \sum_{u \in U} \frac{card(L_u - L_u^*)}{card(L_u)}, \tag{5}$$

where L_u denotes neighbors of the user u, and L_u^* means the reduct-neighbors of user u.

4.2   Comparing CBCF with the Un-reduction Model

We further define Un-CBCF to represent the model without the use of covering reduction. In experiments of both CBCF and Un-CBCF, the top 50% users of the similarity list were selected as the active user's neighbors. To obtain precision, recall, and F1, the number of recommendations were set as 2, 4, 6, 8, 10, and 12.

Table 1. Results of evaluation metrics between CBCF and Un-CBCF

	MAE	RMSE	Reduction Rate	Coverage
CBCF	0.681	0.853	0.795	81.002
Un-CBCF	0.658	0.817	-	88.929

Table 1 illustrates the results of our evaluation metrics between CBCF and Un-CBCF, with values of MAE, RMSE and coverage of Un-CBCF being slightly better than CBCF. For the reduction rate, which only applies to CBCF, on the average, approximately 79.5% of neighbors are removed as redundant neighbors. Given that there are 943 users in the MovieLens dataset and the top 50% users of the similarity list were selected as neighbors, the number of neighbors for Un-CBCF was 471, whereas after making the reduction, the average number of reduct-neighbors for CBCF was only 97.

Figures 1, 2, and 3 show the precision, recall, and F1 metrics, respectively. As shown in the figure, precision had high values for both CBCF and Un-CBCF. The recall and F1 values became higher as the number of recommendations grew. Overall, the precision, recall, and F1 values were almost the same between CBCF and Un-CBCF.

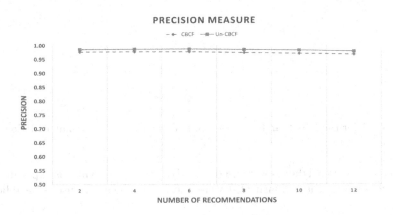

Fig. 1. Precision for the CBCF and Un-CBCF versus the number of recommendations

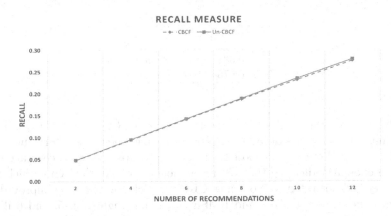

Fig. 2. Recall for the CBCF and Un-CBCF versus the number of recommendations

From these comparative results for CBCF and Un-CBCF, we conclude that our new CBCF model required an average of 97 reduct-neighbors to obtain almost the same results as the Un-CBCF model, which used 471 neighbors; therefore, our new CBCF model would be much more efficient in a RS.

4.3   Comparing CBCF with the Classic CF Model

To further illustrate the performance of our new model, we compared results with the classic CF approach. From the results of reduction in CBCF, on the

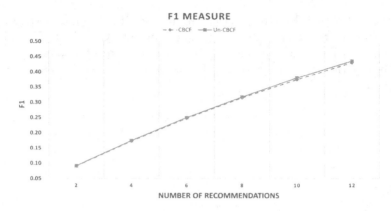

Fig. 3. F1 for the CBCF and Un-CBCF versus the number of recommendations

average, 97 reduct-neighbors were used to make prediction, so in this experiment, the top 97 users of the similarity list were selected as the active user's neighbors for CF.

Table 2. Results of evaluation metrics between CBCF and CF

	MAE	RMSE	Coverage
CBCF	0.681	0.853	81.002
CF	0.675	0.851	50.026

Table 2 shows the results of evaluation metrics for CBCF and CF. As illustrated in the table, values of MAE and RMSE for CF were satisfactory, meaning that the predicted scores were close to the original scores; however, coverage was not good enough, indicating that CF recommended 50% of the items which the active user had not rated. For our new CBCF model, coverage was improved to 81% with the values of MAE and RMSE satisfying. Figures 4, 5, and 6 illustrate the precision, recall, and F1 measures for CBCF versus CF, respectively. From the figures, we note that CBCF was better than CF in terms of precision, decreasing as the number of recommendation grew; however, the values of recall and F1 increased as the number of recommendations grew. Overall, all values were almost the same between CBCF and CF.

Comparative results between CBCF and CF reveal that our new CBCF model could overcome the disadvantage of CF and obtain sufficiently good values of coverage and accuracy.

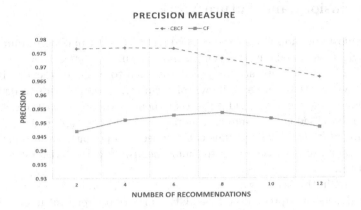

Fig. 4. Precision for CBCF and CF versus the number of recommendations

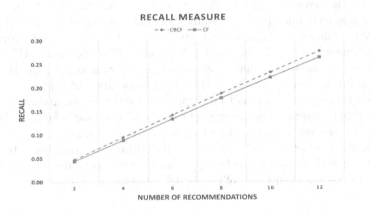

Fig. 5. Recall for CBCF and CF versus the number of recommendations

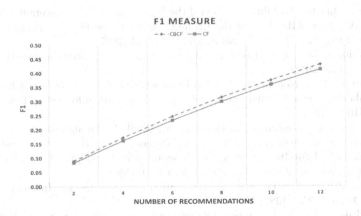

Fig. 6. F1 for CBCF and CF versus the number of recommendations

5   Conclusions and Future Work

CF is the most commonly used and studied technology for making recommendations in RSs. User-based CF has been the most popular recommendation method to date. Generally, we use accuracy and coverage to evaluate the RS, but for user-based CF, if we obtain a high level of accuracy, coverage tends to be unsatisfactory, meaning the RS could only recommend a small set of items. On the contrary, if coverage is outstanding, accuracy tends to decrease, causing the RS to recommend items inexactly, thus causing users to potentially stop using the RS. It is difficult for user-based CF to simultaneously provide satisfying accuracy and coverage.

Therefore, in this study, we presented a new CBCF model based on covering-based rough sets to improve user-based CF, which treats relevant items of every neighbor as a common covering for the active user, and then utilizes covering reduction theory to remove redundant neighbors. The reduct-neighbors are used to predict the score for each candidate, and our approach selects the top N candidates as recommendations. Results of our experiments illustrate that CBCF was able to utilize fewer neighbors to produce almost the same results compared with the model using all neighbors. Moreover, unlike user-based CF, CBCF was able to simultaneously obtain satisfactory accuracy and coverage; therefore, CBCF could be used in practice in the effective operation of RSs.

In our future work, we plan to use upper and lower approximation operators of a covering in covering-based rough set to generate candidates for a RS, because the upper and lower approximation operations are the same between reduction and un-reduction for a covering. We can only utilize reduct-neighbors to obtain the same candidates generated by upper and lower approximation operations, so that accuracy and coverage of the model could be more satisfactory.

References

1. Adomavicius, G., Tuzhilin, A.: Toward the Next Generation of Recommender Systems: A Survey of the State-of-the-Art and Possible Extensions. IEEE Transactions on Knowledge and Data Engineering **17**, 734–749 (2005)
2. Bobadilla, J., Ortega, F.: Recommender System Survey. Knowledge-Based Systems **46**, 109–132 (2013)
3. Herlocker, J.L., Konstan, J.A.: An Empirical Analysis of Design Choices in Neighborhood-based Collaborative Filtering Algorithms. Information Retrieval **5**, 287–310 (2002)
4. Herlocker, J.L., Konstan, J.A., Borchers, A., Riedl, J.: An algorithmic framework for performing collaborative filtering. In: Proceedings of the 22nd Annual International ACM SIGIR Conference on Research and Development in Information Retrieval, pp. 230–237. ACM Press, New York (1999)
5. Pawlak, Z.: Rough Sets. International Journal of Computer and Information Sciences **11**, 341–356 (1982)
6. Pomykala, J.A.: Approximation Operations in Approximation Space. Bull. Pol. Acad. Sci. **35**, 653–662 (1987)

7. Su, X., Khoshgoftaar, T.M.: A Survey of Collaborative Filtering Techniques. Advance in Artificial Intelligence **2009**, 1–19 (2009)
8. Symeonidis, P., Nanopoulos, A., Papadopoulos, A.N., Manolopoulos, Y.: Collaborative Recommender Systems: Combining Effectiveness and Efficiency. Expert Systems with Application **34**, 2995–3013 (2008)
9. Tsang, E., Cheng, D., Lee, J., Yeung, D.: On the upper approximations of covering generalized rough sets. In: Proceedings of the 3rd International Conference Machine Learning and Cybernetics, pp. 4200–4203. IEEE Press, Shanghai (2004)
10. Wang, J., Dai, D., Zhou, Z.: Fuzzy Covering Generalized Rough Sets. Journal of Zhoukou Teachers College **21**, 20–22 (2004)
11. Yang, T.: Li.Q.G.: Reduction about Approximation Spaces of Covering Generalized Rough Sets. International Journal of Approximate Reasoning **51**, 335–345 (2010)
12. Zakowski, W.: Approximations in the Space (u, π). Demonstration Math. **16**, 761–769 (1983)
13. Zhu, W.: Relationship among Basic Concepts in Covering-based Rough Sets. Information Sciences **179**, 2478–2486 (2009)
14. Zhu, W.: Relationship between Generalized Rough Sets Based on Binary Relation. Information Sciences **179**, 210–225 (2009)
15. Zhu, W.: Topological Approached to Covering Rough Sets. Information Sciences **177**, 1499–1508 (2007)
16. Zhu, W., Wang, F.Y.: The Fourth Type of Covering-based Rough Sets. Information Sciences **201**, 80–92 (2012)
17. Zhu, W., Wang, F.Y.: Reduction and Maximization of Covering Generalized Rough Sets. Information Sciences **152**, 217–230 (2003)
18. Zhu, W., Wang, F.Y.: On Three Types of Covering-Based Rough Sets. IEEE Transactions on Knowledge and Data Engineering **19**, 1131–1144 (2007)

Evidence Combination Focusing on Significant Focal Elements for Recommender Systems

Van-Doan Nguyen$^{(\boxtimes)}$ and Van-Nam Huynh

Japan Advanced Institute of Science and Technology (JAIST), Nomi, Japan
nvdoan@jaist.ac.jp

Abstract. In this paper, we develop a solution for evidence combination, called 2-probabilities focused combination, that concentrates on significant focal elements only. Firstly, in the focal set of each mass function, elements with their probabilities in top two highest probabilities are retained; others are considered as noise, which have been generated when assigning probabilities to the mass function and/or by related evidence combination tasks had already been done before, and eliminated. The probabilities of eliminated elements are added to the probability of the whole set element. The achieved mass functions are called 2-probabilities focused mass functions. Secondly, Dempster's rule of combination is used to combine pieces of evidence represented as 2-probabilities focused mass functions. Finally, the combination result is transformed into the corresponding 2-probabilities focused mass function. Actually, the proposed solution can be employed as a useful tool for fusing pieces of evidence in recommender systems using soft ratings based on Dempster-Shafer theory; thus, we also present a way to integrate the proposed solution into these systems. Besides, the experimental results show that the performance of the proposed solution is more effective than a typically alternative solution called 2-points focused combination solution.

Keywords: Information fusion · Uncertain reasoning · Recommender sytem

1 Introduction

For the purpose of increasing sales growth while doing online business in a highly competitive environment, providers try to introduce suitable products or services to each specific customer. On the other hand, while doing shopping online, customers require to share their opinions with one another as well as to be recommended the products or services related to what they are looking for. Thus, over the years, recommender systems have been developed to satisfy both suppliers and customers. Regarding viewpoints of providers, the challenge of these systems is how to generate good recommendations among a large number of products or services being able to recommend whereas evidence of customers' preferences is usually uncertain, imprecise or incomplete.

This research work was supported by JSPS KAKENHI Grant No. 25240049.

V.-N. Huynh et al. (Eds.): IUKM 2015, LNAI 9376, pp. 290–302, 2015.
DOI: 10.1007/978-3-319-25135-6_28

Obviously, recommendation decisions made on the basis of multiple pieces of evidence of customers' preferences should be more effective than the ones based on a single piece of evidence. In addition, it can be seen that Dempster-Shafer (DS) theory [4,9] provides a flexible method for modeling evidence with uncertain, imprecise, and incomplete information. Specially, with this theory, several pieces of evidence can be combined to form more valuable evidence. As a consequence, the theory has been applied in recommender systems using soft ratings [7,8,10]. According to these systems, each rating entry of a user on an item is represented as a mass function as well as considered as a piece of evidence of users' preferences on the item.

Additionally, in a recommender system, the total number of items is usually extremely large, and each user only rates a small subset of items. This issue, known as the sparsity problem, significantly affects quality of recommendations. For the purpose of dealing with the issue, recomender systems using soft ratings employ context information of all users [8,10] or context information extracted from the social network [7] to predict all unrated data. Regarding the predicting unrated data process [7,8,10], pieces of evidence of users' preferences in a specific context are combined together to generate group preferences for each item first; then these group preferences are used for generating unrated data. Assuming that a user U_i has not rated an item I_k, the process to generate the unrated entry of user U_i on item I_k contains several steps such as some relevant group preferences are combined to form concept preferences of user U_i on item I_k, the concept preferences are combined in order to form the context preference of user U_i on item I_k, if the context preference is not vacuous, it is assigned to unrated entry of user U_i on item I_k. Furthermore, in order to generate a recommendation list for an active user, some nearest neighbors of this user are selected; and the rating data of these neighbors are considered as pieces of evidence of the user's preference, thus these pieces of evidence are combined together for predicting the preference of this user. Clearly, evidence combination tasks are used frequently in the recommender systems using soft ratings.

However, in these systems, the method for evidence combination, known as Dempster's rule of combination, is not effective in some cases; for example, when the problem domain contains a large number of elements such as 10 elements in the Flixster data set [7], or when many pieces of evidence have very low probabilities. Therefore, in this paper, we develop a solution, called 2-probabilities focused combination, for the purpose of supporting recommender systems using soft ratings based on DS theory to fuse pieces of evidence. Regarding the proposed solution, pieces of evidence with high probabilities are focused, and the ones with very low probabilities are considered as noise and eliminated.

The remainder of the paper is organized as follows. In the second section, background information about DS theory is provided, and then related work is presented. After that, in the third section, the details of 2-probabilities focused combination solution are described. Next, in the fourth section, we show a way to integrate the proposed solution into recommender systems using soft ratings. In the fifth section, system implementation and discussions are represented. Finally, conclusions are presented in the last section.

2   Background and Related Work

2.1   DS Theory

DS theory [4,9] is a well-known theory of evidence and plausible reasoning; the reason is that it contains a viewpoint on the representation of uncertainty and ignorance, as well as focusing on the fundamental operation of plausible reasoning. In the context of this theory, a problem domain is represented by a finite set $\Theta = \{\theta_1, \theta_2, ..., \theta_L\}$ of mutually exclusive and exhausive hypotheses, called the frame of discernment (FoD) [4]. Each proposition θ_i, with $i = \overline{1, L}$, referred to as a singleton, denotes the lowest level of discernible information in the FoD.

A function $m : 2^{\Theta} \to [0, 1]$ is called a basic probability assignment (BPA) or a mass function if it satisfies $m(\emptyset) = 0$ and $\sum_{A \subseteq \Theta} m(A) = 1$. The quantity $m(A)$ is the measure of the belief that is committed exactly to A. A subset $A \subseteq \Theta$, with $m(A) > 0$, is called a focal element of the mass function m; and the set of all focal elements is called the focal set. A mass function m is considered as vacuous function if $m(\Theta) = 1$ and $\forall A \subset \Theta, m(A) = 0$. When a source of information provides a mass function m and this source has probability of δ of reliability, one may adopt $1 - \delta$ as one's discount rate, resulting in a new mass function m^δ defined by

$$m^\delta(A) = \delta \times m(A), \text{ for } A \subset \Theta;$$
$$m^\delta(\Theta) = \delta \times m(\Theta) + (1 - \delta), \text{ for } A = \Theta,$$

where $\delta \in [0, 1]$ is the degree of trust in mass function m and indicates the reliability of the evidence source.

Based on mass function m, the function $Bel : 2^{\Theta} \to [0, 1]$ defined by

$$Bel(A) = \sum_{\emptyset \neq B \subseteq A} m(B), \text{ for } \forall A \subseteq \Theta$$

is know as a belief function. The quantity of $Bel(A)$ is the measure of the total probability mass constrained to stay somewhere in A. Additionally, a function $Q : 2^{\Theta} \to [0, 1]$ is a commonality function defined by

$$Q(A) = \sum_{\emptyset \neq A \subseteq B} m(B), \text{ for } \forall A \subseteq \Theta.$$

Here, the quantity of $Q(A)$ is the measure of the total probability mass that can move freely to any point in A.

Given a belief function Bel, a function $Dou : 2^{\Theta} \to [0, 1]$, defined by $Dou(A) = Bel(\neg A)$, is called the doubt function. This function measures the probability mass constrained to stay out of A. The function $Pl(A) = 1 - Dou(A) = 1 - Bel(\neg A)$ is called plausibility function which represents the upper probability. The quantity of $Pl(A)$ is the measure of the total probability mass that can move into A, though it is not necessary that it can all move into a single point, hence $Pl(A) = \sum_{A \cap B} m(B)$ is immediate. It follows that $Pl(A) \geq Bel(A)$ since the total mass that can move into A is a super-set of the mass constrained to stay in A.

Let us consider two pieces of evidence on the same frame Θ, represented by two mass functions m_1 and m_2. Dempster's rule of combination, a fundamental operation of evidence reasoning, can be used to combine these two pieces of evidence into a single one denoted by $(m_1 \oplus m_2)$ (also called the orthogonal sum of m_1 and m_2). Formally, this operation is defined as follow

$$(m_1 \oplus m_2)(\emptyset) = 0;$$

$$(m_1 \oplus m_2)(A) = \frac{1}{1-K} \sum_{\{C,D \subseteq \Theta | C \cap D = A\}} m_1(C) \times m_2(D),$$

where $K = \sum_{\{C,D \subseteq \Theta | C \cap D = \emptyset\}} m_1(C) \times m_2(D) \neq 0$, and K represents the basic probability mass associated with conflict. If $K = 1$, then $m_1 \oplus m_2$ does not exist and m_1 and m_2 are said to be totally or flatly contradictory [1].

2.2   Related Work

Assuming that we have N pieces of evidence represented as mass functions which are defined on the same FoD Θ. When combining these pieces of evidence by using Dempster's rule of combination, the computational complexity is dominated by the number of elements in Θ (in the worst case, the complexity of time is $O(|\Theta|^{N-1})$ [3]). As mentioned earlier, in recommender systems using soft ratings, the task of combining multiple pieces of evidence is performed frequently; thus, performances of these systems are not effective because of heavily depending on this task. One way to improve the performances is to reduce the number of pieces of evidence being combined, but possible answers to the question of interest remain [3]. Over the years, some reducing solutions have been developed, such as simple and separable support functions [1], dichotomous function [1], triplet mass function [2,3]. In the rest of this section, we will present briefly these solutions.

Simple and Separable Support Functions. The structure of evidence being taking into account mass function can be significantly simplified. The simplest form of a mass function m can be defined as below

$$m(A) = p;$$
$$m(\Theta) = 1 - p,$$
$$m(B) = 0 \text{ with } B \subset \Theta \text{ and } B \neq A,$$

where p is called the degree of support; the mass function in this form is called a simple support function focused on A [1]. A separable support function is defined as either a simple support function or the orthogonal sum of two or more simple support functions that can be combined [1].

Let us consider a particular case happening frequently. In this case, m_1 and m_2 are simple support functions with their degrees of supports are p_1 and p_2 respectively, and the common focus is $A \subset \Theta$. Let m be the result when combing m_1 and m_2, then we have

$$m(A) = p_1 + p_2 - p_1 \times p_2,$$
$$m(\Theta) = (1 - p_1) \times (1 - p_2),$$
$$m(B) = 0 \text{ with } B \subset \Theta \text{ and } B \neq A.$$

Dichotomous Function. In [1], the author also developed an evidence combination solution based on dichotomous mass functions instead of general mass functions. A mass function m is called a dichotomous function if its focal set, denoted by F, contains only three possible focal elements A, $\Theta \backslash A$, and Θ; in other words, $F = \{A, \Theta \backslash A, \Theta\}$ with $A \subset \Theta$, and $m(A) + m(\Theta \backslash A) + m(\Theta) = 1$. In this case, $m(A)$ is the degree of support for A, $m(\Theta \backslash A)$ is the degree of support for the refutation of A, and $m(\Theta)$ is the degree of the support not assigned for or against the proposition A.

Triplet Mass Function. In [2,3], the authors have introduced a new structure, called focal element triplet, for modeling multiple pieces of evidence, and 2-points focused combination solution for combining pieces of evidence. The focal element triplet originally contains singletons; however, we can extend this structure for representing composites. Formally, let us consider a FoD $\Theta = \{\theta_1, \theta_2, ..., \theta_L\}$, and a mass function $m : 2^\Theta \rightarrow [0, 1]$ with its focal set is $F = \{A_1, A_2, ..., A_n\}$. A focal element triplet is defined as an expression of the form $< X_1, X_2, X_3 >$, where $X_1, X_2 \subseteq \Theta$ and $X_3 = \Theta$. The elements of the triplet are defined as follow

$$X_1 = A_i, \text{ with } m(A_i) = max\{m(A_1), m(A_2), ..., m(A_n)\};$$
$$X_2 = A_j, \text{ with } m(A_j) = max\{m(A_k) \in F \backslash A_i\};$$
$$X_3 = \Theta.$$

The triplet mass function [2,3] associating with this triplet, denoted by $\bar{m}$, is defined as follows

$$\bar{m}(X_1) = m(A_i);$$
$$\bar{m}(X_2) = m(A_j);$$
$$\bar{m}(X_3) = 1 - m(A_i) - m(A_j).$$

Suppose that we are given two triplet mass functions $\bar{m}_1$ and $\bar{m}_2$. When combining $\bar{m}_1$ and $\bar{m}_2$ by using 2-points focused combination solution, the combined

Table 1. Mass function m_1

$m_1(\{1\})$	$= 0.30$
$m_1(\{3\})$	$= 0.30$
$m_1(\{4\})$	$= 0.04$
$m_1(\{5\})$	$= 0.30$
$m_1(\{1,2,3,4,5\})$	$= 0.06$

Table 2. Mass function m_2

$m_2(\{1\})$	$= 0.40$
$m_2(\{2\})$	$= 0.10$
$m_2(\{3\})$	$= 0.07$
$m_2(\{4\})$	$= 0.40$
$m_2(\{1,2,3,4,5\})$	$= 0.03$

Table 3. Mass function $\bar{m}_1^{(1)}$

$\bar{m}_1^{(1)}(\{1\})$	$= 0.30$
$\bar{m}_1^{(1)}(\{3\})$	$= 0.30$
$\bar{m}_1^{(1)}(\{1,2,3,4,5\})$	$= 0.40$

Table 4. Mass function $\bar{m}_1^{(2)}$

$\bar{m}_1^{(2)}(\{1\})$	$= 0.30$
$\bar{m}_1^{(2)}(\{5\})$	$= 0.30$
$\bar{m}_1^{(2)}(\{1,2,3,4,5\})$	$= 0.40$

Table 5. Mass function $\bar{m}_1^{(3)}$

$\bar{m}_1^{(3)}(\{3\})$	$= 0.30$
$\bar{m}_1^{(3)}(\{5\})$	$= 0.30$
$\bar{m}_1^{(3)}(\{1,2,3,4,5\})$	$= 0.40$

result can have three different focal elements (two focal points equal), four different focal elements (one focal point equal), or five different focal elements (totally different focal points); thus this result is transformed into the corresponding triplet mass function [3].

The 2-points focused combination solution is capable of not only helping to distinguish of trivial focal element from significant ones, but also reducing the effective computation time [2,3]. In some cases, however, it is not effective, such as illustrated in Example 1.

Example 1. Let us consider a recommender system using soft ratings with its rating domain $\Theta = \{1,2,3,4,5\}$. Assuming that we are given two ratings represented as two mass functions m_1 and m_2 represented in Tables 1 and 2, respectively. When converting into triplet mass functions, mass function m_1 can be one of three different triplet mass functions, called $\bar{m}_1^{(1)}$, $\bar{m}_1^{(2)}$, and $\bar{m}_1^{(3)}$, as shown in Tables 3, 4 and 5, respectively; and mass function m_2 has an only one triplet mass function, denoted by $\bar{m}_2$, described in Table 6. Regarding three triplet mass function options of mass function m_1, when combing two mass functions m_1 and m_2 using 2-points focused combination solution, we can achieve three possible results as shown in Tables 7, 8, and 9. We can observe that triplet mass function $\bar{m}^{(3)}$ is significantly different from triplet mass functions $\bar{m}^{(1)}$ and $\bar{m}^{(2)}$. Noticeably, the triplet mass function result of the combination of two mass functions m_1 and m_2 depends on the way we choose the triplet mass function regarding mass function m_1; therefore, 2-points focused combination solution is not effective in this case.

3   Proposed Solution

Let us consider a FoD $\Theta = \{\theta_1, \theta_2, ..., \theta_L\}$ and a mass function $m : 2^\Theta \rightarrow [0,1]$. Assuming that the focal set of mass function m, denoted by F, contains n elements. Obviously, the number elements in focal set F is dependent on the number elements in Θ; and the maximum value of n can be $2^{|\Theta|}$. Actually, total

Table 6. Mass function $\bar{m}_2$

$\bar{m}_2(\{1\})$	$= 0.40$
$\bar{m}_2(\{4\})$	$= 0.40$
$\bar{m}_2(\{1,2,3,4,5\})$	$= 0.20$

Table 7. Mass function $\bar{m}^{(1)}$ **Table 8.** Mass function $\bar{m}^{(2)}$ **Table 9.** Mass function $\bar{m}^{(3)}$

$\bar{m}^{(1)}(\{1\})$	$= 0.53$	$\bar{m}^{(2)}(\{1\})$	$= 0.53$	$\bar{m}^{(3)}(\{1\})$	$= 0.31$
$\bar{m}^{(1)}(\{4\})$	$= 0.25$	$\bar{m}^{(2)}(\{4\})$	$= 0.25$	$\bar{m}^{(3)}(\{4\})$	$= 0.31$
$\bar{m}^{(1)}(\{1,2,3,4,5\})$	$= 0.22$	$\bar{m}^{(2)}(\{1,2,3,4,5\})$	$= 0.22$	$\bar{m}^{(3)}(\{1,2,3,4,5\})$	$= 0.38$

elements in focal set F significantly influences performances when combining mass function m with other ones.

It can be seen that, usually, some elements in focal set F have high probabilities, and other elements have very low probabilities. In fact, only the elements with high probabilities significantly influences the results when combining mass function m with other ones. The elements with very low probabilities can be considered as noise, which are generated when assigning probabilities to mass function m and/or by the evidence combination tasks had been done before. As a consequence, in order to improve performances when combining mass function m with other ones, the elements with low probabilities in focal set F should be eliminated.

Regarding this observation, we propose that, in the focal set of mass function m, only elements with their probabilities in top two highest probabilities are retained, other elements are eliminated. The probabilities of the eliminated elements are added to the probability of the whole set element. The achieved mass function is called 2-probabilities focused mass function, denoted by $\bar{m}$. Then, 2-probabilities focused mass function $\bar{m}$ is used instead of mass function m. Formally, assuming that after sorting the elements in focal set F by probabilities, we achieve $F = \{A_1, A_2, ..., A_n\}$, where $A_i \subseteq \Theta$, $m(A_i) = p_i$, and $p_1 \geq p_2 \geq p_3 \geq ... \geq p_n$. Based on mass function m and its sorted focal set F, 2-probabilities focused mass function $\bar{m} : 2^{\Theta} \to [0,1]$ is defined as below

$$\forall A \subset \Theta, \bar{m}(A) = \begin{cases} m(A) \text{ if } m(A) = p_1 \text{ or } m(A) = p_2; \\ 0, \text{ otherwise;} \end{cases}$$

$$\bar{m}(\Theta) = p, \text{ where } p = 1 - \sum_{\{A \subset \Theta | m(A) = p_1\}} m(A) - \sum_{\{B \subset \Theta | m(B) = p_2\}} m(B).$$

Consider now 2-probabilities focused mass functions $\bar{m}_1$ and $\bar{m}_2$ on the same frame Θ. 2-probabilities focused combination solution for fusing these two 2-probabilities focused mass functions, denoted by $\bar{m} = \bar{m}_1 \uplus \bar{m}_2$, containing two steps which are defined as below

- Combining $\bar{m}_1$ and $\bar{m}_2$ by using Dempster's rule of combination. Let m be the result mass function after performing this step, we have $m = \bar{m}_1 \oplus \bar{m}_2$.

Table 10. Mass function $\bar{m}_1$ **Table 11.** Mass function $\bar{m}_2$ **Table 12.** Mass function $\bar{m}$

$\bar{m}_1(\{1\})$	$= 0.30$	$\bar{m}_2(\{1\})$	$= 0.40$	$\bar{m}(\{1\})$	$= 0.60$
$\bar{m}_1(\{3\})$	$= 0.30$	$\bar{m}_2(\{2\})$	$= 0.10$	$\bar{m}(\{4\})$	$= 0.15$
$\bar{m}_1(\{4\})$	$= 0.04$	$\bar{m}_2(\{4\})$	$= 0.40$	$\bar{m}(\{1,2,3,4,5\})$	$= 0.25$
$\bar{m}_1(\{5\})$	$= 0.30$	$\bar{m}_2(\{1,2,3,4,5\})$	$= 0.10$		
$\bar{m}_1(\{1,2,3,4,5\})$	$= 0.06$				

- Converting result mass function m into 2-probabilities focused mass function $\bar{m}$.

With 2-probabilities focused combination solution, the computation time is effective because of elimination trivial focal elements in focal sets. Specially, this solution is capable of overcoming the problem of 2-points focused combination solution [2,3] happening when in the focal set of a mass function contains several elements with similar high probabilities, such as illustrated in Example 2.

Example 2. Let us continue with Example 1. In this case, regarding mass function m_1, there is only one 2-probabilities focused mass function $\bar{m}_1$ as shown in Table 10. The 2-probabilities focused mass function regarding mass function m_2 is represented in Table 11. When combining these two 2-probabilities focused mass functions $\bar{m}_1$ and $\bar{m}_2$ using 2-probabilities focused combination solution, we get only one result as shown in Table 12.

4   Integrating with Recommender Systems Using Soft Ratings

In recommender systems using soft ratings based on DS theory [7,8,10], sets of users and items are defined by $\mathbf{U} = \{U_1, U_2, ..., U_M\}$ and $\mathbf{I} = \{I_1, I_2, ..., I_N\}$, respectively. Additionally, each user rating is represented as a preference mass function spanning over a rating domain $\Theta = \{\theta_1, \theta_2, ..., \theta_L\}$ where $\theta_i < \theta_j$ whenever $i < j$; and all ratings are represented by a DS rating matrix denoted by $\mathbf{R} = \{r_{i,k}\}$ with $r_{i,k}$ being the rating entry of user $U_i \in \mathbf{U}$ on item $I_k \in \mathbf{I}$. Originally, $r_{i,k}$ is represented as general mass function $m_{i,k}$; here, $r_{i,k}$ is represented as 2-probabilities focused mass function $\bar{m}_{i,k}$. Let $^I R_i = \{I_l \in \mathbf{I} \mid \bar{m}_{i,l} \neq vacuous\}$ be the set of items rated by user U_i, and $^U R_k = \{U_l \in \mathbf{U} \mid \bar{m}_{l,k} \neq vacuous\}$ be the set of users already rated item I_k.

In general, the recommendation process of these system can be illustrated in Figure 1; except the system, in [7], of which users are separated into several overlapping communities first and then this recommendation process is applied into each community independently. Regarding the figure, context information is employed for predicting unrated data in rating matrix $\mathbf{R}$; next, user-user similarities are computed by using both provided and predicted ratings; after that, a neighborhood set of an active user is selected; finally, the suitable recommendations for this active user are generated. Note that the task for combining pieces of evidence is used in predicting unrated data and generating recommendation

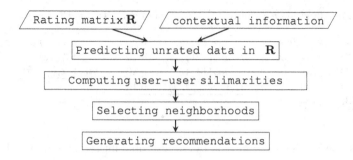

Fig. 1. The recommendation process

steps. Thus, details of these two steps will be represented in the remainder of this section; information about the other ones can be found in [7,8,10].

4.1   Predicting Unrated Data

As mentioned previously, context information from different sources is fused for the purpose of predicting unrated data in the rating matrix. Formally, context information, denoted by $\mathcal{C}$, is defined as set of P concepts, $\mathcal{C} = \{C_1, C_2, ..., C_P\}$; and each concept C_p contains Q_p groups, $C_p = \{G_{p,1}, G_{p,2}, ..., G_{p,Q_p}\}$. In the recommender systems, an item can belong to several groups in a concept; and a user can be interested in some groups in the same concept. Note that users interested in the same group are expected to have similar preferences. Let us consider a concept $C_p \in \mathcal{C}$; the groups to which item I_k belongs are determined by mapping function $g_p : \mathbf{I} \to 2^{C_p} : I_k \mapsto g_p(I_k) \subseteq C_p$ and the groups in which user U_i interests are identified by mapping function $f_p : \mathbf{U} \to 2^{C_p} : U_i \mapsto f_p(U_i) \subseteq C_p$.

Firstly, group preferences of users on each item are generated. Let us consider item $I_k \in \mathbf{I}$, the group preference of this item on group $G_{p,q} \in g_p(I_k)$ is defined by 2-probabilities focused mass function $^{G}\bar{m}_{p,q,k} : 2^{\Theta} \to [0,1]$, computed by combining the provided rating data on item I_k of users interesting in group $G_{p,q}$, as shown below

$$^{G}\bar{m}_{p,q,k} = \biguplus_{\{j | I_k \in {}^{I}R_j, G_{p,q} \in f_p(U_j) \cap g_p(I_k)\}} \bar{m}_{j,k}.$$

Secondly, assuming that user U_i has not rated item I_k, the unrated entry $r_{i,k}$ of user U_i on item I_k is generated by performing three steps as following

– The concept preferences corresponding to user U_i on item I_k, denoted by 2-probabilities focused mass functions $^{C}\bar{m}_{p,i,k} : 2^{\Theta} \to [0,1]$ with $p = \overline{1, P}$, are computed by combining related group preferences of item I_k as follows

$$^{C}\bar{m}_{p,i,k} = \biguplus_{\{q | G_{p,q} \in f_p(U_i) \cap g_p(I_k)\}} {}^{G}\bar{m}_{p,q,k}.$$

- The context preference corresponding to user U_i on item I_k, denoted by 2-probabilities focused mass function $^C\bar{m}_{i,k} : 2^\Theta \to [0,1]$, is achieved by combining all related concept mass functions as below

$$^C\bar{m}_{i,k} = \biguplus_{p=\overline{1,P}} {}^C\bar{m}_{p,i,k}.$$

- Next, if the context preference corresponding to user U_i on item I_k is not vacuous, it is assigned to unrated entry $r_{i,k}$, as follows

$$r_{i,k} = {}^C\bar{m}_{i,k},$$

otherwise, unrated entry $r_{i,k}$ is assigned by the evidence obtained by combining all 2-probabilities focused mass functions of users who have already rated item I_k [8] as below

$$r_{i,k} = \biguplus_{\{j|U_j\in{}^UR_k\}} \bar{m}_{j,k}.$$

It would be worth noting here that, at this point, all unrated data in the rating matrix are completely predicted. Then, predicted as well as provided ratings are used for computing user-user similarities. Assuming that user-user similarities are represented as a matrix $\mathbf{S} = \{s_{i,j}\}$ where $i = \overline{1,M}, j = \overline{1,M}$, and $s_{i,j}$ is referred to as the user-user similarity between users $U_i \in \mathbf{U}$ and $U_j \in \mathbf{U}$.

4.2   Generating Recommendations

Before making a recommendation list for an active user U_i, for each item I_k which has been not rated by user U_i, a K nearest neighborhood set $\mathcal{N}_{i,k}$ containing the users who have already rated item I_k and whose similarities with user U_i are equal or greater than a threshold τ, is selected first. Then, the estimated rating value of user U_i on item I_k is computed as $\hat{r}_{i,k} = \hat{m}_{i,k}$, where $\hat{m}_{i,k} = \bar{m}_{i,k} \uplus \tilde{m}_{i,k}$. Here, $\tilde{m}_{i,k}$ is the 2-probabilities focused mass function corresponding to overall preferences of members in neighborhood set $\mathcal{N}_{i,k}$; $\tilde{m}_{i,k}$ is calculated as below

$$\tilde{m}_{i,k} = \biguplus_{\{j|U_j\in\mathcal{N}_{i,k}\}} \bar{m}_{j,k}^{s_{i,j}}, \text{with } \bar{m}_{j,k}^{s_{i,j}} = \begin{cases} s_{i,j} \times \bar{m}_{j,k}(A), & \text{for } A \subset \Theta; \\ s_{i,j} \times \bar{m}_{j,k}(\Theta) + (1 - s_{i,j}), & \text{for } A = \Theta. \end{cases}$$

After that, estimated rating values of all unrated items are ranked and the suitable recommendation list is generated for user U_i [7,8,10].

5   Implementation and Discussions

To evaluate the proposed solution, we applied it to the recommended system developed in [8], and selected 2-points focused combination solution [2,3] for the purpose of comparison. Since the selected system [8] supports both hard and soft recommendation decisions; in the experiments, we chose some evaluation criteria *MAE* [5], *Precision* [5], *Recall* [5], F_β [5] for evaluating hard decisions and *DS-MAE* [10], *DS-Precision* [6], *DS-Recall* [6], *DS-F$_\beta$* [10] for evaluating soft decisions.

Additionally, we used Movielens data set, MovieLens 100k, in the experiments. This data set contains 100,000 hard ratings from 943 users on 1682 movies with the hard rating value $\theta_l \in \{1, 2, 3, 4, 5\}$. Each user has rated at least 20 movies. Because the selected system [8] requires a domain with soft ratings represented as 2-probabilities focused mass functions, each hard rating entry $\theta_l \in \Theta$ was transformed into the soft rating entry $r_{i,k}$ by the DS modeling function [10] as follows

$$r_{ik} = \bar{m}_{ik} = \begin{cases} \alpha_{ik} \times (1 - \sigma_{ik}), & \text{for } A = \theta_l; \\ \alpha_{ik} \times \sigma_{ik}, & \text{for } A = B; \\ 1 - \alpha_{ik}, & \text{for } A = \Theta; \\ 0, & \text{otherwise,} \end{cases} \text{ with } B = \begin{cases} (\theta_1, \theta_2), & \text{if } l = 1; \\ (\theta_{L-1}, \theta_L), & \text{if } l = L; \\ (\theta_{l-1}, \theta_l, \theta_{l+1}), & \text{otherwise.} \end{cases}$$

Here, $\alpha_{ik} \in [0, 1]$ and σ_{ik} are a trust factor and a dispersion factor, respectively [10].

We adopted the method suggested in [8] for conducting the experiments. Firstly, 10% of the users were randomly selected. Then, for each selected user, we accidentally withheld 5 ratings, the withheld ratings were used as testing data and the remaining ratings were considered as training data. Finally, recommendations were computed for the testing data. We repeated this process for 10 times, and the average results of 10 splits were represented in this section. Note that in all experiments, some parameters were selected as following: $\gamma = 10^{-4}$, $\beta = 1$, $\forall(i,k)\{\alpha_{i,k}, \sigma_{i,k}\} = \{0.9, 2/9\}, w_1 = 0.3, w_2 = 0.1$, and $\tau = 0.7$.

Figures 2 and 3 show overall *MAE* and *DS-MAE* criterion results change with neighborhood size K. In these figures, the smaller values are the better ones. According to Figure 2, the proposed solution is better than 2-points focused mass combination solution in most of cases. Specially, regarding Figure 3, the proposed system is more effective in all selected sizes of K.

Figures 4 and 5 illustrate overall *Precision* and *DS-Precision* criterion results change with neighborhood size K. In these features, the higher values

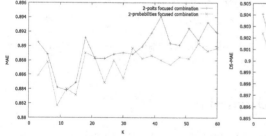

Fig. 2. Overall *MAE* versus K

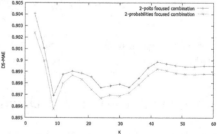

Fig. 3. Overall *DS-MAE* versus K

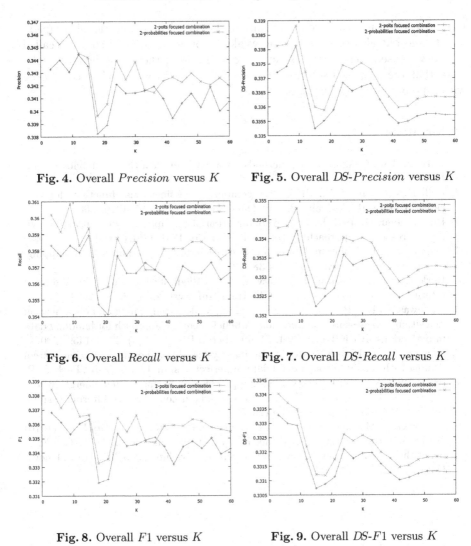

Fig. 4. Overall *Precision* versus *K*   **Fig. 5.** Overall *DS-Precision* versus *K*

Fig. 6. Overall *Recall* versus *K*   **Fig. 7.** Overall *DS-Recall* versus *K*

Fig. 8. Overall *F*1 versus *K*   **Fig. 9.** Overall *DS-F*1 versus *K*

are the better ones. These two figures show that 2-probabilities focused combination solution is better than 2-points focused combination solution. These results are similar the other evaluation criteria such as shown in figures 6, 7, 8, and 9.

6   Conclusions

In this paper, we have developed 2-probabilities focused combination solution for recommender systems using soft ratings based on DS theory. This solution focuses only on significant focal elements, known as the ones with their probabilities in top two highest probabilities, in corresponding focal sets. Specially, the proposed solution is capable of not only improving the computational complexity of time but also overcoming the weakness of an alternative solution called 2-points focused combination solution [2,3]. Besides, the experimental results show that the proposed solution achieves better performance when comparing with 2-points focused combination solution.

References

1. Barnett, J.A.: Computational methods for a mathematical theory of evidence. In: IJCAI 1981, pp. 868–875 (1981)
2. Bell, D.A., Guan, J.W., Bi, Y.: On combining classifier mass functions for text categorization. IEEE Trans. Knowl. Data Eng. **17**(10), 1307–1319 (2005)
3. Bi, Y., Guan, J., Bell, D.A.: The combination of multiple classifiers using an evidential reasoning approach. Artif. Intell. **172**(15), 1731–1751 (2008)
4. Dempster, A.P.: Upper and lower probabilities induced by a multivalued mapping. Annals of Mathematical Statistics **38**, 325–339 (1967)
5. Herlocker, J.L., Konstan, J.A., Terveen, L.G., Riedl, J.: Evaluating collaborative filtering recommender systems. ACM Trans. Inf. Syst. **22**(1), 5–53 (2004)
6. Hewawasam, K.K.R., Premaratne, K., Shyu, M.-L.: Rule mining and classification in a situation assessment application: A belief-theoretic approach for handling data imperfections. IEEE Trans. Syst. Man Cybern., Part B **37**(6), 1446–1459 (2007)
7. Nguyen, V.-D., Huynh, V.-N.: A community-based collaborative filtering system dealing with sparsity problem and data imperfections. In: Pham, D.-N., Park, S.-B. (eds.) PRICAI 2014. LNCS, vol. 8862, pp. 884–890. Springer, Heidelberg (2014)
8. Nguyen, V.-D., Huynh, V.-N.: A reliably weighted collaborative filtering system. In: ECSQARU 2015, pp. 429–439 (2015)
9. Shafer, G.: A Mathematical Theory of Evidence. Princeton University Press (1976)
10. Wickramarathne, T.L., Premaratne, K., Kubat, M., Jayaweera, D.T.: Cofids: A belief-theoretic approach for automated collaborative filtering. IEEE Trans. Knowl. Data Eng. **23**(2), 175–189 (2011)

A Multifaceted Approach to Sentence Similarity

Hien T. Nguyen[1(✉)], Phuc H. Duong[1], and Tuan Q. Le[2]

[1] Faculty of Information Technology, Ton Duc Thang University, Ho Chi Minh City, Vietnam
{hien,duonghuuphuc}@tdt.edu.vn
[2] Ho Chi Minh City University of Transport, Ho Chi Minh City, Vietnam
tuanql@hcmutrans.edu.vn

Abstract. We propose a novel method for measuring semantic similarity between two sentences. The method exploits both syntactic and semantic features to assess the similarity. In our method, words in a sentence are weighted using their information content. The weights of words help differentiate their contribution towards the meaning of the sentence. The originality of this research is that we explore named entities and their coreference relations as important indicators for measuring the similarity. We conduct experiments and evaluate our proposed method on Microsoft Research Paraphrase Corpus. The experiment results show that named entities and their coreference relations improve significantly the performance of paraphrase identification and the proposed method is comparable with state-of-the-art methods for paraphrase identification.

1   Introduction

Sentence similarity is a task of determining the degree of semantic similarity between two irregular sentences. This task has had many applications in natural language processing and related areas such as information retrieval, text mining, question answering, text summarization, plagiarism detection, or assessing the translation quality of automatic translation systems. Due to its applicability, recent literature on sentence similarity has shown abundantly proposed methods, which often focus on identifying paraphrases.

Most of them use corpus-based or knowledge-based word-to-word semantic similarity measures [1], in combination with string matching algorithms [2, 16], syntactic structure information [5], or information exploited from parse trees [8, 9, 10, 13]. More approaches exploit machine translation metrics [7], discourse information [6], graph subsumption [11], or vector space models [15, 17]. However, the research work in literature did not exploit named entities and their coreference relations for measuring sentence similarity.

We propose a novel method to compute semantic similarity of sentence pairs. The method explores named entities and their coreference relations as important indicators for measuring the similarity. It also exploits word-to-word similarity and word-order similarity as proposed by Li *et al.* in [4]. An intuition shows that exploring named entities and their coreference relations is improving the performance of semantic similarity between sentences, especially for those having the same meaning and

© Springer International Publishing Switzerland 2015
V.-N. Huynh et al. (Eds.): IUKM 2015, LNAI 9376, pp. 303–314, 2015.
DOI: 10.1007/978-3-319-25135-6_29

containing named entities but few words in common. For example, with these two sentences *"I am currently working at IBM"* and *"I am a developer at International Business Machines"*, if we only compute the similarity based on words and word-order, the similarity score may not be high as it would be regardless of *IBM* and *International Business Machines* in the contexts of the two sentences have the same referent. The intuition also shows that word-order similarity is a helpful feature. For example, two sentences *"Bob calls the police"* and *"The police calls Bob"* have different meaning but all of their words are common. It is clearly that word order helps distinguish the meaning of these sentences. We evaluate the method on Microsoft Research Paraphrase Corpus (MSRP) [24].

The contributions of this paper are three-fold as follows: (i) we propose a novel method for measuring sentence similarity; (ii) we explore named entities and their coreference relations and prove that they help significantly improve the paraphrase identification performance; and (iii) we investigate several features and evaluate how helpful WordNet and corpora are. The originality of this research is that we explore named entities and their coreference relations as important indicators for measuring sentence similarity.

The rest of this paper is organized as follows. Section 2 presents related work. Section 3 presents our proposed method. Section 4 presents datasets, experiments and results. Finally, we draw conclusion in Section 5.

2   Related Work

Recent literature on sentence similarity has shown abundantly proposed methods [2, 4, 6, 13, 18]. Some of them adapted bag-of-words techniques that are commonly used in Information Retrieval [19] systems such as the method proposed by Sahami and Heilman (2006) [3]. Those bag-of-words techniques rely on assumption that the more similar two sentences are the more same words they share. In [3], the authors proposed a semantic kernel function that takes advantage of web search results to extend context of short texts. Nonetheless, there are several drawbacks of the bag-of-words approach as follows:

- It is not very efficient in computation because sentences are represented as vectors in an *n*-dimension space where *n* is very large compared to the number of words in each sentence. Even though techniques for dimensionality reduction such as Latent Semantic Analysis are applied as presented in [25], the number of dimensions of vectors is still high.
- As concluded in [1], the bag-of-words approach "ignores many of the important relationships in sentence structure" such as syntactic and semantic dependencies between words. Recently, Ştefănescu *et. al.* [13] showed that syntactic dependencies between words are strong indicators for sentence semantic similarity.
- Moreover, it excludes function words such as *the, of, an*, etc. According to Li *et. al.* [4], function words carry out structured information which is helpful in interpreting meaning of sentences. Indeed, as pointed out in [26], function words such

as *because, in spite of, as a result of, according to* are strong discourse markers to extract basic elementary discourse units which are used in [6] for measuring sentence similarity.

Most research work in literature exploited semantic similarity between two words for measuring how similar two input sentences are. Until now, there have been many measures of word-to-word semantic similarity proposed. Notable measures were implemented in [1] where the authors proposed a method - also relies on a bag-of-words technique - that estimates the semantic similarity between two short texts using both corpus-based and knowledge-based similarity measures between two words. The method combines semantic similarity between two words with word specificity. In particular, given two short text segments, the method finds for each word in the first one the most similar matching word in the second one and then similarity between those word pairs will be included in the overall semantic similarity of two text segments. The authors reported that the model combines six different similarity measures between two words gives the best performance, with 81.3% F-measure, on Microsoft paraphrase corpus. The authors report taking word specificity into account, but as argued in [18], advantages of word specificity are not clear.

Li *et. al.* [4] proposed a method that combines word-to-word semantic similarity and word order similarity for measuring sentence similarity. It keeps all function words and weights significance of each word by its information content derived from Brown Corpus. Since methods based on bag-of-words techniques lead to decision similar if two sentences share the same word set, this method takes word order into account to overcome the drawback of bag-of-word techniques. In [2], the authors improve the method of Li *et. al.* [4] by combining word-to-word semantic similarity, word order similarity, and lexical string matching. They used some modified versions of the longest common subsequence algorithm for lexical string matching. Even though these methods is based on bag-of-words techniques, the number of dimensions of vectors representing sentences in a sentence pair is low and equal to the number of words in the joint word set containing all distinct words of two sentences in the pair. The our proposed method in this paper improves the method in [4] by taking named entities and their coreference relations into account with a learning model. Note that this is the first time those information - named entities and their coreference relations - is exploited for sentence similarity.

Some others notable methods proposed in literature are presented in [6], [7], [8], [9], [10], [11], [13], [14], [17], [25]. In [6], in order to identify the paraphrase between two sentences, the authors introduce an approach to exploit elementary discourse units (EDUs). Each input sentence is split into discourse elements, then, semantic similarity between two sentences is computed by using the similarity between these EDUs. This method is quite complexity and overall performance depends on the performance of detecting EDUs. In [7], the authors implemented eight different machine translation metrics for paraphrase detection. Others works exploit latent topics [17, 25], graph structures [11], information from parse trees [8, 9, 10, 13], or vector space models [15, 17]. The method presented in [22] also exploits named entities but is substantially different than ours.

3   Proposed Method

This section presents our proposed method. In particular, Fig. 1 shows the method in details. We will present preprocessing and feature extraction modules respectively in the following sub-sections.

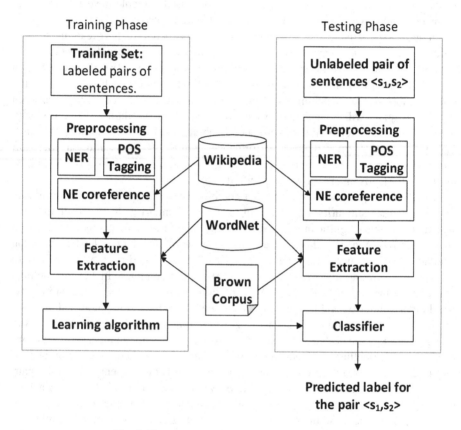

Fig. 1. The proposed method to sentence similarity

3.1   Preprocessing

Given a pair of sentences, we first perform part-of-speech (POS) tagging using Stanford POS tagger[1] to assign parts of speech to each word in the sentences. Since many words are both nouns and verbs vocabulary, it is reasonable to know exactly the POS of each word in the context of a particular sentence to identify the meaning of that word. Let's see the example below:

---

[1] http://nlp.stanford.edu/software/tagger.shtml

S_1: Your help is much appreciated.
S_2: We should help our friends to study well together.

In case of the word *"help"*, its part of speech is noun in first sentence, but verb in second one. It is clear for this case that POS tag is helpful in identifying the meaning of the word *"help"* in a particular context; and this conclusion can generalize for other words as well. Moreover, as argued in [12], all POS information is good indicators for paraphrase recognition. Therefore, we take advantages of POS information in our sentence similarity measure.

We then perform named entity recognition (NER) to identify and classify named entities (NEs) into four classes: PERSON, LOCATION, ORGANIZATION and MISC using Stanford NER[2]. Name-matching rules for NE coreference resolution proposed in [20] are employed to identify if two named entities appearing in a given pair of sentences are coreferent or not. After running the name-matching rules, all named entities in the pair of sentences are grouped in coreference chains, each of which consists of named entities that have the same referent. In order to making easy for processing sentences in next steps, for each of those named entities that do not belong to any coreference chains, we create a new chain for it. Then we assign each coreference chain a unique identifier (ID).

Note that in order to improve the performance of coreference resolution using the name-matching rules, we exploit Wikipedia[3] to build a synonym list and an acronym list. The synonym list consists of pairs of names that can be used to refer to same entity, e.g, *IBM* and *The Big Blue* can be used to refer to International Business Machines Corporation or *Saigon* and *Ho Chi Minh City* can be used to refer to the biggest city in Vietnam. The acronym list consists of pairs of names in which one is acronym of another, e.g., *International Business Machines* and *IBM* or *International Monetary Fund* and *IMF*.

Wikipedia is a free online encyclopedia whose content is contributed by a large number of volunteer users. It consists of a large collection of articles, each of which defines and describes an entity. In reality, each entity may have several names and one name may be used to refer to different entities in different contexts. In Wikipedia, many-to-many correspondence between names and entities can be captured by utilizing *redirect pages*. A redirect page typically contains only a reference to an article. The title of a redirect page is an alternative name of the described entity by that article. We build the synonym and acronym lists based on the titles of Wikipedia articles and the titles of their corresponding redirect pages.

After performing POS tagging, NE recognition and NE coreference resolution, we remove all special characters such as "#", "$", "%", or "." from input sentences. Now we are ready to represent features that we use in our sentence similarity measure.

---

[2] http://nlp.stanford.edu/software/CRF-NER.shtml
[3] https://en.wikipedia.org

3.2   Word-to-Word Similarity and Information Content

We integrate word-to-word similarity into some features used to measure sentence similarity. Therefore, before representing those features in details, we focus on explaining how to compute knowledge-based word-to-word similarity (henceforth *word-sim*) and corpus-based information content of words. In particular, we exploit WordNet[4] to measure *word-sim* between words and Brown Corpus[5] to compute information content of words.

WordNet is a lexical database for English language. Each entry in WordNet is called *synset* - a set of synonyms. The version we used in this paper is WordNet 3.0, which contains 155,287 words organized into 117,659 *synsets*. Each *synset* is a group of synonym words. The definition of words in a *synset* called *gloss text*. Our approach exploits *gloss texts* to measure *word-sim*. Brown Corpus[6] "consists of 1,014,312 English words of running text of edited English prose printed in the United States during the calendar year 1961".

3.3   Word-Sim Measure

Many methods have been proposed for measuring of word-to-word similarity based on WordNet. Some of them were presented in [1]. In our approach, we implement the method proposed by Lesk (1986) [21] for measuring of the similarity between words based on the *gloss texts* derived from WordNet. The reason we use Lesk's method instead of the others is that we identify the meaning of a word based on its POS information, but the taxonomy of synsets in WordNet does not present those POS information explicitly. In particular, we get the appropriate gloss text of a word in a given context based on its POS tag and found that Lesk's method is a reasonable choice. This method measures the similarity between two words based on overlapping degree between two corresponding gloss texts. The higher overlapping degree, the higher similarity. The original formula proposed by Lesk is given as follows:

$$Lesk(w_1, w_2) = |glossText(w_1) \cap glossText(w_2)| \qquad (1)$$

To normalize the similarity score between [0, 1], we suggest an extended formula in Eq. (2) as follows:

$$word - sim(w_1, w_2) = \frac{|glossText(w_1) \cap glossText(w_2)|}{min(|glossText(w_1)|, |glossText(w_2)|)} \qquad (2)$$

3.4   Corpus-Based Information Content

In a sentence, some words may be more important than the others. In order to differentiate the role of different words, we weight each word by its information content

---

[4] https://wordnet.princeton.edu/
[5] http://www.nltk.org/nltk_data/
[6] http://clu.uni.no/icame/brown/bcm.html

(IC). Given a word w, its information content is computed by (this formula is also used in [4]):

$$IC(w) = 1 - \frac{\log(n + 1)}{\log(N + 1)} \tag{3}$$

where N is the total number of words and n is the frequency of the word w in the corpus. The value of IC in Eq. (3) lies between [0, 1]. As above-mentioned, the corpus that we use to compute information content of a word is Brown Corpus.

3.5   Word-Based Cosine Similarity

We present the first feature, namely *word-cosine*, which does not take named entities into account. Given two input sentences, denoted by S_1 and S_2, after go through preprocessing stage, we form a joint word set S that contains all distinct words from S_1 and S_2. Considering the following sentence pair:

S_1: "If we don't march into Tehran, I think we will be in pretty good shape," he said.

S_2: "As long as we don't march on Tehran, I think we are going to be in pretty good shape," he said.

S = {If, we, don't, march, into, tehran, i, think, will, be, in, pretty, good, shape, he, say, as, long, as, on, are, going, to}.

In joint words set S, in the case of *"say"* token, the task of transforming the verb *"said"* to *"say"* is called *stemming*. Unlike the original stemming algorithm - Porter Stemming Algorithm (PSA)[7] - we combine PSA with WordNet to increase the accuracy, e.g., with the verb *"said"*, PSA returns *"said"*; however, in case of combination PSA with WordNet, we get the verb *"say"*. Note that, as above-mentioned, we use POS information to get the semantic of a word; thus, we transform the form of words in both joint word set and the input sentences and keep their POS tags unchanged.

After constructing the set S, we define semantic vectors for the two sentences, denoted by V_1 and V_2. The number of dimensions of each vector is the size of the set S. In order to assign values to all entries of a semantic vector, *word-sim* presented in Eq. (2) is used to determine the values. Taking S_1 as an example:

- Case 1: if a word w in joint word set S appears in S_1, $v_w = 1$.
- Case 2: if w does not appears in S_1, we calculate *word-sim* in turn between w and each word in S_1. If the highest *word-sim* exceeds a preset threshold τ, $v_w = h$, otherwise, $v_w = 0$.

In [4], the authors shown that words appear with high frequency are less important than those appear with low frequency. It means that different words contribute towards the meaning of a sentence with differing degree. We weight each word by its information content to show its significance. The formula presented Eq. (4) shows the weighting scheme where w is the word in joint word set, w' is the word in the

---

[7] http://snowball.tartarus.org/algorithms/porter/stemmer.html

sentence in consideration and having the highest *word-sim* with *w*, and v_w is the entry value of the word *w* in the semantic vector.

$$v_w^{new} = v_w \cdot IC(w) \cdot IC(w') \tag{4}$$

Finally, the semantic similarity between two sentences is computed by cosine coefficient between two semantic vectors:

$$word - cosine(S_1, S_2) = cosine(V_1, V_2) = \frac{V_1 \cdot V_2}{\|V_1\| \cdot \|V_2\|} \tag{5}$$

Note that unlike most of other text similarity methods, as the same way the method proposed in [4] did, we keep all function words in the joint word set, since these words contain syntactic information.

3.6   Word-and-NE-Based Cosine Similarity

We present the second feature, namely *word-and-NE-cosine* which takes named entities and NE coreference relations into account. Given two input sentences, denoted by S_1 and S_2, after go through preprocessing stage, we replace all coreference chains in two sentences by their IDs and achieve sentences S'_1 and S'_2 respectively. Then, we form a joint word set S^{word} that contains all distinct words and a joint coreference chain ID set S^{ID} that contains all distinct IDs from S'_1 and S'_2. We define a semantic vector of a sentence consisting of two parts: $[S^{word} \mid S^{ID}]$. For S^{word}, we compute entry values as doing for *word-cosine* and for S^{ID} we compute entry values as follows: if an ID in S^{ID} appears in S'_1, for instance, $v_{ID} = 1$; otherwise, $v_{ID} = 0$. Considering the following sentence pair:

S_1: In a televised interview on Wednesday[1], ECB[1] President Wim Duisenberg[1] said it was too soon to discuss further interest rate cuts in the 12-nation euro zone.

S_2: European Central Bank[2] President Wim Duisenberg[2] said in a televised interview that it was too soon to discuss further interest rate cuts in the euro zone.

The NE coreference module produces three coreference chains {Wednesday}, {ECB[1], European Central Bank[2]}, and {Wim Duisenberg[1], Wim Duisenberg[2]}. Assuming that we assign ID1, ID2, and ID3 these coreference chains respectively, after replacing coreference chains by their IDs, two sentences will become as follows:

S_1': In a televised interview on ID1, ID2 President ID3 said it was too soon to discuss further interest rate cuts in the 12-nation euro zone.

S_2': ID1 President ID2 said in a televised interview that it was too soon to discuss further interest rate cuts in the euro zone.

After that we construct semantic vectors corresponding to the sentences; and we follow the same way as computing the feature *word-cosine* for *word-and-NE-cosine*. It means that the final formula used to compute the similarity of two semantic vectors is the same as the one presented in Eq. (5).

3.7 Word Order Similarity

In this section, we introduce the third feature, namely *word-order*, in measuring of sentence similarity. This feature takes order of words in a sentence into account. Let's consider two sentences below:

S_1: "Athlete A beats athlete B in the game".
S_2: "Athlete B beats athlete A in the game".

If we only apply *word-cosine* on these sentences, the similarity degree will be 1. The reason is that two sentences consist of the same words, but different in position. Thus, as mentioned in [4], beside word-based cosine similarity, we should exploit the order of words in the two sentences. With example sentence pair, we can draw a joint word set S by taking all the same words. S will be:

$$S = \{ \text{Athlete, A, beats, athlete, B, in, the, game} \}.$$

In order to show the difference of word order, we assign an index number to each word in S_1 and S_2. Then, we form word order vectors, denoted by r_1 and r_2 respectively, for these sentences based on S. To show how to assign entry values for a word order vector, we take S_1 as example:

- Case 1: if a word w in S appears in S_1, we fill its entry value in r_1 with the corresponding index in S_1.
- Case 2: if w does not appears in S_1, we compute *word-sim* between w and in turn words S_1 and keep the word w' having the highest *word-sim* with w; if the highest *word-sim* is greater than a preset threshold, the entry value of w in r_1 is set to the index of w'; otherwise, it is set to zero.

Applying the two cases above, we have the word order vectors of S_1 and S_2 as follows:

$$r_1 = \{1, 2, 3, 4, 5, 6, 7, 8\}.$$
$$r_2 = \{1, 5, 3, 4, 2, 6, 7, 8\}.$$

After all, the word order similarity between two sentences will be measured by the difference of vectors r_1 and r_2, and normalized to vector length. The formula is as followed:

$$word - order(S_1, S_2) = 1 - \frac{\|r_1 - r_2\|}{\|r_1 + r_2\|} \tag{6}$$

4 Evaluation

We tested our method on MSRP. The training set contains 4076 sentence pairs including 2753 true paraphrase pairs and 1323 false paraphrase pairs; the test set contains 1725 sentence pairs including 1147 and 578 pairs, respectively. This corpus is derived from over 9 million sentence pairs of 32.408 news clusters from World Wide Web. We extract a subset of MSRP corpus that consists of all sentence pairs including more than 70% words occurring in WordNet. We call this corpus is *subMSRP*. We obtain the corpus *subMSRP* consisting of 2782 pairs of sentences for training and 1305 pairs of sentences

Table 1. The experimental results using different combinations of features on subMSRP and MSRP in terms of Accuracy (Acc.), precision (P), recall (R) and F-measure (F)

Metric	Brown corpus	WordNet	subMSRP				MSRP			
			Acc.	P	R	F	Acc.	P	R	F
word-cosine	✓		69.6	48.5	69.6	57.2	66.5	44.2	66.5	53.1
		✓	83.5	82.9	83.6	78.8	70.0	66.1	70.1	68.4
	✓	✓	88.5	88.2	88.5	86.4	71.3	66.2	71.1	68.6
word-and-NE-cosine	✓	✓	95.8	95.8	95.9	95.6	79.7	79.3	79.7	79.4
word-and-NE-cosine + word-order	✓	✓	**96.9**	96.9	96.9	**96.8**	**80.7**	80.4	80.8	**80.4**

for testing. The reason we need the *subMSRP* corpus is that we would like to evaluate the role of WordNet in sentence similarity measure. In our experiments, we use Support Vector Machine learning algorithm implemented in WEKA[8] to learn the classifier.

Table 1 presents results. First we evaluate our method in the case the trained classifier using *word-cosine* in three different settings: corpus-based word-cosine (using Brown Corpus), WordNet-based word-cosine, and combination of corpus-based and WordNet-based word-cosine. The experimental results on subMSRP show the combination significant contribute to the similarity in terms of both accuracy and F-measure, in particular 88.5% in comparison with 69.6% and 83.5% accuracy; and 86.4% in comparison with 57.2% and 78.8% F-measure, respectively. However, experimental results on MSRP show the combination does not improve the performance in comparison with the case using only WordNet. It is easy to explain that for this case WordNet cover 64.5% words in MSRP. Therefore, *word-cosine* is set to zero for many words in the corpus.

Table 2. Our results compared to the state-of-the-art Accuracy and F-measure

Model	Accuracy	F
Mihalcea et al. (2006) [1]	70.3	81.3
Qiu et al. (2006) [10]	72.0	81.6
Islam and Inkpen (2007) [2]	72.6	81.3
Rus et al. (2008) [11]	70.6	80.4
Das and Smith (2009) [9]	76.1	82.7
Socher et al. (2011) [8]	76.8	83.6
Madnani et al. (2012) [7]	77.4	84.1
Rus et al. (2013) [17]	73.6	81.8
Ji and Eisenstein (2013) [14]	80.4	85.9
Rus et al. (2014) [13]	74.2	82.1
El-Alfy et al. (2015) [23]	72.1	80.0
This paper	**80.7**	**80.4**

---

[8] http://www.cs.waikato.ac.nz/ml/weka/

We then evaluate our method in the case the trained classifier using *word-and-NE-cosine* and *word-and-NE- cosine + word-order* respectively. In the case of using *word-and-NE- cosine* only, experimental results on MSRP show the performance is improved by ~8% (79.7% in comparison with 71.3%) accuracy and ~ 11% (79.4% in comparison with 68.6%) F-measure. These results prove that named entities and their coreference relations play an important role in the similarity. The experimental results, in the last row of Table 1, on MSRP also show that *word-order* helps improve one point percent in terms of accuracy and F-measure. Table 2 presents our results compared to the state-of-the-art in terms of Accuracy and F-measure. The results show that our method is comparable with state-of-the-art methods for paraphrase identification.

5   Conclusion

This paper introduces a novel method for measuring semantic similarity between two sentences. The originality of our method is that it explores named entities and their coreference relations as important indicators for measuring the similarity. The method exploits WordNet, Wikipedia, and Brown Corpus and learns a classifier using features: semantic similarity between words, semantic similarity between words and named entities, and word-order similarity. The experimental results show that all of those have contributions to the sentence similarity. In comparison with others, our method is straightforward and quite easy to reproduce, while giving state of the art results. Our future work will focus on investigating coreference relations not only between named entities but also between phrases and evaluate it on other data sets.

References

1. Mihalcea, R., Corley, C., Strapparava, C.: Corpus-based and knowledge-based measures of text semantic similarity. In: AAAI, vol. 6, pp. 775–780 (2006)
2. Islam, A., Inkpen, D.: Semantic text similarity using corpus-based word similarity and string similarity. ACM Transactions on Knowledge Discovery from Data 2(2) (2008)
3. Sahami, M., Heilman, T.D.: A web-based kernel function for measuring the similarity of short text snippets. In: Proceedings of the 15th International Conference on World Wide Web, pp. 377–386 (2006)
4. Li, Y., McLean, D., Bandar, Z.A.: O'shea, J. D., Crockett, K.: Sentence similarity based on semantic nets and corpus statistics. IEEE Transactions on Knowledge and Data Engineering 18(8), 1138–1150 (2006)
5. Oliva, J., Serrano, J.I., del Castillo, M.D., Iglesias, Á.: SyMSS: A syntax-based measure for short-text semantic similarity. Data & Knowledge Engineering 70(4), 390–405 (2011)
6. Bach, N.X., Minh, N.L., Shimazu, A.: Exploiting discourse information to identify paraphrases. Expert Systems with Applications 41(6), 2832–2841 (2014)
7. Madnani, N., Tetreault, J., Chodorow, M.: Re-examining machine translation metrics for paraphrase identification. In: Proceedings of 2012 Conference of the North American Chapter of the Association for Computational Linguistics (NAACL), pp. 182–190 (2012)
8. Socher, R., Huang, E.H., Pennington, J., Ng, A.Y., Manning, C.D.: Dynamic pooling and unfolding recursive autoencoders for paraphrase detection. In: NIPS, vol. 24, pp. 801–809 (2011)

9. Das, D., Smith, N.: Paraphrase identification as probabilistic quasi-synchronous recognition. In: Proceedings of the Joint Conference of the 47th Annual Meeting of the ACL and the 4th International Joint Conference on Natural Language Processing of the AFNLP, pp. 468–476 (2009)

10. Qiu, L., Kan, M.Y., Chua, T.S.: Paraphrase recognition via dissimilarity significance classification. In: Proceedings of the 2006 Conference on Empirical Methods in Natural Language Processing (EMNLP 2006), pp. 18–26 (2006)

11. Rus, V., McCarthy, P.M., Lintean, M.C. McNamara, D.S., Graesser, A.C.: Paraphrase identification with lexico-syntactic graph subsumption. In: FLAIRS 2008, pp. 201–206 (2008)

12. Tsatsaronis, G., Varlamis, I., Vazirgiannis, M.: Text relatedness based on a word thesaurus. Journal of Artificial Intelligence Research 37(1), 1–40 (2010)

13. Ştefănescu, D., Banjade, R., Rus, V.: A sentence similarity method based on chunking and information content. In: Gelbukh, A. (ed.) CICLing 2014, Part I. LNCS, vol. 8403, pp. 442–453. Springer, Heidelberg (2014)

14. Ji, Y., Eisenstein, J.: Discriminative improvements to distributional sentence similarity. In: Proceedings of EMNLP, pp. 891–896 (2013)

15. Islam, A., Inkpen, D.: Semantic similarity of short texts. In: Recent Advances in Natural Language Processing V, vol. 309, pp. 227–236 (2009)

16. Kozareva, Z., Montoyo, A.: Paraphrase identification on the basis of supervised machine learning techniques. In: Salakoski, T., Ginter, F., Pyysalo, S., Pahikkala, T. (eds.) FinTAL 2006. LNCS (LNAI), vol. 4139, pp. 524–533. Springer, Heidelberg (2006)

17. Rus, V., Niraula, N., Banjade, R.: Similarity measures based on latent dirichlet allocation. In: Gelbukh, A. (ed.) CICLing 2013, Part I. LNCS, vol. 7816, pp. 459–470. Springer, Heidelberg (2013)

18. Achananuparp, P., Hu, X., Shen, X.: The evaluation of sentence similarity measures. In: Song, I.-Y., Eder, J., Nguyen, T.M. (eds.) DaWaK 2008. LNCS, vol. 5182, pp. 305–316. Springer, Heidelberg (2008)

19. Manning, C.D., Raghavan, P., Schütze, H.: Introduction to information retrieval. Cambridge University Press, Cambridge (2008)

20. Bontcheva, K., Dimitrov, M., Maynard, D., Tablan, V., Cunningham, H.: Shallow methods for named entity coreference resolution. In: Chaînes de références et résolveurs d'anaphores, Workshop TALN (2002)

21. Lesk, M.: Automatic sense disambiguation using machine readable dictionaries: how to tell a pine cone from an ice cream cone. In: Proceedings of the 5th Annual International Conference On Systems Documentation, pp. 24–26 (1986)

22. Nguyen, H.T., Duong, P.H., Vo, V.T.: Vietnamese sentence similarity based on concepts. In: Saeed, K., Snášel, V. (eds.) CISIM 2014. LNCS, vol. 8838, pp. 243–253. Springer, Heidelberg (2014)

23. El-Alfy, E.S.M., Abdel-Aal, R.E., Al-Khatib, W.G., Alvi, F.: Boosting paraphrase detection through textual similarity metrics with abductive networks. Applied Soft Computing 26, 444–453 (2015)

24. Dolan, B., Quirk, C., Brockett, C.: Unsupervised construction of large paraphrase corpora: exploiting massively parallel news sources. In: Proceedings of the 20th International Conference on Computational Linguistics, p. 350. ACL (2004)

25. Rus, V., Niraula, N., Banjade, R.: A study of probabilistic and algebraic methods for semantic similarity. In: Proceedings of the Twenty-Sixth International Florida Artificial Intelligence Research Society Conference (2013)

26. Carlson, L., Marcu, D., Okurowski, M.E.: Building a discourse-tagged corpus in the framework of rhetorical structure theory. In: Current and new directions in discourse and dialogue, pp. 85–112 (2003)

Improving Word Alignment Through Morphological Analysis

Vuong Van Bui[✉], Thanh Trung Tran, Nhat Bich Thi Nguyen,
Tai Dinh Pham, Anh Ngoc Le, and Cuong Anh Le

Computer Science Department, Vietnam National University,
University of Engineering and Technology, 144 Xuan Thuy, Hanoi, Vietnam
{vuongbv-56,cuongla}@vnu.edu.vn

Abstract. Word alignment plays a critical role in statistical machine translation systems. The famous word alignment system, IBM models series, currently operates on only surface forms of words regardless of their linguistic features. This deficiency usually leads to many data sparseness problems. Therefore, we present an extension that enables the integration of morphological analysis into the traditional IBM models. Experiments on English-Vietnamese tasks show that the new model produces better results not only in word alignment but also in final translation performance.

Keywords: Machine translation · Word alignment · IBM models · Morphological analysis

1 Introduction

Most of machine translation approaches nowadays use word alignment as the fundamental material to build their higher models, and make the translation performance highly dependent on the quality of the alignment. Of all word alignment models, IBM models [2], in spite of their ages, are still quite popular and widely used in state-of-the-art systems. They are series of models numbered from 1 to 5 in which each of them is an extension of the previous model. The very first one, IBM Model 1 utilizes only co-occurrence of words in both sentences to train the word translation table. This parameter is used not only to align words between sentences but also to provide reasonable initial parameter estimates for higher models, which use various other parameters involving the order of word (IBM Model 2), the number of words a source word generates (IBM Model 3), etc. A good parameter from Model 1 will efficiently boost the parameter quality of higher models.

Being the model working on words, IBM Model 1 has statistics on only surface forms of words without any utilization of further linguistic features such as part of speech, morphology, etc. Detecting relations between words having the same origin or the same derivation through analyzing linguistic not only

© Springer International Publishing Switzerland 2015
V.-N. Huynh et al. (Eds.): IUKM 2015, LNAI 9376, pp. 315–325, 2015.
DOI: 10.1007/978-3-319-25135-6_30

reduces the sparseness of limited training data but also gives a better explanation of word mapping. However, there are no known general frameworks to utilize these types of information. Each language pair has its own features, that makes the scenario for each one very different. Although there are a number of relating papers published before, they are very specific for some certain language pairs. Unfortunately, as all we know, none of them are about utilizing English-Vietnamese morphological analysis.

In this paper, morphological analysis is used to build a better word alignment for the English-Vietnamese language pair. English is not as rich in morphology as other languages like German, Czech, etc. Each English word usually has less than a dozen of derivatives. However, when being compared to Vietnamese, it is considered to be much richer in morphology. Vietnamese words are really atomic elements, in other words, they are not be able to be divided into any parts, or combined with anyone to make derivatives. For a pair of Vietnamese-English sentences, each Vietnamese word may not only be the translation of an English word, but sometimes actually only a small part of that word. For example, with the translation "*nhngsm_rng*"[1] of "*enlargements*", words "*nhng*", "*s*", "*m_rng*" are respectively actually the translation of smaller parts "*s*", "*ment*" and "*enlarge*". The above example and many other ones are evidences of the fact that while in English, to build a more complex meaning of a word, they combine that word with morphemes to make an extended word, in Vietnamese, additional words with corresponded functions to English morphemes are added surround the main word. In other words, an English word may align to multiple Vietnamese words while most of the time, a Vietnamese word often aligns to no more than one word in English, in many cases, only a part of that word in morphological analysis.

The above property of the language pair plays an important role in our development of the extension. We treat an English word not only in the form of a word, but also in the form after analyzing morphology. The morphemes now can be statistically analyzed, and their correlations with the Vietnamese words which have the same functions can be highlighted. To achieve this, we have a pre-processing step, by which, suitable morphemes will be separated from the original word. After that, the traditional IBM model will be applied to this corpus, and the correspondences between English morphemes and Vietnamese equivalent function words can be shown not only in the probability parameter of the model but also in the most likely alignments the model produces.

Although some of our improvements in result are shown in this paper, our main focus is the motivation of the approach, mostly in the transformation from the specification of the languages to the correspondent difference in models when a few first processing techniques are applied. After relating some previous works, a brief introduction to IBM Model 1 together with its problems when applying to English-Vietnamese copora are presented in the right following section. Next, motivating examples for our method is followed by its details of discription.

---

[1] Vietnamese words is always segmented in our experiments. Various tools are available for this task.

The final result, after experiments on aligning words and translating sentences is placed near the end, before the conclusion of the paper.

2   Related Work

The sparsity problem of IBM models is a well recognized problem. The role of rare words as *garbage collectors*, which strongly affects the alignments of the sentences they appear, was described in [1]. A good smoothing technique for IBM Model 1 is already presented in [7]. However, these problems of rare words are only described in term of surface forms. They treat every word independent of each other regardless of the fact that many groups of them, despite of different appearances, have the same origin or relate to each other in some other way. These relations often depend on the linguistic features of the language. Before trying to solve the general problem of sparsity data, a good idea is attempting to utilize the features of the language to reduce the sparseness first.

Applying morphology analysis is a well known approach to enrich the information of translation models. Most of the time, a corpus is not able to well cover all forms of a lemma. That leads to the situation in which we do not have enough statistics on a derivative word while the statistics for its lemma is quite rich. The traditional IBM models, which works on only surface forms of words, usually find many difficulties when dealing with this kind of sparse data problem. Using morphology information is a natural approach to solutions. Various attempts have been made. The idea of pre-processing the corpus by segmenting words into morphemes and then merging and deleting appropriate morphemes to get the desired morphological and syntactic symmetry for Arabic-English language pair was presented in [6]. Similar ideas of pre-processing can also be found in [11].

A much more general framework is presented in [4] with the integration of additional annotation at word level to the traditional phrase-based models. In the step of establishing word alignment, the system uses the traditional IBM models on the surface forms of words or any other factors. The result was reported to be improved with experiments on lemmas or stems.

3   IBM Model 1

Details of IBM models series are presented in [2]. This paper shows only a brief introduction to IBM model 1, which is the first model in the series to work with word translation probabilities.

3.1   Definition

For a pair of sentences, each word at position j in the target sentence T is aligned to one and only one word at position i in the source sentence, or not aligned to anyone. In the latter case, it is considered to be aligned to a special word *NULL* at position 0 of every source sentence. Denote l, m respectively to be the length of the source sentence and the target sentence, a_j is the position in the source sentence to which the target word j is aligned. The model has a word translation probability tr as its parameter with the meaning of how likely a target word is produced given a known source word. The role of the parameter in the model is described in the following probabilites.

$$P(T, A \mid S) = \epsilon \prod_{j=1}^{m} tr(t_j \mid s_{a_j})$$

$$P(T \mid S) = \epsilon \prod_{j=1}^{m} \sum_{i=0}^{l} tr(t_j \mid s_i)$$

$$P(A \mid T, S) = \prod_{j=1}^{m} \frac{tr(t_j \mid s_{a_j})}{\sum_{i=0}^{l} tr(t_j \mid s_i)}$$

Applying expectation–maximization (EM) algorithm with above equations to a large collection of parallel source-target sentence pairs, we will get the best tr parameter which maximizes the likelihood of this corpus. For a parameter tr we get, the most likely alignment, which is called the Viterbi alignment, is derived as follows.

$$A_{\text{Viterbi}} = \arg\max_A P(A \mid S, T)$$

One point to note here is that the translation of each target word is independent of each other. Therefore, the most likely a_j is the position i which has the highest $tr(t_j \mid s_i)$.

After we have the model, derivations of most likely alignments can be done on sentence pairs of the training corpus, or some other testing copora.

3.2   Some Problems when Applying to English-Vietnamese Corpora

A restriction to IBM models is the requirement of the alignment to be the function of target words. In other words, an target word is aligned to one and only one word in the source sentence (or to *NULL*). The situation may be appropriate when the target language is Vietnamese, in which, each Vietnamese target word most of the time corresponds to no more than one English word. However, in the reversed direction, the scenario is much different. Complex words, which have rich morphology, actually are the translations of two, three, or more Vietnamese words regarded to their complexity in morphology analysis. The restriction that only one Vietnamese word is chosen to align to a complex English word is very unreasonable.

A second problem mentioned here is due to rare words. Assume that in our corpus, there is a rare source word which occurs very few times. Consider a sentence that this source word occurs, this word will play as the role of a *garbage collector*, which makes the EM algorithm to assign very high probabilities in the distribution of that rare word for words in the target sentence to maximize the overall likelihood. This scenario makes many words in the target sentence to be aligned to that rare word. Explanations in detail and an approach to deal with this problem through a smoothing technique can be found in [7]. However, what we want to deal with in this paper is the situation of words which are rare in term of surface form but not in term of morphological form. For example, in a corpus, the word *"enlargements"* may be a rare word, but its morphemes, *"en"*, *"large"*, *"ment"*, *"s"*, in the other way, may be very popular morphemes. By analyzing statistics on smaller parts of the original words, we may highly enrich statistics, and reduce the problem of rare words with popular morphemes.

4   Our Proposal to Extend IBM Model 1

4.1   Motivating Examples

Consider a sample pair of sentences as shown in all three figures: Fig. 1, Fig. 2, Fig. 3.

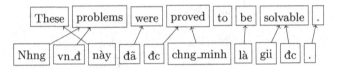

Fig. 1. Alignment with Vietnamese as the target language

The alignment for the Vietnamese target sentence in Fig. 1 is what the model produces after being trained on a *sufficient* corpus. What we mean by *sufficient* is a large enough corpus with no sparsity problems. However, such an ideal corpus is rare. In the case of very few times the word *"solvable"* appears in the corpus (or in a much worse case only one or two times) while other English words are quite common, IBM model 1 will behave very strange when aligning most of words in the Vietnamese sentence to *"solvable"* . Detail explanations can be found in [7]. This is a bad behavior because all wrong alignments of the whole sentence are due to just one rare word. And it is much worse when its lemma, the word *"solve"*, and its suffix *"able"* are very popular in the corpus. In other words, *"solvable"* is not actually a rare word because of its common morphemes. Analyzing statistics on only surface forms regardless of morphology forms has already introduced more and more sparsity problems. On the other hand, analyzing statistics on smaller parts of words can lead to high correlations between Vietnamese words and English morphemes. In our case, particularly, these correlations are between *"solve"* and *"gii"*, *"able"* and *"đc"*, which makes the fact that *"solvable"* is a rare word, is no longer our matter. Denser statistics, especially in our case, seems to produce more reliable decisions.

In the reverse direction, when the target language is English, the alignment the IBM model 1 produces is not sufficient, as shown in Fig. 2.

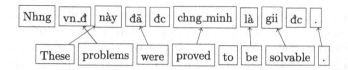

Fig. 2. Alignment with English as the target language

The missing alignments are quite obvious when comparing to the alignment of the other direction in Fig. 1. It is due to the requirement of IBM models that a target word may connect to no more than one word in the source sentence. When these models is applied to our case, complex words like *"problems"*, *"proved"*, *"solvable"*, which are actually the translations of two words in the Vietnamese sentence as in the Fig. 1 of the other direction, make the alignments missing many correct alignments. An important point to note here is that some Vietnamese words actually connect to morphemes of the English words. In the case of *"problems"*, its morphemes *"problem"* and *"s"* respectively connect to *"vn_đ"* and *"nhng"*. The cases for two other words are similar, consider Fig. 3 for details.

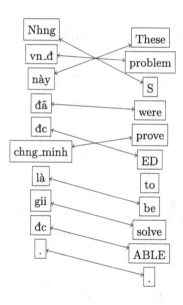

Fig. 3. The symmetric alignment of both directions after breaking words

By an appropriate strategy to break the original English words into parts as in Fig. 3, we can not only enrich the statistics over the corpus but also overcome the matter of aligning one target English word into multiple Vietnamese source words. Therefore, the alignments for both directions when applying this trick tend to be more symmetric and for our example, shown once only in Fig. 3 because of coincidence.

4.2 Our Method

Each English words has its own morphology form, by which we can break it into smaller parts. Each of these parts actually is able to correspond to a Vietnamese word. In other words, one English word is sometimes the translation of multiple Vietnamese words. By breaking the English word into smaller parts, we can assign each individual part to a Vietnamese word.

There are various ways to break an English word into parts. For example, the word "*enlargements*" may be broken into as many parts as "*en+large+ment+s*", but the most suitable one to correspond to its Vietnamese translation "*nhngsm_rng*" is "*enlarge+ment+s*". There is no well known general strategy to figure out which one is best, so in our method, we propose to break on only a very limited set of well known morphological classes, whose morphemes have very high correlations with their translation. Particularly, we focus on investigating the classes including noun+S, verb+ED, verb+ING.

In our method, we will add a pre-processing and a post-processing step to the original model. First, every English word which matches one of the three above mophological forms will be broken into smaller parts. The traditional models will be trained on this pre-processed corpus and produce the Viterbi alignments. After that, the post-processing step will converts these alignments to be compatible with the original corpus. For the case the source language is English, an alignment from a part of an English word means an alignment from that whole word. For the case the source language is Vietnamese, an alignment to any part of an English word means an alignment to that whole word. The post-processing stage is mostly for comparing word alignments produced by different models as it is not appropriate to compare alignments of different corpora.

5 Experiments on Word Alignments

We have our experiments on a corpus of 56000 parallel English-Vietnamese sentence pairs. As usual, this corpus is divided into two parts with the much bigger part of 55000 sentence pairs is for training and the smaller one of 1000 sentence pairs is for testing. For each part, we maintain two versions of the corpus: the original version and the version after the pre-processing stage. The details of each rule in the pre-processing stage and its effects to the translation probability tables after being trained by IBM models are described in the next section. Finally, we manually examine the effects of the rules to the final Viterbi alignments to compare the performances in aligning words between two models.

5.1 The Word Translation Probabilities

We apply morphological analysis in three common classes: noun+S, verb+ED, and verb+ING to pre-process the corpus. Both the original corpus and the pre-processed corpus are trained in totally 20 iterations of IBM models with 5 iterations for each one of models from Model 1 to Model 4. After these training iterations, we examine the translation of new introduced words "PL"[2], "ED" and "ING" in the translation probability tables.

Every plural form of a noun will be separated into two parts: the original noun and the plural notation "PL". For example, "computers" will be broken into two adjacent words "computer" and "PL". The word translation probabilities after being trained IBM models as shown in Table 1 reflect quite well the fact that "PL" usually co-occurs with "nhng", "các", and "nhiu".

Table 1. Probabilities given additional source words after running IBM Models

PL		ED		ING	
các	0.570465	đã	0.841928	đang	0.658824
nhng	0.300285	b	0.152974	vic	0.31915
nhiu	0.0795678	đc	0.000869153	khi	0.0177749
t	0.0321255	vào	0.000844074	bng	0.000399352
đu	0.011416	mc	0,000185142	cách	0.000312531

Every word of the form verb+ING will be divided into two parts: its original verb and the suffix "ING". For example: "running" will be divided into two contiguous words "run" and "ING". The case for "ING" as presented in Table 1 has the same manner as the case of "PL". The highest translation are "dang" for present continuous sentences when the runner-up word "vic" is translated for nouns having verb-ING form.

Every word of the form verb+ED, whatever it is passive form or past form will be split into two parts: the original verb and the suffix "ED". For example: "edited" will become two words "edit" and "ED". The co-occurrence of "ED" in passive form with "b" and "dc", together with its co-occurrence in past form with "dã" are obvious in the word translation probability when these three translation take the top places in the table.

All above results reflect the high correlations between English morphemes and theirs corresponded words in Vietnamese. The estimation produced by IBM Models is nearly like what we expect. They, after all, not only reduce the sparseness in data but also give a clearer explanation for word mappings.

5.2   Improvement in Viterbi Word Alignment

Our pre-processing corpus is not actually compatible with higher IBM models from IBM model 2 because these models employ features like reordering parameters, fertility parameters, etc, whose behaviors are affected by our pre-processing step in an inappropriate way. These facts make the final word alignment produced by higher models quite bad. Therefore, we have our experiments on only IBM model 1 instead. After 20 iterations of IBM model 1, the Viterbi alignment for the testing part will be deduced to check for its correctness.

---

[2] We use notation "PL" instead of "S" to avoid conflicting with original "S" words.

There are many ways to evaluate an alignment model. A popular method is to consider the alignment error rate (AER) [8] as the measurement of the performance. However, in our special case, what we propose, a small modification to IBM Model 1, makes the new word alignment different from what produced by the baseline model in quite few points. Therefore, instead of checking the correctness of every alignment points as the way AER is estimated, we, in our experiments, compares the correctness of alignment points at which two models disagree. For a different point, we credit 1 point for the right model unless both models are wrong. Because these different points throughout the whole testing corpus has a reasonable size, we can definitely check these alignments manually. After all, each model is evaluated on the ratio of times it is correct in this subset of alignment.

After training the two models, one on the original corpus, and the other on the pre-processed corpus, we apply each of these models to get the Viterbi alignments of the testing corpus. The result of evaluating as our method is shown in Table 2. As we can see, our method constitutes about 74% of correct alignments while only 26% is for the original method. The different alignment subset, in our experiments, includes not only points relating to "PL", "ED" and "ING'" but also many other affected cases. In other words, our method has also corrected other alignments not restricted to what is pre-processed.

Table 2. Number of correct alignments in different alignment subset

Original corpus	Pre-processed corpus
173	490

6 Experiments on Translation Performance

We also do some further experiments than comparing the word alignments produced by IBM Model 1. The translation performances of phrase-based machine translation systems built in a traditional way will be another test for our extensions to the baseline.

As usual, each corpus will have its translation model after following the training workflow. First, we use the famous word alignment tool GIZA++ [9], which fully implements the IBM model series to align words for the training part. Together with the word alignment, a language model for the target language, Vietnamese in this case, is also trained by the popular tool IRSTLM [3] on a Vietnamese corpus, particularly the Vietnamese training part in our experiment. Later then, a phrase-based model based on the word alignment and the language model is produced by popular tools in Moses package [5], which actually have some additional actions of extracting phrases and estimating feature scores. Finally, the testing is done with the translation of unseen sentences. The Moses decoder will translate the English testing part of the corpus based on the information the model supply. The result of Vietnamese sentences translated by Moses is evaluated by a BLEU score [10], which is the most popular metric to measure the similarity between the translation of the machine and a reference translation, the Vietnamese testing part in this case. The experiment workflow is done independently for both corpora, and the final BLEU scores retrieved will be the measurement for the translation performance of the two models.

Together with the translation performace, we also want to evaluate the ability to enrich statistics of our new method. Experiments are done on copora of various sizes. The sparsity of a corpus usually increases as the corpus become smaller. We still keep the testing part of 1000 pairs while randomly choosing respectively 10000, 20000, 35000 pairs from the whole training corpus of 55000 pairs for three additional experiments. Increaments in BLEU scores of our method are well recognized in results of total four experiments as shown in Table 3. Not only makes the general translation performace better, our method also demonstrates its ability to reduce the sparseness of data especially when the corpus size is small. The fact that the smaller the corpora get, the farther the distance between BLEU scores is, reflects quite well this point. All of these results are again evidences for the potential of the proposed solution.

Table 3. BLEU scores of two corpora

Size of training part	Original corpus	Pre-processed corpus	Increament
10000	13.85	14.61	5.5%
20000	16.51	16.87	2.2%
35000	18.64	19.07	2.3%
55000	20.49	20.56	0.34%

7   Conclusion and Future Work

We have already presented our approach to employ morphology in building a better word alignment model over the original IBM Model 1. By using the morphological forms of some popular English word classes to pre-process the corpus, we successfully show the high correlations between some Vietnamese words with their corresponded English morphemes. These high correlations are not only reflected in the word translation probability, which is the main parameter of the model, but also in the final Viterbi alignments, and even in the BLEU scores of the baseline phrase-based translation system basing on it.

However, there are still some ways to make our method better. The experiments are tested on quite few classes of words, just a small proportion to the total number of English morphological forms. A broader space of forms may be employed in next improvements. On the other hand, our method should be less manual in choosing morphological forms. We are also looking for an appropriate adaptation for parameters of higher IBM models other than the word translation probability. These additional improvements make our proposed method to be a more general framework, which is actually our target of further development.

Acknowledgments. This paper is partly funded by The Vietnam National Foundation for Science and Technology Development (NAFOSTED) under grant number 102.01-2014.22 and by the project QGTD.12.21 funded by Vietnam National University of Hanoi.

References

1. Brown, P.F., Della Pietra, S.A., Della Pietra, V.J., Goldsmith, M.J., Hajic, J., Mercer, R.L., Mohanty, S.: But dictionaries are data too. In: Proceedings of the Workshop on Human Language Technology, pp. 202–205. Association for Computational Linguistics (1993)
2. Brown, P.F., Pietra, V.J.D., Pietra, S.A.D., Mercer, R.L.: The mathematics of statistical machine translation: Parameter estimation. Computational Linguistics **19**(2), 263–311 (1993)
3. Federico, M., Bertoldi, N., Cettolo, M.: Irstlm: an open source toolkit for handling large scale language models. In: Interspeech, pp. 1618–1621 (2008)
4. Koehn, P., Hoang, H.: Factored translation models. In: EMNLP-CoNLL, pp. 868–876 (2007)
5. Koehn, P., Hoang, H., Birch, A., Callison-Burch, C., Federico, M., Bertoldi, N., Cowan, B., Shen, W., Moran, C., Zens, R., et al.: Moses: open source toolkit for statistical machine translation. In: Proceedings of the 45th Annual Meeting of the ACL on Interactive Poster and Demonstration Sessions, pp. 177–180. Association for Computational Linguistics (2007)
6. Lee, Y.-S.: Morphological analysis for statistical machine translation. In: Proceedings of HLT-NAACL 2004: Short Papers, pp. 57–60. Association for Computational Linguistics (2004)
7. Moore, R.C.: Improving ibm word-alignment model 1. In: Proceedings of the 42nd Annual Meeting on Association for Computational Linguistics, p. 518. Association for Computational Linguistics (2004)
8. Och, F.J.: Minimum error rate training in statistical machine translation. In: Proceedings of the 41st Annual Meeting on Association for Computational Linguistics, vol. 1, pp. 160–167. Association for Computational Linguistics (2003)
9. Och, F.J., Ney, H.: A systematic comparison of various statistical alignment models. Computational Linguistics **29**(1), 19–51 (2003)
10. Papineni, K., Roukos, S., Ward, T., Zhu, W.-J.: Bleu: a method for automatic evaluation of machine translation. In: Proceedings of the 40th Annual Meeting on Association for Computational Linguistics, pp. 311–318. Association for Computational Linguistics (2002)
11. Sadat, F., Habash, N.: Combination of arabic preprocessing schemes for statistical machine translation. In: Proceedings of the 21st International Conference on Computational Linguistics and the 44th Annual Meeting of the Association for Computational Linguistics, pp. 1–8. Association for Computational Linguistics (2006)

Learning Word Alignment Models
for Kazakh-English Machine Translation

Amandyk Kartbayev[✉]

Laboratory of Intelligent Information Systems, Al-Farabi Kazakh National
University, Almaty, Kazakhstan
a.kartbayev@gmail.com

Abstract. In this paper, we address to the most essential challenges
in the word alignment quality. Word alignment is a widely used phenomenon in the field of machine translation. However, a small research
has been dedicated to the revealing of its discrete properties. This paper
presents word segmentation, the probability distributions, and the statistical properties of word alignment in the transparent and a real life
dataset. The result suggests that there is no single best method for alignment evaluation. For Kazakh-English pair we attempted to improve the
phrase tables with the choice of alignment method, which need to be
adapted to the requirements in the specific project. Experimental results
show that the processed parallel data reduced word alignment error rate
and achieved the highest BLEU improvement on the random parallel
corpora.

Keywords: Word alignment · Kazakh morphology · Word segmentation · Machine translation

1 Introduction

In recent years, the several studies were conducted to evaluate the relationships between word alignment and machine translation performance. The phrase
table is the fundamental data structure in phrase-based models, and the training
pipeline of most statistical machine translation (SMT) systems uses a word alignment for limiting the set of the suitable phrases in phrase extraction. Therefore,
the accuracy of the phrase models are highly correlated with the word alignments quality, which are used to learn an accordance between the source and
target words in parallel sentences. However, there is no theoretical support from
the view of providing a formulation to describe the relationship between word
alignments and machine translation performance.

We examine the Kazakh language, which is the majority language in the
Republic of Kazakhstan. Kazakh is part of the Kipchak branch of the Turkic
language family and part of the majority Ural-Altay family, in comparison with
languages like English, is very rich in morphology.

The Kazakh language which words are generated by adding affixes to the root
form is called an agglutinative language. We can derive a new word by adding an

© Springer International Publishing Switzerland 2015
V.-N. Huynh et al. (Eds.): IUKM 2015, LNAI 9376, pp. 326–335, 2015.
DOI: 10.1007/978-3-319-25135-6_31

affix to the root form, then make another word by adding another affix to this new word, and so on. This iteration process may continue several levels. Thus a single word in an agglutinative language may correspond to a phrase made up of several words in a non-agglutinative language[1].

Table 1. An example of Kazakh agglutination

Stem	Plural affixes	Possesive affixes	Case affixes
stem[kol'+]	plural[+der]	1-st pl.[+imiz]	locative[+de]
stem[kol'+]	-	1-st s.[+im]	locative[+de]
stem[kol'+]	-	-	locative[+de]

In this paper, we present a systematic comparison of preprocessing techniques for a Kazakh-English pair. El-Kahlout and Oflazer[2] explored this task for English to Turkish translation, which is an agglutinative language as Kazakh, and they outperformed the baseline results after some morpheme grouping techniques. A research more relevant to this work was done by Bisazza and Federico[3].

Our objective is to produce an word alignment, which can be used to build high quality machine translation systems[4]. These are pretty close to human annotated alignments that often contain m-to-n alignments, where several source words are aligned to several target words and the resulting unit can not be further decomposed. Using segmentation, we describe a new generative model which directly models m-to-n non-consecutive word alignments. The common approaches of word alignment training are IBM Models[5] and hidden Markov model (HMM)[6], which practically use expectation-maximization (EM) algorithm[7]. The EM algorithm finds the parameters that increases the likelihood of the dependent variables. EM transfers the sentences by overlapping the actual parameters, where some rare words align to many words on the opposite sentence pair. The training can be clearly divided into a morphological part of generating the segments and a part modeling the relative features of phrases. Sure, it makes the system is sufficiently general to be applied to other kind of language pair, with the different morphotactics. Potential areas of word segmentation and word alignment problems practical application will be an area, where the incorporation of segments is useful. An application area of this research includes improvements in machine translation, specific experiments with machine learning methods, bioinformatics, and an analysis of knowledge extraction.

Using morphological analysis, and compared to Morfessor tool[8], we out grammatical features of word and can find syntactic structure of input sentence, which further demonstrates the benefit of using this method in machine translation. Previous researches that we explored on our approaches are rule-based morphological analyzers[9], which consist in deep language expertise and a exhaustive process in system development. Unsupervised approaches use actually unlimited supplies of text to cover very few labeled resource and it has been widely studied for a number of languages[10]. However, existing systems are too complicated to extend them with random overlapping dependencies that are crucial to segmentation. For a comprehensive survey of the rule-based morphological analyze we refer a reader to the research by Altenbek[11] and Kairakbay[12].

The paper is structured as follows: Section 2 discusses the proposed model and describes the different segmentation techniques we study. And Section 3 presents our evaluation results.

2   Learning Word Alignment Models

In order to look through this task, we did a series of experiments and found morpheme alignment can be employed to increase the similarity between languages, therefore enhancing the quality of machine translation for Kazakh-English language pair. Our experiments consist of two parts: as the first part of our experiments we morphologically segmented Kazakh input sentences to compute morpheme alignment. For these purposes we used Morfessor, an unsupervised analyzer and Helsinki Finite-State Toolkit (HFST)[13] for the rule-based analyze; finally we conducted a case study of the benefits of morpheme based alignment.

Our study is based on the set of experiments, which have the goal of most properly extraction a phrase table from the word alignment, which assume EM algorithm will be executed for several iterations, and relevant phrase pairs with the word alignment will be extracted. We use the GIZA++[14] tool, as it is, to produce IBM Model 4 word alignment. Most of algorithms usually intersects two word alignments and get alignment points from the union, to produce nearly symmetric results, which leads to higher BLEU scores[15] on average than using them directly. Our studies try to investigate the impact of pruning technique to the overall translation quality by reduction the level of sparse phrases.

2.1   Improving Word Alignment

We suppose a phrase pair is denoted by (F, E) and with an alignment A, if any words f_j in F have a correspondence in a, with the words e_i in E. Formal definition can be described as follows: $\forall e_i \in E : (e_i, f_j) \in a \Rightarrow f_j \in F$ and $\forall f_j \in F : (e_i, f_j) \in a \Rightarrow e_i \in E$, clearly, there are $\exists e_i \in E, f_j \in F : (e_i, f_j) \in A$.

Generally, the phrase-based models are generative models that translate sequences of words in f_j into sequences of words in e_j, in difference from the word-based models that translate single words in isolation.

$$P\left(e_j \mid f_j\right) = \sum_{j=1}^{J} P\left(e_j, a_j \mid f_j\right) \tag{1}$$

Improving translation performance directly would require training the system and decoding each segmentation hypothesis, which is computationally impracticable. That we made various kind of conditional assumptions using a generative model and decomposed the posterior probability. In this notation e_j and f_i point out the two parts of a parallel corpus and a_j marked as the alignment hypothesized for f_i. If $a \mid e \sim ToUniform\,(a; I+1)$, then

$$P\left(e_j^J, a_j^J \mid f_i^I\right) = \frac{f_i}{(I+1)^J} \prod_{j=1}^{J} p\left(e_j \mid f_{a_j}\right) \tag{2}$$

We extend the alignment modeling process of Brown et al. at the following way. We assume the alignment of the target sentence e to the source sentence f is a. Let c be the tag(from Penn Treebank) of f for segmented morphemes. This tag is an information about the word and represents lexeme after a segmentation process. This assumption is used to link the multiple tag sequences as hidden processes, that a tagger generates a context sequence c_j for a word sequence $f_j(3)$.

$$P\left(e_1^I, a_1^I \mid f_1^J\right) = P\left(e_1^I, a_1^I \mid c_1^J, f_1^J\right) \tag{3}$$

Then we can show Model 1 as(4):

$$P\left(e_i^I, a_i^I \mid f_j^J, c_j^J\right) = \frac{1}{(J+1)^I} \prod_{i=1}^{I} p\left(e_i \mid f_{a_i}, c_{a_i}\right) \tag{4}$$

The training is carried out in the tagged Kazakh side and the untagged English side of the parallel text. If we estimate translation probabilities for every possible context of a source word, it will lead to problems with data sparsity and rapid growth of the translation table. We applied EM algorithm to cluster a context of the source sentence using similar probability distributions, avoiding problems with data sparsity and a size of the translation table another case.

We estimate the phrase pairs that are consistent with the word alignments, and then assign probabilities to the obtained phrase pairs. Context information is incorporated by the use of part-of-speech tags in both languages of the parallel text, and the EM algorithm is used to improve estimation of word-to-word translation probabilities. The probability p_k of the word w to the corresponding context k is:

$$p_k\left(w\right) = \frac{p_k f_k\left(w \mid \phi_k\right)}{\sum p_i f_i\left(w \mid \phi_i\right)} \tag{5}$$

Where, ϕ is the covariance matrix, and f are certain component density functions, which evaluated at each cluster. Consecutive word subsequences in the sentence pair are not longer than w words. After we use association measures to filter infrequently occurring phrase pairs by log likelihood ratio r estimation[16]. Because, unaligned words near the word f_j may easily cause random multiple possibilities. For n pairs of the phrases, we can obtain the phrase pairs whose comparative values are larger than a threshold value as follows(6):

$$R\left(f, e\right) = \frac{r\left(f, e\right)}{Max_e\,r\left(f, e\right)} \tag{6}$$

The pruning algorithm may prune all translations of a source phrase that is above a threshold $p(e_i|f_j) > R$, with the probability $p(e_i|f_j)$ of the pair (f_j, e_i) as represented in (6). A pruning threshold is discarded, if $p(e_i|f_j) > Rmax_e p(e_i|f_j)$. Our algorithm, like a middle tier component, processes the input alignment files in a single pass. Current implementation reuses the code from https://github.com/akartbayev/clir that conducts the extraction of phrase pairs and filters out low frequency items. After the processing all valid phrases will be stored in the phrase table and be passed further. This algorithm proposes refinement by adding morphological constraints between the direct and the reverse directions of the alignment, which may improve the final word alignments.

2.2   Morphological Segmentation

Our preprocessing job usually starts from morphological segmentation, which includes running Morfessor tool and HFST to each entry of the corpus dictionary. The first step of word segmentation aims to get suffixes and roots from a vocabulary consisting of 1500k unique word forms taken from Kazakh Wikipedia dump[17]. Accordingly, we take surface forms of the words and generate their all possible lexical forms. Also we use the lexicon to label the initial states as the root words by parts of speech such as noun, verb, etc. The final states represent a lexeme created by affixing morphemes in each further states.

The schemes presented below are different combinations of outputs determining the removal of affixes from the analyzed words. The baseline approach is not perfect since a scheme includes several suffixes incorrectly segmented. In this case, we mainly focused on detection a few techniques for the segmentation of such word forms. In order to find an effective rule set we tested several segmentation schemes named S[1..5], some of which have described in the following Table 2.

Table 2. The segmentation schemes

Id	Schema	Examples	Translation
S1	stem	el	state
S2	stem+case	el + ge	state + dative
S3	stem+num+case	el + der + den	state + num + ablativ
S4	stem+poss+	el + in	state + poss2sing
S5	stem+poss+case	el + i +ne	state + poss3sing + dative

Kazakh has 7 noun cases: nominative, accusative, dative, ablative, genitive, locative and instrumental. Kazakh verbs take suffixes to show their tense and mood. Verbs can have a morphological past, present and future tense. Kazakh doesn't need morphological agreement between words, for example noun and adjective agreement. Nominal cases that are expected to have an English counterpart are split off from words: these are namely dative, ablative, locative and instrumental, often aligning with the English prepositions 'to', 'from', 'in' and 'with/by'. The remaining case affixes nominative, accusative and genitive are not have English counterparts. After treating case affixes we split of possessive suffixes from nouns of all persons except the first singular, which doesn't need remove.

There are large amount of verbs presenting ambiguity during segmentation, which do not take personal endings, but follow conjugated main verbs. During the process, we hardly determined the border between stems and inflectional affixes, especially when the word and the suffix matches entire word in the language. In fact, there are lack of syntactic information we cannot easily distinguish among similar cases.

In order to solve the problems represented above, we have to split up Kazakh words into the morphemes and some tags which represent the morphological information expressed on the suffixation. Splitting Kazakh words in this way, we expect to reduce the sparseness produced by the agglutination being of Kazakh and the drought of training data. Anyway, the segmentation model takes into account the several segmentation options of both sides of the parallel corpus while looking for the optimal segmentation. In order to define the most convenient segmentation for our Kazakh-English system, we checked most of the segmentation options and have measured their impact on the translation quality.

While GIZA++ tool produces a competitive alignment between words, the Kazakh sentences must be segmented as we already have in the first step. Therefore our method looks like an word sequence labeling problem, the contexts can be presented as POS tags for the word pairs.

Table 3. Part of Speech tag patterns

Tag	Sample	Tag	Sample
NN (Noun)	"el"-"state"	JJS (Adjective, super.)	"tym"-"most"
NNP (Proper noun)	"biz"-"we"	VB (Verb, base form)	"bar"-"go"
JJ (Adjective)	"jasyl"-"green"	VBD (Verb, past tense)	"bardy"-"went"
JJR (Adj, comp.)	"ulkenirek"-"bigger"	VBG (Verb, gerund)	"baru"-"to go"
RB (Adverb)	"jildam"-"speedy"	CC (Conjunction)	"jane"-"and"

The underlying concept of our POS tags approach is to take the parts of the words (and closed class items) into consideration during the selection process. Most of the text contains short and frequent words that will be selected by frequency-based approach. Because Kazakh derivational suffixes cannot occur freely, only in conjunction with a word stem, so each input word was reduced to its lemma and POS tagged word parts.

Present application of morphological processing aims to find several best splitting options that the each Kazakh word ideally corresponds to one English words, so the deep analysis is more desirable.

3   Evaluation

For evaluation the system, three samples of text data were processed with 50k sentences each one, which were used in raw form and with special segmentation. The expert decisions about a segmentation quality were defined by our university undergraduate students. The data samples were stored randomly into a training set and a test set had one sample for each of the phrase-based Moses[18] system run. After the most of the samples were found processed correctly, which means the same interpretation of data was selected as acceptable by the experts, we decided the system was trained well, and that is a good result.

Our corpora consists of the legal documents from http://adilet.zan.kz, a content of http://akorda.kz, and Multilingual Bible texts, and the target-side language models were trained on the MultiUN[19] corpora. We conduct all experiments on a single PC, which runs the 64-bit version of Ubuntu 14.10 server edition on a 4Core Intel i7 processor with 32 GB of RAM in total. All experiment files were processed on a locally mounted hard disk. Also we expect the more significant benefits from a larger training corpora, therefore we are in the process of its construction.

We did not have a gold standard for phrase alignments, so we had to refine the obtained phrase alignments to word alignments in order to compare them with our word alignment techniques. We measure the accuracy of the alignment using precision, recall, and F-measure, as given in the equations below; here, A represents the reference alignment; T, the output alignment; A and T intersection, the correct alignments.

$$pr = \frac{|A \cap T|}{|T|}, re = \frac{|A \cap T|}{|A|}, F - measure = \frac{2 \times pr \times re}{pr + re} \qquad (7)$$

Table 4. Best performance scores

System	Precision	Recall	F-score	AER	BLEU	METEOR	TER
Baseline	57.18	28.35	38.32	36.22	30.47	47.01	49.88
Morfessor	71.12	28.31	42.49	20.19	31.90	47.34	49.37
Rule-based	89.62	29.64	45.58	09.17	33.89	49.22	48.04

The alignment error rate (AER) values for the trained system show distinct tendencies which were consistent through the iteration of different training parameters. The values show completely the higher rates for raw lexeme than for segmented one, which seems suitable for an alignment task. Another tendency is that the differences of context receive smaller impact than the precision of segmentation. This was not clear since removing or normalization causes a change in word structure. A problem in interpreting these training results depend on the scaling of the morpheme probability, which can be of different variation, and the scale needs to be appropriate to the text domain and segmentation schemes. We assume that phrase alignment connects word classes rather than words. Consequently, the phrase translation table has to be learned directly from phrase alignment models, and an estimation of phrase distribution probability is internally part of the process.

The system parameters were optimized with the minimum error rate training (MERT) algorithm [20], and evaluated on the out-of and in-domain test sets. All 5-gram language models were trained with the IRSTLM toolkit[21] and then were converted to binary form using KenLM for a faster execution[22]. The translation performance scores were computed using the MultEval[23]: BLEU, TER[24] and METEOR[25]; and we ran Moses several times per experiment setting, and report the best BLEU/AER combinations obtained. Our survey shows that translation quality measured by BLEU metrics is not strictly related with lower AER.

4 Conclusions

We have described an word alignment model for SMT that improves the previous state-of-the-art models by incorporating a morpheme information to the alignment relation between the sentence pair. We extracted good phrase pairs from word alignment and we discussed the essential steps in establishing a phrase extracting procedure to create a pool of necessary phrase pairs and constructing a phrase translation table. A method has been proposed for evaluating alignments and phrase relation rating of internal sequences based on analyze in two levels, phrase pairs also could be found by pattern detection algorithms from parallel sentences through their possible co-occurrence. A phrase does not need to be a sequence of consecutive words, the gaps between the words and the subphrases are allowed. After the work of the component no particular phrase reordering training is required. By using a knowledge about morphemes in combination with the machine learning procedure, the task can be implemented with the shown encouraging results.

The improved model performs at slightly the same speed as the previous one, and gives an increase of about 3 BLEU over baseline translation. This is a pretty modest improvement, but on the other hand it is achieved by adding only middle-tier component to the baseline pipeline. I think that it is a demonstration of the potential of word alignments for SMT quality, and we plan to investigate more complicated methods in the future researches, possibly adding the new alignment features to the model.

References

1. Bekbulatov, E., Kartbayev, A.: A study of certain morphological structures of Kazakh and their impact on the machine translation quality. In: IEEE 8th International Conference on Application of Information and Communication Technologies, Astana, pp. 1–5 (2014)

2. Oflazer, K., El-Kahlout, D.: Exploring different representational units in English-to-Turkish statistical machine translation. In: 2nd Workshop on Statistical Machine Translation, Prague, pp. 25–32 (2007)
3. Bisazza, A., Federico, M.: Morphological pre-processing for Turkish to English statistical machine translation. In: International Workshop on Spoken Language Translation 2009, Tokyo, pp. 129–135 (2009)
4. Moore, R.: Improving IBM word alignment model 1. In: 42nd Annual Meeting on Association for Computational Linguistics, Barcelona, pp. 518–525 (2004)
5. Brown, P.F., Della Pietra, V.J., Della Pietra, S.A., Mercer, R.L.: The mathematics of statistical machine translation: Parameter estimation. Computational Linguistics **19**, 263–311 (1993). MIT Press Cambridge, MA
6. Vogel, S., Ney, H., Tillmann, C.: HMM-based word alignment in statistical translation. In: 16th International Conference on Computational Linguistics, Copenhagen, pp. 836–841 (1996)
7. Dempster, A., Laird, N., Rubin, D.: Maximum likelihood from incomplete data via the EM algorithm. Journal of the Royal Statistical Society. Series B **39**, 1–38 (1977). Wiley-Blackwell, UK
8. Creutz, M., Lagus, K.: Unsupervised models for morpheme segmentation and morphology learning. ACM Transactions on Speech and Language Processing **4**, article 3. Association for Computing Machinery, New York (2007)
9. Beesley, K.R., Karttunen, L.: Finite State Morphology. CSLI Publications, Palo Alto (2003)
10. Goldsmith, J.: Unsupervised learning of the morphology of a natural language. Computational Linguistics **27**, 153–198 (2001). MIT Press Cambridge, MA
11. Altenbek, G., Xiao-Long, W.: Kazakh segmentation system of inflectional affixes. In: CIPS-SIGHAN Joint Conference on Chinese Language Processing, Beijing, pp. 183–190 (2010)
12. Kairakbay, B.: A nominal paradigm of the Kazakh language. In: 11th International Conference on Finite State Methods and Natural Language Processing, St. Andrews, pp. 108–112 (2013)
13. Lindén, K., Axelson, E., Hardwick, S., Pirinen, T.A., Silfverberg, M.: HFST—framework for compiling and applying morphologies. In: Mahlow, C., Piotrowski, M. (eds.) SFCM 2011. CCIS, vol. 100, pp. 67–85. Springer, Heidelberg (2011)
14. Och, F.J., Ney, H.: A Systematic Comparison of Various Statistical Alignment Models. Computational Linguistics **29**, 19–51 (2003). MIT Press Cambridge, MA
15. Papineni, K., Roukos, S., Ward, T., Zhu, W.: BLEU: a method for automatic evaluation of machine translation. In: 40th Annual Meeting of the Association for Computational Linguistics, Philadephia, pp. 311–318 (2002)
16. Dunning, T.: Accurate methods for the statistics of surprise and coincidence. Computational Linguistics **19**, 61–64 (1993). MIT Press Cambridge, MA
17. Gabrilovich, E., Markovitch, S.: Computing semantic relatedness using wikipedia-based explicit semantic analysis. In: 20th International Joint Conference on Artificial Intelligence, Hyderabad, pp. 1606–1611 (2007)
18. Koehn, P., Hoang, H., Birch, A., Callison-Burch, C., Federico, M., Bertoldi, N., Cowan, B., Shen, W., Moran, C., Zens, R., Dyer, C., Bojar, O., Constantin, A., Herbst, E.: Moses: open source toolkit for statistical machine translation. In: 45th Annual Meeting of the Association for Computational Linguistics, Prague, pp. 177–180 (2007)

19. Tapias, D., Rosner, M., Piperidis, S., Odjik, J., Mariani, J., Maegaard, B., Choukri, K., Calzolari, N.: MultiUN: a multilingual corpus from united nation documents. In: Seventh Conference on International Language Resources and Evaluation, La Valletta, pp. 868–872 (2010)
20. Och, F.J.: Minimum error rate training in statistical machine translation. In: 41st Annual Meeting of the Association for Computational Linguistics, Sapporo, pp. 160–167 (2003)
21. Federico, M., Bertoldi, N., Cettolo, M.: IRSTLM: an open source toolkit for handling large scale language models. In: Interspeech 2008, Brisbane, pp. 1618–1621 (2008)
22. Heafield, K.: Kenlm: faster and smaller language model queries. In: Sixth Workshop on Statistical Machine Translation, Edinburgh, pp. 187–197 (2011)
23. Clark, J.H., Dyer, C., Lavie, A., Smith, N.A.: Better hypothesis testing for statistical machine translation: controlling for optimizer instability. In: 49th Annual Meeting of the Association for Computational Linguistics, Portland, pp. 176–181 (2011)
24. Snover, M., Dorr, B., Schwartz, R., Micciulla, L., Makhoul, J.: A study of translation edit rate with targeted human annotation. In: Association for Machine Translation in the Americas, Cambridge, pp. 223–231 (2006)
25. Denkowski, M., Lavie, A.: Meteor 1.3: automatic metric for reliable optimization and evaluation of machine translation systems. In: Workshop on Statistical Machine Translation EMNLP 2011, Edinburgh, pp. 85–91 (2011)

Application of Uncertainty Modeling Frameworks to Uncertain Isosurface Extraction

Mahsa Mirzargar, Yanyan He[✉], and Robert M. Kirby

Scientific Computing and Imaging Institute, University of Utah, Salt Lake City, USA
yhe@sci.utah.edu

Abstract. Proper characterization of uncertainty is a challenging task. Depending on the sources of uncertainty, various uncertainty modeling frameworks have been proposed and studied in the uncertainty quantification literature. This paper applies various uncertainty modeling frameworks, namely possibility theory, Dempster-Shafer theory and probability theory to isosurface extraction from uncertain scalar fields. It proposes an uncertainty-based marching cubes template as an abstraction of the conventional marching cubes algorithm with a flexible uncertainty measure. The applicability of the template is demonstrated using 2D simulation data in weather forecasting and computational fluid dynamics and a synthetic 3D dataset.

Keywords: Uncertainty quantification · Isosurface visualization · Possibility theory · Dempster-Shafer theory

1 Introduction

Uncertainty is an inevitable component of predictive simulations based on computational models since the models are often imperfect or might involve uncertain parameters. As a result, understanding and quantifying the uncertainty in model output (data) is of critical importance.

To account for the uncertainty in data, an integral component of data analysis — visualization — has been combined with uncertainty modeling frameworks to constitute a special topic in the visualization community called uncertainty visualization. Uncertainty visualization is not a new topic and various uncertainty visualization techniques have been defined and studied [9,13]. Most relevant one to the current work is isosurface extraction in the presence of uncertainty and hence, we focus only on visualization of uncertain isosurfaces. In order to quantify and visualize the uncertainty in isosurfaces extracted from uncertain scalar fields, parametric probabilistic models have been used to approximate Level-Crossing Probabilities (LCP) [15,16]. The concept of level-crossing probabilities has been deployed to extend the conventional marching cubes algorithm, the predominant isosurface visualization scheme, for probabilistic modeling of uncertainty in scalar fields [17,18].

Probabilistic modeling is a well-developed approach for uncertainty when its source is a stochastic/random process (called aleatory uncertainty). However, it

© Springer International Publishing Switzerland 2015
V.-N. Huynh et al. (Eds.): IUKM 2015, LNAI 9376, pp. 336–349, 2015.
DOI: 10.1007/978-3-319-25135-6_32

requires complete probability information for the random process, which could be quite difficult, if not impossible. Modeling *epistemic uncertainty*, referring to the uncertainty due to lack of knowledge [20], requires the exploration of the alternatives of probabilistic approaches. In the past few decades, possibility theory [6] and Dempster-Shafer (DS) theory [21], have been explored and studied in the uncertainty quantification literature for a better representation of the epistemic uncertainty. Yager [22] has developed a unified theoretical structure for uncertainty representation using various models.

In this work, we apply modern uncertainty modeling frameworks to the isosurface extraction from uncertain scalar fields, and propose an abstraction of the conventional marching cubes algorithm in terms of a template infrastructure that is flexible enough to incorporate various uncertainty modeling frameworks to model uncertainty in the spatial location of isosurfaces extracted from uncertain scalar fields. Our uncertainty-based marching cubes template can be considered as the extension of level-crossing probability to a general uncertainty measure.

The contributions of this work can be summarized as follows:

- We apply various uncertainty modeling frameworks to the isosurface extraction from uncertain scalar fields.
- We provide an abstraction of the marching cubes algorithm that we call *uncertainty-based marching cubes (UMC) template*. This template is amenable to deploy different mathematical frameworks for uncertainty modeling.
- We demonstrate the effectiveness and applicability of the template in a few examples of uncertain isosurface extraction from an ensemble of scalar fields.

The rest of the paper proceeds as follows. Section 2 is devoted to a brief introduction to the various uncertainty modeling frameworks. In Section 3, we propose the uncertainty-based marching cubes template. Section 4 demonstrates the effectiveness of our template by presenting the results for several examples. We conclude our work in Section 5 and provide some avenues for future investigation.

2   Mathematical Frameworks for Uncertainty Modeling

Let $Y \in U \subseteq \mathbb{R}$ denote a random quantity of interest. We assume that the probability distribution of Y is unknown, instead, we have a finite number of possible realizations of Y (i.e., $\{y^i\}_{i=1}^N$) available. In such a situation, it is suggested that the random variable Y should be represented "as a mixture of natural variability (aleatory) and estimation errors (epistemic)" since " a finite number of samples from a population leads to epistemic uncertainty [8]." The goal is to model the mixed uncertainty from the ensemble $\{y^i\}_{i=1}^N$ using uncertainty modeling frameworks. We consider the propositions in the form of "the true value of Y is in A" for any subset $A \subseteq U$, and adopt the measures from the uncertainty modeling frameworks to quantify the strength of the support from the evidence (i.e., the ensemble $\{y^i\}_{i=1}^N$) for proposition A. For example, the probability measure quantifies the chance of proposition A being true.

In this section we introduce the mathematical notation used to express the various frameworks of fuzzy measures. Note: we skip the basics of probability theory due to the space limitations. Interested readers can consult [11].

2.1 Possibility Theory

Instead of a single measure in probability theory, possibility theory defines a dual measure (possibility and necessity) as [7]

$$Poss(A) = \max_{y \in A} \pi_Y(y), \quad Nec(A) = 1 - Poss(A^c), \tag{1}$$

where $\pi_Y : U \to [0,1]$ is a *possibility distribution*, and A^c is the complement of A. The possibility distribution describes the degree to which it is *possible* that the element $y \in U$ is the true value of Y.

The possibility function $Poss(A)$ measures the maximum possible support from the evidence for proposition A whereas the necessity function $Nec(A)$ quantifies the minimum support for proposition A. The length/distance between $Nec(A)$ and $Poss(A)$ indicates the epistemic uncertainty regarding proposition A.

Assume that the Y_js are independent variables associated with possibility distribution $\pi_{Y_j}(y_j)$, the joint distribution is defined using $\min(\cdot, \cdot)$ as the joint operator [10,23]

$$\pi_Y(y) = \min(\min(\dots \min(\pi_{Y_1}(y_1), \pi_{Y_2}(y_2)), \dots), \pi_{Y_M}(y_M)). \tag{2}$$

Although there have been a few attempts to discover the correlation between uncertain variables in possibility theory [3], it is still an open problem to construct joint distribution for dependent variables from ensembles. This topic lies outside the scope of the current work, but remains an active area of research in the field of uncertainty quantification.

Construction of Possibility Distribution: Here, we provide two examples of constructing a possibility distribution.

1. With assumption of triangular shape distribution (referred as parametric technique): We construct a possibility distribution for the variable Y based on a modified version of a triangular shape:

$$\pi_Y(y) = \begin{cases} 1 - \frac{(y - y_{mean})(1 - p^+)}{y_{max} - y_{mean}} & \text{if } y_{mean} \le y \le y_{max} \\ p^+ + \frac{(y - y_{min})(1 - p^+)}{y_{mean} - y_{min}} & \text{if } y_{min} \le y < y_{mean} \\ p^+ & \text{otherwise} \end{cases} \tag{3}$$

where y_{min}, y_{mean} and y_{max} are the minimum, sample mean and the maximum of the ensemble data, respectively; p^+ is the upper bound of the probability of the true value of Y falling outside the range of the ensemble data, which is estimated using Goodman's simultaneous confidence interval [4] as follows

$$p^+ = \frac{a + 2n + \sqrt{D}}{2(N+a)}, \quad D = a(a + \frac{4n(N-n)}{N}), \tag{4}$$

where a is the quartile of order $1 - \alpha$ of the chi-square distribution with one degree of freedom and N is the size of the ensemble. There is no data that supports the value of Y falling outside of the ensemble, and hence, $n = 0$. The probability that *the chance of "the value of Y falls outside of the ensemble range" is less than p^+* is no less than $1 - \alpha$, i.e., $Prob(0 \leq Prob(Y \notin [y_{\min}, y_{\max}]) \leq p^+) \geq 1 - \alpha$ holds. Therefore, it is reasonable to assign p^+ to the values outside the ensemble range in the possibility distribution. In the current work, we take one of the usual probability levels $\alpha = 0.025$.

2. Without assumption of shapes for distribution (referred as nonparametric technique): We construct a possibility distribution using the combination of histogram and the probability-possibility transformation proposed by Dubois et al. [5] as follows. The transformation $Prob \rightarrow \pi$ is based on the principle of maximum specificity, which aims at finding the most informative possibility distribution [5].

We first construct the probability distribution $Prob$ from the histogram of the ensemble data $\{y^i\}_{i=1}^{N}$, where we fix the sample mean at the boundary of one of the bins of the histogram. Let $\{x_j\}_{j=1}^{l}$ be the bins and the probability values be $p_j = Prob(x_j)$. If the probability values are ordered, i.e., $p_1 \geq p_2 \geq \ldots \geq p_l$, then a possibility distribution can be obtained using the transformation as follows:

$$\pi_1 = 1, \quad \pi_j = \sum_{k=j}^{l} p_k. \tag{5}$$

If there exist j such that $p_j = p_{j+1}$, there will be different possibility distributions $\pi^{(t)}$ obtained for each permutation of the equal probability values. Then we choose the one that minimizes the possibility values as $\pi_j = \min_t \pi_j^{(t)}$. In order to provide smooth transition between the values in adjacent bins, we use a Gaussian filtering at the end for smoothing.

2.2 Dempster-Shafer Theory

Analogous to possibility theory, Dempster-Shafer (DS) theory also defines a dual measure (plausibility and belief), for $\forall A \subseteq U$, as

$$Pl(A) = \sum_{B \cap A \neq \emptyset} m(B), \quad Bel(A) = \sum_{B \subseteq A} m(B), \tag{6}$$

where $m : 2^U \rightarrow [0,1]$ is a *basic belief assignment* (BBA), also called *m-function*. An m-function satisfies the following two conditions:

$$m(\emptyset) = 0, \quad \sum_{A \subseteq \mathbf{U}} m(A) = 1. \tag{7}$$

The plausibility function $Pl(A)$ and the belief function $Bel(A)$ quantify the maximum and minimum strength of the evidence that supports the proposition, respectively. The length between $Bel(A)$ and $Pl(A)$ also indicates the epistemic uncertainty regarding proposition A.

Let Y_js be independent variables associated with basic belief assignments $m_{Y_j}(A_j)$ $(A_j \in 2^{U_j})$, the joint basic belief assignment (m-function) can be calculated by taking the product (joint operator) over all the components of $\mathbf{Y} = \{Y_j\}_{j=1}^{M}$ as [12]

$$m(\mathbf{A}) = m_1(A_1)m_2(A_2)\ldots m_M(A_M), \tag{8}$$

where the hypercube $\mathbf{A}$ is the Cartesian product, i.e., $\mathbf{A} = A_1 \times A_2 \times \ldots \times A_M$.

Due to space limitations, we do not introduce further concepts for the dependent case, but refer the interested reader to [19].

Construction of Basic Belief Assignment: We construct belief/plausibility functions using the method proposed by Denœux [4] as follows. Consider two ordered consecutive intervals: $A_1 = \{Y \leq \theta\}$ and $A_2 = \{Y > \theta\}$ (the universal set becomes $\{A_1, A_2\}$), and "ordered" means that the elements in A_i are no larger than the elements in A_j if $i < j$. Let n_k be the number of samples falling inside $\{A_k\}_{k=1}^{2}$ and

$$P_k^- = \frac{a + 2n_k - \sqrt{D_k}}{2(N + a)}, \quad P_k^+ = \frac{a + 2n_k + \sqrt{D_k}}{2(N + a)}, \tag{9}$$

where a and D_k are computed using the relation in Eq. 4 with $n = n_k$. Let $A_{k,j}$ $(k \leq j)$ denote the union $A_k \cup A_{k+1} \cup \ldots \cup A_j$. Then the m-function is constructed as

$$\begin{aligned}
m(A_{k,j}) &= P_k^-, \quad \text{if } j = k, \\
m(A_{k,j}) &= P^-(A_{k,j}) - P^-(A_{k+1,j}) - P^-(A_{k,j-1}), \quad \text{if } j = k + 1, \\
m(A_{k,j}) &= P^-(A_{k,j}) - P^-(A_{k+1,j}) - P^-(A_{k,j-1}) + P^-(A_{k+1,j-1}), \\
&\qquad \text{if } j > k + 1,
\end{aligned} \tag{10}$$

where

$$P^-(B) = \max\Big(\sum_{A_k \subset B} P_k^-, 1 - \sum_{A_k \not\subset B} P_k^+\Big), \quad \forall B \neq \emptyset. \tag{11}$$

3 Application of Uncertainty Modeling Frameworks to Isosurface Extraction

In this section, we apply the introduced uncertainty modeling frameworks to isosurface extraction and introduce our uncertainty-based marching cubes (UMC) template. We first recall the fundamentals of the concept of level crossing in the (deterministic) marching cubes algorithm.

3.1   Deterministic Marching Cubes Algorithm

In the absence of uncertainty, a deterministic scalar field can be considered as a discrete representation of a continuous multivariate function $g(\cdot)$ using a set of *deterministic* scalar values $\{y_j\}_{j=1}^M$ on a grid, where M denotes the resolution of the grid. An isosurface of $g(\cdot)$ associated with a given isovalue θ is defined as: $C = \{x \in \mathbb{R}^d, \; g(x) = \theta\}$, where d is the embedding dimension. The goal of the marching cubes algorithm (in 3D) is to extract an approximation of the isosurface based on the trilinear approximation (i.e., tensor product of linear interpolation in the univariate case) of the underlying continuous function giving rise to the scalar field [14].

The local nature of the trilinear approximation simplifies the isosurface extraction significantly. Trilinear approximation requires the information only at the corners of a cell (e.g., a cube on a 3D Cartesian lattice). Therefore, the presence of a level crossing inside each cell (i.e., cell crossing) is *locally* determined based on the values of the scalar field at the corners of the cell. A cell crossing happens if at least one sign change occurs in the set of differences $\{y_j - \theta\}_{j=1}^M$ for the scalar values y_j at the corners of the cell under question. In the presence of a cell crossing, the values at the corners of the cell also determine the (approximate) polygonal tessellation and the spatial location of the isosurface.

In the presence of uncertainty or error associated with the scalar field, the scalar values are no longer known deterministically. Consequently the spatial location of the isosurface from an uncertain scalar field becomes uncertain. Therefore, the conventional marching cubes algorithm must be extended to incorporate the uncertainty to provide reliable information about the presence or absence of an isosurface inside a cell.

3.2   Uncertain Cell Crossing and UMC Template

We propose an uncertainty-based marching cubes (UMC) template that encompasses the essential concepts from the deterministic and probabilistic marching cubes algorithm [18] and is flexible enough to be adopted for various uncertainty modeling frameworks. Similar to the conventional marching cubes algorithm, the UMC template (algorithm) proceeds through the uncertain scalar field, and at each cell it quantifies how much the available information at the corners supports the incidence of a cell crossing.

To construct an abstract template, which accommodates various uncertainty modeling frameworks, we adopt a generic and flexible (uncertainty) measure that we call *U-Measure* to indicate the presence of cell crossing for each constituent cell of an uncertain scalar field (i.e., uncertain cell crossing). The U-Measure lends itself to various uncertainty modeling frameworks, and hence, the cell-crossing U-Measure value can be computed based on the axioms of a chosen uncertainty modeling framework.

For what follows, let us define a *cell-crossing* proposition:

$$C: \quad \text{"there exists a level-crossing in a cell"}. \tag{12}$$

The goal is to evaluate the cell-crossing U-Measure values for proposition (12) (i.e., U-Measure(C)) for each cell in an uncertain scalar field. Computation of the U-Measure(C) can be broken down into a few steps.

For every cell, the first step in computing U-Measure (C) is to construct a **joint distribution function** based on the ensemble about the uncertain scalar values at the corners. As discussed in Section 2, construction of a joint distribution function can be carried out either by

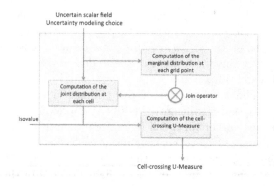

Fig. 1. Schematic illustration of the uncertainty-based marching cubes template.

- first constructing a **marginal distribution function** at each grid point and then using a **joint operator** to define the joint distribution. For example, use Eqs. (3) or (5) in possibility theory, and Eqs. (10) and (8) in Dempster-Shafer theory.
- or in special cases, such as in the probabilistic case [18], directly from the available information at the corners.

The second step is to **compute the cell-crossing U-Measure** value(s) for proposition C for a cell after construction of its joint distribution function. The value(s) of U-Measure(C) for a cell can be evaluated using Eq. 1 for possibility theory or Eq. 6 for DS theory or Eq. (14) for probability theory. These steps constitute the building blocks of the uncertainty-based marching cubes template that has also been demonstrated schematically in Fig. 1.

Without loss of generality, we illustrate the idea for a 1D example (similar to the example presented in [18]) while the concepts extend to higher dimensions.

Consider two adjacent points as corners of an edge with associated (uncertain) scalar variables y_1 and y_2, respectively. Proposition C is equivalent to considering the incidence of the values (y_1, y_2) belonging to

$$A = \{(y_1, y_2) : (y_1 \geq \theta \, \& \, y_2 \leq \theta) \mid (y_1 \leq \theta \, \& \, y_2 \geq \theta)\}. \tag{13}$$

Now let us consider an ensemble of tuple values (y_1, y_2) as the uncertain scalar values at the endpoints of an edge. A tuple from the ensemble results in an edge crossing if the linear interpolant connecting its endpoints crosses value θ

(Fig. 2 (a)). Fig. 2 (b) illustrates another representation (scatterplot) of the set of (tuple) scalar values in Fig. 2 (a). Fig. 2 (b) makes it more clear that if the values of a tuple belong to one of the subregions marked as $A_2 = \{y_1 \geq \theta \ \& \ y_2 \leq \theta\}$ or $A_4 = \{y_1 \leq \theta \ \& \ y_2 \geq \theta\}$, the linear interpolant associated with them will result in an edge crossing. To evaluate the edge-crossing U-Measure values (i.e., the measure of the support for proposition C from the available data), one needs to construct a joint distribution function (e.g., $f_Y(y)$ in Fig. 2 (c)) based on a chosen uncertainty modeling framework. The edge-crossing U-Measure values can then be computed as discussed earlier. For instance, in probabilistic modeling [18], probability measure is used to compute the U-Measure(C) values as

$$\text{U-Measure}(C) = Prob(C) = \int_{A_2 \cup A_4} f_Y(y), \quad y = (y_1, y_2). \tag{14}$$

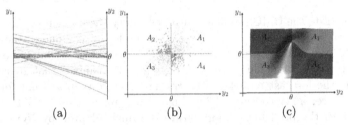

(a)   (b)   (c)

Fig. 2. (a) An ensemble of linear interpolants in 1D: the linear interpolants that correspond to the presence of an edge crossing are highlighted. (b) The values at the end points of each linear interpolant have been demonstrated as points in $\mathbb{R}^2$. The highlighted points correspond to tuples that result in presence of edge crossing in (a). (c) A joint distribution function $f_Y(y)$ inferred by fitting a multivariate Gaussian function to the data in (b). The regions corresponding to the presence of edge crossing are highlighted in dark blue.

Note: i) The UMC template maintains a linear computational cost in terms of the number of the cells and takes advantage of the local nature of the concept of cell crossing similar to the deterministic version. However, it is important to note that the overall computational cost of instances (instantiations) of the UMC template for different uncertainty modeling frameworks is higher than the deterministic version due to the cost associated with the construction of the distribution function and computation of the U-Measure. ii) We focus only on the construction of the joint distributions that use an independent assumption (as examples) even though our proposed framework can easily incorporate dependence.

4   Results and Applications

In this section, we demonstrate the utility of the UMC template with three experimental examples from each uncertainty modeling framework. In all the examples, the colormap has been scaled so that the highest level-crossing U-Measure value is assigned to blue and the minimum level-crossing U-Measure value is assigned to white. The scaling of the colormap helps to provide better color contrast.

Note: The goal of the current work is to demonstrate the applicability of the UMC template with different mathematical uncertainty modeling frameworks. The comparisons among the mathematical modeling frameworks and among the corresponding isocontour/isosurface extraction results are beyond the scope of the current manuscript; they represent present and future work within the uncertainty quantification field which may in part facilitated by our template. Therefore, we present each example with specific mathematical modeling framework and the corresponding visualization results.

4.1   Temperature Forecast Example Using Possibility Theory

We demonstrate the UMC template with possibility theory using a dataset from the weather forecast application. For this example, we use one of the publicly available weather forecast ensembles called SREF-CONUS (40km) temperature ensemble by NOAA [1]. This ensemble consists of 21 members that are generated by varying the forecast model and the initial conditions to account for various sources of uncertainty (both model and parameter uncertainty). We have chosen to use one of the predefined temperature isovalues adopted by NOAA that is $-15C$ at 500mb.

For this example, we use both parametric (Eq. 3) and nonparametric (Eq. 5) techniques to construct possibility distributions. First two rows of Fig. 3 provides the visualization of the level-crossing possibility and necessity values in both parametric and nonparametric settings. The possibility values suggest the maximum/optimistic estimation for the chance of the presence of level-crossing at each cell, and the region near the mean isocontour (visualized in black) has a relatively higher maximum chance of the presence of level crossing. The necessity values indicate the minimum/conservative estimation of the chance of the presence of level crossing. Note that the colormap for necessity values has been flipped to make them more visible. The region with parametric nonzero necessity values is coincident with the mean isocontour, which is due to (a) the degree of possibility $\pi(y) = 1$ is assigned to the mean ensemble y_{mean} when we construct the possibility distribution; and (2) the relation $Nec(A) \neq 0$ if $Poss(A) = 1$.

Note that the lack of smoothness of the computed possibility/necessity values in the nonparametric case is due to the oscillatory nature of the possibility distribution function in the nonparametric setting. Instances of the distribution functions constructed from an ensemble at one grid location are shown in Fig. 3 (c) and Fig. 3 (f). The parametric technique of constructing possibility distribution has fewer degrees of freedom compared to the nonparametric technique; therefore the resulting distribution function from the parametric technique is less oscillatory.

4.2   Computational Fluid Dynamics Example Using DS Theory

We use DS theory as the underlying mathematical framework for instantiation
of the uncertainty-based marching cubes template, and demonstrate the results
for a fluid simulation example, which is motivated by the use of ensembles in
computational fluid dynamics to study structures such as vortices.

For this example, we use the simulation of flow past a circular cylinder.
When the fluid passes an obstacle, the eddies or vortices are formed in a periodic
fashion and move in the direction of the flow field shown in Fig. 4 (a). Studying
the pressure of a flow field is among the simplest approaches to study vortex
structures. The center of a vortex typically corresponds to minimum pressure
values. Therefore, isocontours of the pressure field can be used to approximate

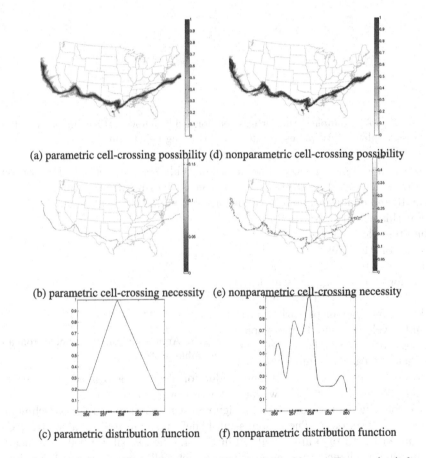

(a) parametric cell-crossing possibility (d) nonparametric cell-crossing possibility

(b) parametric cell-crossing necessity   (e) nonparametric cell-crossing necessity

(c) parametric distribution function   (f) nonparametric distribution function

Fig. 3. The UMC template instantiated with possibility theory. Figures (a-c) demonstrate the results for parametric and figures (d-f) demonstrate the results for nonparametric technique. The possibility distribution demonstrated in (c) and (f) are
constructed from an ensemble at a single point in the dataset (the ensemble values are
represented with dots).

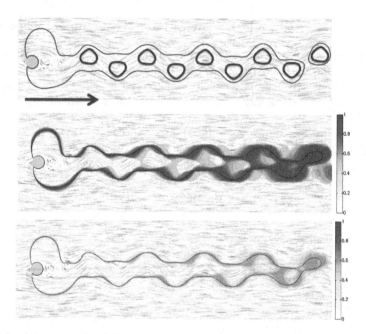

Fig. 4. (top) An example of the vortex street formed by a flow past a cylinder, (middle) cell-crossing plausibility values, (bottom) cell-crossing belief values.

the position and size of the vortices in a flow field. The number and the position of vortices generated is affected by variation of the simulation parameters such as the Reynolds number, initial conditions and boundary conditions.

For this example, we used the 2D incompressible Navier-Stokes solver as part of the Nektar++ software package [2] to generate simulation results for fluid passing our stationary obstacle. We generated our ensemble of size 40 by random perturbation of the inlet velocity and the Reynolds number. After normalizing the pressure field of each ensemble member

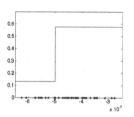

Fig. 5. An m-function constructed from an ensemble (dots).

based on the average of the pressure value for a unique and fixed point inside the field behind the cylinder, we have chosen isovalue=-0.005.

We constructed the basic belief assignment (m-function) using the technique discussed in Section 2. The cell-crossing belief and plausibility values (i.e., U-Measure in DS theory) are visualized in Fig. 4. The plausibility values indicate the maximum chance of level crossing at each cell whereas the belief values indicate the minimum chance. As an example, the m-function constructed from the ensemble at one grid location is shown in Fig. 5 and is constructed as

$$m(\{y <= \theta\}) = 0.1306, \ m(\{y > \theta\}) = 0.5748, \qquad (15)$$

and $m(A_1 \cup A_2) = 1 - m(A_1) - m(A_2)$ where $A_1 = \{y <= \theta\}$, $A_2 = \{y > \theta\}$. As Fig. 5 demonstrates, the constructed m-function is not oscillatory and hence the resultant U-Measures values have smooth transitions.

4.3   Synthetic Example Using Probability Theory

In the last example, we show that our UMC template is applicable to a high-dimensional dataset. Specifically, we provide the result of our UMC template on a 3D synthetic example given in [18] through volume rendering of the probability cell-crossing values (i.e., the U-Measure in probability theory). The synthetic data contains an ensemble of 45 volumetric data using the analytical formula: $\mu(x, y, z) = (cos(7x) + cos(7y) + cos(7z))exp(ar)$ where $r = \sqrt{x^2 + y^2 + z^2}$ and $a = -4.5$. Each ensemble member is a volumetric data of size $300 \times 300 \times 250$ where a has been perturbed by adding normally distributed noise values. For this example, we use our UMC template with the probability-theory-based parametric modeling assumption. In this case, the instantiation of our template is equivalent to the probabilistic marching cubes algorithm [18] and the U-Measure in this case is the chance of the presence of level crossing (i.e., probability values).

Fig. 6. 3D Uncertainty-based marching cubes: with a parametric modeling assumption, the UMC (template) translates into probabilistic marching cubes [18].

Fig. 6 shows that probability values are much smaller around the corners of the volumetric data compared to the values at the center.

5   Summary and Conclusions

This paper applies possibility theory and Dempster-Shafer (DS) theory as alternative uncertainty modeling frameworks of probability theory to the isosurface extraction from uncertain scalar fields. It proposes an uncertainty-based marching cubes template as an abstraction of the conventional marching cubes algorithm with a flexible uncertainty measure. The effectiveness and applicability of the template are demonstrated in a few examples of uncertain isosurface extraction from an ensemble of scalar fields.

In the future, the proposed framework can be used to further study the choice of different mathematical frameworks for the representation and quantification of specific types of uncertainty (aleatoric, epistemic or their mixture) in different data formats. Some of the limitations of the current work also suggest interesting avenues for future research. For example, designing techniques to infer dependence structure suitable for possibility and Dempster-Shafer theories is an interesting avenue of future research that can potentially result in improving the quality and accuracy of uncertainty modeling using these theories; and decision making based on visualization of the results of dual measures could be an interesting future research direction as well.

Acknowledgement. The first and third authors acknowledge the support of National Science Foundation (NSF) grant IIS-1212806. The second author acknowledges support by the Army Research Laboratory under Cooperative Agreement Number W911 NF-12-2-0023. The views and conclusions contained in this document are those of the authors and should not be interpreted as representing the official policies, either expressed or implied, of the Army Research Laboratory or the U.S. Government. The U.S. Government is authorized to reproduce and distribute reprints for Government purposes notwithstanding any copyright notation herein.

References

1. National oceanic and atmospheric administration (2015). http://www.noaa.gov/
2. Nektar++ (2015). http://www.nektar.info
3. Carlsson, C., Fullér, R., Majlender, P.: On possibilistic correlation. Fuzzy Sets and Systems **155**(3), 425–445 (2005)
4. Denœux, T.: Constructing belief function from sample data using multinomial confidence regions. International Journal of Approximate Reasoning **42**, 228–252 (2006)
5. Dubois, D., Foulloy, L., Mauris, G., Prade, H.: Probability-possibility transformations, triangular fuzzy sets, and probabilistic inequalities. Reliable Computing **10**, 273–297 (2004)
6. Dubois, D., Prade, H.: Possibility theory. Plenum Press, New York (1988)
7. Dubois, D.: Possibility theory and statistical reasoning. Computational Statistics & Data Analysis **51**(1), 47–69 (2006)
8. Iaccarino, G.: Uncertainty quantification in computational science (2011). http://web.stanford.edu/group/uq/events/pdfs/lecture_2.pdf
9. Johnson, C., Sanderson, A.: A next step: Visualizing errors and uncertainty. IEEE Computer Graphics and Applications **23** (2003)
10. Klir, G., Yuan, B.: Fuzzy Sets and Fuzzy Logic: Theory and Applications. Prentice-Hall Inc. (1995)
11. Kolmogorov, A.: Foundations of the Theory of Probability. Chelsea Pub. Co., June 1960
12. Kriegler, E., Held, H.: Utilizing belief functions for the estimation of future climate change. International Journal of Approximate Reasoning **39**, 185–209 (2005)
13. Lodha, S., Wilson, C., Sheehan, R.: Listen: sounding uncertainty visualization. In: Proceedings of IEEE Visualization, pp. 189–195 (1996)
14. Lorensen, W.E., Cline, H.E.: Marching cubes: A high resolution 3d surface construction algorithm. Computer Graphics **21**(4), 163–169 (1987)
15. Pfaffelmoser, T., Reitinger, M., Westermann, R.: Visualizing the positional and geometrical variability of isosurfaces in uncertain scalar fields. Computer Graphics Forum **30**(3), 951–960 (2011)

16. Pöthkow, K., Hege, H.C.: Positional uncertainty of isocontours: Condition analysis and probabilistic measures. IEEE Trans. on Visualization and Computer Graphics **17**(10), 1393–1406 (2011)
17. Pöthkow, K., Petz, C., Hege, H.C.: Approximate level-crossing probabilities for interactive visualization of uncertain isocontours. International Journal for Uncertainty Quantification **3**(2), 101–117 (2013)
18. Pöthkow, K., Weber, B., Hege, H.C.: Probabilistic marching cubes. Computer Graphics Forum **30**(3), 931–940 (2011)
19. Regan, H., Ferson, S., Berlent, D.: Equivalence of methods for uncertainty propagation of real-valued random variables. International Journal of Approximate Reasoning **36**, 1–30 (2004)
20. Roy, C.J., Oberkampf, W.L.: A comprehensive framework for verification, validation, and uncertainty quantification in scientific computing. Computer Methods in Applied Mechanics and Engineering **200**, 2131–2144 (2011)
21. Shafer, G.: A Mathematical Theory of Evidence. Princeton University Press (1976)
22. Yager, R.R.: A general approach to uncertainty representation using fuzzy measures. In: FLAIRS Conference, pp. 619–623. AAAI Press (2001)
23. Zadeh, L.A.: Fuzzy sets as a basis for a theory of possibility. Fuzzy Sets and Systems **1**, 3–28 (1978)

On Customer Satisfaction of Battery Electric Vehicles Based on Kano Model: A Case Study in Shanghai

Yanping Yang, Hong-Bin Yan$^{(\boxtimes)}$, and Tieju Ma

School of Business, East China University of Science and Technology,
Meilong Road 130, Shanghai 200237, People's Republic of China
hbyan@ecust.edu.cn

Abstract. Due to the greenhouse effect and limited energy resources, more and more countries and firms have put more attention to clean energy so as to reduce pollution emissions. The development of battery electric vehicle (BEV) becomes crucial to meet the government and society's demands. As one new product with immature technology, there are many factors affecting the wide utilization of BEV. It is necessary to study customer satisfaction of BEV so as to distinguish customer needs, help find the way to improve customer satisfaction, and identify critical factors. Considering the non-linear relationship between product performance and customer satisfaction, the Kano model is used to analyze customer needs for the BEV so as to promote the adoption of BEV in Shanghai. Four approaches to Kano model are used to categorize the BEV attributes as must-be quality, one-dimensional quality, attractive quality and indifferent quality. According to the strategic rule $M > O > A > I$, the priorities of efforts towards promoting the adoption of BEV is identified, i.e., the government and vehicle firms have to fulfill all the must-be requirements. They should make great improvement of one-dimensional qualities to make the BEV competitive to the traditional motor vehicles. Finally, the customers will be very satisfied if the attractive requirements are fulfilled.

Keywords: Battery electric vehicle · Customer satisfaction · Priorities · Kano model

1 Introduction

With the rapid economic development over the past three decades, the total energy consumption in China has increased greatly from 57144 to 375000 (10000 tons of SCE), and the crude oil consumption surges more than 5 times from 12971.6 to 69000 (10000 tons of SCE) [11]. Such a large amount of energy consumption has created heavy energy emissions, which leads to quite serious air pollution in China. As one major source of pollution emissions in China, the road transportation has accounted for about 8% of pollution emissions [1].

© Springer International Publishing Switzerland 2015
V.-N. Huynh et al. (Eds.): IUKM 2015, LNAI 9376, pp. 350–361, 2015.
DOI: 10.1007/978-3-319-25135-6_33

With the rapid economic development, the quantity of motor vehicles in China has reached 137 millions until 2013. Unfortunately, the majority meet relatively low emission standards.

Consequently, Chinese government has been actively exploring and developing effective solutions to reduce exhaust emissions brought by the popularization of motor vehicles so as to reduce the crude oil consumption and improve the air quality. The promotion of new energy vehicles, especially battery electric vehicle (BEV), is one effective way to reduce air pollution [4]. The BEV, one energy-saving and environment-friendly technology, is completely powered by rechargeable batteries (such as lead-acid batteries, nickel cadmium batteries), and excels in lower driving cost, comfort, and quiet driving performance [5]. By replacing conventional motor vehicles, the BEV will save the oil consumption and reduce pollution emissions. It's essential for the government to support the development of BEV technologies. On the other hand, the firms have to continually improve BEV's performance to meet customers' needs by R&D activities.

Quite different from the internal-combustion engine in conventional motor vehicles, the BEV is characterized by limited driving range, new form of fuel refilling, and other attributes, which may create customers' resistance to utilize the BEV in practice. Many studies have focused on the factors affecting the purchasing decision of BEV to enhance customer satisfaction, which is considered to be important for product design and development to succeed in the market place [10]. For example, Sierzchula et al. [13] discussed the relationship between the utilization rate of electric vehicles and the fiscal stimulus, and other social factors by collecting more than thirty countries' data. It was concluded that the fiscal stimulus, number of charging piles, and the presence of local electric vehicle production base are associated with the utilization of pure electric vehicles; however, the fiscal stimulus or sufficient charging infrastructures cannot grantee high utilization rate of electric vehicles. Mau et al. [9] concluded that customers' preferences on price, fuel costs, government subsidies, driving range, charging facilities, service and maintenance have great effects on the utilization rate of electric vehicles. Park et al. [12] predicted the impact on the fuel cell vehicles caused by price change rate and charging stations, based on Bass diffusion model and dynamics.

In summary, there are many factors affecting the adoption of BEV in practice. Most researchers only focus on the analysis of barriers and policies [3,6,14], little research has focused on the development orientation and the priority of factors to improve the BEV. It is important for the improvement of BEV's performance to understand customers' preferences on the adoption of BEV. It is necessary to study customer satisfaction of BEV, so as to distinguish customer' needs, help find the way to improve customer satisfaction, and identify critical factors. Due to the variety of customers' preferences, it is difficult for the firms and government to determine what they should focus on and how to be more targeted. In general, analysis of customer needs involves three aspects [15]: (1) knowing what the customers want and prefer to, (2) the product functional requirements prioritization, and (3) the classification of the functional requirements.

Traditional customer need analysis assumes a linear relationship between customer satisfaction and the performance of product. Taking a different perspective, Kano model [7] combines with hygiene-motivational factors, which is based on nonlinear relationship and takes into account the psychology of customers. The Kano model can identify and classify customer needs by the common surveys.

Toward this end, the main focus of this research is to identify and classify the customer needs of BEV based on the Kano model. By this way, we may use a systematic approach to distinguish consumer needs, identify and prioritize the key factors, and find ways to improve customer satisfaction of the BEV. This research will offer a guidance to the firms to make a trade-off of customer satisfaction and R&D costs. The government can also put forward relevant policies to accelerate the promotion of BEV. The reminder of this paper is as follows. Section 2 reviews four approaches to Kano model. The methodology of customer satisfaction based on Kano model is presented in Section 3. In Section 4, the proposed methodology is applied to BEV in Shanghai by means of the four approaches to Kano model. The results are also compared with each other. Finally, this paper is concluded in Section 5.

2   The Kano Model

Inspired by Herzberg's two-factor theory, Noriaki Kano [7] proposed the Kano model in 1984 to establish the cognitive dimension about fulfillment of product quality characteristics and customer satisfaction. In general, Kano model divides the product attributes into six types:

- Must-be attribute (M). It's the product's basic requirement and essential to the product or service. If well fulfilled, the satisfaction of customers will not be improved; but if not fulfilled, customers will be extremely dissatisfied with the product.
- One-dimensional attribute (O). If well fulfilled, the customer satisfaction will be improved; if not fulfilled, customers will be dissatisfied with the product. Such a type of attributes has a a linear relationship with customer satisfaction.
- Attractive attribute (A). Such a type of attribute will surprise the customer and cause satisfaction. But if it doesn't exist, it will not cause dissatisfaction.
- Indifferent attribute (I). The attribute doesn't have significant influence on the satisfaction. Customers will not pay attention to this type of attributes.
- Reverse attribute (R). The customer doesn't expect this attribute. Its presence will cause dissatisfaction.
- Questionable attribute (Q). The customer gives conflicting answers to this type of attributes.

Fig. 1 shows the relationship between product performance and customer satisfaction. The horizontal axis represents the state of fulfillment of the product

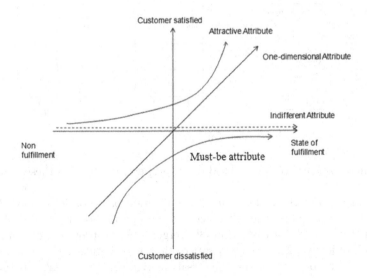

Fig. 1. The Kano model.

performance and the vertical axis represents the customer satisfaction [7]. In this axis, only four attributes are described.

There are several possible analytical methods to rank customer needs. The simple way is to divide the attributes by the frequencies of responses, known as "Frequency-based Attributes Category". The attribute will belong to the category which is the mostly frequently occurring dimension. Such a method is the traditional way to categorize these attributes based on the mode statistic, described as follows:

$$\text{Grade} = \max\{M, O, A, I, R, Q\}. \tag{1}$$

The frequency based method can increase the "noise level" to a point where all "requirements" are considered indifferent. For example, if 18 answerers classify a function as one-dimensional, 19 as attractive, 18 as must-be, 20 as indifferent, 2 as reverse, and 3 as questionable, then the mode statistic classifies this function as indifferent even though 57 out of 82 people answering say that they need this function in one way or the other. One way to decrease the noise level is the comparison-based method, which modifies the mode statistic as follows:

$$\text{Grade} = \begin{cases} \max\{M, O, A\}, & \text{if } (M + O + A) > (I + R + Q); \\ \max\{I, R, Q\}, & \text{otherwise.} \end{cases} \tag{2}$$

which indicates that if the sum of One-dimensional, Attractive, and Must-be is larger than sum of Indifferent, Reverse and Questionable, then the attribution will fall into the maximum of the first three. According to the statistic mode above, the attributes can be arranged into groups as the following order:

$$M > O > A > I > R > Q.$$

M:must-be attribution	I:indifferent attribution	Dysfunctional				
O:one-dimensional	R:reverse attribution	Like	Must-be	Neutral	Live with	Dislike
A:attractive attribution	Q:questionable answer					
Functional	Like	Q	A	A	A	O
	Must-be	R	I	I	I	M
	Neutral	R	I	I	I	M
	Live with	R	I	I	I	M
	Dislike	R	R	R	R	Q

Fig. 2. The Kano evaluation table.

The evaluation rule above can be used to set proper order to fulfil these requirements.

The above two methods define final classification based on frequency. However, it is difficult to reflect the difference in the properties for the impact of customer satisfaction and dissatisfaction [8]. Berger et al. [2] proposed two indexes to reach this objective. The customer satisfaction index indicates that the degree of satisfaction can be created and the dissatisfaction can be prevented by meeting the function. The customer requirements can be classified by the method proposed in [2]. When the attribute is equipped or not, the assignment of the satisfaction and dissatisfaction level is proposed. The attributes can fall into two-dimensional diagram based on satisfaction index and dissatisfaction index, defined as follows:

$$\text{SatIndex} = \frac{A+O}{A+O+M+I}$$
$$\text{DisSatIndex} = \frac{M+O}{A+O+M+I} \tag{3}$$

The satisfaction index is between 0 and 1, when it is close to 1 means the attribute will make highly influence on customer satisfaction, while it is close to 0 means little influence on customer satisfaction. Similarly, value of dissatisfaction index is greater means the impact is greater on customer dissatisfaction. For each attribute, different coordinate value corresponds to different location. Each attribute means one point, every point scatters in the corresponding location.

- If satisfaction index SatIndex < 0.5, dissatisfaction index DisSatIndex < 0.5, the attribute is indifferent.
- If satisfaction index SatIndex < 0.5, dissatisfaction index DisSatIndex ≥ 0.5, the attribute is must-be.
- If satisfaction index SatIndex ≥ 0.5 and dissatisfaction index DisSatIndex ≥ 0.5, it is one-dimensional.
- If satisfaction index SatIndex ≥ 0.5 and dissatisfaction index DisSatIndex < 0.5, the attribute is attractive.

The $M > O > A > I$ rule is be used to organize the importance of these qualities. The preliminary category of these three methods above is determined by means of the Kano evaluation table in 2.

Table 1. Scales of the functional/dysfunctional attributes

Answers	I like	It must be	I am neutral	I can live with	I dislike
If the attribute is provided.	1	0.5	0	-0.25	-0.5
If the attribute isn't provided.	-0.5	-0.25	0	0.5	1

In order to further classify the attributes quantitatively, the analytical Kano model [15] is put forward to analyze customer needs. If the product is composed of I attributes, the product attributes can be identified as $F = \{f_i | i = 1, 2, \ldots, I\}$, where f_i represents the i-th attribute. Each respondent evaluates the functional and disfunctional attributes as $e_{ij} = (x_{ij}, y_{ij}, w_{ij})$, where x_{ij} is the evaluation of the jth respondent to the product without the i-th attribute or function, y_{ij} is the evaluation of the jth respondent to the product with the i-th attribute or function, w_{ij} is the importance of the j-th respondent to the i-th attribute. Next, for each f_i, the average level of satisfaction for the dysfunctional form question is defined as $\overline{X}_i$, and the average level of satisfaction for the functional form question is defined as $\overline{Y}_i$, i.e.,

$$\overline{X}_i = \frac{1}{J} \sum_{j=1}^{J} w_{ij} \cdot x_{ij}, \quad \overline{Y}_i = \frac{1}{J} \sum_{j=1}^{J} w_{ij} \cdot y_{ij} \tag{4}$$

The value pair $(\overline{X}_i, \overline{Y}_i)$ can be plotted in a two-dimensional diagram, where the horizontal axis indicates the dissatisfaction score and the vertical axis stands for the satisfaction score. The analytical Kano takes into account of the customer self-stated importance and the score of the importance falls into the interval $[0, 1]$. This method designs the scoring scale of satisfactions and dissatisfactions, which are asymmetric and view positive answers are stronger than the negative ones, as shown in Table 1.

3 Research Methodology

In this study, the Kano model is applied to analyze the factors affecting customer's purchasing decision and prioritize these customer requirements to improve the performance targeted. The steps to apply Kano model into analysis of BEV are as follows.

- **1) Collect Customer Needs.** There are many attributes affecting the development of BEV. It is necessary to distinguish key BEV attributes. The common method is to confirm potential customer requirements which constitutes questionnaire by a vast amount of literature collecting and summarizing.
- **2) Develop the Kano Questionnaire.** To construct the questionnaire, we formulate a pair of questions for each potential customer need for which you desire customer feedback. The first question is how a customer feels if the attribute/function is provided, and the other is that how the customer

feels if the attribute/function isn't provided. For each question, customer is required to select one from five answers: "I like it that way", "It must be that way", "I am neutral", "I can live with it that way" and "I dislike it that way". Crossing the answers of each pair of question, customer needs can be evaluated into different dimensions (M: must-be attribute, O: one-dimensional attribute, A: attractive attribute, I: indifferent attribute, R: reverse attribute, Q: questionable attribute).

- **3) Test the Questionnaire and Revise if Necessary.** When a questionnaire is to be sent to many customers, it is important that it be understandable. This is especially true of a Kano questionnaire, since it is unfamiliar to most people asked to fill it out. Therefore a test run will help us identify unclear wording, typographical errors, or confusing instructions. To do so, firstly try to predict the interviewees' response and guess the questions the customer may not understand; secondly, select some students to answer the questionnaire, analyze the results to check the problems which may exist in the it, revise the questions and retest; finally comprehend the interviewees' feedback and revise the questions if necessary.

- **4) Administer the Questionnaires to Customers.** Select customers to fill in the questionnaires. This research is about behavior of buying cars in Shanghai. So the customers working or settling in Shanghai are the ones who can afford the cars or be familiar with cars.

- **5) Process and Analyze the Results.** Based on the data collected, we will analyze the customer needs of BEV by means of the four methods reviewed in Section 2. Consequently, BEV attributes are divided into four categories, which will support the BEV firms and government to make appropriate decisions and policy suggestions to promote faster development of the BEV sector.

4 Case Study: The BEV in Shanghai

In this section, the methodology introduced in Section 3 will be used to analyze the customer needs of BEV in Shanghai. As one new product with immature technology, the attributes affecting the customers' purchasing decision of BEV are quite different from the traditional motor vehicles. Therefore, it is necessary to gain insight into the attributes with respect to the customer satisfaction and firm's capacity. By summarizing a large number of relevant literature and taking into account of the environment and policy issues in Shanghai, finally 20 attributes of BEV are identified, as shown in Table 2.

Our Kano questionnaire is composed of two parts: the first one is the demographic information of respondents, e.g., gender, education; the second one consists of 20 pairs of questions with respect to the 20 attributes in Table 2. Taking attribute f_2 "government subsidy" as an example, the questions are designed as shown in Fig. 3. The respondents who have cars or want to buy new cars in Shanghai were asked to provide their answers with respect to the 20 pairs of questions via two means: face-to-face survey and e-mail. Overall, 103 questionnaires

Table 2. BEV attributes

Attributes	Description of BEV attributes	Benefits provided to customers
f_1	Relatively low price	Beneficial
f_2	Government gives subsidy	Beneficial
f_3	Low fuel costs	Beneficial
f_4	Low maintenance costs	Beneficial
f_5	Free license plate	Beneficial
f_6	Sufficient charging station	Convenient
f_7	High level after-sale service	Convenient, cheerful
f_8	Pre-sale consulting service	Convenient, cheerful
f_9	Complete charging in 15 minutes	Fast, convenient
f_{10}	Battery cycle life is longer than 5 years	Beneficial
f_{11}	Maximum speed is over 120km/h	Fast
f_{12}	Driving range exceeds 120km	Convenient
f_{13}	A wide variety of BEV types	Cheerful
f_{14}	Attractive vehicle's appearance	Cheerful
f_{15}	Good reputation	Cheerful
f_{16}	Various brands	Cheerful
f_{17}	Operational convenience	Convenient, safe
f_{18}	Good acceleration	Fast, safe
f_{19}	Comfortable	Cheerful
f_{20}	Safe	Safe

If you will be given government subsidies when buying the BEV, how do you feel?	I like it that way. It must be that way. I am neutral. I can live with it that way. I dislike it that way.
If you won't be given government subsidies when buying the BEV, how do you feel?	I like it that way. It must be that way. I am neutral. I can live with it that way. I dislike it that way.

Fig. 3. Example of Kano questionnaire: Government subsidy.

were distributed to customers. Since some respondents gave apparently para-doxical answers, the questionnaires answered by them were regarded as invalid. Finally a total of 77 respondents' answers were viewed as reasonable, i.e. the effective response rate is 74.8%.

With the Kano evaluation data collected, we now use the four Kano methods to process and analyze the customer needs of BEV in Shanghai. With the principle of frequency-based and comparison-based category, the classification of the attributes can be obtained, as shown in Columns 2-3 of Table 3. Based on the index-based and analysis-based method, classifications of the BEV attributes are shown in Columns 4-5 of Table 3 and plotted in Fig. 4. These two methods

Table 3. Kano categorization of BEV attributes based on four methods.

Attributes	Frequency-based	Comparison-based	Index-based	Analysis-based	Category
f_1	I	A	A	O	A
f_2	A	A	A	O	A
f_3	A	A	A	O	A
f_4	O	O	O	O	O
f_5	O	O	O	O	O
f_6	M	M	M	O	M
f_7	M	M	M	O	M
f_8	I	M	I	O	I
f_9	A	A	A	O	A
f_{10}	M	M	M	O	M
f_{11}	M	M	M	O	M
f_{12}	M	M	M	O	M
f_{13}	I	I	I	A	I
f_{14}	I	A	A	A	A
f_{15}	A	A	A	O	A
f_{16}	A	A	A	A	A
f_{17}	M	M	M	O	M
f_{18}	I	M	I	O	I
f_{19}	M	M	M	O	M
f_{20}	M	M	M	O	M

are based on the satisfaction and dissatisfaction indexes to categorize the BEV attributes. It should be noted that in the analysis-based method, the importance values of BEV attributes are set to be 1 so as to make the four methods be compared reasonably with each other.

It is clearly seen from Table 3 that with the same data set, different results can be obtained by means of the four Kano methods. The classification results of the first three methods are very similar with each other. With the analysis-based method, the attributes are located close to each other in the coordinate axis (Fig. 4), in other words, the classification results by the analysis-based methods is not so good. Therefore, we only compare the first three methods. In addition, the "majority rule" is used to obtain the final categorizations of the BEV attributes, as shown in Column 6 of Table 3.

It is derived that the attributes "f_6: Sufficient charging station", "f_7: High level after-sale service", "f_{10}: Battery cycle life is longer than 5 years", "f_{11}: Maximum speed is over 120km/h", "f_{12}: Driving range exceeds 120km", "f_{17}: Operational convenience", "f_{19}: Comfortable", and "f_{20}: Safe" are categorized as must-be type. It means that without these functions, the customers will be very dissatisfied; however, these functions can not increase customer satisfaction. Taking the attribute "Sufficient charging station" as an example, as a new tech-

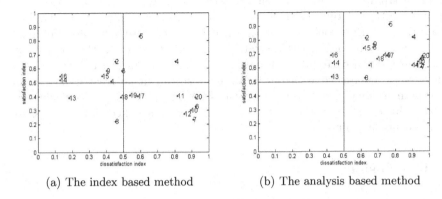

(a) The index based method   (b) The analysis based method

Fig. 4. Attributes classification of BEV

nology, the customers are worry about the charging stations of BEV. If there are fewer stations, the customers are very dissatisfied. In this sense, "sufficient charging station" is a prerequisite for the customers to use the BEV in practice.

The attributes "f_4: Low maintenance costs" and "f_5: Free license plate" are viewed as one-dimensional quality. It means that with these functions, the customers will be satisfied; and without these functions the customers will be dissatisfied. The Shanghai government has put forward "vehicle license auction policy" to limit the quantity of traditional motor vehicles. In April 2015, more than 150 thousands customers bid the license plate in Shanghai, and the auction price of one license plate is quite expense as high as the license plate in Shanghai can cost as high as 80, 600 RMB (about 13,000 US \$). Compared with the traditional motor vehicles, free license plate is a great advantage and will promote the adoption of BEV. In Shanghai, many customers choose to buy one BEV just because the license plate is free and bid is not needed. The license plate can make influence on customer's satisfaction and dissatisfaction. Therefore, the government should keep the policy of free license plate until the BEV technology has been greatly improved. Lower maintenance cost will increase customer satisfaction and higher maintenance cost will increase customer dissatisfaction.

The attributes "f_1: Relatively low price", "f_2: Government gives subsidy", "f_3: Low fuel costs", "f_9: Complete charging in 15 minutes", "f_{14}: Attractive vehicle's appearance", "f_{15}: Good reputation", and "f_{16}: Various brands" are classified as attractive quality. We can conclude that "high price" has little effect on customer dissatisfaction of BEV. Customers who tend to buy BEV already know the prices of BEV are more expensive than the one of traditional motor vehicles; due to the immature technology and lack of large-scale production, high price of the BEV means it doesn't create customer dissatisfaction. However, lower price will significantly increase customer satisfaction. As for the government subsidy, the Shanghai government now provides subsidy to customers who buy BEV. This policy has greatly promoted the adoption of BEV.

The attributes "f_8: Pre-sale consulting service", "f_{13}: A wide variety of BEV types" and "f_{18}: Good acceleration" are categorized as indifferent quality. This means that the firms don't need pay much attention to these functions at present.

In summary, according to the strategic rule $M > O > A > I$, the government and vehicle firms have to fulfill all the must-be requirements, otherwise customers will be very dissatisfied with the BEV. They should make great improvement of one-dimensional qualities to make the BEV competitive to the traditional motor vehicles. Finally, the customers will be very satisfied if the attractive requirements are fulfilled.

5   Conclusions

As one energy-saving and environment-friendly technology, there are many obstacles to the wide adoption of BEV in practice due to the immature technology. It is important to identify and classify the customer needs of BEV so as to help government and firms to pay attention to the attributes which affects customers' purchasing decision of BEV. In this study, customer need analysis for the BEV is investigated based on Kano model to promote the adoption of BEV in Shanghai. To do so, a total of 20 BEV's attributes are firstly determined by summarizing the previous researches and combining current advantages of policy. Secondly, it takes into the customer's psychology account, and combines the frequency-based, comparison-based, index-based and analysis-based methods to categorize the BEV attributes. The BEV attributes are classified as must-be quality, one-dimensional quality, attractive quality and indifferent quality. According to the strategic rule $M > O > A > I$, the priorities of efforts towards promoting the adoption of BEV are identified, i.e., the government and vehicle firms have to fulfill all the must-be requirements, otherwise customers will be very dissatisfied with the BEV. They should make great improvement of one-dimensional qualities to make the BEV competitive to the traditional motor vehicles. Finally, the customers will be very satisfied if the attractive requirements are fulfilled.

Acknowledgement. We would like to appreciate constructive comments and valuable suggestions from the three anonymous referees, which have helped us efficiently improve the quality of this paper. This study was partly supported by the National Natural Sciences Foundation of China (NSFC) under grant nos. 71471063 and 71125002; sponsored by the Innovation Program of Shanghai Municipal Education Commission under grant no. 14ZS060, the Shanghai Pujiang Program, and the Fundamental Research Funds for the Central Universities in China under grant no. WN1424004.

References

1. CO$_2$ emissions from fuel combustion highlights. Tech. rep., International Energy Agency, Paris (2013)
2. Berger, C., Blauth, R., Boger, D., Bolster, C., Burchill, G., DuMouchel, W., Pouliot, F., Richter, R., Rubinoff, A., Shen, D., Timko, M., Walden, D.: Kano's methods for understanding customer-defined quality. Center for Quality Management Journal **2**(4), 3–35 (1993)
3. Caperello, N.D., Kurani, K.S.: Households' stories of their encounters with a plug-in hybrid electric vehicle. Environment and Behavior **44**(4), 493–508 (2012)
4. Duvall, M., Knipping, E., Alexander, M.: Environmental assessment of plug-in hybrid electric vehicles. volume 1: Nationwide greenhouse gas emissions. Tech. rep., Electric Power Research Institute, Palo Alto, CA (2007)
5. Granovskii, M., Dincer, I., Rosen, M.A.: Economic and environmental comparison of conventional, hybrid, electric and hydrogen fuel cell vehicles. Journal of Power Sources **159**(2), 1186–1193 (2006)
6. Hidrue, M.K., Parsons, G.R., Kempton, W., Gardner, M.P.: Willingness to pay for electric vehicles and their attributes. Resource and Energy Economics **33**(3), 686–705 (2011)
7. Kano, N., Seraku, N., Takahashi, F., Tsuji, S.: Attractive quality and must-be quality. Journal of the Japanese Society for Quality Control **14**(2), 147–156 (1984)
8. Lee, M.C., Newcomb, J.: Applying the kano methodology to meet customer rquirements: Nasa's microgravity science program. Quality Management Journal **4**, 95–106 (1997)
9. Mau, P., Eyzaguirre, J., Jaccard, M., Collins-Dodd, C., Tiedemann, K.: The 'neighbor effect': Simulating dynamics in consumer preferences for new vehicle technologies. Ecological Economics **68**(1), 504–516 (2008)
10. McKay, A., de Pennington, A., Baxter, J.: Requirements management: a representation scheme for product specifications. Computer-Aided Design **33**(7), 511–520 (2001)
11. National Bureau of Statistics of China: China Statistical Yearbook 2013. China Statistics Press, Beijing (2013)
12. Park, S.Y., Kim, J.W., Lee, D.H.: Development of a market penetration forecasting model for hydrogen fuel cell vehicles considering infrastructure and cost reduction effects. Energy Policy **39**(6), 3307–3315 (2011)
13. Sierzchula, W., Bakker, S., Maat, K., van Wee, B.: The influence of financial incentives and other socio-economic factors on electric vehicle adoption. Energy Policy **68**, 183–194 (2014)
14. Skippon, S.M.: How consumer drivers construe vehicle performance: Implications for electric vehicles. Transportation Research Part F: Traffic Psychology and Behaviour **23**, 15–31 (2014)
15. Xu, Q., Jiao, R.J., Yang, X., Helander, M.: An analytical kano model for customer need analysis. Design Studies **30**(1), 87–110 (2009)

Co-Movement and Dependency Between New York Stock Exchange, London Stock Exchange, Tokyo Stock Exchange, Oil Price, and Gold Price

Pathairat Pastpipatkul[✉], Woraphon Yamaka, and Songsak Sriboonchitta

Faculty of Economics, Chiang Mai University, Chiang Mai, Thailand
ppthairat@hotmail.com, woraphon.econ@gmail.com

Abstract. This paper aims to analyze the co-movement and dependence of three stock markets, oil market, and gold market. These are gold prices as measured by gold future, crude oil prices as measured by Brent, and stock prices as measured by three developed stock markets comprising the U.S. Dow Jones Industrial Average, the London Stock Exchange, and the Japanese Nikkei 225 index. To capture the correlation and dependence, we employed the application of C-vine copula and D-vine copula. The results demonstrate that the C-vine copula is a structure more appropriate than the D-vine copula. In addition, we found positive dependency between the London Stock Exchange and the other markets; however, we also obtained complicated results when the London Stock Exchange, the Dow Jones Industrial Average, and Brent were given as the conditions. Finally, we found that gold might be a safe haven in this portfolios.

Keywords: C-D vine copula · London stock exchange · Tokyo stock exchange · Oil price · Gold price

1   Introduction

In the present day, stock markets, oil markets, and gold markets play an important role in world economy. These markets are of considerable interest to investors who need to diversify their portfolio. In mid-2014, the New York Stock Exchange (Dow Jones Industrial Average: DJIA), the London Stock Exchange (FTSE 100), and the Tokyo Stock Exchange (Nikkei) were the three stock exchange markets that handled the largest stock exchange in the world in both market capitalization and trade value. The total market capitalization of these markets amounted to more than $20,000 billion, whereas the total trade value was almost $40,000 billion. These figures indicate that these stock markets attract many investors. However, as mentioned earlier, investors need to allocate their investment; thus, the gold market and the oil market become lucrative alternative-investment choices for investors.

Oil markets and gold markets are markets popular to investors. They play an important role in the world economy because these are, respectively, the most traded raw material and precious metal. The study conducted by Miller and Ratti [7] revealed that oil, the main input in production, can influence the price of the production.

© Springer International Publishing Switzerland 2015
V.-N. Huynh et al. (Eds.): IUKM 2015, LNAI 9376, pp. 362–373, 2015.
DOI: 10.1007/978-3-319-25135-6_34

Additionally, a study carried out by Samanta and Zadeh [10] confirmed that the high volatility of oil price has serious impact on other macroeconomic variables including gold and stock price. In the case of the gold market, the role of the gold market is important to the investor, especially during periods of turbulence. The gold market is a safe haven for investment when the economy fares worse than expected [8]. Volatility in the stock or the oil market leads the investor face with higher risks. Moreover, global economic crises, especially the Euro zone and the United States crises, are other factors that cause higher volatility in global economy. Therefore, investors choose to reduce the risk involved by allocating their investment to the gold market.

Although investors could reduce their portfolio risk by allocating their investment to various other markets, the movements of these markets, sometimes, are highly volatile, and they effect considerable impact on each other. Therefore, understanding the co-movement and dependency between stock prices, oil prices, and gold prices has become important enough to be of interest for financial analysis and portfolio management, especially in periods of turbulence.

There are many methods that examine the relationship between two random variables; however, these methods cannot provide any information about the variation of variables across their distributions [8]. Thus, one way to analyze the co-movement and dependences between these three markets is to use the copula approach. In the last decade, copula modeling has become a crucial tool in financial economics. The copula theory was introduced in the form of Sklar's theorem and developed by Joe and Hu [5] who originally proposed the pair-copula construction (PCC). Then, Bedford and Cooke [3] extended PCC and proposed copula models, called regular vine copulas, which are flexible for use even in high dimensions and allow the mixing of several families in high dimension copulas. As a result, vine copulas have been mostly employed in financial time series. In recent times, it was Aas, Czado, Frigessi, and Bakken [1] who first applied vine copulas and introduced the canonical vine (C-Vine) and the drawable vine (D-vine) copulas, which provide a starting point for high dimensions and allow the employment of different dependency structures between the different pair copulas [12]. Therefore, in this study, we propose a C-vine and D-vine copula approach to estimate the co-movement of and the dependence between three stock markets, the oil market, and the gold market. These are gold prices as measured by gold future, crude oil price as measured by Brent, and stock prices as measured by the three developed stock markets comprising the U.S. Dow Jones Industrial Average, the London Stock Exchange, and the Japanese Nikkei 225 index. The vine copula has been used in many studies which include Ayusuk and Sriboonchitta [2], Puarattanaarunkorn and Sriboonchitta [9], Sriboonchitta, Liu, Kreinovich, and Nguyen [12,13], Liu, Sriboonchitta, Nguyen and Kreinovich [6], Sriboonchitta, Liu and Wiboonpongse [14], Tang, Sriboonchitta and Yuan [15], and Yuan, Sriboonchitta and Tang [16].

The remainder of this study is constructed as follows. In section 2, we describes the C-vine and the D-vine copulas, and the estimation strategy. Section 3 presents the data used. Section 4 provides the empirical results and discussion. In the last section, the conclusion is provided.

2   Methodology

2.1   Basic Concepts of Copula

According to Sklar's theorem [11], an n-dimensional copula $C(u_1,....,u_n)$ is a multivariate distribution function in $[0,1]^n$ whose marginal distribution (u) is uniform in the $[0,1]$ interval. In addition, Skalar [11] showed a link between multivariate distribution functions and their marginal distribution functions, and presented a basic concept for any joint distribution $H(x_1,...,x_n)$ with marginal distribution $F_1(x_1),...,F_n(x_n)$, as follows:

$$H(x_1,...,x_n) = C(F_1(x_1),...,F_n(x_n)) \tag{1}$$

The copula density C is obtained by differentiating Eq. (1); thus, we get

$$c(F_1,(x_1),...,F_n(x_n)) = \frac{h(F_1^{(-1)}(u_1), F_2^{(-1)}(u_2))}{\prod_{i=1}^{2} f_i(F_i^{(-1)}(u_i))} \tag{2}$$

where h is the density function associated to H; f_i is the density function of each marginal distribution; and C is the copula density

2.2   GARCH Models for Univariate Distributions

In this study, we propose a univariate ARMA(p,q)-GARCH(m,n) specification which is often chosen to model the marginal distribution of data. It can be described as

$$x_t = \phi_0 + \sum_{i=1}^{p} \phi_t x_{t-1} + \alpha_t - \sum_{j=1}^{q} \theta_j \alpha_{t-j} = u_t + \alpha_t \tag{3}$$

$$\alpha_t = h_t^{1/2} \varepsilon_t \tag{4}$$

$$h_t = \alpha_0 + \sum_{i=1}^{m} \alpha_i \alpha_{t-i}^2 + \sum_{j=1}^{n} \beta_j h_{t-j} \tag{5}$$

where u_t and h_t are, respectively, the conditional mean and variance equation, given past information. α_t is the residual term which consists of the standard variance, h_t, and the standardized residual, ε_t, which is assumed to have a Gaussuan distribution, a Student-t distribution, and a skewed-t distribution. Then, the best-fit marginal distribution provides a standardized ARMA(p,q)-GARCH(m,n) residual which is transformed into a uniform distribution in $[0,1]$. This step is the first step of the estimation procedure; thus, it is necessary to choose the best-fit ARMA(p,q)-GARCH(m,n) to obtain standardized residuals since they are joint distribution functions defined over $u_{i,t} = F_i(x_{i,t} | u_{i,t}, h_{i,t})$ and

$u_{i,t} = F_{i+1}(x_{i+1,t} | u_{i+1,t}, h_{i+1,t})$, $i = 1, \ldots n-1$, where $u_{i,t}$ $u_{i+1,t}$ are uniform in $(0,1)$. Finally, the cumulative distributions of the standardized residuals are plugged into the model comprising the vine copulas further.

2.3   C-Vine and D-Vine Copulas

In this study, we estimated the co-movement between three developed stock markets, the gold price, and the oil price based on C-vine and D-vine copulas, as introduced by Aas et al. [1]. In this section, we provides general form of C-vine and D-vine copulas, and proposed a starting point for the joint probability density function of d-dimension, as follows:

$$f(x_1, \ldots, x_d) = f(x_1) f(x_2 | x_1) f(x_3 | x_1, x_2) \ldots f(x_d | x_1, x_2, \ldots, x_{d-1}) \quad (6)$$

According to Sklar's theorem, as far as the examples of a three-dimensional random vector case are concerned, we know that

$$f(x_2 | x_1) = c_{12}(F_1(x_1), F_2(x_2)) f_2(x_2)$$

$$f(x_3 | x_1, x_2) = c_{23|1}(F_{2|1}(x_2 | x_1), F_{3|1}(x_3 | x_1)) \cdot c_{13}(F_1(x_1), F_3(x_3) \cdot f_3(x_3))$$

thus, we get

$$f(x_1, x_2, x_3) = c_{23|1}(F_{2|1}(x_2, x_1), F_{3,1}(x_3 | x_1)) \, c_{12}(F_1(x_1), F_2(x_2))$$
$$c_{13}(F_1(x_1), F_3(x_3)) f_1(x_1) f_2(x_2) f_3(x_3) \quad ,$$

where c_{12}, c_{13}, and $c_{23|1}$ are the pair copulas; $F_i(\cdot)$ is the cumulative distribution function (cdf) of x_i.

In the vine copula approach, there are several possible decompositions of the conditional distributions; therefore, we proposed the application of C-vine and D-vine copulas which are classes of the vine copula approach. As far as Bedford and Cooke [3] are concerned, they introduced a graphical model called vine and proposed two structure vines called C-vine tree and D-vine tree. In the C-vine tree, each tree has a unique node, the first root node, which connects with all the other nodes. Thus, the joint probability density function of d-dimension for C-vine can be form as

$$f(x_1, \ldots x_d) = \prod_{k=1}^{d} f_k(x_k) \prod_{j=1}^{d-1} \prod_{i=1}^{d-j} c_{j,j+i|1,\ldots,j-1}(F(x_j | x_1, \ldots x_{j-1}), F(x_{j+i} | x_1, \ldots, x_{j-1})) (7)$$

By contrast, in the D-vine tree, each tree is not connected to more than two nodes. Thus, the joint probability density function of d-dimension for D-vine can be form as

$$f(x_1,....x_d) = \prod_{k=1}^{d} f_k(x_k) \prod_{j=1}^{d-1} \prod_{i=1}^{d-j} c_{i,i+j|(i+1),...,(i+j-1)} (F(x_i|x_{i+1},....,x_{i+j-1}), F(x_{j+i}|x_{i+1},....,x_{i+j-1})) \ (8)$$

where $f_k(x_k)$ denotes the marginal density of x_k, $k=1,....,d$; $c_{j,j+j|1,...,j-1}$ and $c_{i,i+j|(i+1),...,(i+j-1)}$ are the bivariate copula densities of each pair copula in C-vine and D-vine, respectively. Lastly, the conditional distribution function for the d-dimensional vector V can be written as

$$F(x|v,\theta) = F(x|v) = \frac{\partial C_{xv_j|v_{-j}} (F(x|v_{-j}), F(v_j|v_{-j}))}{\partial F(v_j|v_{-j})}$$

where the vector v_j is an arbitrary component of vector V and the vector v_{-j} is vector V excluding v_j [11]. Further, $C_{xv_i|v_{-j}}$ is a bivariate conditional distribution of x and v_j is conditioned on v_{-j} with parameters (θ) specified in tree d.

The crucial question regarding the structures of C-vine and D-vine copulas is how to order the sequences of variables in the C-vine and the D-vine models. Czado, Schepsmeier, and Min [4] suggested that ordering may be given by choosing the strongest correlation in terms of absolute empirical values of pairwise Kendall's $\tau's$ as the first node. For the D-vine copula, the ordering may be given by the order that has the biggest value of the sum of empirical Kendall's $\tau's$.

3   Estimation

In the first step of the estimation procedure, the ARMA-GARCH filter is employed to obtain the univariate parameters and then the data are transformed using empirical cumulative distribution (ecdf). To test whether the transformed standardized residuals have uniform distribution in [0,1], we use the Kolmogorov–Smirnov test for goodness of fit. In addition, the Ljung–Box test is employed to test autocorrelation in the standardized residual.

Previously, we provided general expressions of C-vine and D-vine copulas for five variables comprising three developed stock markets, the oil price, and the gold price, and presented a starting point for the joint probability density function of five dimensions as $X = x_1, x_2, x_3, x_4, x_5$ with the marginal distribution function F_1, F_2, F_3, F_4, F_5. By recursive conditioning, we can write the joint density function as

$$f(x_1, x_2, x_3, x_4, x_5) = f(x_1)f(x_2|x_1)f(x_3|x_1,x_2)f(x_4|x_1,x_2,x_3)f(x_5|x_1,x_2,x_3,x_4) \ (9)$$

In the process of parameter estimation in the C-vine and the D-vine copulas, by following Czedo et al. [4], firstly, the sequential maximum likelihood estimation was conducted in order to obtain the initial value for the C-vine copula and the D-vine copula. Note that, in this study, the Gaussian copula, T copula, Clayton copula, Frank copula, Gumbel copula, Joe copula, BB1 copula, BB6 copula, BB7 copula, BB8 copula, and rotate copulas are bivariate copula families which are selected for each conditional pair of variables in the sequential estimation. Nevertheless, an appropriate bivariate copula family was chosen by taking into consideration the lowest Akaike information criteria (AIC) and Bayesian information criteria (BIC). Secondly, the maximum likelihood estimation is applied to estimate the final parameters of the pair copulas, by taking the initial value from the first step. Finally, the C-vine copula and the D-vine copula were compared in order to select the fitting vine copula structure, using the Clarke test and the Vuong test.

4   Empirical Results

In this study, we use the daily log returns of gold future (spot gold), Brent crude oil index (Brent), Dow Jones Industrial Average (DJIA), the London Stock Exchange (FTSE 100), and Nikkei 225 index (N225) over the period from January 3, 2009, to October 20, 2014, totaling 1774 observations.

Table 1. Summary Statistics

	Brent	DJIA	FTSE100	N225	Spot Gold
Mean	−2.59E-05	5.19E-05	−7.31E-06	−3.16E-06	9.81E-05
Median	7.67E-05	0.000135	0	0	0
Maximum	0.058649	0.045637	0.040756	0.057477	0.029712
Minimum	−0.04833	−0.03561	−0.04024	−0.0526	−0.04168
Std. Dev.	0.00849	0.005702	0.005786	0.007371	0.005665
Asymmetry	0.172085	−0.04088	−0.09187	−0.57206	−0.41045
Kurtosis	8.1534	12.7549	10.9659	11.2121	8.141602
Jarque–Bera	1971.833	7034.222	4692.899	5081.54	2003.876
Probability	0.000	0.000	0.000	0.000	0.000
Sum	−0.0459	0.092134	−0.01296	−0.00561	0.173959
Sum Sq. Dev.	0.127811	0.057638	0.059346	0.096329	0.056902

Source: Calculation.

Table 1 provides the descriptive statistics of our data. We can see that all the variables have a kurtosis above 8, and that their asymmetry coefficients are almost negative except in the case of Brent. This means that the marginal distribution of our data has a heavy tail to the left rather than to the right. Furthermore, the normality of these marginal distributions is strongly rejected by the Jarque–Bera test, prob.=0. Thus, these findings indicate that normal distribution might not be appropriate for our data. Consequently, we assume that the skewed Student-t distribution and the skewed GED are appropriate marginal distributions for our data.

Table 2. Results of Marginal Distribution: ARMA(2,2)-GARCH(1,1)

	Brent	DJIA	FTSE100	N225	Spot Gold
Mean Equation					
ω_i	0.0001	0.0002***	0.0001**	0.0001	0.0001***
AR(2)	−0.760***	0.531**	−0.796***	−0.806***	0.708***
MA(2)	0.780***	−0.529**	0.795***	0.814***	−0.713***
Variance Equation					
β_i	0.0001	0.0001***	0.0001***	0.0001	0.0001***
v_i	0.039***	0.116***	0.093***	0.102***	0.038***
h_i	0.958***	0.876***	0.902***	0.020***	0.953***
KS test (prob.)	0.991	0.993	0.993	0.993	0.993
$Q^2(10)$ (prob.)	0.368	0.264	0.118	0.301	0.968
AIC (skewed Student-t)	−7.112	−8.138	−7.968	−7.347	−7.787
AIC (skewed GED)	−7.119	−8.157	−7.952	−7.362	−7.800

Note: "*," "**," and "***" denote rejections of the null hypothesis at the 10%, 5%, and 1% significance levels, respectively.
Source: Calculation.

Table 2 shows the results of the marginal distribution of the skewed Student-t distribution and the skewed generalized error distribution (GED) with the ARMA(2,2)-GARCH(1,1) model for all the variables. We choose the best specifications for the marginal base on the AIC. The result of the AIC test show that ARMA(2,2)-GARCH(1,1) with the skewed GED is the best specification for the gold future, Brent crude oil index, Dow Jones Industrial Average, and Nikkei 225 index, and that ARMA(2,2)-GARCH(1,1) with the skewed Student-t distribution is the best specification for the London Stock Exchange.

In addition, the Kolmogorov–Smirnov (KS) test is used as the uniform test for the transformed marginal distribution functions of these residuals. The result shows that none of the KS test accepts the null hypothesis. Therefore, it is evident that all the marginal distributions are uniform on $[0,1]$. Moreover, the Ljung–Box test, which is used as the autocorrelation test on standardized residuals, confirms that there is no rejection of the null hypothesis, i.e. no autocorrelation in any of the series.

Table 3. Sum of Empirical Kendall's $\tau's$

	DJIA	Brent	N225	Spot Gold	FTSE100
DJIA	1	0.162936	0.071576	0.02072	0.414646
Brent	0.162936	1	0.096928	0.195815	0.272034
N225	0.071576	0.096928	1	0.01741	0.158285
Spot Gold	0.02072	0.195815	0.01741	1	0.074468
FTSE100	0.414646	0.272034	0.158285	0.074468	1
Sum	1.669877	1.727713	1.344198	1.308413	1.919433

Source: Calculation.

Table 4. Results of C-vine Copulas

Copula family	Parameter	Lower and upper tail dependence	AIC	BIC
$C_{1,2}$ BB1	0.4036***	0.2986,0.3711	−859.172	−848.214
$C_{1,3}$ Student-t	0.4119***	0.126495	−366.372	−355.413
$C_{1,4}$ Clayton	0.3246***	0.118248,0	−123.872	−118.392
$C_{1,5}$ Student-t	0.1220***	0.051111	−58.5388	−47.5802
$C_{2,3|1}$ Student-t	−0.0055	0.001087	-3.391	7.567093
$C_{2,4|1}$ BB8 (270 degrees)	−1.1288	0	−8.70948	2.249118
$C_{2,5|1}$ Student-t	−0.0503	0.0028	−11.7295	−0.7709
$C_{3,4|12}$ Frank	0.3536**	0	−4.1207	1.358578
$C_{3,5|1,2}$ BB8 (180 degrees)	1.8977***	0	−138.096	−127.137
$C_{4,5|123}$ Clayton (270 degrees)	−0.0412	0	−0.76	4.719291
Log-likelihood	804.5391		−1574.76	−1481.61
D-Vine copula				
Log-likelihood	788.6617		5078.634	5160.826

Note: "*," "**," and "***" denote rejections of the null hypothesis at the 10%, 5%, and 1% significance levels, respectively.

Source: Calculation.

Table 3 provides the result of the sum of empirical Kendall's $\tau's$ and shows that FTSE100 has the strongest dependency in terms of the empirical value of pairwise Kendall's tau; thus, we determine FTSE100 as the first root node, and the order should be the following: FTSE100 (order 1), DJIA (order 2), Brent (order 3), N225 (order 4), spot gold (order 5) for C-vine. For D-vine, we find that the following order of spot gold, Brent, FTSE100, DJIA, N225 is the biggest value of the sum of Kendall's tau. We can see that C-vine and D-vine have different structures of pair copulas; therefore, it is reasonable to calculate both C-vine and D-vine in order to find the best-fit structure for this analysis.

According Table 4, it is evident that the C-vine structure for these five markets is more appropriate than the D-Vine one because the sum values of AIC and BIC are the smallest for D-vine. In addition, the Vuong test and the Clarke test, the results of which are presented in Table 5 provide the small p-values, confirming that the C-vine and the D-vine copula models for these five market data sets could be statistically distinguished. Thus, in this study, we choose C-vine to analyze the co-movement and dependency between these five markets.

As for the dependence result of C-vine copula, which is given in Table 4, It can be observed that BB1, Student-t, rotated BB8 270 degrees, Frank, BB8 180 degrees, and Clayton 270 degrees are the best pair-copula families in terms of the lowest AIC and BIC values for C-vine. We find that there is significant co-movement and tail dependence in some market pairs. Among the five market pairs, it is seen that the pair FTSE100 and DJIA accounts for the largest, including its upper tail (0.2986) and lower tail (0.3711), followed by FTSE100 and Brent (lower tail and upper tail, 0.1264), FTSE100 and N225 (lower tail, 0.1182), and FTSE100 and spot gold (lower tail and upper tail, 0.0511).

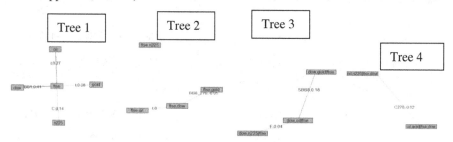

Fig. 1. The C-vine tree plot.

Table 5. Results of Comparison between Two Non-nested Parametric Models

Test	statistic	Statistic Akaike	Statistic Schwarz	p-value	p-value Akaike	p-value Schwarz
Vuong Test	1.8395	1.6078	0.9730	0.0658	0.1078	0.3305
Clarke Test	975	968	953	0.0000	0.0000	0.0014

Source: Calculation.

Upon observing Figure 1 it can be seen that the tree plot shows us the four layers of the C-vine copula model, comprising tree 1, tree 2, tree 3, and tree 4. Each tree presents a combination of the dependency of the market pairs of these five markets, and their Kendall's Tau, as follows:

The first tree shows that the Kendall's tau of $C_{1,2}$, $C_{1,3}$, $C_{1,4}$, and $C_{1,5}$ are 0.41, 0.27, 0.14, and 0.08. The second tree demonstrates three conditional copula models, which include the Kendall's tau of $C_{2,3|1}$, $C_{2,4|1}$, $C_{2,5|1}$. We found dependence between DJIA and spot gold conditional on FTSE100, which presented the Kendall's tau as −0.05. This means that if the FTSE100 is given as the condition of DJIA and spot gold dependence, the Kendall's tau becomes negitive. For the pair DJIA and Brent and the pair DJIA and N225, if the FTSE100 is given as the condition, the Kendall's tau becomes independent. The third tree presents the two conditional copulas, which include $C_{3,4|12}$ and $C_{3,5|1,2}$. So, if FTSE100 and DJIA are given as the condition, the Kendall's tau for Brent and N225 falls approximately by 5.69%, while the Kendall's tau for Brent and spot gold increases by 1.581%. Finally, the fourth tree presents the three conditional copulas for $C_{4,5|123}$; thus, if the FTSE100, DJIA, and Brent are given as the condition, the Kendall's tau becomes negative.

To summarize the above result, there exists positive interdependency between gold future, Brent crude oil index, Dow Jones Industrial Average, the London Stock Exchange, and Nikkei 225 index, according to Kendall's tau, however, some complicated relationships arise between these five markets if there are given conditions by other markets. The London Stock Exchange, Dow Jones Industrial Average, and Brent have effective impact on the dependence of market pairs; hence, investors in each market should closely monitor these markets, especially the London Stock Exchange. Moreover, there are positive and significant average dependence between these five markets. However, this study also found a negative, dependency between gold market and two markets, namely Dow Jones Industrial Average and Nikkei 225 markets if the FTSE100, DJIA, and Brent are given as the condition. This indicates that gold can act as safe haven against extreme those price movements.

Table 6. Portfolios Value at Risk based C-Vine

Portfolio		Risk Value		
		1%	5%	10%
Portfolios	VaR	−2.68%	−1.57%	−1.12%
	ES	−3.46%	−2.28%	−1.80%
Portfolios	VaR	−3.22%	−1.82%	−1.26%
without Gold	ES	−4.42%	−2.75%	−2.13%

*Source:*Calculation.

According to our finding that gold serve as a safe haven for Dow Jones Industrial Average and Nikkei 225 markets if the FTSE100, DJIA, and Brent are given as the condition. To confirm our finding, we, then, computed the Value at Risk (VaR) and Expected Shortfall (ES) to measure the risk of two portfolios namely; Brent, DJIA, FTSE 100, N225 and gold portfolios (BDFNG-Portfolios) and Brent, DJIA, FTSE 100, N225 portfolios (BDFN-Portfolios). Following Sriboonchitta, Liu, Kreinovich, and Nguyen[13], we make use of the Monte Carlo simulation by simulated 10,000 jointly-dependent uniform variates from the estimated C-vine copulas.to calculate the VaR and ES of equally weighted portfolio since C-vine copulas since this structure provide an appropriate than D-vine (lowest AIC and BIC). Table 6, reveal the calculated average of 1%, 5%, and 10% VaR and ES of two portfolios. We can observe the average loss between the BDFNG-Portfolios and the BDFN-Portfolios and find that gold could decrease the average loss in the BDFNG-Portfolios by around 0.14–0.96%. Therefore, we can conclude that gold might be a safe haven in periods of markets stress.

5   Conclusion

This paper employed the C-vine and D-vine copula approach to estimate the co-movement and dependence between gold future, Brent crude oil, the U.S. Dow Jones Industrial Average, the London Stock Exchange, and the Japanese Nikkei 225 index. The empirical results confirm that the C-vine structure is more appropriate than the D-vine structure. As for the estimated C-vine results, we find that bivariate C-vine copulas tend to have both a symmetric structure and a hierarchal structure. Additionally, it is found that there exists a positive correlation between the London Stock Exchange and the other markets; however, it is observed that if the London Stock Exchange, the Dow Jones Industrial Average, and Brent are given as the condition for the other market pairs, it will have an effect on the dependency of those market pairs. In addition, the VaR and ES confirmed that gold might be a safe haven in this portfolios.

As far as further studies are concerned, the paper can expand the scope of analysis of the co-movement of other countries with factors such as bond yield and exchange rates, as well as try to extend the vine copula approach to the Markov-switching model with time-varying parameters in order to provide a new insight into the dependence dynamics of regimes with high dimension.

Acknowledgement. We are grateful for financial support from Puay Ungpakorn Centre of Excellence in Econometrics, Faculty of Economics, Chiang Mai University. We thanks Nguyen, H. T. and Kreinovich, V. for valuable comment to improve this paper.

References

1. Aas, K., Czado, C., Frigessi, A., Bakken, H.: Pair-copula constructions of multiple dependence. Insurance: Mathematics and Economics **44**(2), 182–198 (2009)
2. Ayusuk, A., Sriboonchitta, S.: Risk Analysis in Asian Emerging Markets using Canonical Vine Copula and Extreme Value Theory. Thai Journal of Mathematics, 59–72 (2014)
3. Bedford, T., Cooke, R.M.: Vines: A new graphical model for dependent random variables. Annals of Statistics, 1031–1068 (2002)
4. Czado, C., Schepsmeier, U., Min, A.: Maximum likelihood estimation of mixed C-vines with application to exchange rates. Statistical Modelling **12**(3), 229–255 (2012)
5. Joe, H., Hu, T.: Multivariate distributions from mixtures of max-infinitely divisible distributions. Journal of Multivariate Analysis **57**(2), 240–265 (1996)
6. Liu, J., Sriboonchitta, S., Nguyen, H.T., Kreinovich, V.: Studying volatility and dependency of chinese outbound tourism demand in Singapore, Malaysia, and Thailand: a vine copula approach. In: Huynh, V.-N., Kreinovich, V., Sriboonchitta, S. (eds.) Modeling Dependence in Econometrics. AISC, vol. 251, pp. 295–311. Springer, Heidelberg (2014)
7. Miller, J.I., Ratti, R.A.: Crude oil and stock markets: Stability, instability, and bubbles. Energy Economics **31**(4), 559–568 (2009)
8. Najafabadi, A.T.P., Qazvini, M., Ofoghi, R.: The Impact of Oil and Gold Prices' Shock on Tehran Stock Exchange: A Copula Approach. Iranian Journal of Economic Studies **1**(2), 23–47 (2012)

9. Puarattanaarunkorn, O., Sriboonchitta, S.: Modeling dependency in tourist arrivals to Thailand from China, Korea, and Japan using vine copulas. In: Huynh, V.-N., Kreinovich, V., Sriboonchitta, S. (eds.) Modeling Dependence in Econometrics. AISC, vol. 251, pp. 433–448. Springer, Heidelberg (2014)
10. Samanta, S.K., Zadeh, A.H.: Co-Movements of Oil, Gold, the US Dollar, and Stocks. Modern Economy **3**, 111 (2012)
11. Sklar, M.: Fonctions de repartition 'an dimensions et leursmarges. Publ. Inst. Statist. Univ. Paris **8**, 229–231 (1959)
12. Sriboonchitta, S., Liu, J., Kreinovich, V., Nguyen, H.T.: Vine copulas as a way to describe and analyze multi-variate dependence in econometrics: computational motivation and comparison with bayesian networks and fuzzy approaches. In: Huynh, V.-N., Kreinovich, V., Sriboonchitta, S. (eds.) Modeling Dependence in Econometrics. AISC, vol. 251, pp. 205–219. Springer, Heidelberg (2014)
13. Sriboonchitta, S., Liu, J., Kreinovich, V., Nguyen, H.T.: A vine copula approach for analyzing financial risk and co-movement of the indonesian, Philippine and Thailand stock markets. In: Huynh, V.-N., Kreinovich, V., Sriboonchitta, S. (eds.) Modeling Dependence in Econometrics. AISC, vol. 251, pp. 281–294. Springer, Heidelberg (2014)
14. Sriboonchitta, S., Liu, J., Wiboonpongse, A.: Vine copula-cross entropy evaluation of dependence structure and financial risk in agricultural commodity index returns. In: Huynh, V.-N., Kreinovich, V., Sriboonchitta, S. (eds.) Modeling Dependence in Econometrics. AISC, vol. 251, pp. 313–326. Springer, Heidelberg (2014)
15. Tang, J., Sriboonchitta, S., Yuan, X.: A Mixture of Canonical Vine Copula GARCH Approach for Modeling Dependence of European Electricity Markets. Thai Journal of Mathematics, 165–180 (2014)
16. Yuan, X., Sriboonchitta, S., Tang, J.: Analysis of International Trade, Exchange Rate and Crude Oil Price on Economic Development of Yunnan Province: A GARCH-Vine Copula Model Approach. Thai Journal of Mathematics, 145–163 (2014)

Spillovers of Quantitative Easing on Financial Markets of Thailand, Indonesia, and the Philippines

Pathairat Pastpipatkul[1(✉)], Woraphon Yamaka[1], Aree Wiboonpongse[2,3], and Songsak Sriboonchitta[1]

[1] Faculty of Economics, Chiang Mai University, Chiang Mai, Thailand
ppthairat@hotmail.com, woraphon.econ@gmail.com
[2] Faculty of Economics, Prince of Songkla University, Hat Yai, Thailand
[3] Department of Agricultural Economics and Institute for Sufficiency Economy Research and Promotion, Faculty of Agriculture, Prince of Songkla University, Hat Yai, Thailand

Abstract. This paper provides the results of the effectiveness of the quantitative easing (QE) policy, including purchasing mortgage-backed securities, treasury securities, and other assets in the United States, on the financial markets of Thailand, Indonesia, and the Philippines (TIP) in the post-QE introduction period. In this study, we focused on three different financial markets, which include the exchange rate market, stock market, and bond market. We employed a Bayesian Markov-switching VAR model to study the transmission mechanisms of QE shocks between periods of expansion in the QE policy and turmoil with extraordinarily negative events in the financial markets and the global economy. We found that QE may have a direct substantial effect on the TIP financial markets. Therefore, if the Federal Reserve withdraws the QE policy, the move might also have an effect on the TIP financial market. In particular, purchasing the mortgage-backed securities (MBS) program is more likely to affect the TIP financial markets than purchasing the other programs.

Keywords: Markov switching · Bayesian · QE · TIP financial market

1   Introduction

After quantitative easing (QA) was announced, the amount of private security holds decreased while the purchases increased the price of the asset, thus lowering its yield. The expected return on the security had to fall. Therefore, the private investors, both resident and non-resident, were willing to adjust their portfolio. Moreover, the U.S. public debt had risen from 70% of GDP to 100% of GDP since QE was first announced. This can affect portfolio decisions and asset price by altering the confidence of investors who understand that the economic conditions are worse than expected, and this makes them lean toward safety as a precaution to any heavy financial loss. These effects would certainly spill over to other assets that are similar in attribute, and the fact is that investors are willing to substitute one asset for the other [12]. As a result, these effects contributed tremendously to the sharp rebound of capital inflow to the rest of the world, especially to the Asia-Pacific region, because the investors

© Springer International Publishing Switzerland 2015
V.-N. Huynh et al. (Eds.): IUKM 2015, LNAI 9376, pp. 374–388, 2015.
DOI: 10.1007/978-3-319-25135-6_35

wanted to reduce tail risks and avoid the lowering bond yield rates in the U.S., which had redirected capital inflow toward Asia [13].

There are several channels of QE operations which could have an effect on the economy. Joyce, Lasaosa, Stevens, and Tong [8] suggested three main channels which might influence the economy. The first is the signaling channel (policy news), which comes from the announcement of QE, information economic indicator, and the reaction of FOMC. Krishnamurthy and Jorgensen [12] found that the signaling channel could lower the yield of intermediate and long-term bonds. Bauer and Rudebusch [1] found that signaling the channel also lowers rates in all fixed income markets because all interest rates depend on the expected future path of policy rates. The second is the rebalancing channel, which comes from the purchasing of long-term securities so that the bond yield could be lowered. However, we found that there was no evidence of decline in longer-term rates as the term premiums come from the portfolio balance channel. In contrast, Cho and Rhee [3] isolated a portfolio rebalancing channel of QE and found that QE has effects in terms of lowering the bond yield, and raising output and inflation. The third is the liquidity channel, which results from the purchasing of the securities program of central bank so that it reduces the premium on most liquid bonds as well as the interest rate. Herrenbrueck [7] found that QE can reduce yield across the board and simulate investment because when the central bank employs a monetary expansion, the domestic interest rate does decrease, thereby attracting investment from domestic investors.

At present, however, the Fed has begun tapering the QE by winding down its bond purchasing at $20,000 million per month since November 2013. Also, they aim to end the program if the economic condition is expected to make a recovery. So this QE tapering will have an impact on the TIP currencies, bond markets, and stock markets. Although the Fed has not yet ended the QE program, its actions have already affected the TIP's stock, bond, and currency markets and have resulted in capital outflows from the markets. The impact on the TIP markets is likely to be very severe. Sharp fall in the TIP currencies, bond trade volume, and stock index has been found as occurring since the Fed announced it in May 2013. By January 3, 2014, the Thai, Indonesian, and Philippine currencies had depreciated by 12%, 25%, and 8%, respectively. In the stock markets, SET, JKSE, and PSEi indexes were found to have fallen by 19%, 12%, and 13%, respectively. In the bond markets, 10-year government bond yields in Thailand were observed to have shot up to 62 basis points (bps) since May 1, 2013; the corresponding rise in Indonesia and the Philippines were about 300 bps and 170 bps, respectively.

In this paper, we particularly focus on how QE affects the performance of TIP financial markets. Therefore, we would like to examine the QE effect on the TIP financial markets. Although this recent research area provides a better theoretical understanding, there is a scarcity in the amount of empirical research attempting to analyze how financial markets react to QE shock over time. Moreover, most of the empirical research in this area employs linear models such as linear structural vector autoregressions (SVARs) and linear regression, which might not be compatible for these data that have extreme volatility and have not corresponded to the real economy with regime shifts. Thus, this paper aims to employ the Markov-switching vector autoregressive

(MS-VAR) models as an innovative tool. The work asserts that the methods are more appropriate to the financial time series in order to capture the discrete changes in the financial time series data, for dating the effect of volatility in each period as well as identifying the factors that lead to TIP financial markets switching from one state to another, and also signal ahead a turbulent regime as an early warning system. In addition, the Bayesian estimator has been adopted in order to deal with the over-parameterization problem in the MS-VAR model which is estimated by maximum likelihood estimation (MLE) [16]. In general, the model which is estimated by the maximum likelihood estimation has complicated estimation and bias. MLE has been found to be employed in many econometric fields, such as the studies of Curdia, Ferrero, and Chen [5], Sriboonchita, Wong, Dhompongsa, and Nguyen [22], Do, Mcaleer and Sriboonchitta [6], Chinnakum, Sriboonchitta, and Pastpipatkul [4], Liu and Sriboonchitta [13], Jintranun, Calkins and Sriboonchitta [9], Kiatmanaroch and Sriboonchitta [10,11], and Praprom and Sriboonchitta [15]. These papers may not provide accurate results; thus, the purpose of this study is to employ a Bayesian estimator which is more informative, flexible, and efficient than MLE. Brandt and Freeman [2] suggested the use of Bayesian prior on the coefficient matrix of the model to reduce the estimation uncertainty and to obtain accurately the inference and the forecast.

The next section briefly summarizes some of the literature concerning the QE effects. In section 3, we outline the methods and procedures for estimating MS-BVARs. The results are reported in section 4. The economic implications are presented in section 5. Section 6 provides the conclusion of the work.

2   Methods and Procedures

2.1   Markov-Switching Bayesian VAR

The model is estimated by using a block EM algorithm where the blocks are Bayesian vector autoregressive (BVAR) regression coefficients for each regime (separating for intercepts, AR coefficients, and error covariance) and the transition matrix [14].

Sims, Waggoner, and Zha [16] provided the estimation MS-BVAR models of lag length k, as follows:

$$Y_t^{'} A_0(s_t) = \sum_{j=1}^{k} Y_{t-j}^{'} A_j(s_t) + \alpha_j C(s_t) + \varepsilon_t^{'} \Gamma^{-1}(s_t), \quad t = 1, ..., T, \quad (1)$$

where $Y_t^{'}$ is n-dimensional column vector of endogenous variables; A_0 n×n non singular matrix; s_t h dimension vector of regimes ; h the finite set of integers H; A_j is n×n matrix; C is vector of intercept terms; ε_t is the vector of n unobserved shocks; Γ is n×n diagonal matrix of variance of ε_t ; and s_t is state or regime.

Sims, Waggoner, and Zha [16] provide a distributional assumption with densities of the MS-BVAR disturbances, as follows:

$$P(\varepsilon_t | Z_{t-1}, S_t, \omega, \Theta) = N(\varepsilon_t | 0_{n \times 1}, I_n) \tag{2}$$

where Z_{t-1} is the lag coefficient

$$P(Y_t | Y_{t-1}, S_t, \omega, \Theta) = N(Y_t | u_z(s_t), \Sigma_z(s_t)), \tag{3}$$

where $Y_t = [Y_1', Y_2', \dots Y_t']$, $S_t = [s_0', s_1' \dots s_t']$, $u_z(s_t)$ is an error term of each equation, and ω is the vector of probabilities which is estimated by the Markov chain

$$\Theta = [A_0(1)A_0(2)\dots A_0(h), A(1)A(2)\dots A(h), \alpha_j]' \tag{4}$$

$$\Sigma_z(\cdot) = [\Gamma(\cdot)]^{-1} \tag{5}$$

$$Q = \begin{bmatrix} \rho_{11} & \cdots & \cdots & \rho_{1h} \\ \vdots & \rho_{22} & \cdots & \vdots \\ \vdots & \vdots & \ddots & \vdots \\ \rho_{n1} & \cdots & \cdots & \rho_{nh} \end{bmatrix} \tag{6}$$

The matrix Q given in Eq. (6) contains a set of transition probability dynamics which has been estimated by the MS-BVAR model, prior, and data. Suppose $h = 3$; the set of transition probability events would occur in the early, middle, and later parts of the sample.

The likelihood of the MS-BVAR model is built by Y_t and the assumptions Eq. (2), Eq. (3), Eq. (4), and Eq. (5).

$$\ln P(Y_t | \alpha_t, \omega, \Theta) = \sum_{t=1}^{T} \ln[\sum_{s_t \in H} P(Y_t | \alpha_t, \omega, \Theta) P(S_t | \alpha_t, \omega, \Theta)], \tag{7}$$

where $P(Y_t | \alpha_t, \omega, \Theta)$ is the density used to sample the probability that S_t is in regime l, given $S_{t-1} = j$. Sims, Waggoner, and Zha [16] also proposed Gibb sampling methods to construct the log-likelihood along with the conditional densities of $\Theta, P(\Theta | \alpha_{t-1}, S_t, \omega)$, and $\omega, \omega, P(\Theta | \alpha_{t-1}, S_t, \Theta)$, where the vector of regimes, S_T, is integrated out of the log-likelihood.

The estimation of the MS-BVAR model Eq. (1) depends on the joint posterior distribution of Θ and ω. This posterior is calculated in the MS-BVAR model Eq. (1) using Bayes' rule, which gives

$$P(\omega, \Theta | Y_T, \alpha_T, \omega, \Theta) \propto P(Y_t | \alpha_t, \omega, \Theta) P(\omega, \Theta) \tag{8}$$

where $P(\omega, \Theta)$ denotes the prior of ω and Θ.

2.2   Data, Prior, and Procedures

The raw monthly data, Fed's total asset (TA), U.S.'s mortgage bank securities (MBS), U.S.'s Treasury securities (TS), Stock Exchange of Thailand index (SET), Jakarta composite index (JKSE), the Philippine Stock Exchange composite index (PSEi), Singapore Stock Exchange (SGX), THB/USD (Exth), IDR/USD (Exind), SGD/USD (Exsin), PHP/USD (Exphp), government bond yield (THY), Indonesia government bond yield (INDY), and the Philippine government bond yield (PHY) were collected from Thomson Reuters DataStream; Financial Investment Center (FIC), Faculty of Economics, Chiang Mai University; and www.federalreserve.gov for the period from December 18, 2002, to March 19, 2014. However, all of these observations have been transformed to the first difference form (R Y_t).

Before estimating the parameters, we separate the parameters into three groups namely Thai (TH) group consisting of RTA, RTS, RMBS, RSET, RExth, and RTHY, Indonesia (IND) group consisting of RTA, RTS, RMBS, RJKSE, RExind, and RIN-DYand Philippines (PH) group consisting of RTA, RTS, RMBS, RPsei, RExphp, and RPHY.

We, then, use the following techniques to estimate the relationship between the parameters in each model using the MS-BVAR method, as follows:

1) The determination of the lag length (k) in the MS-BVAR model is carried out using AIC and BIC. In comparing models, the model with the lowest value of AIC is preferred.

2) The estimation of MS-BVAR(p), purposed by Nason and Tallman [14], using the multi-step procedure is as follows:

3.1 Setting the random walk, smoothness, and duration prior to the MS-BVAR(p).

Sims, Waggoner, and Zha [16] and Sims and Zha [17, 18] suggested the prior hyper-parameters that we believe are about the conditional mean of the coefficients and the lagged effects. In this study, we propose normal-Wishart prior, normal-flat prior, and flat-flat prior as the priors in model.

3.2 Estimate the initial parameters for MS-VAR model using the maximum likelihood estimator. Then the obtained parameters, $A = A_0, A_1, ..., A_h$, $\alpha = \alpha_{j,h}$, and $\Gamma = \Gamma_0, \Gamma_1, ..., \Gamma_h$, become three arbitrary starting values of $\tilde{A}$, $\tilde{\alpha}$, and $\tilde{\Gamma}$ for sampling the data. To estimate each parameter, the block optimization has been employed. In addition, to sample the initial values, Gibbs sampler is preferred and the procedure is as follows: 1) Draw $\tilde{A}_1$ from $f_1(A_1 | A_{2,0}, A_{3,0}, Y_t)$; 2) Draw $\tilde{\Gamma}_1$ from $f_2(\Gamma_2 | \Gamma_{3,0}, \Gamma_{1,1}, Y_t)$, and 3) Draw $\tilde{\alpha}_1$ from $f_3(\alpha_3 | \alpha_{1,1}, \alpha_{2,1}, Y_t)$.

This completes a Gibb burn-in by 2,000 times, and these parameters become $\tilde{A}_1$, $\tilde{\Gamma}_1$, and $\tilde{\alpha}_1$.Thereafter, these new parameters are used as starting values in order to repeat the prior iteration 10,000 times to obtain a final of random draws.

The obtained sample parameters, in each block, are summed up and divided by the number of samples in order to get the mean of the parameters. Additionally, we aim to detect the number of regimes by estimating and comparing multiple change point models. We estimate two models of change point case, including the model with one change point and the model with two change points.

3) Rerun the MS-BVAR(p) models that achieve the best fit to the data with the highest value of log marginal likelihood to produce the transition probabilities (Q).

4) The estimated MS-BVAR(p) model produces probabilities of the regime $j, j = 1, ..., 3$ on date t in order to separate the data of the different regimes.

6) Construction of the impulse response.

3 Empirical Results

3.1 Model Fit

In this section, we compare the various types of models with the following specification. For each model, the normal-Wishart prior, normal-flat prior, and flat-flat prior estimations have been conducted. Additionally, we compared the number of change points, as well. We learned that among the trial runs of several alternative prior specifications for each model, the results provide evidence that the normal-Wishart prior, which has the highest MDDS, has the best fit among all the three groups of models. This prior is common to all countries and appropriates for these sample periods. However, there are different numbers of change points in the different models. The result confirms that the data of the TH and the PH groups support one change point, while the data of the IND group support two change points.

Table 1. Estimates of transition matrices

Transition Matrices							
Q: TH Group		**Q: IND Group**			**Q:PH Group**		
0.920	0.090	0.9227	0.0387	0.0385	0.9482	0.0517	
0.080	0.910	0.0386	0.9221	0.0391	0.0528	0.9477	
		0.0387	0.0392	0.9224			
Duration				**Duration**		**Duration**	
Regime 1	12.5	Regime 1	12.9300			Regime 1	18.93
Regime 2	11.11	Regime 2	2.83			Regime 2	19.12
		Regime 3	12.8800				

Source: Calculation

3.2 Estimation of MS-BVAR(1) TH group

Regime 1 (St=1)

$$
\begin{bmatrix} RTA \\ RTS \\ RMBS \\ REXTH \\ RSET \\ RTHY \end{bmatrix} = \begin{bmatrix} 7.31^{***} \\ 1.213 \\ -0.411^{***} \\ -0.223 \\ 2.756 \\ -0.738 \end{bmatrix} + \begin{bmatrix} 0.068 & -0.11 & -0.307 & 3.37 & 0.127 & -0.135^{***} \\ 0.745 & -0.031 & -0.131 & -1.055 & -0.049 & -0.008 \\ -0.128 & 0.628^{***} & -0.011 & -0.293 & -0.077 & 0.003 \\ -0.297 & 0.003 & 1.052^{***} & 0.775 & -0.001 & -0.0002 \\ 0.338 & 0.183 & 0.089 & 0.014 & 0.368 & -0.019 \\ 0.526 & -0.013 & -0.267 & -1.581^{***} & 0.315^{**} & -0.007 \end{bmatrix} \cdot \begin{bmatrix} RTA \\ RTS \\ RMBS \\ REXTH \\ RSET \\ RTHY \end{bmatrix}_{t-1} + \begin{bmatrix} \varepsilon_{RTA} \\ \varepsilon_{RTS} \\ \varepsilon_{RMBS} \\ \varepsilon_{REXTH} \\ \varepsilon_{RSET} \\ \varepsilon_{RTHY} \end{bmatrix}
\tag{9}
$$

Regime 2 (St=2)

$$
\begin{bmatrix} RTA \\ RTS \\ RMBS \\ REXTH \\ RSET \\ RTHY \end{bmatrix} = \begin{bmatrix} 2.282 \\ 1.1569 \\ -0.612^{***} \\ 0.340 \\ 2.029 \\ -0.574 \end{bmatrix} + \begin{bmatrix} -2.506 & -0.052 & 0.600 & -2.984 & -0.348 & -0.009 \\ -0.440 & -0.016 & -0.385 & 0.488 & -0.788^{*} & -0.002 \\ -0.016 & 0.624^{***} & -0.023 & 0.135 & -0.027 & 0.002 \\ 0.624^{***} & -0.057 & 0.665^{***} & 1.202 & -0.163 & 0.036^{**} \\ 0.386 & -0.09 & 0.386 & 4.028 & -1.354^{**} & 0.044 \\ 0.398^{**} & 0.001 & 0.398^{**} & -0.001 & 0.337^{**} & -0.035^{***} \end{bmatrix} \cdot \begin{bmatrix} RTA \\ RTS \\ RMBS \\ REXTH \\ RSET \\ RTHY \end{bmatrix}_{t-1} + \begin{bmatrix} \varepsilon_{RTA} \\ \varepsilon_{RTS} \\ \varepsilon_{RMBS} \\ \varepsilon_{REXTH} \\ \varepsilon_{RSET} \\ \varepsilon_{RTHY} \end{bmatrix}
\tag{10}
$$

Eq. (9) and Eq. (10) show that the estimated means (intercept) of the MS-BVAR(1) model for each of the two regimes seem to have an economic interpretation. The first regime indicates that most of the values of mean in each equation are larger than those of the second regime. Thus, this indicates that regime 1 is among the high growth regimes, or there is an expansion of the QE programs, while regime 2 is among the low growth regimes, or the QE programs remain in the same level. Furthermore, considering RSET, RExth, and RTHY equations in the high growth regime (regime 1), we can see that RExth seems to be significantly driven by RMBS, whereas the reaction of RExth is positive in the first lag period of mortgage-backed securities (MBS). It can be seen, however, that the shock coefficients of RSET and RTHY are basically not statistically significant among RTA, RTS, and RMBS. For the low growth regime (regime 2), the reactions of the RTHY and REXTH are positive in the first lag period of RMBS and RTA, and the shock coefficients are statistically significant. But the shock coefficient of RSET is not statistically significant among RTA, RTS, and RMBS. Therefore, these results indicate that mortgage-backed securities (MBS) and expansion of Fed's total asset (TA) are QE programs that have significance to the Thai currency and bond markets, while Thailand's financial market has not been affected by any program in either of the regimes.

3.3 Estimation of MS-BVAR(1) IND Group

Regime 1 (St=1)

$$
\begin{bmatrix} RTA \\ RTS \\ RMBS \\ REXIND \\ RIDX \\ RIDY \end{bmatrix} = \begin{bmatrix} 3.712 \\ -0.326 \\ -0.365^{***} \\ -0.765 \\ 0.361 \\ 0.294 \end{bmatrix} + \begin{bmatrix} -0.645 & 0.145 & 0.180 & 0.219 & 0.152 & 0.006 \\ -0.158 & 0.065 & 0.372 & -1.215 & 0.149 & 0.012 \\ 0.129 & 0.750^{***} & -0.066 & -0.152 & 0.026 & -0.004 \\ 0.970^{**} & 0.017 & 1.041^{***} & 0.546 & 0.168 & -0.031 \\ -1.401 & -0.049 & -0.090 & 1.683 & 0.206 & -0.020 \\ 0.009 & -0.036 & 0.209 & -0.554 & 0.812 & -0.016 \end{bmatrix} \cdot \begin{bmatrix} RTA \\ RTS \\ RMBS \\ REXIND \\ RIDX \\ RIDY \end{bmatrix}_{t-1} + \begin{bmatrix} \varepsilon_{RTA} \\ \varepsilon_{RTS} \\ \varepsilon_{RMBS} \\ \varepsilon_{REXIND} \\ \varepsilon_{RIDX} \\ \varepsilon_{RIDY} \end{bmatrix}
\tag{11}
$$

Regime 2 (St=2)

$$
\begin{bmatrix} RTA \\ RTS \\ RMBS \\ REXIND \\ RIDX \\ RIDY \end{bmatrix} = \begin{bmatrix} 3.326 \\ 0.507 \\ -0.304^{**} \\ -0.390 \\ -1.919^{**} \\ 0.147 \end{bmatrix} + \begin{bmatrix} 0.967 & 0.087 & 0.272 & -1.465 & -0.469 & -0.071 \\ -0.665 & 0.079 & 0.060 & 1.066 & -0.878^{**} & 0.005 \\ -0.071 & 0.733^{***} & -0.129^{***} & 0.049 & -0.132^{***} & -0.003 \\ -0.512 & -0.027 & 0.864^{***} & -0.418 & -0.100 & -0.004 \\ -0.711 & 0.116 & 0.992 & 1.423 & 0.373 & -0.065 \\ -0.430 & 0.001 & 0.086 & -0.891^{**} & 0.477^{***} & -0.014 \end{bmatrix} \begin{bmatrix} RTA \\ RTS \\ RMBS \\ REXIND \\ RIDX \\ RIDY \end{bmatrix}_{t-1} + \begin{bmatrix} \varepsilon_{RTA} \\ \varepsilon_{RTS} \\ \varepsilon_{RMBS} \\ \varepsilon_{REXIND} \\ \varepsilon_{RIDX} \\ \varepsilon_{RIDY} \end{bmatrix} \quad (12)
$$

Regime 3 (St=3)

$$
\begin{bmatrix} RTA \\ RTS \\ RMBS \\ REXIND \\ RIDX \\ RIDY \end{bmatrix} = \begin{bmatrix} -3.303 \\ -0.622 \\ -0.339^{***} \\ -0.672 \\ 3.533^{**} \\ -0.123 \end{bmatrix} + \begin{bmatrix} 0.006 & 0.094 & -0.868 & -3.078 & -0.815 & -0.130^{**} \\ 0.991 & 0.079 & -0.649 & 0.882 & -0.400 & -0.038 \\ 0.014 & 0.729^{***} & -0.018 & 0.098 & 0.061 & -0.004 \\ 0.0004 & 0.097^{***} & 1.015^{***} & 1.145^{**} & 0.333 & 0.033 \\ -1.373 & -0.0007 & 0.851 & 0.790 & 1.101^{***} & 0.048 \\ -0.707^{**} & 0.064^{**} & 0.308 & -0.273 & 0.563^{***} & -0.012 \end{bmatrix} \begin{bmatrix} RTA \\ RTS \\ RMBS \\ REXIND \\ RIDX \\ RIDY \end{bmatrix}_{t-1} + \begin{bmatrix} \varepsilon_{RTA} \\ \varepsilon_{RTS} \\ \varepsilon_{RMBS} \\ \varepsilon_{REXIND} \\ \varepsilon_{RIDX} \\ \varepsilon_{RIDY} \end{bmatrix} \quad (13)
$$

Eq. (11), Eq. (12), and Eq. (13) show that there are three different regimes in the IND group model. The first regime and the second regime do not seem to have a lot of difference in the means of the equations. Therefore, we interpreted the regime by considering the coefficients of the variables of interest. It can be seen that the RIDX equation mostly has a positive sign in regime 1, while it mostly has a negative sign in regime 2. This indicates that regime 1 is a high growth regime, while regime 2 is an intermediate growth regime, or in the small QE program expansion. As for the third regime, it captures the low growth regime with a negative sign of means in almost all the equations. The results also report the parameters that have an estimate of three regimes MS-BVAR(1). The coefficients of RIDX, RExind, and RINDY equations demonstrate that there exists some relationship between the QE programs and the financial markets of Indonesia. RMBS shows a significant effect on RIDX in every regime, while RTA only has an effect on RIDX in the high growth regime (regime 1). In addition, RTA and RTS seem to influence RINDY only in the low growth regime (regime 3).

3.4   Estimation of MS-BVAR(1) PH group

Regime 1 (St=1)

$$
\begin{bmatrix} RTA \\ RTS \\ RMBS \\ REXPH \\ RPHEi \\ RPHY \end{bmatrix} = \begin{bmatrix} 9.558 \\ -17.04^{**} \\ -0.905 \\ 1.136 \\ 6.030 \\ 6.901^{***} \end{bmatrix} + \begin{bmatrix} -14.07^{**} & 0.925 & 0.596 & -14.181 & -2.636 & 0.087 \\ -7.214 & 0.249 & 0.344 & 14.120 & -0.198 & 0.060 \\ 0.289 & 0.525 & -0.117 & 1.174 & -0.218 & 0.021^{**} \\ 0.102 & 0.317 & -0.610 & -3.182 & 1.883^{***} & 0.066^{**} \\ 5.532 & 0.592 & 2.515 & -7.768 & 1.290 & 0.034 \\ -1.376 & 0.072 & 0.574 & -1.575 & 1.555^{***} & 0.023 \end{bmatrix} \begin{bmatrix} RTA \\ RTS \\ RMBS \\ REXPH \\ RPHEi \\ RPHY \end{bmatrix}_{t-1} + \begin{bmatrix} \varepsilon_{RTA} \\ \varepsilon_{RTS} \\ \varepsilon_{RMBS} \\ \varepsilon_{REXPH} \\ \varepsilon_{RPHEi} \\ \varepsilon_{RPHY} \end{bmatrix} \quad (14)
$$

Regime 2 (St=2)

$$
\begin{bmatrix} RTA \\ RTS \\ RMBS \\ REXPH \\ RPHEi \\ RPHY \end{bmatrix} = \begin{bmatrix} -19.37^{**} \\ 2.121 \\ -1.097 \\ 3.129 \\ 12.098 \\ 4.902^{**} \end{bmatrix} + \begin{bmatrix} 3.834 & -0.021 & -1.458 & -10.873 & 1.996 & -0.073 \\ 0.232 & 0.509 & -6.164^{***} & -11.766 & -3.758^{***} & 0.127 \\ -0.277 & 0.528^{***} & -0.008 & 0.546 & -0.094 & -0.009 \\ 1.496 & -0.230 & 2.603^{***} & -1.976 & 0.279 & 0.051 \\ 5.582 & -1.364^{**} & -3.384 & -5.086 & -1.808 & 0.201 \\ -0.100 & -0.111 & -1.273 & 4.008 & 0.936 & -0.015 \end{bmatrix} \cdot \begin{bmatrix} RTA \\ RTS \\ RMBS \\ REXPH \\ RPHEi \\ RPHY \end{bmatrix}_{-1} + \begin{bmatrix} \varepsilon_{RTA} \\ \varepsilon_{RTS} \\ \varepsilon_{RMBS} \\ \varepsilon_{REXPH} \\ \varepsilon_{RPHEi} \\ \varepsilon_{RPHY} \end{bmatrix} \quad (15)
$$

Eq. (14) and Eq. (15) provide the parameter estimations of the two regimes. Similar to the IND group model, the first regime and the second regime seem not to have much of a difference in the means in the equations; therefore, we interpreted the regime by considering it in terms of the coefficients of the variables of interest. In the RPHEi equation, the RExph equation, and the RPHY equation, there are mostly positive signs in regime 1 while there are mostly negative signs in regime 2. Thus, we interpreted regime 1 as the high growth rate regime and regime 2 as the low growth rate regime. Furthermore, Eq. 15 provides the results for the estimated coefficients in the first lag term. It has only RTS and RMBS which seem to significantly influence Rexph and RPHEi, respectively, in the low growth regime. This indicates that the PHEi index and the Philippines currency have been driven by some QE programs only in the low growth regime, while it has no significant influence at all on the Philippines' financial markets in the first regime.

3.5   Regime Probabilities

We plot the probabilities for the MS-BVAR model TH group, a single MS chain of two regimes, in Figure 1. Figure 2 presents the probabilities for the MS-BVAR model IND group, which is a single MS chain of three regimes. Lastly, the MS-BVAR model PH group chain of two regimes is presented in Figure 3.

Figure 1 shows the smooth probability plots; smooth probability is the probability of staying in either regime 1 or regime 2, between the periods of 2009 and 2014. It shows that the TH group model is consistent with the hypothesis that high growth and low growth represent different financial outcomes. Regime 1 of the TH group model is plotted in the top panel of Figure 1. We interpreted this regime as the era of the expansion in the QE programs. The bottom panel contains regime 2, which consists of a severe political event, a flooding disaster, and the speculative shock in the financial markets and the QE tapering. A distinguishing feature of regime 1 and regime 2 is the stark difference in the impacts of the QE programs. In addition, regime 1 seems to have higher probabilities than regime 2. This result indicates that announcements of the different QE programs are captured by the MS-BVAR(1) model since the first announcement of the QE program. In Table 1, it can be seen that the estimation of the transition matrix Q for the TH group shows that its regime 1 and regime 2 are persistent because the probability of staying in regime 1 and regime 2 is 92 percent, while the probability of moving between these regimes is only 8 percent. This indicates that only an extreme event can switch the series to change from regime 1 to regime 2. Moreover, the result also shows that the high growth regime has a duration of approximately 12.15 months, while the low growth regime has a duration of 11.11 months. This result indicates that Thai financial markets have high volatility because the duration of each regime corresponds to a short period of time.

Figure 2 shows the smooth probability of regime 1, regime 2, and regime 3 of the IND group model. Regime 1, regime 2, and regime 3 are plotted in the top panel, middle panel, and bottom panel, respectively. Regime 1 and regime 2, respectively, are interpreted as high growth regime and low growth regime. They are detected by the MS-BVAR(1) model many times in this long period; however, the probability of staying in each period is quite short. As far as regime 2 is concerned, we interpreted this regime as the era of normal economy. The probability of staying in this regime is longer than the probability of staying in the other regimes. This result indicates that the Indonesian financial markets mostly stay in the normal economy and that QE programs have little impact on that country's financial market. The transition matrices of the IND group are presented in Table 1. The result shows that regime 1, regime 2, and regime 3 are persistent because the probability of staying in each of these regimes is more than 92 percent, while the probability of moving between these regimes is only nearly 4 percent. Whereas the high growth regime has a duration of approximately 12.93 months, the intermediate growth regime and the low growth regime have durations of 12.83 months and 12.88 months, respectively. The results are also similar to the TH group result that it is only an extreme event that can switch the series moving between the regimes. In addition, we observed high financial market volatility during the QE operations.

Figure 3 is similar to Figure 1 which has two regimes. However, the probability of staying in regime 2 is greater than that in regime 1. Thus, this indicates that the Philippine financial markets remain in the low growth regime more than in the high growth regime. In addition, the transition matrix arises for the PH group, as demonstrated in Table 1, which implies a probability of 5.2 percent of the series moving between these two regimes. Conversely, the probability of staying in its own regime is nearly 95 percent; whereas the high growth regime has a duration of approximately 18.93 months, the low growth regime has a duration of 19.12 months. This result indicates that only an extreme event can switch the series to change between these two regimes. However, the Philippine financial markets are less volatile than the Thai and the Indonesian financial markets.

4   Economic Implication

4.1   Impulse Response

The first two panels in Figure 4 reports the impulse responses for the TH group model. Each panel displays the deviation in percent for the series entered in difference for every endogenous variable. In the first panel, it displays the impulse response in the high growth regime, while the second panel of the figure displays the low growth regime. The feedback of the QE programs differs considerably between regimes. In the high growth regime, the shock in QE has a great and persistent negative effect on

RExth (appreciation value of Thai baht). It then falls sharply and reaches the steady state within 4 months. However, QE creates a positive sharp-shaped response in RSET, and RTHY dies out in about 2 months. In this regime, RMBS is more likely to affect the Thai financial markets than any other programs, followed by RTA and RTS, respectively. As far as the low growth regime is concerned, the QE programs seem to have a different effect on the Thai financial markets. After a positive innovation in RMBS, the Thai currency falls sharply and reaches its minimum after about 3 months. Conversely, it has a positive effect on RSET and RTHY. RTA and RTS seem to have a similar effect on the Thai financial markets. The result shows that a shock to TA and TS produces a positive response to Exth, but the same creates a negative response to RSET and RTHY; at the same time, RTA has a larger impact than RTS.

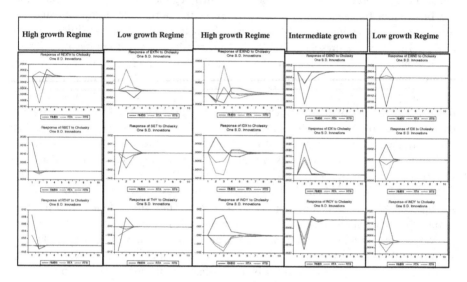

Fig. 1. The impulse responses for the TH and IND group model.

In Figure 4 also present the impulse response function for the changes in Indonesia's financial markets to a shock of QE in the high growth regime, the intermediate growth regime, and the low growth regime. The third panel of Figure 4, depicting the high growth regime, illustrates that a shock to RMBS causes RExind to fall after about 3 months, following which it begins to increase and eventually overshoots, thereby leading to an increase in RExind about 3–4 months later. Moreover, it creates a positive sharp-shaped response in IDX and INDY in about 2 months and 3 months, respectively, following which it falls to the equilibrium. Furthermore, shocks of TA and TS are likely to establish similar responses from the Indonesian financial markets, which would be an initial negative response and increase after that. In the middle panel, the intermediate growth regime shows that shocks to RMBS, RTA, and RTS decrease RExind and RINDY. Although they reach the peak at 2 months before returning to equilibrium within 3–4 months, the event causes a rise in the RIDX, which peaks at 2 months before returning to equilibrium within 3–5 months. In this regime, RMBS is more likely to affect the Indonesian financial markets than any other programs, followed by RTA and RTS, respectively. As for the fifth panel of Figure 4, the low growth regime shows that a shock to RMBS creates a negative sharp-shaped response in RExind, which then rises to equilibrium in 3 months, while increasing the RIDX and the RINDY peaks at 2 months before returning to equilibrium within 3 months. Conversely, shocks to RTA and RTS make the Indonesian financial markets respond in the opposite direction.

In Figure 5, we present the impulse response function for the changes in the Philippines' financial markets to a shock of QE in the high growth regime and the low growth regime. While no significant change takes place in the financial markets in response to a shock in the QE programs in the high growth regime (see Eq. 15), a shock in the QE programs is observed to have some positive effects on the Philippine financial markets. However, it only has RMBS, which has some negative effects on RExph.

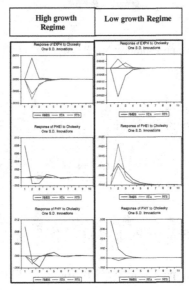

Fig. 2. The impulse responses for the PH group model.

5   Conclusion

Based on the results, we suggest that QE may have a direct substantial effect on the TIP financial markets. Therefore, if the Fed withdraws the QE programs, the move might have an effect on the TIP financial markets. In particular, purchasing the mortgage-backed securities (MBS) program is more likely to have an effect on the TIP financial markets than purchasing any other programs, so the TIP government, TIP central banks, and investors should emphasize on the MBS program and consider matters with extreme caution when a shock occurs. Additionally, it has to be borne in mind that the Thai financial markets and the Philippine financial markets have high volatility because the duration of each of their regimes is longer than that of the Indonesia.

Acknowledgement. We are grateful for financial support from Puay Ungpakorn Centre of Excellence in Econometrics, Faculty of Economics, Chiang Mai University. We thanks Nguyen, H. T. and Kreinovich, V. for valuable comment to improve this paper.

A   Appendix

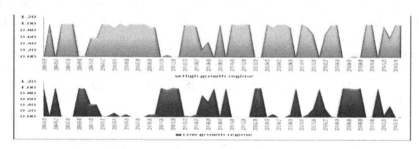

Fig. 3. Two regime probabilities: MS-BVAR (2) model TH group, 2009–2014.

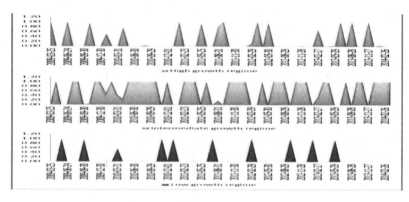

Fig. 4. Three regime probabilities: MS-BVAR (3) model IND group, 2009–2014.

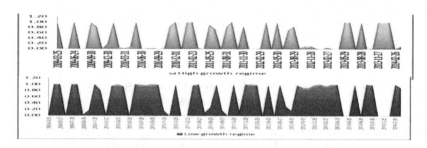

Fig. 5. Two regime probabilities: MS-BVAR (2) model PH group, 2009–2014

References

1. Bauer, M.D., Rudebusch, G.D.: The Signaling Channel for Federal Reserve Bond Purchases (2013)
2. Brandt, P.T., Freeman, J.R.: Advances in Bayesian time series modeling and the study of politics: Theory testing, forecasting, and policy analysis. Political Analysis **14**(1), 1–36 (2006)
3. Cho, D., Rhee, C.: Effects of Quantitative Easing on Asia: Capital Flows andFinancial Markets (ADB Working Papers 350) (2013)
4. Chinnakum, W., Sriboonchitta, S., Pastpipatkul, P.: Factors affecting economic output in developed countries: A copula approach to sample selection with panel data. Int. J. Approximate Reasoning **54**(6), 809–824 (2013)
5. Curdia, V., Ferrero, A., Chen, H.: The Macroeconomic Effects of Large-Scale Asset Purchase Programs (SED Working Papers 372). Saint Louis, MO: Society for Economic Dynamics (2012)
6. Do, G.Q., Mcaleer, M., Sriboonchitta, S.: Effects of international gold market on stock exchange volatility: evidence from asean emerging stock markets. Economics Bulletin **29**(2), 599–610 (2009)
7. Herrenbrueck, L.: Quantitative Easing, Frictional Asset Markets, and the Liquidity Channel of Monetary Policy (2013)
8. Joyce, M., Lasaosa, A., Stevens, I., Tong, M.: The financial market impact of quantitative easing in the United Kingdom. International Journal of Central Banking **7**(3), 113–161 (2011)
9. Jintranun, J., Calkins, P., Sriboonchitta, S.: Charitable giving behavior in northeast Thailand and Mukdaharn province: multivariate tobit models. In: Huynh, V.N., Kreinovich, V., Sriboonchitta, S., Suriya, K. (eds.) TES 2013. AISC, vol. 200, pp. 271–284. Springer, Heidelberg (2013)
10. Kiatmanaroch, T., Sriboonchitta, S.: Relationship between exchange rates, palm oil prices, and crude oil prices: a vine copula based GARCH approach. In: Huynh, V.-N., Kreinovich, V., Sriboonchitta, S. (eds.) Modeling Dependence in Econometrics. AISC, vol. 251, pp. 449–463. Springer, Heidelberg (2014)
11. Kiatmanaroch, T., Sriboonchitta, S.: Dependence Structure between World Crude Oil Prices: Evidence from NYMEX, ICE, and DME Markets. Thai Journal of Mathematics, 181–198 (2014)
12. Krishnamurthy, A., Vissing-Jorgensen, A.: The effects of quantitative easing on interest rates: channels and implications for policy (NBER Working Papers 17555). National Bureau of Economic Research, Cambridge, MA (2011)
13. Liu, J., Sriboonchitta, S.: Analysis of volatility and dependence between the tourist arrivals from China to Thailand and Singapore: a copula-based GARCH Approach: implications for risk management. In: Huynh, V.N., Kreinovich, V., Sriboonchitta, S., Suriya, K. (eds.) Uncertainty Analysis in Econometrics with Applications. Intelligent Systems and Computing, pp. 283–294. Springer, Heidenberg (2014)
14. Nason, J.M., Tallman, E.W.: Business cycles and financial crises: the roles of credit supply and demand shocks (FRBP Working Papers 1221). Philadelphia, PA: Federal Reserve Bank of Philadelphia (2012)

15. Praprom, C., Sriboonchitta, S.: Dependence analysis of exchange rate and international trade of Thailand: application of vine copulas. In: Huynh, V.-N., Kreinovich, V., Sriboonchitta, S. (eds.) Modeling Dependence in Econometrics. AISC, vol. 251, pp. 265–279. Springer, Heidelberg (2014)
16. Sims, C.A., Waggoner, D.F., Zha, T.: Methods for inference in large multiple-equation Markov-switching models. Journal of Econometrics **146**(2), 255–274 (2008)
17. Sims, C.A., Zha, T.: Bayesian methods for dynamic multivariate models. International Economic Review **39**(4), 949–968 (1998)
18. Sims, C.A., Zha, T.: Error bands for impulse responses. Econometrica **67**(5), 1113–1155 (1999)

Impacts of Quantitative Easing Policy of United States of America on Thai Economy by MS-SFABVAR

Pathairat Pastpipatkul[1]([✉]), Warawut Ruankham[1], Aree Wiboonpongse[2,3], and Songsak Sriboonchitta[1]

[1] Faculty of Economics, Chiang Mai University, Chiang Mai, Thailand
ppthairat@hotmail.com
[2] Faculty of Economics, Prince of Songkla, Songkhla, Thailand
[3] Faculty of Agriculture, Chiang Mai University, Chiang Mai, Thailand

Abstract. This paper provides new empirical evidences by combining the advantage of principal component analysis (PCA) with Markov-switching Bayesian VAR (MS-BVAR) to examine the durations and impacts of quantitative easing (QE) policy on the Thai economy. The results claimed that QE policy created monetary shock to the Thai economy around 4–5 months in each cycle before getting back to equilibrium. The result from foreign direct investment (FDI) was similar to foreign portfolio investment (FPI) channel that when QE was announced, excess capital stocks were injected into the emerging economies, including the Thailand stock market. Excess liquidity as a result of the QE policy pushed up the SET index of Thailand to reach the highest point in 2012. The booming stock market generated more real output (GDP), and greater levels of employment, private consumption, and policy interest rates. Also, it produced shocks for just 4–5 months for one cycle of QE. On the other hand, excess liquidity from QE caused the Thai Baht to appreciate significantly, and this affected Thailand's trade balance negatively. Impulse response and filtered probability yielded similar results that the QE had the impact on the Thai economy seasonally, and that the impact was around 4–5 months in each cycle.

Keywords: Quantitative easing · Markov-switching · Principal component analysis · MS-BVAR

1 Introduction

Under the liberalization of trade and finance since 1980, world financial market has been connected together through trading and capital mobility. Thus, it is very much possible that a financial crisis that once happened in one country impacts extensively on the world financial market. Subprime crisis, known as the "hamburger crisis" was one of exam that existed in the United States of America in 2008. The crisis caused the U.S. economy to fall into severe recession. The Federal Bank of America (Fed), headed by Ben Bernanke at that time, the chairman, needed to stimulate the economy by reducing the policy interest rate (Fed funds rate) to almost 0 percent. However, just lowering the policy interest rate could not resolve the crisis; the situation did not turn

© Springer International Publishing Switzerland 2015
V.-N. Huynh et al. (Eds.): IUKM 2015, LNAI 9376, pp. 389–402, 2015.
DOI: 10.1007/978-3-319-25135-6_36

for the better until an unconventional monetary policy, or quantitative easing (QE), has been launched. As a result, in recent years, the federal banks of many countries recovered their economies by relying heavily on quantitative policy.

A series of unconventional monetary policies, or quantitative easing (QE) was, accordingly, used by the U.S. Federal Bank to stimulate the economy to come out of the crisis. Starting in the period Q4/2008–Q3/2010, QE1 was announced by the Fed as meant for taking measures to strengthen liquidity in the banking system and to stabilize real estate prices by providing short-term loans to financial institutions through Term Auction Facilities amounting to over 4.3 billion USD and purchasing securities or mortgage loans (Mortgage-Backed Securities) through Term Asset-Backed Loan Facilities (TALF). This brought about a change in its balance sheet and it becomes more than double, and the money supply (High Power Money) was increased by more than 1.2 trillion USD, which was pumped into the economy system. Unfortunately, the outcome of QE1 was not as effective as it should have been. The U.S. GDP growth was found to have increased in only small amounts, while the unemployment rate climbed up to 9.6%. Therefore, phase 2 of the QE program was implemented by providing long-term U.S. government bonds (Long-term Treasury) in amounts of more than 6 billion USD to the U.S. economy (Fiscal policy office, 2010).

The QE policy of the United States plays a major role in causing shock to many countries. The excess liquidity resulting from the QE policy has been added into the world financial market through banking systems and multinational firms who have distributed their investment and disseminated their speculation to the emerging markets, including Southeast Asia and Thailand where the growth rates and the interest rates are higher.

In the case of Thailand, under the open financial market, the transmission of impact from QE relies heavily on the capital inflow. The key factors that bring the impact of QE from the U.S. into Thai economy, as considered in this study, are FDI and FPI during QE1 and QE2. Therefore, in order to measure impacts of capital inflow as a result of the quantitative easing policy on the Thai economy system, the study provided a new empirical finding by combining the advantage of principal component analysis (PCA) in collaboration with Markov-switching Bayesian VAR (MS-BVAR) to measure empirical impacts and its duration on Thai economy by adding macro variables (134 variables) after classifying them into nine sections by using the concept of factor analysis so as to avoid price puzzle and degree of freedom problems, as discussed in the study of Bernanke , Bovin, and Eliasz [5]. Moreover, we found that there were numerous studies which are employ a Maximum likelihood estimator (MLE) in the macroeconomic fields and seem to lack of the efficient and accurate result. These studies include Shibamoto [23], Bellone [4], Chang, Sriboonchitta, and Wiboonpongse [8], Do, Mcaleer and Sriboonchitta [11], Chinnakum, Sriboonchitta, and Pastpipatkul [9], Liu and Sriboonchitta [17], Autchariyapanitkul, Chanaim and Sriboonchitta, [3], Ayusuk and Sriboonchitta [2], Boonyanuphong and Sriboonchitta [6,7], Jintranun, Calkins and Sriboonchitta [14], Kiatmanaroch and Sriboonchitta [15,16], Praprom and Sriboonchitta [18,19,20], and Puarattanaarunkorn and Sriboonchitta [21,22]. Therefore, the paper aims to use the Bayesian estimator rather than MLE.

Thereafter, the duration of the impact was calculated by analyzing the filtered probability change in each of the regimes (both the inflow and the outflow regimes) from the MS-BVAR modeling. The combination method applied in this study is called Markov-switching Structural Factor Augmented Bayesian VAR (MS-SFABVAR), according to the study of Girardin and Moussa [13]. The study measures 134 monthly time series data on Thai macro economy variables during January 2008 to July 2013, and then classifies the data into nine sectors. We focused on the transmission of the QE policy under FDI and FPI channels only. Series were estimated using BVAR model to identify shocks, and using MS-BVAR model to calculate probability and its duration during each regime. Regimes were set by applying concept of business cycle that switching regimes creates a cycle of economy. Therefore, regimes were classified into the inflow regime (regime 1), or period that FDI and FPI were invested in Thai economy, and the outflow regime (regime 2), or period that FDI and FPI were flowing back to the U.S. economy.

2   Methodology

2.1   Bayesian Vector Auto Regressions (BVAR)

The Sims–Zha prior reduced form Bayesian VAR (B-SVAR) model was described by Sims and Zha [10] and Sims [24]. It is based on the dynamic simultaneous equation developed from the regular VAR model. The prior is constructed for the structural parameters. The basic SVAR model has the following form, as given by

$$y_t' A_0 = \sum_{\ell=1}^{p} Y_{t-\ell}' A_\ell + z_t' D + \varepsilon_t' , t = 1,..., T, \tag{1}$$

where A_i is an $m \times m$ parameter matrix for contemporaneous and its lag within the endogenous variable, $D_{\,}$ is the $h \times m$ parameter matrix for the exogenous variable; intercept, Y_t, is the endogenous variable with the dimension $m \times 1$, z_t is the $h \times 1$ vector for the exogenous and its intercept, and ε_t is the structural shock with dimension $m \times 1$( ). The mean and the variance of the shock can be seen to be the following:

$$E\left[\varepsilon_t \middle| y_1,..., y_{t-1}, z_1,..., z_{t-1}\right] = 0 \tag{2}$$

$$E\left[\varepsilon_t \varepsilon_t' \middle| y_1,..., y_{t-1}, z_1,..., z_{t-1}\right] = I \tag{3}$$

The reduced form representation of SVAR model can be written as

$$y_t' A_0 A_0^{-1} = \sum_{\ell=1}^{p} Y_{t-\ell}' A_\ell A_0^{-1} + z_t' D A_0^{-1} + \varepsilon_t' A_0^{-1} \tag{4}$$

$$y'_t = \sum_{\ell=1}^{p} Y'_{t-\ell} B_\ell + z'_t T + \varepsilon'_t A_0^{-1} \tag{5}$$

The reduced form error covariance matrix is

$$\Sigma = E\left[(\varepsilon_t A_0^{-1})(\varepsilon'_t A_0^{-1})'\right] = \left[A_0 A'_0\right]^{-1}, \tag{6}$$

where A_0 is the identity matrix of shocks (hit) in each of the equations[10]

2.2   Impulse Response Function

The objective of the Impulse Response Function is to find out the variations regarding how endogenous variables, in terms of one standard deviation, impact the exogenous variables in the meantime and in future, or to investigate the shock during each period, and predict when it will return to equilibrium [1]. The Vector Moving Average (VMA) before calculating the IRF can be written as

$$y_t = \bar{y} + \sum_{i=0}^{\infty} \phi_i u_{t-1} \quad ; \quad \phi = A_i B^{-1} \tag{7}$$

2.3   Markov-Switching Bayesian VAR (MS-BVAR)

In this section, we carry out the Markov-switching Bayesian reduced form vector autoregression model setup and posterior mode estimation. The MS-BVAR model is estimated using block EM algorithm where the blocks are 1) the BVAR regression coefficients for each regime (separating optimum for intercepts, AR coefficients, and error covariance) and 2) the transition matrix. Considering the autoregressive model order ρ for K dimension of time series vector with general form [12],

$$y_t = (y_{t1}, ..., y_{Kt})', t = 1, ..., T \tag{8}$$

$$y_t = v + A_1 y_{t-1} + ... + A_p y_{t-p} + \mu_t \tag{9}$$

where $\mu_t \sim IID(0, \Sigma)$ and $y_0, ..., y_{t-p}$ are constantly fixed.

The mean adjusted of VAR

$$y_t - \mu = A_1(y_{t-1} - \mu) + A_P(y_{t-P} - \mu) + \mu_t \tag{10}$$

where $\mu = (I_K - \sum_{j=1}^{P} A_j)^{-1} v$ is the mean of y_t with $(K \times 1)$ dimension.

A VAR model with changes in regimes will have a constant parameter that does not vary over time. Consider, for example, the unobservable variable s_t. These unobservable variables are not controlled and might cause an error in the model. Therefore, Markov-switching VAR (MS-VAR) is considered to be a structural change in regime-switching. MS-VAR is based on the specification of the underlying data-generating process of time series data where y_t is the observed variable that relies on the unobserved regime s_t. Here, the probability differentiated at different regimes (shift regime) in terms of transition probabilities is given as follows

$$P_{ij} = \Pr(s_{t+1} = j | s_t = i), \ \sum_{j=1}^{M} P_{ij} = 1 \ \forall_{i,j} \in \{1,...,M\} \tag{11}$$

3 Empirical Results of Research

3.1 Component Analysis

According to the study carried out by Bernanke, Bovin, and Eliasz [5], a shortage in the number of variables creates a technical problem called price puzzle and outcome of estimated model under VAR can be misunderstood from the theory and concept of reality. For that reason, this study tried to collect as many monetary variables as possible including all macroeconomic variables in order to avoid this problem. Study set up experiment by rotating Varimax and rotating component to find 1 representative factor that exemplified its group. See table 1 for representative factor from each group.

Table 1. Rotated Component Matrix

Factor	Variables
FA1: Real output	Industrial Production, Capacity Utilization Rate, Agricultural Production, Mining Production
FA2: Employment	Unemployment rate, Labor force, Seasonally inactive labor force
FA3: Private consumption	Private Consumption Index, Sales of Benzene and Gasohol, Sales of NGV, Passenger car sales, Real imports of consumer goods, Commercial car sales, Sales of LPG, Electricity consumption, Motorcycle Sales, Sales of Diesel
FA4: Stock market	SET index, SET50, Market capitalized, Dividend yield, Price–Earnings ratios
FA5: Exchange rate	(REER), (NEER), Forex: Thai Baht to U.S. Dollar
FA6: Interest rate	Prime rate: (MLR), T-BILL1M (28) days, Interbank overnight lending rates, Repurchase rate, Prime rate: (MOR), Government bond yield
FA7: Money supply	Monetary Base (MB), Money Supply: Narrow-Money
FA8: Price Index	Export Price, CPI,PPI,Import Price, Housing Price Index,
FA9: Trade balance	Export volume Index, Export Value (f.o.b.), Import Value (f.o.b.),Import volume Index, Trade Balance, Term of trend

3.2 VAR Lag Order Selection Criteria Test

Table 2 displays the various computed values for all lags. The minimum value of the criterion shows the appropriate lags. Minimized Schwarz information criterion (SC) can identify the optimum lag length. The result shows that lag 2 with an SC value of 1.3595 is the appropriate lag length. So we will use lag 2 to estimate the MS-BVAR model.

Table 2. Lag Order Selection Criteria

Lag	AIC	SC	HQ
0	2.027217	2.125389	2.053262
1	1.494270	1.788783	1.572404
2	**0.868683***	**1.359539***	**0.998907***
3	1.107649	1.794847	1.289963

Source: Calculation.

3.3 Estimated VAR/BVAR and Its Impulse Response Function

Because the study is designed to measure the impact of the QE policy of the United States on the Thai economy system, so as to clarify the results in deeper detail we estimate the model in pairs (binary estimation). This is performed by starting from the transmission of QE through FDI to the nine sectors of Thai economy, followed by FPI, in that order. Then, the results between the VAR and the Sims–Zha prior reduced form Bayesian VAR (BVAR) models will be computed to confirm the long-run relationship between FDI and nine sectors of Thai economy and FPI and nine sectors of Thai economy. Additionally, we will find the Impulse Response Function to identify the duration of the monetary shock from the QE policy to the Thai economy system and its returning point to the equilibrium.

Impact of QE Through FDI Channel

Table 3 illustrates comparison result of the VAR and BVAR models. Results said QE policy transmission under FDI channel affects Thai economy (in all nine sectors). Negative and positive coefficient represents impacts. In the first period, capital inflow in terms of FDI is observed to positively affect Thailand's real output factor (FA1), push up the employment factor (FA2), and stimulate private consumption (FA3), raise the stock market factor (FA4), increase money supply (FA7), and increase price index (FA8). More importantly, QE affected Thai exchange rate to appreciate and negatively affects Thailand trade balance.

Table 3. Estimates VAR and BVAR (FDI and nine sectors of Thai economy)

Model	VAR			BVAR		
	Coeff: t(-1)	Coeff: t(-2)	Constant	Coeff: t(-1)	Coeff: t(-2)	Constant
FDI & (FA1)	0.2490	0.6874	0.2762	0.0107	−0.0025	0.1176
FDI & (FA2)	−0.1500	−0.1563	0.2758	0.0022	−0.0139	0.1317
FDI & (FA3)	0.4846	−0.0504	0.2730	0.0091	−0.1384	−0.0062
FDI & (FA4)	0.4712	−0.0344	0.2928	0.0107	−0.0025	0.1176
FDI & (FA5)	0.4797	−0.0583	0.2854	−0.0189	−0.5729	−0.0310
FDI & (FA6)	−0.1829	−0.0885	0.3693	−0.0501	−0.4264	−0.0287
FDI & (FA7)	0.5101	−0.0685	0.2659	0.0355	−0.1170	−0.0267
FDI & (FA8)	0.4707	−0.0496	0.2703	0.0341	1.1280	0.0248
FDI & (FA9)	−1.5876	−0.7361	0.3234	−0.0045	−0.0006	0.0134

Source: Calculation.

Fig. 1 provides confirmations for the impulse responses to the monetary shocks based on nine binary VAR and BVAR models. In the Figure, the inner line indicates the median impulse response from VAR and Bayesian VAR with 5000 draws. The monetary shock from VAR seemed to be more effective as it described that the shock had existed since the 1st and the 2nd months before getting back to equilibrium in the period of months 4th–5th. QE shock is temporary or takes time for 4–5 months approximately. However, it is noteworthy that the impulse response from the BVAR model indicates the explosive concept, that the shock is permanent and not getting back to equilibrium.

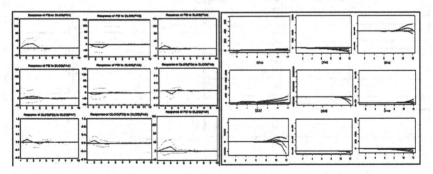

Fig. 1. Impulse Response Function (IRF) from VAR (left) and BVAR (right)

Impact of QE through FPI Channel

Foreign Portfolio Investment (FPI) is another channel bringing on QE's impact to the Thai economy. In this study, we measured FPI as the excess capital from the U.S. economy that flowed into the Thai stock market only. It is excluded from the capital that is invested in the real sector. In Fig. 1, it can be seen that the computed coefficient values significantly claimed results equivalent to the FDI channel in that FPI in the first period affected the Thai exchange rate (FA5) and caused it to appreciate besides negatively reducing Thailand's trade balance (FA9). Also, FPI is seen to have increased the real output factor (FA1), employment (FA2), stock market (FA4), interest rate (FA6), and money supply (FA7). Additionally, the empirical results from BVAR demonstrate that when capital stocks are invested in the stock market, the price index (FA8) keeps rising. This is because the consumer believes that when the stock market index increases, it is a sign of the economy booming. Consequently, they decide to spend more, and this causes more inflation in the next period.

Table 4. Estimates VAR and BVAR (FPI and nine sectors of Thai economy)

Model	VAR			BVAR		
	Coeff: t(-1)	Coeff: t(-2)	Constant	Coeff: t(-1)	Coeff: t(-2)	Constant
FPI& (FA1)	−0.3365	−0.1038	0.0912	0.0012	−0.0006	0.0012
FPI& (FA2)	−0.3178	−0.0989	0.0549	0.0051	0.0153	0.0051
FPI& (FA3)	−0.3455	−0.1262	0.8194	−0.0065	0.1387	−0.0065
FPI& (FA4)	−0.3315	−1.1212	0.6165	0.0164	−0.0852	0.0164
FPI& (FA5)	−0.2994	−0.1532	−3.1895	−0.0040	−0.5745	−0.0040
FPI& (FA6)	−0.3237	−0.0994	−1.2142	0.0778	0.4274	0.0778
FPI& (FA7)	−0.3682	−0.1132	−1.1182	0.0189	−0.1157	0.0189
FPI& (FA8)	−0.3275	−0.1214	1.5484	0.3535	−1.0568	0.3535
FPI& (FA9)	0.4482	−0.0754	0.8615	−0.0097	0.0407	−0.0097

Source: Calculation.

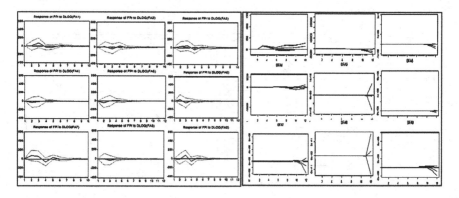

Fig. 2. Impulse Response Function (IRF) from VAR (left) and BVAR (right)

From Fig. 2, it can be concluded that the impulse response function yielded results equivalent to the FPI channel that IRF from the VAR model is more effective than that from BVAR. In the Figure on the left-hand side, the plotting of the IRF of the nine binary models points to the fact that monetary shock existed in the period of 1st–2nd month since QE1 and 2 are used. Later, the shock is observed to increase during the months 3–4 before adjusting back to equilibrium during the months 5–6. The monetary shock stayed on when the QE policy was announced, and it existed for 5–6 months in each cycle.

4   Estimated MS-BVAR

The advantage of Markov-switching model is that it can express the probability of switching regimes. The regime set in this study is separated into two states. The first state is inflow regime (regime 1) where excess capital flows into the Thai economy system via FDI and FPI. The second state is outflow regime (regime 2) where excess capital flows out of the Thai economy. The study was conducted to find out the fluctuating probability in each of the regimes and its duration of stay in each regime. After that, we plot the filtered probability diagram to demonstrate the change and its duration during QE1 and QE2. The calculated results are shown separately, starting from the FDI channel up to the FPI channel, as follows:

Duration of QE's Shock (FDI Channel)

Table 5 displays the transition probability in each of the regimes and its duration. The computed probability from the nine models gave a similar result that transition probabilities staying in the inflow regime (regime 1) are higher than those in the outflow regime (regime 2). The average duration of the inflow regime is about 6–7 months. This is because foreign direct investments made in Thailand are long-term investments in the real sector; for example, investments in the form of building factories, buying new machines, employing long-term employees, etc. For this reason, FDI stays longer in regime 1. On the other hand, transition probability and its duration in the outflow regime are smaller. The average time taken in the outflow regime is approximately 1–2 months before it turns back to the inflow regime. This means that

Table 5. Transition Probabilities P[i,i]

Model	Prob. Regime 1	Duration	Prob. Regime 2	Duration
(FDI) & (FA1)	0.9622	**2.00**	0.4976	**1.99**
(FDI) & (FA2)	0.8720	**7.81**	1.00E-04	**1.10**
(FDI) & (FA3)	0.8177	**5.48**	1.00E-01	**1.00**
(FDI) & (FA4)	0.8901	**9.10**	0.0663	**1.07**
(FDI) & (FA5)	0.7715	**4.37**	2.28E-01	**1.3**
(FDI) & (FA6)	0.7999	**5. 00**	0.3745	**1.60**
(FDI) & (FA7)	0.6933	**3.26**	1.00E-01	**1.00**
(FDI) & (FA8)	0.8540	**6.84**	1.46E-01	**1.17**
(FDI) & (FA9)	0.7827	**4.60**	2.17E-01	**1.27**

Source: Calculation.

when QE is used appropriately, the investors perceive it as a good sign that the U.S. economy is on the road to recovery. Yet, the real GDP growth of the U.S. is not on an increase, as employment rate is still high in the U.S. in comparison with the emerging countries. Thus, the capital is bound to flow back into the market with higher growth.

In addition, upon plotting the filtered probability diagram from the nine models, the outcomes are found to be comparable. Fig. 3 displays the diagram of filtered probability, where the vertical axis represents the probability from 0–1 where "1" is 100% probability to stay in the regime and "0" is 0% probability to stay in the regime, and the horizontal axis represents the time period from 2008 to 2013. The upper line denotes the inflow regime (regime 1) and the median line denotes the outflow regime (regime 2). Diagrams from the nine models suggest that Foreign Direct Investment (FDI) received from America switches within regimes seasonally with the cycle of 1–2 months of changes. In the middle right figure, starting from the beginning of 2008, the capital stayed in the inflow regime until the end of 2008 when the QE1 policy was announced on 25th November 2008. Speculators expected that the U.S. economy would recover through the QE1 policy; hence the change in capital between the inflow regime and the outflow regime for 1–2 months. Thereafter, when the U.S. economy did not reach the recovery as expected, all the capital was brought back to the emerging countries, including Thailand, where the growth rate and policy interest rate are higher and which keep staying in the inflow-regime for 3–4 months.

Later, QE2 was announced on 3th November 2010. The flowing capital shifted from the inflow regime (regime 1) to the outflow regime (regime 2) due to the expectation that the U.S. economy will be boosted. The capital flowing out to the U.S. economy took a duration of 2–3 months before moving back to Thailand in the beginning and the middle of 2011. The cycle of the shifting regime relied on the period of the QE policy announcement and the growth number of the U.S. economic indicator adjustment.

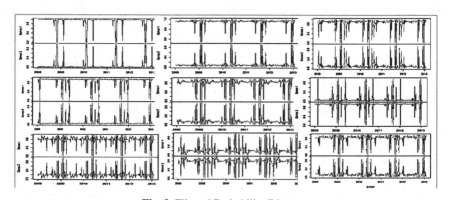

Fig. 3. Filtered Probability Diagram

Duration of QE's Shock (FPI channel)

Under Foreign Portfolio Investment (FPI), the transition probability from the MS-BVAR model gave the same result that the probability as well as the duration of staying in the inflow regime (regime 1) is higher. At the same time, the probability in the outflow regime is less and the average time of staying in the outflow regime is approximately 1–2 months. The duration of the stay depended on the expectation according to the QE announcement as well as the expectation of the speculators in the stock market.

Table 6. Transition Probabilities P[i,i]

Model	Prob. Regime 1	Duration	Prob. Regime 2	Duration
(FPI) & (FA1)	0.9621	26	0.6882	3.20
(FPI) & (FA2)	0.9822	22.12	0.71811	3.54
(FPI) & (FA3)	0.9577	23.60	0.04965	1.05
(FPI) & (FA4)	0.9452	18.20	0.09916	1.11
(FPI) & (FA5)	0.9506	10.27	1.00E-04	1.00
(FPI) & (FA6)	0.9615	25.9	1.00E-04	1.00
(FPI) & (FA7)	0.9747	29.53	1.00E-04	1.00
(FPI) & (FA8)	0.24492	1.32	0.9283	13.96
(FPI) & (FA9)	0.9506	20.27	1.00E-04	1.00

Source: Calculation.

Fig. 4 provides confirmations regarding the filtered probability diagrams from nine binary models under the FPI channel. The probability is seen to switch within the regime seasonally. Each of the cycles changing from the inflow regime to the outflow regime took about 1 month. This is because the investors in the stock market can rotate their portfolios faster than investors in the real sectors. Thus, when QE1 and QE2 were announced, it made the shock to the stock market quicker than that to the real sector in the economy. According to the Table, the first shock existed from the end of

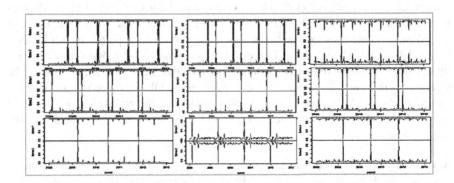

Fig. 4. Filtered Probability Diagram

2008 to the beginning of 2009, which was the same period when QE1 was used, and it took around 1 month to get back to the inflow regime. The second shock existed from the middle to the end of 2010, the same as the period when QE2 was used. The capital inflow that was invested in Thailand moved out to the U.S. in the expectation that the U.S. economy was booming again. The movement cycle depended on the time of announcement of the QE policy.

5   Conclusion

The study demonstrates that the influence of the quantitative easing (QE) policy of the United States of America depended on the capital inflow that moved from the U.S. economy to Thailand. Transmission of the QE policy to the Thai economy system has two channels: the Foreign Direct Investment (FDI) channel and the Foreign Portfolio Investment (FPI) channel. Under the FDI channel, when QE was used, excess liquidity was introduced into the Thai economy. This stimulated the real output factor and employment. Also, foreign investment served to motivate the private sector to consume more and caused the interest rate to increase. At the same time, the QE policy caused the Thai Baht to appreciate, and this affected the trade balance negatively. The QE policy caused monetary shock to the Thai economy system for around 4–5 months in each cycle before getting back to equilibrium. As for the FPI channel, the result was similar to that of the FDI channel. When QE was announced, excess capital was injected into emerging economies, including the Thailand stock market. Excess liquidity as a result of the QE policy pushed up the SET index of Thailand to reach the highest point in 2012. The booming stock market caused increases in the real output (GDP), level of employment, and policy interest rate. Yet, it created shock just for 4–5 months for 1 cycle of QE. On the other hand, excess liquidity from QE caused the Thai Baht to appreciate significantly, and this affected Thailand's trade balance negatively.

Consequently, the Bank of Thailand (BOT), as the monetary policy maker, should be concerned more about flowing liquidity that comes from outside as it makes its impact felt on several sectors of our economy, especially with regard to the fluctuation of the Thai currency. Monetary policy should be more concerned about the time period of its content as to whether it should be a short-term policy or a long-term policy. QE created shocks to the Thai economy seasonally (4–5 months in one cycle); thus, evidently, unconventional monetary policies that deal with shocks should be effective in the short term. In addition, impacts of the QE policy or shocks from the currency war should be reported and published to reach the private sector as quickly as possible so that the private sector is able to set suitable strategies to deal with the outside shock.

Acknowledgement. We are grateful for financial support from Puay Ungpakorn Centre of Excellence in Econometrics, Faculty of Economics, Chiang Mai University. We thanks Nguyen, H. T. and Kreinovich, V. for valuable comment to improve this paper.

References

1. Alejandro, R.G., Jesus, R.G.: Structural changes in the transmission mechanism of monetary policy in Mexico: a non-linear VAR approach. The Working Papers series of Bank of Mexico (2006)
2. Ayusuk, A., Sriboonchitta, S.: Risk Analysis in Asian Emerging Markets using Canonical Vine Copula and Extreme Value Theory. Thai Journal of Mathematics 59–72 (2014)
3. Autchariyapanitkul, K., Chanaim, S., Sriboonchitta, S.: Portfolio optimization of stock returns in high-dimensions: A copula-based approach. Thai Journal of Mathematics 11–23 (2014)
4. Bellone, B.: Classical Estimation of Multivariate Markov-Switching Models using MSVARlib (2005)
5. Bernanke, B., Bovin, J., Eliasz, P.: Measuring of Monetary Policy: A Factor-Augmented Vector Autoregressive (FAVAR) Approach. NBER working paper National Bureau of Economic Research Massachusetts, Cambridge, USA (2004)
6. Boonyanuphong, P., Sriboonchitta, S.: The Impact of Trading Activity on Volatility Transmission and Interdependence among Agricultural Commodity Markets. Thai Journal of Mathematics 211–227 (2014)
7. Boonyanuphong, P., Sriboonchitta, S.: An analysis of volatility and dependence between rubber spot and futures prices using copula-extreme value theory. In: Huynh, V.-N., Kreinovich, V., Sriboonchitta, S. (eds.) Modeling Dependence in Econometrics. AISC, vol. 251, pp. 431–444. Springer, Heidelberg (2014)
8. Chang, C.L., Sriboonchitta, S., Wiboonpongse, A.: Modelling and forecasting tourism from East Asia to Thailand under temporal and spatial aggregation. Mathematics and Computers in Simulation 79(5), 1730–1744 (2009)
9. Chinnakum, W., Sriboonchitta, S., Pastpipatkul, P.: Factors affecting economic output in developed countries: A copula approach to sample selection with panel data. International Journal of Approximate Reasoning 54(6), 809–824 (2013)
10. Sims, C.A., Zha, T.: Bayesian Methods For Dynamic Multivariate Models. Yale University, Federal Reserve Bank of Atlanta, U.S.A. (1998)
11. Do, G.Q., Mcaleer, M., Sriboonchitta, S.: Effects of international gold market on stock exchange volatility: evidence from ASEAN emerging stock markets. Economics Bulletin 29(2), 599–610 (2009)
12. George, H., Evangelia, P.: Stock returns and inflation in Greece: A markov switching Approach. Economic Research Department, Bank of Greece, Athens (2005)
13. Girardin, E., Moussa, Z.: Quantitative easing works: Lessons from the unique experience in Japan 2001–2006. Grouping of Economic Research Universities of Aix-Marseille (2010)
14. Jintranun, J., Calkins, P., Sriboonchitta, S.: Charitable giving behavior in Northeast Thailand and Mukdaharn Province: multivariate tobit models. In: Huynh, V.N., Kreinovich, V., Sriboonchitta, S., Suriya, K. (eds.) TES'2013. AISC, vol. 200, pp. 269–281. Springer, Heidelberg (2013)
15. Kiatmanaroch, T., Sriboonchitta, S.: Relationship between exchange rates, palm oil prices, and crude oil prices: a vine copula based GARCH approach. In: Huynh, V.-N., Kreinovich, V., Sriboonchitta, S. (eds.) Modeling Dependence in Econometrics. AISC, vol. 251, pp. 399–413. Springer, Heidelberg (2014)
16. Kiatmanaroch, T., Sriboonchitta, S.: Dependence Structure between World Crude Oil Prices: Evidence from NYMEX, ICE, and DME Markets. Thai Journal of Mathematics 181–198 (2014)

17. Liu, J., Sriboonchitta, S.: Analysis of volatility and dependence between the tourist arrivals from China to Thailand and Singapore: a copula-based GARCH approach. In: Huynh, V.N., Kreinovich, V., Sriboonchitta, S., Suriya, K. (eds.) TES'2013. AISC, vol. 200, pp. 283–294. Springer, Heidelberg (2013)
18. Praprom, C., Sriboonchitta, S.: Dependence analysis of exchange rate and international trade of Thailand: application of vine copulas. In: Huynh, V.-N., Kreinovich, V., Sriboonchitta, S. (eds.) Modeling Dependence in Econometrics. AISC, vol. 251, pp. 229–243. Springer, Heidelberg (2014)
19. Praprom, C., Sriboonchitta, S.: Extreme value copula analysis of dependences between exchange rates and exports of Thailand. In: Huynh, V.-N., Kreinovich, V., Sriboonchitta, S. (eds.) Modeling Dependence in Econometrics. AISC, vol. 251, pp. 187–199. Springer, Heidelberg (2014)
20. Praprom, C., Sriboonchitta, S.: Investigation of the Dependence Structure Between Imports and Manufacturing Production Index of Thailand using Copula-Based GARCH Model. Thai Journal of Mathematics 73–90 (2014)
21. Puarattanaarunkorn, O., Sriboonchitta, S.: Copula based GARCH dependence model of Chinese and Korean tourist arrivals to Thailand: implications for risk management. In: Huynh, V.-N., Kreinovich, V., Sriboonchitta, S. (eds.) Modeling Dependence in Econometrics. AISC, vol. 251, pp. 343–365. Springer, Heidelberg (2014)
22. Puarattanaarunkorn, O., Sriboonchitta, S.: Modeling dependency in tourist arrivals to Thailand from China, Korea, and Japan using vine copulas. In: Huynh, V.-N., Kreinovich, V., Sriboonchitta, S. (eds.) Modeling Dependence in Econometrics. AISC, vol. 251, pp. 383–398. Springer, Heidelberg (2014)
23. Shibamoto, M.: An analysis of monetary policy shocks in Japan: a factor augmented vector autoregressive approach. Discussion paper Osaka University, Osaka, Japan (2005)
24. Sims, C.A.: Interpreting Macroeconomics Time Series Facts: The Effect of Monetary Policy. Paper presented at International Seminar on Macroeconomics, Madrid (1992)

Volatility and Dependence for Systemic Risk Measurement of the International Financial System

Jianxu Liu[1], Songsak Sriboonchitta[1(✉)], Panisara Phochanachan[1], and Jiechen Tang[1,2]

[1] Faculty of Economics, Chiang Mai University, Chiang Mai, Thailand
[2] Faculty of Management and Economics, Kun Ming University of Science and Technology, Kun Ming, China
songsakecon@gmail.com

Abstract. In the context of existing downside correlations, we proposed multi-dimensional elliptical and asymmetric copula with CES models to measure the dependence of G7 stock market returns and forecast their systemic risk. Our analysis firstly used several GARCH families with asymmetric distribution to fit G7 stock returns, and selected the best to our marginal distributions in terms of AIC and BIC. Second, the multivariate copulas were used to measure dependence structures of G7 stock returns. Last, the best modeling copula with CES was used to examine systemic risk of G7 stock markets. By comparison, we find the mixed C-vine copula has the best performance among all multivariate copulas. Moreover, the pre-crisis period features lower levels of risk contribution, while risk contribution increases gradually while the crisis unfolds, and the contribution of each stock market to the aggregate financial risk is not invariant.

Keywords: Vine copulas · Component expected shortfall · GARCH · G7

1 Introduction

It is well known the acceleration in global integration has brought many benefits. It has also created fragility because a country's financial instability poses a potential threat to the whole financial system. For example, the 2007-2009 financial crises happened in USA, but it swept across the global financial markets, such as England, Europe, Japan, etc. Systemic risk can be magnified by forces of risk-contagion and correlation of financial markets. As a result, accurate modeling and forecasting of the dependence structures and the contribution of a financial market to systemic risk are of considerable interest to financial researchers, financial regulators, and investors.

This paper aims at modelling the dependence structures of G7 stock markets, forecasting the systemic risks of G7 countries and the contribution of each

© Springer International Publishing Switzerland 2015
V.-N. Huynh et al. (Eds.): IUKM 2015, LNAI 9376, pp. 403–414, 2015.
DOI: 10.1007/978-3-319-25135-6_37

stock market to the entire financial system. Multivariate copulas and vine copulas were used to investigate the dependence structures of G7, while measuring the systemic risk was implemented by applying a new approach which is called component expected shortfall (CES). Banulescu and Dumitrescu [4] proposed the CES method that not only calculates systemic risk, but also measures the institutions absolute contribution to the ES of financial system. It overcomes the main drawbacks of MES (Acharya et al. [2]) and SRISK (Brownlees and Engle [5]), such as no assumption for the liabilities of institutions. However, Banulescu and Dumitrescu [4] and Brownlees and Engle [5] used a bivariate GARCH or DCC-GARCH model to fit the asset returns. This implies that the correlation between asset returns is linear, and there exists symmetric tail dependency. Obviously, linear correlation is not invariant under non-linear strictly increasing transformation. In reality, finance asset return has the presence of heavy tails and asymmetry as well. Embrechts et al.[7] showed that linear correlation breaks down and leads to a number of fallacies in the non-elliptical world. Therefore, the problem raised from normality could lead to an inadequate CES estimate.

In order to overcome these drawbacks, we used copula functions to construct a flexible multivariate distribution with different marginals and different dependent structures. Copula functions allow us to capture symmetric and asymmetric non-linear correlation structures depending on the copula chosen. To more appropriately fit asset returns, many GARCH family models were chosen to capture the volatilities of G7 stock returns. In addition, multivariate copulas, e.g., multivariate Gaussian, T, Clayton, Frank and Gumbel copulas, and several vine copulas were performed to capture the dependencies of G7 stock returns. In the last decade, there are a few studies that apply multivariate copulas, such as multivariate Archimedean copulas, Clayton canonical vine copula (CVC), T canonical vine copula, and drawable vine copula, to capture asymmetric dependence and measure risk. Patton [15], Wu and Liang [18] and Ba [3] showed that bivariate or trivariate copulas can produce significant gains for the investor with no short-sales constraints. Low et al. [14] showed that Clayton CVC has better performance than multivariate normal distribution or standard Clayton copula for portfolio and risk management. Moreover, we also have several studies in copulas models, such as Sriboonchitta et al. [17], Liu and Sriboonchitta [12] and Liu et al. [13] , etc.

The novelty of our contribution is that we incorporated CES with copula-GARCH models, which allows for higher scalability for capturing asymmetric dependence. Moreover, we made a comparison between various kinds of multivariate copulas in terms of AIC. It is crucial to capture asymmetric dependence and measure systemic risk. For vine copulas, we built a cascade of bivariate copulas among vine copulas that would be conducive to the flexibility of multivariate copulas. In this study, the Gaussian copula, T copula, Clayton copula, Frank copula, Gumbel copula, Joe copula, BB1 copula, BB6 copula, BB7 copula, BB8 copula, survival copulas, and rotated copulas were candidates in the selection of the best one by using the model comparison criteria, AIC and BIC. This is beneficial in improving the flexibility of vine copulas. Last, we selected the best GARCH model from the 8 kinds of GARCH models with asymmetric distributions in terms of AIC and BIC for the stock returns of G7, thereby incorporating asymmetries and volatility within forecasting process.

The paper is organized as follows. In Section 2, the methodology are presented, which include the family GARCH models for marginals, multivariate copula constructions, and CES model. Section 3 analyzes the empirical results. Lastly, Section 4 concludes the paper.

2   Methodology

2.1   The Family GARCH Model

Hentschel [11] proposed a family of GARCH models that include 8 kinds of the most popular GARCH models, such as GARCH, TGARCH, GJR-GARCH, AVGARCH, NGARCH, NAGARCH, APARCH and ALLGARCH. Following Ghalanos [10], the family ARMA (p, q)-GARCH (k, l) model can be formulated as:

$$r_t = c + \sum_{j-1}^{p} \phi_j r_{t-j} + \sum_{j=1}^{q} \psi_j \epsilon_{t-j} + \epsilon_t, \tag{1}$$

$$\epsilon_t | \Psi_{t-1} = z_t \sigma_t, \tag{2}$$

$$\sigma_t^\lambda = \omega + \sum_{j=1}^{k} \alpha_j \sigma_{t-j}^\lambda \left(|z_{t-j} - \eta_{2j}| - \eta_{1j}(z_{t-j} - \eta_{2j}) \right)^\delta + \sum_{j=1}^{l} \beta_j \sigma_{t-j}^\lambda \tag{3}$$

where $\sum_{i=1}^{p} \phi_i < 1$. The z_t represents the standardized residuals, and Ψ_{t-1} denotes past information. The parameters λ, δ, η_1 and η_2 are used to decide the types of submodel. Various submodels are shown as follows: when $\lambda = \delta = 2$ and $\eta_{1,j} = \eta_{2,j} = 0$, it is GARCH model; when $\lambda = \delta = 1$ and $|\eta_{1j}| \leq 1$, the AVGARCH model is generated; when $\lambda = \delta = 2$ and $\eta_{2,j} = 0$, the GJR-GARCH model is performed; when $\lambda = \delta = 1$, $|\eta_{ij}| \leq 1$ and $\eta_{2,j} = 0$, the submodel is TGARCH model; when $\lambda = \delta$ and $\eta_{1,j} = \eta_{2,j} = 0$, the model becomes NARCH; when $\lambda = \delta = 2$ and $\eta_{1,j} = 0$, the NAGARCH is represented; when $\lambda = \delta$, $\eta_{2,j} = 0$ and $|\eta_{1k}| \leq 1$, it becomes APARCH; the full GARCH model is performed when $\lambda = \delta$.

To describe the possibly asymmetric and heavy-tailed characteristics of the stock index returns, the error term z_t is assumed to be a skewed student-t distribution (SSTD) or skewed generalized error distribution (SGED). The submodels of the family ARMA-GARCH model are estimated by maximum quasi-likelihood method. The best performance model with SSTD or SGED is selected in terms of most information criteria.

2.2   Multivariate Copulas

Sklar [16] first gave the definition of a copula as follows: Let $x = x_1, x_2, ..., x_n$ be a random vector with the joint distribution function H and the marginal distribution $F_1, F_2, ..., F_n$. Then, there exists a function C called copula function such that

$$F(x_1, x_2, ..., x_n) = C(F_1(x_1), F_2(x_2), ..., F_n(x_n)). \qquad (4)$$

Compared to bivariate copula function with some classical families of bivariate distributions such as bivariate normal and Student-t distributions, many copula models can capture asymmetric tail dependences and rank correlation except elliptical copulas. Elliptical copulas are simply the copulas of elliptically contoured distributions. Gaussian and T copulas are constructed from elliptical distributions via Sklar's Theorem [16]. Multivariate elliptical copulas are very popular in risk management due to their easy implementation. The linear correlation coefficients ρ in multivariate elliptical copulas have different structures. Normally, there are four kinds of parameter structures for elliptical copulas: autoregressive of order 1 (ar1), exchangeable (ex), Toeplitz (toep), and unstructured (un) (Yan [19]). The parameter structures of a three-dimension copula can be expressed as:

$$\begin{pmatrix} 1 & \rho_1 & \rho_1^2 \\ \rho_1 & 1 & \rho_1 \\ \rho_1^2 & \rho_1 & 1 \end{pmatrix}, \begin{pmatrix} 1 & \rho_1 & \rho_1 \\ \rho_1 & 1 & \rho_1 \\ \rho_1 & \rho_1 & 1 \end{pmatrix}, \begin{pmatrix} 1 & \rho_1 & \rho_2 \\ \rho_1 & 1 & \rho_1 \\ \rho_2 & \rho_1 & 1 \end{pmatrix}, \begin{pmatrix} 1 & \rho_1 & \rho_2 \\ \rho_1 & 1 & \rho_3 \\ \rho_2 & \rho_3 & 1 \end{pmatrix},$$

respectively. When modelling the joint distribution of multiple assets, such limited parameter models are unlikely to adequately capture the dependence structure. This means the multivariate elliptical copulas are inflexible in high dimensions. Fortunately, vine copulas can overcome these drawbacks. Vine copulas are multivariate modelling tool. They construct the multivariate distribution by building bivariate copulas of conditional distributions. A d-dimensional vine copula are built by $d(d-1)$ bivariate copulas in a $(d-1)$-level tree form. There are different ways to construct a copula tree. C-vines and D-vines are the selected tree types in this paper. If C-vines or D-vines are constructed by different pair-copula families, then this is called mixed C-vine or D-vine copula. Following Aas et al. [1], the d-dimensional densities of C-vine and D-vine copulas are given

$$f(x_1, x_2, ..., x_d) = \prod_{k=1}^{d} f(x_k) \times \prod_{j=1}^{d-1} \prod_{i=1}^{d-j} c_{j,i+j|1,L,j-1}(F(x_j|x_{1:j-1}), F(x_{i+j}|x_{i+j|1:j-1})), \qquad (5)$$

$$f(x_1, x_2, ..., x_d) = \prod_{k=1}^{d} f(x_k) \times \prod_{j=1}^{d-1} \prod_{i=1}^{d-j} c_{i,i+j|i+1,K,1+j-1}(F(x_i|x_{i+1:i+j-1}), F(x_{i+j}|x_{i+j1|i+1:i+j-1})),$$
$$(6)$$

respectively. The conditional distribution functions in C-vine and D-vine copulas can be gotten by the following equation

$$F(r|v) = \frac{\partial C_{r,v_j|v_{-j}}(F(r|v_{-j}), F(v_j, v_{-j}))}{\partial F(v_j|v_{-j}}, \qquad (7)$$

where v denotes all the conditional variables. Aas et al. [1] and Czado et al. [6] proposed maximum likelihood method to estimate vine copulas. The estimation can be explained by three steps. Firstly, the variables are ordered according to their influence that depends on the empirical Kendalls tau. Secondly, all bivariate copula families are individually selected in terms of AIC. Last, we follow the estimation method of Czado [6] who used sequential estimates as starting values and then estimate the vine copulas through the maximum likelihood estimation method, again.

2.3 Component Expected Shortfall

Banulescu and Dumitrescu [4] defined CES as that which measures the absolute contribution of a firm to the risk of the financial system; the systemic risk can be measured by linearly aggregating the component losses, as well. They are given by

$$CES_{it} = w_{it} \frac{\partial ES_{m,t-1}(VaR_\beta(W))}{\partial w_{it}} \tag{8}$$

$$CES\%_{it}(VaR_\beta(W)) = \frac{CES_{it}(VaR_\beta(W))}{SES_{t-1}(VaR_\beta(W))} \times 100 \tag{9}$$

and

$$SES_{t-1}(VaR_\beta(W)) = \sum_{i=1}^{n} CES_{it}(VaR_\beta(W)). \tag{10}$$

where $VaR_\beta(W)$ is the VaR under the β confidence level and the W portfolio allocations, and the ES is expected shortfall. This systemic risk measure by CES allows us to identify pockets of risk concentrations by directly ranking the institutions in term of their riskiness. The larger $CES\%_{it}$ demonstrates that the institution i is the more systemically important. With these formulas, the absolute contribution of a firm and the systemic risk can be measured under given weights. In this study, each index has equal weight. For out-of-sample systemic risk, Banulescu and Dumitrescu [4] show that it may be solved using Equation (11):

$$CES\%_{i,T+1:T+h}(C\%) = \frac{CES_{i,T+1:T+h}(C\%)}{\sum_{i=1}^{n} CES_{i,T+1:T+h}(C\%)} \tag{11}$$

where $C\%$ is set to be the out-of-sample VaR-HS for cumulative market returns at coverage rates of 1%, 2%, and 5%, and h represents a forecasting horizon. The multi-period ahead forecasts of CES are implemented by combining rolling window and Monte Carlo simulation methods with the C-vine copula-GARCH model. We used the mixed C-vine copula with the estimated parameters to generate random numbers 10,000. Thereafter, the inverse functions of the corresponding marginal distribution were employed to calculate standardized residuals of each variable. Then the predicted values of each stock return at one-ahead period can be obtained by using the preferable GARCH families. Thus, the CES and CES% can be obtained by equations (10) and (11).

3   Empirical Results

3.1   The Data

Our data set consists of G7 weekly returns on 7 indices including US S&P 500, Canada GSPTSE, England FTSE, France CAC40, Germany DAX, Italy FTSEMIB, and Japan NKY225. The sample period extends from 3 January 2000 to 29 December 2014, yielding 780 observations in total. All data are obtained from Thomson Reuters ECOWIN. We use the daily logarithmic returns defined as , where Pt is the close price of G7 stock indexes at time t. We partition the data into two parts: in sample and out of sample. The in-sample data from 3 January 2000, to 31 December, 2004, with 260 observations is used to estimate the parameters of the marginal models and the multivariate copulas. Thereafter, the 520 observations in the out of sample are used to estimate the multi-period forecasts of the CES% by using the principle of the weekly rolling window forecasting of returns. The Kendalls tau displayed in Table 1 reveals that the returns for each pair of indices are closely related to each other, with the biggest dependence being that between the France and Germany stock returns, and the smallest dependence is between Canada and Japan.

Table 1. The Kendalls tau of G7 stock returns

Countries	USA	Canada	England	France	Germany	Italy	Japan
USA	1.0000	0.5289	0.5987	0.6220	0.6081	0.5408	0.3467
Canada	0.5289	1.0000	0.4961	0.4855	0.4628	0.4333	0.3240
England	0.5987	0.4961	1.0000	0.6950	0.6240	0.5818	0.3510
France	0.6220	0.4855	0.6950	1.0000	0.7398	0.7092	0.3761
Germany	0.6081	0.4628	0.6240	0.7398	1.0000	0.6354	0.3768
Italy	0.5408	0.4333	0.5818	0.7092	0.6354	1.0000	0.3482
Japan	0.3467	0.3240	0.3510	0.3761	0.3768	0.3482	1.0000

3.2   Results for Copulas

Before we conducted the copula analysis, we had to consider marginal distribution for adjusting the return distribution because the return series has volatility clustering. The preferable model was selected from the 8 kinds of GARCH families for each stock return in terms of AIC and BIC. The two asymmetric distributions, SSTD and SGED, were performed with GARCH families. Table 2 shows that NAGARCH with SSTD model fits the US and England stock returns very well, while the SGED has the better performance than SSTD for others. Also it is obvious that ALLGARCH model is selected for France and Italy stock returns, and NAGARCH model is used to fit Germany and Japan stock returns, and the best fitting GARCH family of Canada stock return is NGARCH. Therefore, there exists asymmetric effect in the stock returns of G7.

After having the estimated parameters of marginal distributions, we turn to estimate the DCC model and the multivariate copulas, e.g., Gaussian, T,

Clayton, Frank, Gumble, mixed C-vine, Clayton C-vine, T C-vine, mixed D-vine, Clayton D-vine and T D-vine. The log-likelihood and AIC of each multivariate copula are shown in Table 3. According to the log-likelihood and AIC, we found that the performance of the multivariate Clayton copula was the worst. This could be because the data contained lower tail dependence which could not be measured properly by the Clayton copula. The values of all the criteria for both multivariate Gaussian and T copula were also not good. This implies that a purely elliptical approach may not be appropriate to measure the systemic interdependency of the series, especially when the tail dependence is asymmetric. It is no surprise that the mixed C-vine copula has the best performance among all the multivariate copulas. Although Low et al. [14] showed that Clayton C-vine copula has a good performance in managing portfolios of high dimensions and measuring VaR, the mixed C-vine copula should be more flexible corresponding to capture asymmetric dependences. Also, we use the preferable GARCH families to derive standard residuals, and input them into the DCC model (see Engle [8][9]). The values of log-likelihood and AIC of the DCC model are 119.1192 and −242.2384, respectively. This implies that the DCC model is worse than the copula models in terms of the log-likelihood and AIC.Therefore, this paper estimates the parameters of the mixed C-vine copula using maximum likelihood method, and combines mixed C-vine copula with CES to forecast the systemic risk and the contribution of each stock market to the entire financial system.

Table 4 reports the estimated parameters of the mixed C-vine copulas, the Kendalltau, the upper tail dependence coefficient and the lower tail dependence coefficient for G7 stock returns. "1", "2", "3", "4", "5", "6", and "7" represent the stock returns of US, Canada, England, France, Germany, Italy and Japan, respectively. The second column shows the best copula family of each pair in terms of AIC and BIC. "SBB1", "SJoe", "RJoe", and "SClayton" represent survival BB1, survival Joe, rotate Joe by 270 degrees and survival Clayton copulas, respectively. First, the mixed C-vine copula has been selected to be different from their special cases, the Gaussian copula and T copula. This indicates that a purely elliptical approach to measuring systemic interdependencies falls short of adequately capturing all relevant dependence characteristics, in particular the asymmetric tail behavior. Second, with the increasing of the amount of conditional variables, the conditional dependences become very tiny. Third, there does exist obvious asymmetric dependences between the stock returns of Germany and US, Germany and Canada. The estimated coefficient of lower tail dependence between Germany and US equals to 0.5528, which indicates the dependence between the stock market returns during bear markets is stronger than the dependence during bull markets. While the lower and upper dependency between Germany and Canada are about 0.4, which reflects there exists the strong dependency of Germany and Canada stock market returns during booms and crashes. Analogously, the symmetric dependencies between Germany and England, Germany and Italy also yield similar status. Last, the estimated conditional dependencies during some stock returns are linear and symmetric.

Table 2. The Selection of the GARCH Families

	GARCH	TGARCH	...	NGARCH	NAGARCH	ALLGARCH
US					SSTD	
CANADA			SGED			
ENGLAND				SSTD		
FRANCE						SGED
GERMANY				SGED		
ITALY						SGED
JAPAN					SGED	

Table 3. Model Selection for Different Copulas

Copulas	Num.paras	LogL.	AIC
Gaussian(ar1)	1	539.3931	-1076.79
Gaussian(ex)	1	587.5398	-1173.08
Gaussian(toep)	6	624.023	-1236.05
Gaussian(un)	21	798.7267	-1555.45
T(ar1)	2	582.4287	-1160.86
T(ex)	2	635.3522	-1266.7
T(toep)	7	664.3147	-1314.63
T(un)	22	823.1099	-1602.22
Clayton	1	457.9531	-913.906
Frank	1	515.6621	-1029.32
Gumbel	1	492.1471	-982.294
mixed C-vine	28	841.5398	-1627.08
Clayton C-vine	21	650.4606	-1258.92
T C-vine	42	834.5952	-1585.19
mixed D-vine	27	833.6134	-1613.23
Clayton D-vine	21	642.3388	-1242.68
T D-vine	42	836.0883	-1588.18

This may imply conditional variables have profound effect on the tail dependence of each pair.

3.3   Results for Systemic Risk

Figure 1 displays the evolution of CES from January, 2005 to December, 2014. We found that the pre-crisis period features lower levels of risk contribution, while risk contribution increased gradually while the crisis unfolds. The predicted CES peaked at the beginning of 2009 and then decayed slowly to a certain level, but it was still higher than pre-crisis period. This finding is consistent with the financial events that hit the stock market at that time. The two-month period from January 1-February 27 in 2009 represented the worst start to a year in the history of the S&P 500 with a drop in value of 18.62%. For the first quarter of 2009, the annualized rate of decline in GDP was 14.4% in Germany, 15.2% in Japan, 7.4% in the UK. In the spring of 2009, the CES reached extreme value

Table 4. Estimated Results of the Mixed C-vine Copula

Pairs	Copulas	Parameters	Estimators	Kendall's tau	Upper tail	Lower tail
51	SBB1	θ	0.24012***	0.5239	0.0044	0.5528
		δ	1.8754***			
52	BB1	θ	0.4729***	0.4479	0.395	0.3677
		δ	1.4651***			
53	T	ρ	0.8235***	0.616	0.4245	0.4245
		v	6.3822***			
54	Frank	θ	10.2376***	0.672	0	0
56	T	ρ	0.8641***	0.6643	0.3654	0.3654
		v	11.1301*			
57	Frank	θ	2.6674***	0.2775	0	0
12|5	Gaussian	ρ	0.5246***	0.3516	0	0
13|5	BB7	θ	1.158***	0.1601	0.1805	0.0325
			0.2023			
14|5	SJoe	θ	1.1215***	0.0652	0	0.1446
16|5	T	ρ	0.3741***	0.2441	0.0137	0.0137
		v	15.6637			
17|5	Frank	θ	0.7851**	0.0867	0	0
23|15	RJoe	θ	1.036***	-0.0204	0	0
24|15	SClayton	θ	0.0598	0.029	0	0
26|15	T	ρ	0.0974	0.0621	0.043	0.043
		v	6.2776**			
27|15	Gaussian	ρ	0.1794***	0.1148	0	0
34|125	Joe	θ	1.0923***	0.0504	0.1138	0
36|125	Clayton	θ	0.1734***	0.0797	0	0.0183
37|125	Clayton	θ	0.0476***	0.0232	0	0
46|1235	Gaussian	ρ	0.3334	0.2164	0	0
47|1235	Gaussian	ρ	0.0999***	0.0637	0	0
67|12345	Clayton	θ	0.1646**	0.076	0	0.0148

Note: *, ** and *** represent significance at levels 10%, 5%, and 1%, respectively.

again. This should be ascribed to Greece's sovereign credit rating that was downgraded to junk. And four day after the activation of a 45-billion EU-IMF bailout was performed thereby triggering the decline of stock markets worldwide and of the Euro's value, and furthering a European sovereign debt crisis. Moreover, the Dow Jones Industrial Average suffered its worst intra-day point loss, dropping nearly 1,000 points on 6th May, 2010. The series of predicted CES obtained a high level from August of 2011 to the beginning of 2012, which can be explained by the sharp drop in stock prices of US, Europe, and Asia in August 2011. Considering the slow economic growth of the United States and contagion of the European sovereign debt crisis to Spain and Italy, a severe volatility of stock market indexes continued for the rest of the year 2011 and the first quarter of 2012. Figure 2 displays the multi-period forecasts of CES% at coverage rates of 1%, 2%, and 5% from January 2005 to December 2014. Taking it by and large, the contribution of each stock market to the aggregate financial risk is

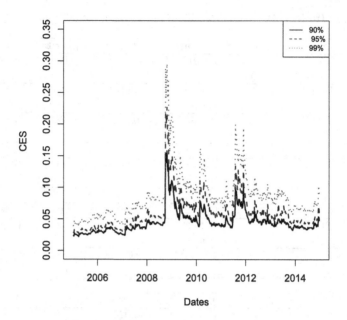

Fig. 1. The Component Expected Shortfall From 2005 to 2014

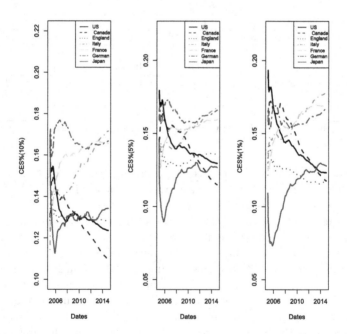

Fig. 2. The Contribution of Each Stock Market to the Entire Financial System

not invariant. Germany, Italy, and France have the greater contribution to the systemic risk than other stock markets. This is maybe because the crisis rapidly developed and spread into a global economic shock, resulting in a number of European bank failures, and declined in various stock indexes. Especially, the economic system in Germany was deeply hit by the financial crisis. In 2008, the annual economic growth rate fell to 1% and in 2009 it even became negative at -4.7%. The contribution of US stock market to systemic risk is not consistent with what we imagined. This means that the strong contagion caused a severe effect to the Europe an union, and the Europe debt crisis undoubtedly worsened the already grave situation.

4   Conclusions

In this paper, we extended existing CES-based bivariate GARCH models for measuring systemic risk by proposing a model that incorporates multivariate copula-GARCH and CES. The model is used to gauge the systemic risk of G7 stock markets and their dependences. In the dependence analysis we found the mixed C-vine copula has better performance than other multivariate copulas, which gives evidence of non-elliptical structures, especially of asymmetric tail behavior. It is crucial to take into account in systemic risk. In systemic risk analysis, we found that the pre-crisis period featured lower levels of risk contribution, while risk contribution increased gradually as the crisis unfolded. Also, the systemic risk of G7 obtained a high level after financial crisis due to the Europe debt crisis. The present article focuses on studying systemic risk and dependences of G7 stock market returns. Further research might examine the financial contagion and portfolio analysis for G7.

Acknowledgment. The financial support from the Puay Ungphakorn Centre of Excellence in Econometrics is greatly acknowledged. The authors are very grateful to Professor Vladik Kreinovich and Professor Hung T. Nguyen for their comments.

References

1. Aas, K., Czado, C., Frigessi, A., Bakken, H.: Pair-copula constructions of multiple dependence. Insurance: Mathematics and Economics **44**(2), 182–198 (2009)
2. Acharya, V., Pedersen, L., Philippe, T., Richardson, M.: Measuring Systemic Risk. Technical Report, Department of Finance, NYU (2010)
3. Ba, C.: Recovering copulas from limited information and an application to asset allocation. Journal of Banking & Finance **35**(7), 1824–1842 (2011)
4. Banulescu, G.D., Dumitrescu, E.I.: Which are the SIFIs? A Component Expected Shortfall (CES) Approach to Systemic Risk. Journal of Banking & Finance **50**, 575–588 (2015)
5. Brownlees, T.C., Engle, R.F.: Volatility, Correlation and Tails for Systemic Risk Measurement, Working Paper, NYU-Stern (2012)
6. Czado, C., Schepsmeier, U., Min, A.: Maximum likelihood estimation of mixed C-vines with application to exchange rates. Statistical Modelling **12**, 229–255 (2012)

 7. Embrechts, P., McNeil, A., Straumann, D.: Correlation and dependence in risk management: properties and pitfalls. In: Dempster, M.A.H. (ed.) Risk Management: Value at Risk and Beyond. Cambridge University Press, Cambridge (2002)
 8. Engle, R.: Dynamic Conditional Correlation: A Simple Class of Multivariate Generalized Autoregressive Conditional Heteroskedasticity Models. Journal of Business and Economic Statistics **20**(3), 339–350 (2002)
 9. Engle, R.: Anticipating Correlations: A New Paradigm for Risk Management. Princeton University Press, Princeton (2009)
10. Ghalanos, A.: Introduction to the rugarch package (2012). http://cran.rproject. org/web/packages/rugarch Version 1.0-11
11. Hentschel, L.: All in the family nesting symmetric and asymmetric garch models. Journal of Financial Economics **39**(1), 71–104 (1995)
12. Liu, J., Sriboonchitta, S.: Analysis of volatility and dependence between the tourist arrivals from China to Thailand and Singapore: a copula-based GARCH approach. In: Huynh, V.N., Kreinovich, V., Sriboonchitta, S., Suriya, K. (eds.) Uncertainty Analysis in Econometrics with Applications. AISC, vol. 200, pp. 283–294. Springer, Heidelberg (2013)
13. Liu, J., Sriboonchitta, S., Denœux, T.. Economic Forecasting Based on Copula Quantile Curves and Beliefs. Thai Journal of Mathematics, 25–38 (2014)
14. Low, R.K.Y., Alcock, J., Faff, R., Brailsford, T.: Canonical vine copulas in the context of modern portfolio management: Are they worth it? Journal of Banking & Finance **37**, 3085–3099 (2013)
15. Patton, A.J.: On the out-of-sample importance of skewness and asymmetric dependence for asset allocation. Journal of Financial Econometrics **2**(1), 130–168 (2004)
16. Sklar, M.: Fonctions de répartition à n dimensions et leurs marges, vol. 8, pp. 229–231. Publications de l'Institut de Statistique de L'Université de Paris (1959)
17. Sriboonchitta, S., Nguyen, H.T., Wiboonpongse, A., Liu, J.: Modeling volatility and dependency of agricultural price and production indices of Thailand: Static versus time-varying copulas. International Journal of Approximate Reasoning **54**(6), 793–808 (2013)
18. Wu, C.C., Liang, S.S.: The economic value of range-based covariance between stock and bond returns with dynamic copulas. Journal of Empirical Finance **18**, 711–727 (2011)
19. Yan, J.: Enjoy the Joy of Copulas: With a Package copula. Journal of Statistical Software **21**(4), 1–21 (2007)

Business Cycle of International Tourism Demand in Thailand: A Markov-Switching Bayesian Vector Error Correction Model

Woraphon Yamaka[✉], Pathairat Pastpipatkul, and Songsak Sriboonchitta

Faculty of Economics, Chiang Mai University, Chiang Mai, Thailand
woraphon.econ@gmail.com, ppthairat@hotmail.com

Abstract. This paper uses the Markov-switching Bayesian Vector Error Correction model (MS-BVECM) model to estimate the long-run and short-run relation for Thailand's tourism demand from five major countries, namely Japan, Korea, China, Russia, and Malaysia. The empirical findings of this study indicate that there exist a long-run and some short-run relationships between these five countries. Additionally, we analyses the business cycle in a set of five major tourism sources of Thailand and find two different regimes, namely high tourist arrival regime and low tourist arrival regime. Secondary monthly data results of forecasting were used to forecast the tourism demand from five major countries from November 2014 to August 2015 based on the period from January 1997 to October 2014. The results show that Chinese, Russian, and Malaysian tourists played an important role in Thailand's tourism industry during the period from May 2010 to October 2014.

Keywords: MS-VECM · Bayesian · Thailand's tourism demand · Forecasting

1 Introduction

Thailand was ranked 10th in the list of world international tourism destinations in 2013. It is a famous destination in South-east Asia because it has many attractive places such as historical sites, both old traditional and new traditional, beaches, and scenic mountains. Despite severe political conflicts, Thailand's tourism industry has grown rapidly in the past decade to become one of the industries most important to Thai economy. According to a report by the WTTC, the total contribution of travel and tourism to the gross domestic product (GDP) in Thailand is accounted to be between 9% and 10% in 2013. In addition, the council forecasts that the total contribution to the GDP will rise to 22.7% of the GDP by 2024. However, Thailand's tourism industry is faced with high competition from all the major tourism markets in Southeast Asia because these markets are of similar nature and they propose promotions to attract tourists.

Recently, the total volume of international tourist arrival in Thailand was found to have reached 26 million. Chinese tourists retain their position of 1st rank and account for the most arrivals in Thailand in terms of nationality, with 17.6% shares of the total inbound tourists, while Malaysian tourists hold the 2nd rank with 11.2%, Japanese

© Springer International Publishing Switzerland 2015
V.-N. Huynh et al. (Eds.): IUKM 2015, LNAI 9376, pp. 415–427, 2015.
DOI: 10.1007/978-3-319-25135-6_38

tourists, the 3rd rank, with 5.8%, Korean tourists, the 4th rank, with 4.9%, and Russian tourists, the 5th rank, with 3.5%, according to the Tourism Authority of Thailand (TAT). This indicates that these five countries can be considered as priority markets for Thailand. Thus, it is important to study the movement, both short-run and long-run, and forecast the tourism demand from these five counties in order to provide an accurate number of tourists in each regime and provide the necessary information for policy makers as well as private and government tourism industry sectors. Additionally, this study can help private and government tourism industry sectors to adjust and adapt their strategies for sustainable tourism industry development.

In this study, we aim to construct an econometric model to forecast and estimate the long-run and short-run dynamics for five major markets of Thailand's tourism industry using the cointegration and error correction approach. Wong [16] studied the business cycle in in-ternational tourist arrival and found that there may be a cyclical trend in the international tourist arrival. Song and Li [11] and Witt.S add Witt.C [15] reviewed the published studies on tourism demand modeling and forecasting and found that there exist a tourism cycle and seasonality that is discussed in many studies. In addition, Chaitip and Chaiboonsri [1] captured the two regimes, high season period and low season period, in the number of international tourist arrivals in Thailand. These findings indicate that Thailand seems to have a tourism season; hence, we extend the model to a nonlinear process, and employ the Markov-switching Bayesian Vector error correction model (MS-BVECM). We found that this model was more flexible to use than the conventional MS-VECM because the expectation–maximization (EM) algorithm used to estimate the unobserved Markov state variables is not conditional on the value of the cointegration vector and other parameters. Furthermore, we found a numerous studies which were related to this study such as Mazumder, Al-Mamun, Al-Amin, and Mohiuddin [10] Harcombe [4], Chang, Sriboonchitta and Wiboonpongse [2], Chaiboonsri, Sriwichailamphan, Chaitip and Sriboonchitta [3], and Liu and Sriboonchitta [7].They employed the classical estimation, Maximum likelihood estimator (MLE), to estimate the parameters which are treat as a constant and do not allow putting a prior belief about the likely values of estimated parameter. Thus, the estimated results are not accurately. In this study, the Bayesian Estimation is preferred rather than MLE. The model outperform the MLE because we can give the prior on the estimated parameter and lead the estimation to gain more efficiency.

The rest of the paper is organized as follows. Section 2 discusses the MS-BVECM model. The data description and the estimation results are presented in section 3 and 4. Finally, section 5 summarizes and concludes the paper.

2   Methodology

2.1   Markov Vector Error Correction Model

Following MS-VECM of Jochmann and Koop [5], the model which allows the intercept term, the adjustment term, the cointegrating term, the autoregressive term, and the variance–covariance matrix subject to the unobserved state variable can be written as

$$\Delta y_t = c_{S_t} + \sum_{i=1}^{k} \Gamma_{iS_t} \Delta y_{t-i} + \Pi_{S_t} y_{t-i} + \varepsilon_t \tag{1}$$

where Δy_t denotes an $I(1)$ vector of the n-dimensional time series data, c is a vector of the state-dependent intercept term, Γ_i stands for the state-dependent $n \times n$ autoregressive parameter matrices of the vector Δy_{t-i}, Π_{S_t} is the state-dependent error correction term, the long-run impact matrices, which are defined by $\Pi_{S_t} = \alpha_{S_t} \beta'$, α_{S_t} is the state-dependent adjustment term, the state-dependent β is the $r \times n$ matrix of the cointegrating vectors, and ε_t is the error term which is assumed to be $N(0, \Sigma(S_t))$. Thus, Eq. (1) can be rewritten as follows:

$$\Delta y_t = c_{S_t} + \sum_{i=1}^{k} \Gamma_{iS_t} \Delta y_{t-i} + \alpha_{S_t} \beta' y_{t-1} + \varepsilon_t \tag{2}$$

In this study, the state variable with the first order Markov process is assumed. Thus, the N state variables ($S_t = 1, \ldots, N$) and the transition probabilities can be defined as $P_{ij}(s_t = i | s_{t-1} = j)$, $i, j = 1, \ldots, N$

$$P = \begin{bmatrix} p_{11} & p_{21} & \cdots & p_{N1} \\ p_{12} & p_{22} & \cdots & p_{N2} \\ \vdots & \vdots & \cdots & \vdots \\ p_{1N} & p_{2N} & \cdots & p_{NN} \end{bmatrix}, \tag{3}$$

where p_{ij} is the probability that regime i is followed by regime j.

2.2 Prior Distributions

We select flat prior density for the intercept term (c), the autoregressive term (Γ), the adjustment term (α), and the cointegrating vector (β); Inverted Wishart prior for the variance–covariance matrix (Σ), and Dirichlet prior for the transition probabilities (p_{ij}).

For a prior for c, Γ, α, and β, we assume these parameters as Least Informative Priors, that is, flat priors, where the prior is simply a constant.

$$p(\theta) = k = \frac{1}{b-a} \quad for \ a \leq \theta \leq b \tag{4}$$

with a flat prior, the posterior is just a constant times the likelihood,

$$p(\theta|x) = C\ell(\theta|x) \tag{5}$$

where θ contains c, Γ, α, and β which have a uniform distribution from negative infinity to positive infinity. This allows the posterior distribution to be mainly affected by data rather than prior information.

With regard to the prior for Σ, we assume the variance–covariance matrix to be with the degree of freedom h

$$\Sigma_i \sim IW(\Phi_i, h_i) \quad i = 1, \dots, N \tag{6}$$

where $\Phi_i \in R^{n \times n}$, as Inverse Wishart prior for the transition probabilities $p_{ij}; i, j$ $= 1, \dots, N$, in this study. We assume that the two regimes are the high tourism arrival regime and the low tourism arrival regime; thus, we assign a beta distribution, assuming $N = 2$. To summarize, the likelihood function for c, Γ, β, Σ, and $\tilde{s}_t$ is given by

$$\xi(B, \beta, \Sigma_1, \dots, \Sigma_n, \tilde{s}_T \Big| Y) \propto \left(\prod_{i=1}^{n} |\Sigma_i|^{-t_i/2} \right) \exp\left(-\frac{1}{2} tr\left[\sum_{i=1}^{n} \{ \Sigma_i^{-1}(Y_i - W_i B)'(Y_i - W_i B) \} \right] \right) \tag{7}$$

$$= \xi(B, \beta, \Sigma_1, \dots, \Sigma_n, \tilde{s}_T \Big| Y) \propto \left(\prod_{i=1}^{n} |\Sigma_i|^{-t_i/2} \right) \exp\left(-\frac{1}{2} tr\left[\sum_{i=1}^{n} \{ vec(Y_i - W_i B)'(\Sigma_i \otimes I_\tau)(vec(Y_i - W_i B)) \} \right] \right) \tag{8}$$

where $B = \{c, \Gamma\}$ and $W = (I_1 Z\beta, \dots, I_{1n} Z\beta)$; $Z = (Y_{p-1}, \dots, Y_{T-1})'$, and t_i is the number of observations when $s_t = i, i = 1, 2$.

2.3   Posterior Specifications

The posterior densities were obtained from the priors and the likelihood functions. Sugita [13] proposed two steps for posterior estimation via Gibbs sampling. First, by using the multi-move Gibbs sampling method, the state variable $\tilde{s}_\tau = \{ s_\tau = 1, s_\tau = 2 \}$ should be estimated, and then the posterior densities for the intercept term, the adjustment term, the cointegrating term, the autoregressive term, and the variance–covariance matrix should be estimated.

To sample the state (or regime) variable ($\tilde{s}_t$), Kim and Nelson [8] proposed the multi-move Gibbs sampling to simulate the state variable $\tilde{s}_t$ as a block from the following conditional distribution

$$p(s_\tau|s_{\tau+1}, \Theta, Y) = \frac{p(s_{\tau+1}|s_\tau, \Theta, Y) p(s_\tau|\Theta, Y)}{p(s_{\tau+1}|\Theta, Y)} \alpha p(s_{\tau+1}|s_\tau) p(s_\tau|\Theta, Y) \tag{9}$$

where $p(s_\tau | s_{\tau+1})$ is the transition probability, $\Theta = \{B, \beta, \Sigma_1, \Sigma_2, p_{11}, p_{22}\}$, and $p(s_\tau | \Theta, Y)$ is the probability of staying at t conditional on the vector of variables and other parameters. We can compute it from the Hamilton filter.

After drawing $\tilde{s}_\tau$, we generate the transition probabilities, p11 and p22; this takes a draw from the Dirichlet posterior by multiplying Eq. (7) and Eq. (8) by the likelihood function, Eq. (13). We get

$$p(p_{11}, p_{22} | \bar{s}_t) \alpha \, p_{22}^{u_{22}+n_{22}-1} (1-p_{22})^{u_{21}+n_{21}-1} \, p_{11}^{u_{11}+n_{11}-1} (1-p_{11}^{u_{12}+n_{12}-1}) \qquad (10)$$

Then, to estimate Σ, B, and β, Gibbs sampling can be used to generate sample draws from the staring values for p_{ik}^0, B^0, β^0, and Σ^0 to generate p_{ik}^{j+1}, β^{j+1}, Σ_i^{j+1}, and B^{j+1}, and then repeat this previous iteration for N times, $j = 1, \ldots, N$. As a result, we obtain the posterior means and the standard deviations of these updated parameters. Note that the number of burn-in iterations N_0 for the Gibbs sampling should be large enough for handling the non-stationary effect of the initial values. this study, we construct the posterior of an MS-BVECM by using a pilot drawing 10,000 times from the basic algorithm for Gibbs sampling and discarding the first 2000 as burn-in.

3   Empirical Results

3.1   Results of Unit Root Test and Seasonal Unit Root Test

In this study of international tourism demand to Thailand from five major countries comprising Japan, Korea, China, Russia, and Malaysia, the number of tourist arrivals from those origins is used to represent the tourism demand to Thailand. The data are monthly time series data for the period from January 1997 to October 2014, which is collected from Thomson Reuters DataStream, from Financial Investment Center (FIC), Faculty of Economics, Chiangmai University. Additionally, we transform these variables into logarithms before estimating. In this study, we employ the ADF test statistics to analyze the order of integration of our variables. The result of the ADF test statistics provide that logarithm of Japan, Korea, China, Russia, and Malaysia are *I(1)*.

3.2   Lag Length Selection

In this section, we have to specify the lag length for the MS-BVECM model. We employ the vector error correction lag length criteria to find the appropriate number of lag lengths. For the VECM lag length criteria, the result reveals that the AIC and the

HQIC values for p=2 are lower than those for p=1 and p=3, while the BIC value is the lowest for p=1. The different results are obtained from these different information criteria. However, in this study, we prefer the results of the BIC tests, which point to the lag length, p=1.

Table 1. VECM Lag Length Criteria

Lag	AIC	HQIC	BIC
0	2.3597	2.3925	2.4408
1	−5.0143	−4.8176	−4.5281*
2	−5.2171*	−4.8565*	−4.3256

Source: Calculation.

3.3   Cointegration Test

To determine the rank or the number of cointegration vectors, trace tests are carried out, as shown in Table 3; we test the rank of the long-run relationship using Johansen's trace test which is obtained from BVECM with a Minnesota prior. In this study, we specify a tightness parameter, a decay parameter, and a parameter for the lags of the variables as 0.10, 0.10, and 0.50, respectively. Based on the trace statistics test, the null of one or fewer cointegrating vectors is rejected at the 5% significance level. This indicates that the model has one cointegrating vectors; therefore, the study chooses r=1.

Table 2. Cointegration Test

Bayesian Error Correction Model Johansen MLE estimates				
NULL	Trace Statistics	Crit. 90%	Crit. 95%	Crit. 99%
r <= 0	90.186	65.82	69.819	77.82
r <= 1	45.753	44.493	47.855	54.682
r <= 2	21.128	27.067	29.796	35.463

Source: Calculation.

Table 3. Transition matrix of MS(2)-VECM(1)

	p_{1t}	p_{2t}	Duration	Observation	BIC
Regime 1	0.968	0.019	52.63	210	27.7525
Regime 2	0.032	0.981	31.25	4	

Source: Calculation; *Note:* () is a standard error and "*," "**," and "***" denote rejections of the null hypothesis at the 10%, 5%, and 1% significance levels, respectively.

3.4   Estimates of MS(2)-VECM(1)

Regime 1 Equation A

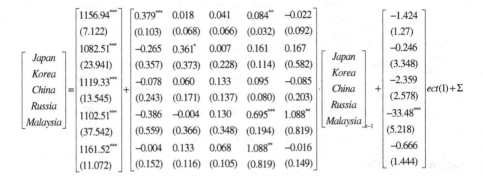

Regime 2 Equation B

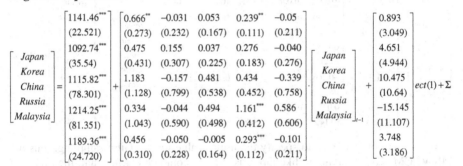

Table 4. Transition matrix of MS(2)-BVECM(1)

	p₁ₜ	p₂ₜ	Duration	Observation	BIC
Regime 1	0.961	0.026	25.64	68	21.8271
Regime 2	0.039	0.974	38.46	146	

Source: Calculation. *Note:* () is a standard error and "*," "**," and "***" denote rejections of the null hypothesis at the 10%, 5%, and 1% significance levels, respectively.

3.5   Estimates of MS(2)-BVECM(1)

Regime 1 Equation C

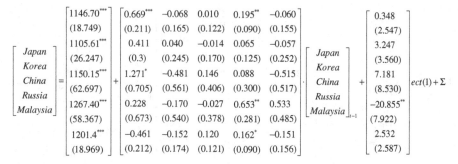

Regime 2 Equation D

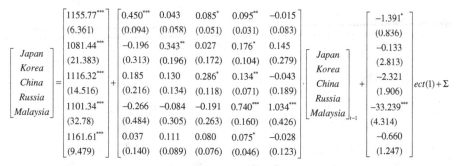

In this section, we need to specify the initial value for the Gibb sampling, thus the MS-VECM based Maximum likelihood is, firstly, computed. After that, the obtained parameters in the first step are used as a starting value to compute the posterior. The MS-VECM and MS-BVECM have been computed and shown in Equation A, B, C and D. Before, we interpret the results, table 4 and 5 provide the interesting result that the BIC value of MS-BVECM is lower than the BIC value of MS-VECM. Thus, it confirms that the Bayesian estimation in our model is outperforming the conventional Maximum likelihood estimation.

According to the estimated result of the final MS(2)-BVECM(1). We can see that the intercept term seems to have an economic interpretation.[1] The value of the intercept term in the first regime is less than that in the second regime. Therefore, we interpret that regime 1 is the high tourist arrival regime, while regime 2 is the low tourist arrival regime. To capture the short-run equilibrium relation, we consider the short-run coefficients for all the variables, which are presented in the regime 1 equation, and find that there exist some short-run relationships between these five countries, especially in the low tourist arrival regime. For the long-run relationship, we take into consideration the coefficients of the error correction terms (ECT1), which show the speed of adjustment of long-run equilibrium in the period of the study. The

---

[1] VARGAS III [14] has interpreted the regime based on either the intercept term or the variance term (for more details, see VARGAS III, 2009).

results in regime 1 show that the adjustment terms mostly have positive signs, except for Russia. We found that ECT1 of the Russian equation shows a negative sign with statistical significance and this indicates that there is a long-run relationship for tourism demand to Thailand in regime 1. For regime 2, there is also only one coefficient of adjustment that is statistically significant with a negative sign in ECT1 thereby indicating the existence of the long-run relationship in this regime. In addition, Table 5 provides the results of transition probabilities which show that their regime 1 and regime 2 are persistent because the probability of staying in regime 1 is 96.1 percent and the probability of staying in regime 2 is 97.4 percent, while the probability of moving between these regimes is less than 5 percent. This indicates that only an extreme event can switch the series to change from regime 2 to regime 1. Moreover, the result also shows that the high tourism arrival regime has a duration of approximately 68 months, while the low tourism arrival regime has a duration of 146 months.

The estimated MS-BVECM model also produces smoothed probabilities which can be understood as the optimal inference on the regime using the full-sample information [9]. The time paths of the smoothed probability for the high tourist arrival regime, between the periods of 1997 and 2014, are plotted in Figure 1. We observe that the low tourist arrival regime is spread over a long period between 1997 and late 2007. As for the high tourist arrival regime, it has been captured during the period since 2008 to the present; however, the recession period was found with very short duration in the mid of 2009. This result seems to correspond to the event in which the National United Front of Democracy against Dictatorship (UDD) stormed the Fourth East Asia Summit in Pattaya, resulting in the declaration of a state of emergency in Bangkok and five neighboring provinces. Moreover, we also found that Japan, China, Malaysia, Russia, and Korea had warned their citizens to avoid traveling to Thailand during those periods. However, the trend of tourist arrivals was observed to switch to the high tourist arrival regime again after that. We found that the Ministry of Tourism and Sport proposed tourism policies such as regional road shows, partnerships with airlines and tour agencies, and hard-sell marketing events in 2010. Additionally, the Suvarnabhumi Airport has increased the capacity to accommodate up to a 100 million passengers per year thereof. As a result, Thailand has the potential, and it does attract a large number of tourists.

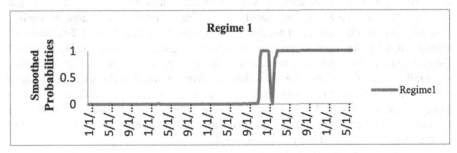

Fig. 1. The high tourist arrival regime of MS-BVECM[2]

---

[2] Note: The low tourist arrival regime of MS-BVECM is the inverted Figure 1.

3.6   Forecasting of International Tourism Demand in Thailand

In this section, we extend our model to forecast and predict Japanese, Korean, Chinese, Russian, and Malaysian tourist arrivals in Thailand in the two different regimes during the period from November 2014 to August 2015 without transform any variables series since we aim to forecast the real number of tourist arrivals in Thailand. This model become more flexible than other model such as ARIMA and regression since these model need transform the data to be stationary before estimating. For the forecasting process, firstly, we separate the observation into two regimes using smoothed probability. For a case of two regimes, Krolzig [9] suggested assigning the observation to the first regime if the smooth probability of the observation is more than 0.5 and assigning the observation to the second regime if the smoothed probability of the observation is less than 0.5. Secondly, we employ a BVECM with a Minnesota prior and specify a tightness parameter, a decay parameter, and a parameter for the lags of the variables as 0.10, 0.10, and 0.50, respectively, and employ first order contiguity of the states as weights, as in Lesage [6] to forecast the number of tourist arrivals in Thailand in each regime. Thirdly, the two regimes forecast values at time $t = (1,...,T)$ which obtained from the previos step are ,then, conducted to construct the MS-BVECM forecasting at time $t = (1,...,T)$ using weighted average of forecast value in each regime[3].   Finally, For the prediction period from November 2014 to August 2015, we tried to make the prediction without weighting the forecasting of two regime in order to provide a prediction of the number of tourist arrival in 2 regimes.

The results presented in Figure 2, Figure 3, Figure 4,   5 and Figure 6 show a comparison between the forecasting performance accuracy and the actual number of tourist arrivals in Thailand during the period from May 2010 to October 2014. In addtion, the model also plots a prediction of the number of tourist arrivals in Thailand for the period from November 2014 to August 2015. Upon observing the results, it is evident that the forecasting trends for tourists during the comparison periods exceed the actual numbers. It is surprising to see that the forecasting trend in every country exceeds the actual tourism number. Sookmark [12] suggests that tourists are sensitive to shocks in the destination country, such as an unstable political situation in Thailand, and so, such shock may, in consequence, become the cause of an overestimation. For the prediction period, we are not restrict the   we find that Chinese, Russian, and Malaysian tourists play an important role in Thailand's tourism industry because the number of tourist arrivals from these countries to Thailand will continue to grow for the next ten years in both the regimes. By contrast, Japanese and Korean tourists are found to have a low number of tourist arrivals in Thailand, especially Korean tourists. We find that the Korean tourism demand has a large difference between regime 1 and regime 2; thus, this indicates that if Thailand has an extremely severe event, for example, a severe political event, Korean tourists are certain to be not confident enough to travel to Thailand, which will result in a large drop in the Korean tourist arrivals in Thailand.

---

[3] To weight the forecast value at time $t = (1,...,T)$, we use the filter probabilities to weight the forecast of 2 regimes in order to generate average forecasting of the MS-BVECM .

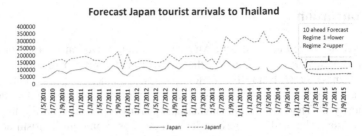

Fig. 2. The forecast of Japanese tourist arrivals in Thailand.

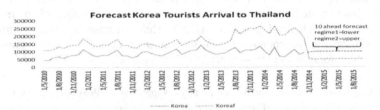

Fig. 3. The forecast of Korean tourist arrivals in Thailand.

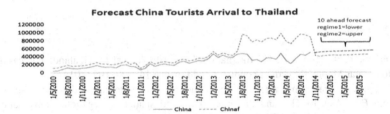

Fig. 4. The forecast of Chinese tourist arrivals in Thailand.

Fig. 5. The forecast of Russian tourist arrivals to Thailand.

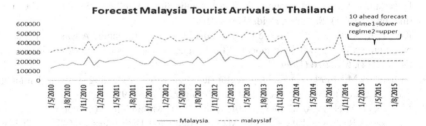

Fig. 6. The forecast of Malaysian tourist arrivals in Thailand.

4   Conclusion

In this study, we aimed to construct a Markov-switching Bayesian Vector Autoregressive model (MS-BVAR) to estimates the long-run and short-run relation for Thailand's tourism demand from five major countries, namely Japan, Korea, China, Russia, and Malaysia. Based on the above results, it might be suggested that the Tourism Authority of Thailand (TAT) might support further research on the impacts of factors affecting Thailand's   tourism demand such as promoting tourism activities, improving safety and security in tourism, reducing or controlling the cost of living of tourists, and improving the standard of service and quality as well as controlling the political situation in the countiy in order to increase the number of international tourist arrivals in Thailand. Additionally, the Thai government should control the politcal situation in the country and ensure that an atmosphere of peace prevails in order to attract foreign tourists — especially, Korean tourists — to Thailand.

Acknowledgement. We are grateful for financial support from Puay Ungpakorn Centre of Excellence in Econometrics, Faculty of Economics, Chiang Mai University. We thanks Nguyen, H. T. and Kreinovich, V. for valuable comment to improve this paper.

References

1. Chaitip, P., Chaiboonsri, C.: International Tourists Arrival to Thailand: Forecasting by Non-linear Model. Procedia Economics and Finance **14**, 100–109 (2014)
2. Chang, C.L., Sriboonchitta, S., Wiboonpongse, A.: Modelling and forecasting tourism from East Asia to Thailand under temporal and spatial aggregation. Mathematics and Computers in Simulation **79**(5), 1730–1744 (2009)
3. Chaiboonsri, C., Sriboonjit, J., Sriwichailamphan, T., Chaitip, P., Sriboonchitta, S.: A panel cointegration analysis: an application to international tourism demand of Thailand. Annals of the University of Petrosani, Economics **69** (2010)
4. Harcombe, D.: The economic impacts of tourism. ABAC Journal **19**(2) (1999)
5. Jochmann, M., Koop, G.: Regime-switching cointegration. Studies in Nonlinear Dynamics and Econometrics (2011)
6. LeSage, J.P.: Spatial econometrics. Regional Research Institute, West Virginia University (1999)
7. Liu, J., Sriboonchitta, S.: Analysis of volatility and dependence between the tourist arrivals from China to Thailand and Singapore: a copula-based GARCH approach: implications for risk management. In: Huynh, V.N., Kreinovich, V., Sriboonchitta, S., Suriya, K. (eds.) Uncertainty Analysis in Econometrics with Applications. Intelligent Systems and Computing, pp. 283–294. Springer, Heidenberg (2014)
8. Kim, C.J., Nelson, C.R.: Has the US economy become more stable? A Bayesian approach based on a Markov-switching model of the business cycle. Review of Economics and Statistics **81**(4), 608–616 (1999)
9. Krolzig, H.M.: Markov switching vector autoregressions: modeling, statistical inference and application to business cycle analysis. Springer, Berlin (1997)
10. Mazumder, M.N.H., Al-Mamun, A., Al-Amin, A.Q., Mohiuddin, M.: Economic Impact of Tourism–A Review of Literatures on Methodologies and Their Uses: 1969–2011 (2012)

11. Song, H., Li, G.: Tourism demand modelling and forecasting—A review of recent research. Tourism Management **29**(2), 203–220 (2008)
12. Sookmark, S.: An Analysis of International Tourism Demand in Thailand. Doctor of Philosophy' Thesis, School of development Economics, National Institute of Development Administration (2011)
13. Sugita, K.: Bayesian Analysis of Markov Switching Vector Error Correction Model (No. 2006-13). Graduate School of Economics, Hitotsubashi University (2006)
14. Vargas, G.A., I., II.: Markov switching var model of speculative pressure: An application to the asian financial crisis. (Order No. 1483224, Singapore Management University (Singapore)) (2009)
15. Witt, S.F., Witt, C.A.: Forecasting tourism demand: A review of empirical research. International Journal of Forecasting **11**(3), 447–475 (1995)
16. Wong, K.K.: The relevance of business cycles in forecasting international tourist arrivals. Tourism Management **18**(8), 581–586 (1997)

Volatility Linkages Between Price Returns of Crude Oil and Crude Palm Oil in the ASEAN Region: A Copula Based GARCH Approach

Teera Kiatmanaroch[1], Ornanong Puarattanaarunkorn[2],
Kittawit Autchariyapanitkul[3], and Songsak Sriboonchitta[4]([✉])

[1] Department of Agricultural Economics, Faculty of Agriculture,
Khon Kaen University, Khon Kaen 40002, Thailand
teera@kku.ac.th
[2] Department of Economics, Faculty of Management Sciences,
Khon Kaen University, Khon Kaen 40002, Thailand
pornan@kku.ac.th
[3] Faculty of Economics, Maejo University, Chiang Mai 50200, Thailand
kittawit_autchariya@cmu.ac.th
[4] Faculty of Economics, Chiang Mai University, Chiang Mai 50200, Thailand
songsakecon@gmail.com

Abstract. This paper used the copula based ARMA-GARCH to examine the dependence structure between the weekly prices of two commodities, namely Crude oil and Crude palm oil. We found evidence of a weak positive dependence between two commodities prices. These findings suggest that the crude oil market of the Middle East and the crude palm oil market of Malaysia are linked together. This information is useful for decision making in various area, such as the risk management in financial field and the international trade in agricultural commodities.

Keywords: Price volatility · Crude oil · Palm oil · Copula · ARMA-GARCH

1 Introduction

Crude oil (CO) and crude palm oil (CPO) are commodities related to energy security and food security, respectively because they are used as a primary input factor in the supply chain of necessary goods and services in our daily lives. It is well known that CO is a primary input factor for energy industries and petrochemical industries. CPO can be modified as cooking oil, margarine, ice-cream, soap, and as an ingredient for animal feedstock. Also, it can be used to produce alternative energy such as the biodiesel type, Palm Methyl Ester (PME).

According to the information mentioned above CPO can be seen as a valuable agricultural commodity. The major producer countries of CPO are located in the ASEAN region. USDA [1] states that the top three palm oil producing countries are Indonesia, Malaysia, and Thailand, respectively. The total CPO production

© Springer International Publishing Switzerland 2015
V.-N. Huynh et al. (Eds.): IUKM 2015, LNAI 9376, pp. 428–439, 2015.
DOI: 10.1007/978-3-319-25135-6_39

output of Indonesia and Malaysia is about 90 percent of CPO produced globally per annum [2].

Over the past years, one of the reasons that make the food prices and the agricultural prices rise and having volatility, is due to an increase and volatility of oil prices in the world market. There are many studies found evidence of the relationship between crude oil prices and food prices and agricultural commodity prices. For example, Baffes [3] showed that the change in oil prices had an impact on the change in food commodity prices and fertilizers. Similarly, Balcombe [4] found that oil price volatility had an influence on the commodity prices. Moreover, Nazlioglu and Soytas [5] found that a change in the world oil prices had an impact on the agricultural commodity prices.

The high oil prices have an effect on various operating cost such as production and transportation [6]. Continuous increase in oil prices have contributed directly and indirectly to a rise in the agricultural production cost, e.g. increasing the price of fertilizer and fuel [7]. So, increases of oil prices can have an effect towards the cost of producing CPO. Moreover, the demand for using palm oil for biodiesel production will increase from the consequences of a rise in crude oil price, and followed by the rise of CPO prices [2]. Thus, we can see how two commodities prices are linked, where the increase or decrease in CPO prices are related to the change of CO prices.

Figure 1 shows the time series data of CPO prices and CO prices during 24th December 2004 – 26th December 2014, which is an average weekly prices data of Malaysia and Dubai markets, respectively. The figure shows that the change in the prices of CPO and CO seems to be consistent in both upward and downward directions. Thus, it is interesting to examine the magnitude of the relationship between the two commodity prices. That's because both commodities are important to people in terms of energy security and food security in the ASEAN region.

Fig. 1. The crude palm oil prices of Malaysia and crude oil prices of Dubai during 24th December 2004 – 26th December 2014 (sources: Thomson Reuters Datastream database (2015)).

The simultaneous increase in these commodity prices can have an adversely affect on the consumer; the purchasing power dwindles as some goods begin to have an increase in price. However, the increase in CO prices would be good for farmers who cultivate oil palm tree in the ASEAN region. Since the demand for using CPO for biodiesel production is increasing, it will result in an increase in the farmers' income [8]. Moreover, if there is an appropriate policy in energy security, including a promotion for production and use of biodiesel from palm oil, it can cause a positive effect on the industries that are involved in this commercial process. This would create some benefits for the economic system, social livelihood, and environment [9] [10]. For the future sustainability of biodiesel production, Mukherjee [9] suggested that the government should support the development of new technology to produce biodiesel, in order to increase production efficiency: this includes using feedstock from several sources for biodiesel production. The aim of this method is to reduce the dependence between CPO and CO.

To investigate the dependence between two commodities, prices dependence is one of the measurements that can explain the relationship between the prices of CPO and CO. We can employ this measurement for decision making in various area. For example, it will be useful for government sectors in being able to monitor the changes in the crude oil prices for risk prevention that is in the context of food security and energy security. This is regarded as information and knowledge for helping farmers, who cultivate palm oil tree and are consumers themselves, to become immune towards an unforeseen crisis. Moreover, it is useful for trading companies dealing with policy formulation on agricultural products, as well as for risk management of agricultural futures exchanges and commodity market. The effective investment will also have a positive affect on the agricultural sector as well.

Therefore, this research is interested in studying the relationship between the prices of two commodities. To find the relationship, this study uses the measurement of the dependence between the volatility return of Crude oil prices and the volatility return of Crude palm oil prices. Analyzing the dependence between two volatility returns, we use copula model. Since the copula model offers more flexibility than the traditional approach because the copula can cross over the restriction of normal distribution and linear assumption. Another advantage of copula is that we can find out the dependence without actually knowing the real marginal distributions of the variables. Because the copula provides many forms of multivariate distribution functions, or copula families, which can allow us to know the dependence structure between variables. Thus, the copula model leads us to find more accurate relationship and prediction e.g. when the oil prices have changed then we are able to predict the changes in the palm oil prices from the magnitude of dependence and the dependence structure. Moreover, the empirical results of our method can be extended to other models such as the value at risk (VaR) and expected short fall (ES).

The purpose of the study are the following: 1) To analyze the dependence between prices of crude palm oil and crude oil, 2) To apply the copula function to valuate the value at risk (VaR) and Expected Short fall (ES).

The observations of the two data series were obtained from the Thomson Reuters Datastream during the period from December 2004 to December 2014. For the prices of crude palm oil and crude oil, we used the average weekly prices. Since Malaysia is a world exporter of crude palm oil [1], we used crude palm oil prices that are related with the Malaysian market. With ASEAN relying on crude oil imports from the Middle East [11] so the crude oil price of the Arab Gulf Dubai was used in this study: its price is highly related to the crude oil prices of North America, Europe and Maya [12] [13] corresponding to Adelman [14] [15] assumption, "world oil market, like the world ocean, is one great pool". Both commodity prices are related to the international commodity market and ASEAN.

The remainder of the article is organized as followed: In Section 2, we introduce the ARMA-GARCH model and describe copula function procedures in the empirical analysis. In Section 3, we describe the data that were used in this study, and the empirical results. In Section 4, we provide the application of copula in VaR and ES valuation. Finally, Section 5 presents the conclusion and policy implication.

2   Methodology

The copula is an effective tool that is used to measure the dependence structure in many studies. For example, it used for analyzing the dependence structure of the data series in the stock prices, the asset prices, and the commodity prices (see Sriboonchitta et al. [16]; Kiatmanaroch and Sriboonchitta [17]; Puarattanaarunkorn et al. [18] ; Sriboonchitta et al. [19].

Therefore, to answer the research questions, the Copula based ARMA-GARCH model was used in this study. First, the ARMA(p,q)-GARCH(1,1) model is adopted to find out the marginal distributions of each data series, $(F(x), G(y))$, which is used as input data for the copula model. Since this model can capture the volatility of the price returns. Next, the copula model is used to estimate the dependence between two marginal distributions, which we obtained from the ARMA-GARCH model.

2.1   ARMA-GARCH Model for Marginal Distributions

The appropriate marginal distributions for copula model are required. This study applied the ARMA(p,q)-GARCH(1,1) model to estimate the marginal distributions.

$$y_t = a_0 + \sum_{i=1}^{p} a_i y_{t-i} + \sum_{i=1}^{q} b_i \varepsilon_{t-i} + \varepsilon_t \tag{1}$$

$$\varepsilon_t = z_t \sqrt{h_t}, z_t \sim Dist. \tag{2}$$

$$h_t = \omega_t + \alpha\varepsilon_{t-1}^2 + \beta h_{t-1} \tag{3}$$

An ARMA(p,q) model is shown in equation (1) where y_{t-i} is an autoregressive term of y_t, ε_t is an error term and p is the lag order of the autoregressive and q is the lag order of the moving average. Equation (2) then define this error term as the product between conditional variance h_t and a residual z_t. A residual z_t will be assumed to follow an appropriate distribution. Equation (3) presents GARCH(1,1) process [20] where $\omega_t > 0$, $\alpha \geq 0$, $\beta \geq 0$ are sufficient to ensure that the conditional variance $h_t > 0$. The $\alpha\varepsilon_{t-1}^2$ represent the ARCH term and α refers to the short run persistence of shocks, while βh_{t-1} represent the GARCH term and β refers to the contribution of shocks to long run persistence $(\alpha + \beta)$.

2.2 Copula Model

One approach of modelling the multivariate dependence is the copula. The copula is a function which joins a multivariate distribution function to form a joint distribution function. The fundamental theorem of copula is Sklar's theorem by Sklar [21]. A description of the copula theory has presented in Nelson [22] as the following:

Sklar's Theorem. Let H be a joint distribution function with marginal distributions F and G. Then there exists a copula C for all x, y in real line,

$$H(x,y) = C(F(x), G(y)) \tag{4}$$

If F and G are continuous, then C is unique; otherwise, C is uniquely determined on [0,1]. Conversely, if C is a copula and F and G are univariate distribution functions, then the above function H in (4) is a joint distribution function with marginal distributions F and G.

Given $F(x) = u, G(y) = v$, for u, v in $[0,1]^2$ then we obtain $x = F^{-1}(u)$, $y = G^{-1}(v)$. If H is known, the copula is an equation (4) that one can get from the form,

$$C(u,v) = F(F^{-1}(u), G^{-1}(v)) \tag{5}$$

where F^{-1} and G^{-1} are the quantile functions of the marginal distributions.

This study used various copula families such as the Gaussian copula; $C(u, v; \rho)$ $= \Phi_G(\Phi^{-1}(u), \Phi^{-1}(v); \rho) = \int_{-\infty}^{\phi^{-1}(u)} \int_{-\infty}^{\phi^{-1}(v)} \frac{1}{2\Pi\sqrt{(1-\rho^2)}} \times [\frac{-(s^2 - 2\rho st + t^2)}{2(1-\rho^2)}]dsdt$, the Student's t copula; $C^t(u, v; \rho, \nu) = \int_{-\infty}^{t_\nu^{-1}(u)} \int_{-\infty}^{t_\nu^{-1}(v)} \frac{1}{2\Pi\sqrt{(1-\rho^2)}} \times [1 + \frac{(s^2 - 2\rho st + t^2)}{\nu(1-\rho^2)}]^{-(\frac{\nu+2}{2})}dsdt$, the BB8 (Joe-Frank) copula; $C(u, v; \theta, \delta) = \delta^{-1}[1 - \{1 - [1 - (1-\delta)^\theta]^{-1}[1 - (1-\delta u)^\theta][1 - (1-\delta v)^\theta]\}^{1/\theta}$ and the Rotated BB8 (Joe-Frank) $180°$ copula; $C(u, v; \theta, \delta) = u + v - 1 + \delta^{-1}[1 - \{1 - [1 - (1-\delta)^\theta]^{-1}[1 - (1-\delta(1-u))^\theta][1 - (1-\delta(1-v))^\theta]\}^{1/\theta}$ (see Joe [23], Trivedi and Zimmer [24], Nelson [22]). Each copula family has different characteristics of dependence between the data series. It can be described as followed: the Gaussian (Normal) copula allows for equal degrees of positive and negative dependence and has parameter range $\rho \in (-1, 1)$, the Student's t copula exhibits equal tails (upper and lower) dependence and has parameter ranges $\rho \in (-1, 1)$, $\nu > 2$. For BB8 copula, it consists of the Joe copula and the Frank copula, and it has parameter ranges $\theta \in [1, \infty)$, $\delta \in (0, 1]$. The Joe copula is an extreme value copula, which has upper tail dependence that allows for positive dependence, and the Frank copula accounts for strong positive or negative dependence. Trivedi and Zimmer [24] said that the dependence in the tails of this copula are relatively weak, and the strongest dependence in the center of the distribution that is most appropriate for data showing weak tail dependence. For selection, one has to consider which copula family is the most appropriate to describe the dependence structure between two price returns of the crude oil and the crude palm oil. This study used Akaike Information Criterion (AIC) and a goodness-of-fit test based on Kendall's tau by Genest and Rivest [25].

To estimate dependence parameter of copula, we used the method of maximum pseudo-log likelihood by Genest et al. [26]. Since the marginal distribution functions F and G are not observable, we use the pseudo-observations $(F_n(X_i), G_n(Y_i))$, $i = 1, 2, ..., n$ to form an approximate likelihood function for parameter θ.

Under the assumption that the marginal distributions F and G are continuous, the copula C_θ is a bivariate distribution with density c_θ and pseudo-observations $F_n(X_i)$ and $G_n(Y_i)$, $i = 1, 2, ..., n$. The pseudo-log likelihood function of θ can be given as

$$L(\theta) = \Sigma_{i=1}^{n} \log[c_\theta(F_n(X_i), G_n(Y_i))]. \tag{6}$$

where the empirical distributions are $F_n(x) = \frac{1}{n+1}\Sigma_{i=1}^{n} 1(X_i \leq x)$, $G_n(x) = \frac{1}{n+1}\Sigma_{i=1}^{n} 1(Y_i \leq y)$ and the density function of copula is $c_\theta = \frac{\partial^2 C_\theta(F_n(x), G_n(y))}{\partial x \partial y}$.

3 Data and Empirical Result

3.1 Data

This study used data of the average weekly prices of the crude oil and the crude palm oil from 24th December 2004 to 26th March 2014, from the Thomson Reuters Datastream database. Each price data series is transformed into the log-difference, $\ln \frac{P_t}{P_{t-1}}$, or the price returns before are used for analyzing.

The descriptive statistics of two data series are presented in Table 1. Both data series show the negative skewness, which means that two price returns have downside risk or have a probability of a negative return. The kurtosis of two data series is greater than three. This means that the price returns of the crude oil and the crude palm oil have a probability distribution function with heavy tails. The Jarque-Bera tests show that two data series do not follow a normal distribution. In addition, the result of the Augmented-Dickey-Fuller test shows that both data series are stationary at p-value 0.01.

Table 1. Data Descriptive Statistics of Price Returns of Crude oil and Crude Palm Oil.

	Crude oil	Crude palm oil
Mean	0.0010	0.0011
Median	0.0042	0.0026
Maximum	0.1495	0.1284
Minimum	−0.1926	−0.1704
Std. Dev.	0.0404	0.0351
Skewness	−0.6027	−0.6507
Kurtosis	5.4619	5.4193
p-value of Jarque-Bera	< 2.2e-16	< 2.2e-16
p-value of Augmented-Dickey -Fuller test	0.01	0.01
Observations	522	522

3.2 Results of ARMA-GARCH Model

The ARMA(p,q)-GARCH(1,1) model is used to estimate the appropriate marginal distributions for copula model. The empirical findings show that ARMA(1,1)-GARCH(1,1) with skewed normal distribution of residual fits to the price return of Crude oil, and ARMA(1,1)-GARCH(1,1) with a normal distribution of residual fits to the price return of Crude palm oil. Identification of the optimal models was based on the Akaike information criterion (AIC). The results of the estimated parameters of ARMA-GARCH models are shown in Table 2. All the parameters of the model are significant at levels 0.001, 0.01, 0.05 except that the constant parameter (mu) of both data series and parameter ω of Crude oil are insignificant at level 0.1.

The values $\alpha + \beta$ in GARCH term for Crude oil and Crude palm oil are 0.996 and 0.963, respectively; this implies that their volatilities have a long-run persistence.

Table 2. Results of ARMA(1,1)-GARCH(1,1) with Skewed Normal Distribution of Residual for Crude oil Data and ARMA(1,1)-GARCH(1,1) with Normal Distribution of Residual for Crude palm oil Data.

	Crude oil	Std. error	p-value	Crude palm oil	Std. error	p-value
mu	−8.636e-05	3.634e-04	0.81213	8.335e-04	7.211e-04	0.24775
ar1	7.711e-01	2.880e-01	0.00743**	5.628e-01	1.359e-01	3.47e-05***
ma1	−7.302e-01	2.830e-01	0.00987**	−3.952e-01	1.505e-01	0.00862**
ω	9.757e-06	7.267e-06	0.17938	4.224e-05	1.809e-05	0.01955*
α	8.233e-02	1.879e-02	1.18e-05***	1.622e-01	3.939e-02	3.84e-05***
β	9.141e-01	1.775e-02	<2e-16***	8.003e-01	4.581e-02	<2e-16***
λ (skewness)	8.007e-01	4.765e-02	<2e-16***	-	-	-
Log-likelihood	1009.809	-	-	1101.508	-	-
AIC	−2005.618	-	-	−2191.0161	-	-
K-S test	-	-	1	-	-	1
Box-Ljung test						
1st moment	-	-	0.3212	-	-	0.3980
2st moment	-	-	0.0944	-	-	0.3550
3st moment	-	-	0.1250	-	-	0.4667
4st moment	-	-	0.4096	-	-	0.4829

Note: Significant codes: 0 *** 0.001 ** 0.01 * 0.05.

For the short run persistence, the values α are considered. Therefore, they have the values of 0.082 and 0.162, and a small impaction for volatility.

After estimating ARMA-GARCH model with an appropriate residual distribution of each data series, we transformed the standardized residuals from the ARMA-GARCH model into the uniform [0,1] by using the empirical distribution function to get the pseudo-observation as the input for copula model. The transformed data are used in the Kolmogorov-Smirnov (K-S) test for uniform [0,1] and the Box-Ljung test for serial correlation, which are the test for the marginal distribution that has resulted in the uniform (0,1) and i.i.d., respectively. This is a precondition for using the copula model. The results of K-S tests show that the marginal are uniform by accepting the null hypothesis, and at 5% level, the results of Box-Ljung tests show that the marginal distributions are i.i.d. by accepting the null hypothesis that there is no serial correlation, as shown in Table 2.

3.3   Results of Copula Model

The results of copula estimation are shown in Table 3. By considering the AIC value, the rotated BB8 (Joe-Frank) 180 degrees copula is selected to describe the dependence structure between Crude oil and Crude palm oil. To ensure that this chosen copula family is appropriate for describing the dependence structure, the goodness-of-fit test (GoF) based on Kendall's tau process is used. The GoF offers the p-values of Cramér-von Mises (CvM) test statistic and Kolmogorov-Smirnov (KS) test statistic, which are 0.69 and 0.84, respectively. Thus, it indicates that the rotated BB8 (Joe-Frank) 180 degrees copula is fitted by accepting the null hypothesis at level 0.05. The rotated BB8 (Joe-Frank) 180 degrees copula has two copula parameters, $\theta = 1.9356$ and $\delta = 0.8827$, which are significant at level 0.001 when we convert the copula parameters to the Kendall's tau correlation, which is in the range of [-1,1]. The Kendall's tau correlation of the rotated BB8 (Joe-Frank) 180 degrees copula is 0.2354, so there exists a weak positive dependence. Moreover, the results show no lower (T^L) and upper tail (T^U) dependences.

According to the results of copula parameters ($\theta = 1.9356$ and $\delta = 0.8827$) and also a Kendall's tau correlation (0.2354) of the rotated BB8 (Joe-Frank) 180 degrees, it means that the price returns of Crude oil and Crude palm oil have co-movement that is both upward and downward, but with a weak dependence. The change of the Crude oil price is slightly correlated by the change of the Crude palm oil price.

4   Application of Copula in VaR and ES Valuation

In order to extend the practical applicability of our results, this study shows a simple example of applying the copula function to valuate the value at risk (VaR) and the expected shortfall (ES) of equally weighted portfolio with two price returns of Crude oil and Crude palm oil at period $t + 1$. This study used the equally weighted portfolio because it is easy to use formulae, and it is just an

Table 3. Results of Copula Model

Copula	Parameter	Std. error (p-value)	Kendall's tau	T^L (lower tail dep.)	T^U (upper tail dep.)	AIC	p-value of CvM	p-value of KS
Gaussian	$\theta = 0.3558$	0.0361 (0.0000)	0.2316	0	0	-68.3700	0.19	0.28
Student's t	$\theta = 0.3590$	0.0373 (0.0000)	0.2338	0.0013	0.0013	-67.2600	NA	NA
	$\nu = 26.2258$	9.2997 (0.0281)						
BB8	$\theta = 6.0000$	2.5980 (0.0107)	0.2276	0	0	-61.5675	0	0
	$\delta = 0.3322$	0.1290 (0.0052)						
Rotated	$\theta = 1.9356$	0.3589 (5.273E-08)	0.2354	0	0	-74.9714	0.69	0.84
BB8 180°	$\delta = 0.8827$	0.0898 (0.0000)						

example to show that the copula model can improve VaR forecast in portfolio risk management, as mentioned in Embrechts et al. [27], and Cherubini et al. [28]. In the economics and financial field, the VaR and the ES are a tool of risk measurement. The VaR is the maximum loss on investment in securities or a portfolio for a period of time at a given confidence level. Artzner et al. [29] introduced the ES for a risk measurement. The ES is the expected value of a portfolio loss given a value at risk exceedance has occurred [30].

This study follows the steps of estimation as presented in Sriboonchitta et al. [31] and Ouyang et al. [32]. After we got an appropriate ARMA-GARCH model with fit distribution of residual, and the copula parameter from an appropriate family (the rotated BB8 180 degrees copula): (1) we predicted one-ahead of conditional mean and conditional variance from the appropriate ARMA-GARCH model of each data series; (2) we generated 10,000 random variable u, v with uniform distribution [0,1] using the copula function, by Monte Carlo simulation; (3) The simulated standardized residuals are obtained by using the inverse functions of the skew normal distribution for Crude oil and the normal distribution for Crude palm oil that is the marginal distribution from the ARMA-GARCH model; (4) the simulated price returns are obtained by using the standardized residuals in (3) and the predicted conditional mean and conditional variance in (1); (5) we distribute equal weights to each price return, and then we get the returns after the weighting; (6) the VaR and ES can be measured at the 5% and 1% levels. We repeat the above steps 1000 times to get the convergence values.

Table 4. VaR and ES of Equally Weighted at 5% and 1% level

	5% level	1% level
VaR	−0.0525	−0.0770
ES	−0.0671	−0.0911

Table 4 provides the results of VaR and ES at levels of 5% and 1% with equally weighted. The 5%VaR is -0.0525, which means that the chance of the maximum loss on investment in this portfolio does not exceed 5.25% at 95% confidence level, or the loss will exceed to 5.25% with 5% risk. The 1%VaR is -0.0770, which means that the chance of the maximum loss on investment in this portfolio does not exceed 7.70% at 99% confidence level, or the loss will exceed to 7.70% with 1% risk. Furthermore, the 5%ES is -0.0671, which means that the expected average value of a portfolio loss given a value-at-risk exceedance is 6.71. Also, the 1%ES is -0.0911, which means that the expected value of a portfolio loss given a value-at-risk exceedance is 9.11%.

However, for the portfolio risk management, we should use VaR if single individual commodity. We cannot use VaR for portfolio because it doesn't satisfy subadditivity. VaR is criticized for not being diversified risk measure [33]. Therefore, we recommend to use ES for a portfolio analysis because it has the desirable subadditivity property.

5   Conclusion and Policy Implication

We used the copula based ARMA-GARCH model to examine the dependence between the volatility of price returns of two data series, namely the Crude oil and the Crude palm oil. The empirical results show that the volatility of these price returns has a co-movement that are both upward and downward, but with a weak positive dependence. This draws a conclusion that the crude oil market of the Middle East and the crude palm oil market of Malaysia are linked together.

Moreover, we applied the copula function to estimate the value at risk (VaR) and the expected shortfall (ES) of an equal weight portfolio with two price returns of Crude oil and Crude palm oil. The results show that the copula function can be used for valuation the VaR and the ES, and there exist different values in different confidence levels, which is presented in the previous section. This information is useful for investors who are deciding on making an investment, and to be aware of a loss in these commodities.

In the past there had been a dramatic rise of crude oil prices on the world market. It caused an increasing demand for biodiesel, eventually leading to a rise in CPO prices. One of the advantages of a rise in crude palm oil prices is to increase income of farmers, but on the other hand, it will have an impact to consumers due to the rise of commodities and food prices. Serra [34] studied the price volatility transmission between ethanol and corn markets, the results show that there existed an evidence of price volatility transmission between ethanol and corn markets. Moreover, they also suggested that stock management can reduce corn price fluctuations. Just as with crude oil and crude palm oil, to reduce the link between their volatility price returns there should be some appropriate plans on the stock of crude palm oil both for the adoption of renewable energy and on the other side. The appropriate renewable energy policy from crude palm oil can help stabilize the prices of agricultural products. It can also provide a benefit to the farmers who plant palm oil tree. For a situation where crude oil prices are rising, the renewable energy such as biodiesel from crude palm oil can help relieve the energy problems for consumers.

References

1. USDA: Table 11: Palm Oil: World Supply and Distribution. United States Department of Agriculture (2015). https://apps.fas.usda.gov/psdonline/circulars/oilseeds.pdf/ (accessed April 16, 2015)
2. Sheil, D., et al.: The impacts and opportunities of oil palm in Southeast Asia: What do we know and what do we need to know? Occasional paper no. 51. CIFOR, Bogor, Indonesia (2009)
3. Baffes, J.: Oil spills on other commodities. Resources Policy 32, 126–134 (2007)
4. Balcombe, K.: The nature and determinants of volatility in agricultural prices: an empirical study. In: Prakash, A. (ed.) Safeguarding Food Security in Volatile Global Markets, pp. 85–106. FAO, Rome (2011)
5. Nazlioglu, S., Soytas, U.: Oil Price, agricultural commodity prices, and the dollar: a panel cointegration and causality analysis. Energy Economics 34, 1008–1104 (2012)
6. Asian Development Bank: Global food price inflation and developing Asia. Asian Development Bank (2011). http://www.adb.org/publications/global-food-price-inflation-and-developing-asia (accessed May 23, 2013)
7. Arshad, F.M., Abdel Hameed, A.A.: Crude Oil, Palm Oil Stock and Prices: How They Link. Review of Economics & Finance 3, 48–57 (2012)
8. Kochaphum, C., et al.: Does biodiesel demand affect palm oil prices in Thailand? Energy for Sustainable Development 17, 658–670 (2013)
9. Mukherjee, I., Sovacool, B.K.: Palm oil-based biofuels and sustainability in southeast Asia: A review of Indonesia, Malaysia, and Thailand. Renewable and Sustainable Energy Reviews 37, 1–12 (2014)
10. Johari, A., et al.: The challenges and prospects of palm oil based biodiesel in Malaysia. Energy 81, 255–261 (2015)
11. Speed, P.A.: ASEAN. The 45 Year Evolution of a Regional Institution. POLINARES working paper no. 61, University of Westminster (2012)
12. Reboredo, J.C.: How do crude oil prices co-move? A copula approach. Energy Economics 33, 948–955 (2011)
13. Kiatmanaroch, T., Sriboonchitta, S.: Dependence Structure between World Crude Oil Prices: Evidence from NYMEX, ICE, and DME Markets. Thai Journal of Mathematics, Special Issue on Copula Mathematics and Econometrics 181–198 (2014)
14. Adelman, M.A.: International oil agreements. The Energy Journal 5, 1–9 (1984)
15. Adelman, M.A.: Is the world oil market "One Great Pool"?-Comment. The Energy Journal 13, 95–107 (1992)
16. Sriboonchitta, S., et al.: Modeling volatility and dependency of agricultural price and production indices of Thailand: Static versus time-varying copulas. International Journal of Approximate Reasoning 54(6), 793–808 (2013)
17. Kiatmanaroch, T., Sriboonchitta, S.: Relationship between exchange rates, palm oil prices, and crude oil prices: a vine copula based GARCH approach. In: Huynh, V.-N., Kreinovich, V., Sriboonchitta, S. (eds.) Modeling Dependence in Econometrics. AISC, vol. 251, pp. 399–413. Springer, Heidelberg (2014)

18. Puarattanaarunkorn, O., et al.: Dependence Structure between TOURISM and TRANS Sector Indices of the Stock Exchange of Thailand. Thai Journal of Mathematics. Special Issue on: Copula Mathematics and Econometrics 199–210 (2014)

19. Sriboonchitta, S., Liu, J., Wiboonpongse, A.: Vine copula-cross entropy evaluation of dependence structure and financial risk in agricultural commodity index returns. In: Huynh, V.-N., Kreinovich, V., Sriboonchitta, S. (eds.) Modeling Dependence in Econometrics. AISC, vol. 251, pp. 275–287. Springer, Heidelberg (2014)

20. Bollerslev, T.: Generalized Autoregressive Conditional Heteroskedasticity. Journal of Econometrics **31**, 307–327 (1986)

21. Sklar, A.: Fonctions de rpartition n dimensions et leurs marges. Publications de l'Institut de Statistique de L'Université de Paris **8**, 229–231 (1959)

22. Nelson, R.B.: An Introduction to Copulas, 2nd edn. Springer, New York (2006)

23. Joe, H.: Multivariate Models and Dependence Concepts. Chapman and Hall, London (1997)

24. Trivedi, P.K., Zimmer, D.M.: Copula Modeling: An Introduction for Practitioners. Foundations and Trends in Econometrics **1**(1), 1–111 (2005)

25. Genest, C., Rivest, L.P.: Statistical Inference Procedures for Bivariate Archimedean Copulas. Journal of the American Statistical Association **88**(423), 1034–1043 (1993)

26. Genest, C., et al.: A Semiparametric Estimation Procedure of Dependence Parameters in Multivariate Families of Distributions. Biometrika **82**, 543–552 (1995)

27. Embrechts, P., Lindskog, F., McNeil, A.J.: Modelling dependence with copulas and application to risk management. In: Rachev, S.T. (ed.) Handbook of heavy tailed distribution in finance. Elsevier, Amsterdam (2003)

28. Cherubini, U., Luciano, E., Vecchiato, W.: Copula methods in finance. Wiley, London (2004)

29. Artzner, P., et al.: Coherent measures of risk. Mathematical Finance **9**(3), 203–228 (1999)

30. Sheppard, K.: Financial Econometrics Notes. University of Oxford (2013)

31. Sriboonchitta, S., Liu, J., Kreinovich, V., Nguyen, H.T.: A vine copula approach for analyzing financial risk and co-movement of the Indonesian, Philippine and Thailand stock markets. In: Huynh, V.-N., Kreinovich, V., Sriboonchitta, S. (eds.) Modeling Dependence in Econometrics. AISC, vol. 251, pp. 245–257. Springer, Heidelberg (2014)

32. Ouyang, Z., Liao, H., Yang, X.: Modeling dependence based on mixture copulas and its application in risk management. Appl. Math. J. Chinese Univ. **24**(4), 393–401 (2009)

33. Autchariyapanitkul, K. et al.: Portfolio optimization of stock returns in high-dimensions: A copula-based approach. Thai Journal of Mathematics. Special Issue on: Copula Mathematics and Econometrics 11–23 (2014)

34. Serra, T., Zilberman, D.: Biofuel-related price transmission literature: A review. Energy Economics **37**, 141–151 (2013)

The Economic Evaluation of Volatility Timing on Commodity Futures Using Periodic GARCH-Copula Model

Xue Gong$^{(\boxtimes)}$, Songsak Sriboonchitta, and Jianxu Liu

Faculty of Economics, Chiang Mai University, Chiang Mai 50200, Thailand
snowswirlin@gmail.com

Abstract. Corn is rapidly emerging used as an energy crop. As such, it strengthen the corn-ethanol-crude oil price relationship. In addition, both corn price and crude oil price have been shown to have seasonal changes and also exhibit an asymmetric or tail dependence structure. Hence, this paper uses a periodic GARCH Copula model to explore the volatility and dependence structure between the corn and oil price. More importantly, an asset-allocation strategy is adopted to measure the economic value of the periodic GARCH Copula models. The out-of-sample forecasts show that periodic GARCH copula model performs better than other parametric models as well as a non-parametric model. This result is important since the copula-based GARCH not only statistically improved the traditional method, but has economic benefit to its application. The in-sample and out-of-sample results both show that a risk-averse investor should be willing to switch from non-parametric method, DCC model to Copula based Model.

Keywords: Corn · Crude oil · Periodic GARCH-Copula · Economic evaluation

1 Introduction

Volatility modeling, which includes estimating variance and covariance, is important for risk management, portfolio allocation and hedging in commodity futures. Variance and covariance represents risk in a single stock and portfolio, respectively, and risk is the important factor when investors selected the stock and portfolio. A group of assets return volatility and correlation models have been developed over the last decade. For the volatility models, there are Autoregressive Conditional Heteroskedasticity Model (ARCH), GARCH, and Exponential GARCH models (Engle, 1982 [1]; Bollerslev, 1986 [2]). For the covariance part, it dates from constant correlation models, such as CCC (Bollerslev, 1990 [3]), VECH (Bollerslev et al., 1988 [4]), and BEKK to the time varying DCC models. Recently, the copula based GARCH model takes a significant step forward in

---

X. Gong—Thank you for the research funding from Research Center, Chiang Mai University

V.-N. Huynh et al. (Eds.): IUKM 2015, LNAI 9376, pp. 440–451, 2015.
DOI: 10.1007/978-3-319-25135-6_40

correlation analysis. Chen and Fan (2005)[5] and Lee and Long (2009)[6] also verified that the model is statistically better than traditional model.

Although these works show that the copula-based GARCH model outperforms others, the evidence is statistical. The question is whether the copula-based GARCH model has economic benefit, that is, whether the gains in the precision are sufficient to have an effect on better asset allocation. The point is that there is even an accuracy increase in copula-based GARCH model; the traditional model may work as well in the real situation. Whether the investors are willing to switch from copula-based GARCH model to the traditional one is unknown.

This study evaluates the economic benefit of the copula-based GARCH model in the framework of the mean-variance analysis. This framework is first invented by Markowitz (1959)[7] and later developed by Fleming, Kirby, and Ostdiek (2001, 2003)[8,9]. The core of this framework is first to find out the efficient frontier of the portfolio and give an optimal choice to allocate the portfolio. Then a risk averse investor is introduced to balance the return and risk, and get economic value of different models.

The results of empirical analysis show that the economic benefits of Copula-Periodic-GARCH models (hereafter CPG model) is large, when applying it to the commodity future return forecast. We estimate that different risk adverse investors are willing to pay 6 to 180 basis points per year and to switch from the benchmark method.

The paper will be conducted as follows. Section 2 describes the properties of data used and evaluates the performance of the different strategies. Section 3 introduces the asset allocation methodology, economic value measurement, and the copula-based and the traditional multivariate model. Section 4 presents our out-of-sample forecasting results. Finally, the conclusion is shown in section 4.

2   Data

We investigate two commodity futures returns: Corn No.2 Yellow futures (Chicago Board of Trade), WTI Crude Oil (New York Mercantile Exchange). The sample span is from 12th June 1996 to 25th May 2011, totally 15 years, 781 observations. We obtain closing prices on Friday from Datastream as weekly prices. If there were any two futures prices missing, we use the data from Thursday instead. To construct the weekly return series, we generally use the nearby contract in each market. However, we switch to the second nearby contract when the first nearby contract is in the delivery month. We use the 3-month Treasury bill rate to substitute for the risk free rate, which can be downloaded from the Federal Reserve Board. The mean of the 3-month Treasury bill rate is around 0.06% weekly (3.2% annually).

2.1   The Framework of a Minimum Variance Strategy

To carry out this study, the framework of a minimum variance strategy was adopted, which was conductive to determining the accuracy of the time-varying

Table 1. Summary Statistics

	Corn	Oil
Mean	0.064	0.208
Median	0.236	0.409
Maximum	17.020	30.300
Minimum	-18.230	-23.260
Std. Dev.	4.414	5.286
Skewness	-0.158	-0.105
Excess Kurtosis	2.013	2.126
Jarque-Bera	136.968	150.439
	(***)	(***)
Correlation	0.166	

covariances. Let the returns of assets denote as $R = (r_{1t}, ..., r_{nt})'$, where r_{it} is the weekly returns ($\log P_t - \log P_{t-1}$) of risky asset, t is the week, $i = 1, 2, ..., N - 1$. The first conditional moment (conditional mean) and second conditional moment (conditional variance) can be expressed as $E(R_t|I_{t-1}) = (\mu_{t,1}, \mu_{t,2}, ..., \mu_{t,N-1})' = \mu$, which is the $(N-1) \times 1$ vector of conditional means and $\sum_t = E[(R_t - \mu_t)(R_t - \mu_t)']$, which is $(N - 1) \times (N - 1)$ conditional covariance matrix, respectively, where I_{t-1} is the $t - 1$ period information set (Bai et al., 2009)[10].

The mean of the portfolio represents the mean return, while the variance of the portfolio represents the risk. The mean-variance problem proposed by Markowitz (1952) is to choose a proper portfolio weights to minimize the variance of the portfolio return with the constraint of the target expected return of the portfolio. That is, an investor will choose certain portfolio weights $(w_{t,1}, w_{t,2}, ... , w_{t,N-1})'$ to minimize the portfolio risk between $(N - 1)$ risky assets and the weight in risk-free asset has given. The mathematical notation is as the following:

$$min \quad w_t' \sum_t w_t \tag{1}$$

$$s.t. \quad E(w_t' R_t) = w_t' \mu = \mu_0 \tag{2}$$

Following Merton (1972), we solve the optimization problem by Lagrange multipliers, and get the solutions of optimal weights:

$$w_t = \frac{\mu_0 \sum_t^{-1} \mu_t}{\mu_t' \sum_t^{-1} \mu_t} \tag{3}$$

where μ_0 is the target return of portfolio, it should be noted that since we study the commodity futures, eq (3) stated that the number of contracts the investor hold. It also implied that the return on this portfolio should closely approximate the excess return of underlying spot assets. With the relation of cost of carry, the futures return is the spot return minus the interest rate. Therefore, the investor can give the same weights to the spot asset and invest 100% of investor's funds in the risk-free assets. We also assume that μ_t are constant, only the covariance matrix can be time-varying.

2.2   Measuring the Performance Gains

We use an investor utility function to measure the value of performance gains given estimator of the conditional covariance matrix by different multivariate GARCH models. We consider an investor's utility function as:

$$U(W_t) = W_t R_{p,t} - \frac{\alpha W_t^2}{2} R_{p,t}^2 \qquad (4)$$

Equation (4) is a second-order approximation to the investor's true utility function (Markowitz, 1991)[11], where W_t is the investor's wealth at t and α is his absolute risk aversion. $R_{p,t}$ is the portfolio return at time t. As the initial wealth W_0 and risk aversion α are given, the utility function is a trade-off between the expected return ($R_{p,t}$) and the variance ($R_{p,t} - \mu_0$)^2. To compare within the portfolios, we give an assumption that αW_t is certain, which to say that with more wealth, the less risk-averse the investor is. However, the investor's relative risk aversion,

$$\gamma_t = U''(W_t)/U'(W_t) = \alpha W_t/(1 - \alpha W_t) \qquad (5)$$

Therefore, it implies that γ_t is also constant ($\gamma_t = \gamma$). Then we solve equation (5) to get γ and plug into the utility function (4). It implies that the expected utility is linearly related to wealth. With this assumption, the average realized utility $\bar{U}(\cdot)$ can be used in estimating the expected utility with a given initial wealth W_0.

$$\bar{U}(\cdot) = W_0 \sum_{t=1}^{T} [R_{p,t} - \frac{\gamma}{2(1+\gamma)} R_{p,t}^2] \qquad (6)$$

where W_0 is the initial wealth.

Therefore, the value of volatility timing calculated by equating the average utilities for two alternative portfolios is expressed as:

$$\sum_{t=1}^{T} [(R_{b,t} - \delta) - \frac{\gamma}{2(1+\gamma)} (R_{b,t} - \delta)^2] = \sum_{t=1}^{T} [R_{a,t} - \frac{\gamma}{2(1+\gamma)} (R_{a,t})^2] \qquad (7)$$

where δ is the maximum expense that an investor would be willing to pay to switch from the strategy a to the strategy b. $R_{a,t}$ and $R_{b,t}$ are the gross returns of the portfolios from the strategy a and b, which is using different estimators of the conditional covariance matrix.

$$R_{p,t} = 1 + r_{p,t} = 1 + (1 - w_t'1)r_f + w_t' r_t, \quad p = a, b \qquad (8)$$

If the expense δ is a positive value, it means the strategy b is more valuable than the strategy a. In our empirical study, we reported δ as an annualized expense with three risk aversion levels of $\gamma = 1, 5$, and 10.

3   The Econometric Methods

In this section, we adopt several methods to estimate the conditional covariance matrix in commodity futures, they are, multivariate GARCH Copula, multivariate GARCH model, nonparametric method. We will focus on introducing parametric model: the Periodic GARCH copula model. And the nonparametric model will be introduced in the empirical study.

3.1   Conventional Parametric MGARCH Models

Suppose the return series r_t of commodities futures follows the stochastic process:

$$r_t|F_{t-1} \sim P(\mu_t, H_t; \theta) \tag{9}$$

where $r_t = (r_{1,t}, r_{2,t})'$, F_{t-1} is the information set at time t-1, P is the joint distribution of r_t and θ is the unknown parameters, and the first and second conditional moment of the return data are:

$$E(r_t|F_{t-1}) = \mu_t \ \ and \ \ E(r_t r_t'|F_{t-1}) = H_t \tag{10}$$

We transform the data to make sure that the conditional μ_t is zero. Then we standardized return data to get the standardized error $e_t = H^{-1/2} r_t$, the respective first and second moment is

$$E(e_t|F_{t-1}) = 0 \ \ and \ \ E(e_t e_t'|F_{t-1}) = I \tag{11}$$

Here, the conditional covariance matrix H_t could be decomposed as $D_t R_t D_t$, where R_t is the conditional correlation matrix between r_t, and $D_t = diag(H_t)^{(1/2)}$. We assume that e_t follows normal distribution such that $e_t \sim N(0, I)$ and the conditional correlation between $r_{1,t}$ and $r_{2,t}$ is equal to the conditional covariance between the standardized variables $\epsilon_t = (\epsilon_{1,t}, \epsilon_{2,t})' = D_t^{-1} r_t$:

$$\rho_{1,2} = corr(r_{1,t}, r_{2,t}|F_{t-1}) = cov(\epsilon_{1,t}, \epsilon_{2,t}|F_{t-1}) \tag{12}$$

The key to the multivariate GARCH model is modeling the conditional covariance H_t. There are three problems of modeling H_t: (1) there are too many parameters to be estimated, and (2) the conditional covariance matrix H_t may not be positive definite. (3) The assumption of constant correlation among the financial commodity markets is doubted.

To overcome these problems in the previous studies (such as Du et al., 2011 [12]; Nazlioglu et al., 2013 [13]), a dynamic correlation model (DCC) was proposed by Engle (2002) [14]. The formula for DCC model are:

$$h_{i,t} = w_i + a_i h_{i,t-1} + b_i e_{i,t-1}^2, \ \ i = 1, 2, \ \ \epsilon_t = r_t/\sqrt{h_t} \tag{13}$$

$$Q_t = (1 - \alpha - \beta)Q + \beta(\epsilon_{t-1}\epsilon_{t-1}') \tag{14}$$

where Q is sample covariance matrix of $\hat{\epsilon}_t$, with the stationary condition that $\alpha > 0$, $\beta > 0$ and $\alpha + \beta < 1$. The second challenge of non- positive definiteness of the conditional covariance matrix was overcome by the transformation of $R_t = diag\{Q_t\}^{-1} Q_t diag\{Q_t\}^{-1}$ Engle and Sheppard (2001)[15].

3.2   Copula GARCH Model

Copula. The copula is a powerful tool to link different margins into a multivariate distribution. It is much more flexible than the conventional multivariate joint distribution (Chinnakum, Sriboonchitta, and Pastpipatkul,2013 [17]; Puarattanaarunkorn and Sriboonchitta, 2014 [18];Wichian and Sriboonchitta, 2014 a,b [19,20]).

Let's introduce the Sklar theorem first: suppose P be a joint distribution function with margins F and G. Then, there exists a copula C such that for all η_1, η_2.

$$P(\eta_1, \eta_2) = C(F(\eta_1), F(\eta_2)) = C(u, v) \tag{15}$$

If F and G are continuous, then C is unique; otherwise, C is uniquely given by $C(u, v) = P(F^{-1}(u), G^{-1}(v))$ for $(u, v) \times [0, 1]$ where $F^{-1}(u) = inf\{\eta_1 : F(\eta_1) \geq u\}$ and $G^{-1}(v) = inf\{\eta_2 : G(\eta_2) \geq v\}$. Conversely, if C is a copula and F and G are distribution functions, then the function C defined above is a joint distribution function with marginal F and G. The characteristic of copula function is its ability to decompose the joint distribution into two parts: margins and dependence. Any changes in margin or dependence can lead to a different joint distribution. That is the reason why we called it more flexible than any other existed joint distribution. The details of the Copula can be found in Joe (1997)[21] and Rakonczai and Tajvidi (2010)[22].

To adapt copula into the multivariate GARCH, we assume that we get the marginal conditional probability density as:

$$u_{1,t} = F_{1,t}(y_{1,t}|\psi_{t-1}) \tag{16}$$

and

$$u_{2,t} = F_{2,t}(y_{2,t}|\psi_{t-1}) \tag{17}$$

And then we link the two margins by the following time-varying copula. Afterwards, we get bivariate conditional CDF:

$$C_t(u_{1,t}, u_{2,t}|\psi_{t-1}) = F(y_{1,t}, y_{2,t}|\psi_{t-1}) \tag{18}$$

Here, we used the time-varying correlation structure: that is, the correlation function or covariance changes with time. we learn from the previous literature, and adopt the following correlation:

$$\rho_t = \lambda(\alpha + \beta\rho(t-1) + \gamma(u_{1,t-1} - u_{2,t-1})^2) \tag{19}$$

where λ is the logistic transformation, which assures the correlation is in the range of [-1,1].

4   Candidates for the Margins

4.1   The Period GARCH (P-GARCH) Model

Since the multivariate GARCH model can be separated into two steps, the first step is to estimate the volatility of each commodity futures. The GARCH models model the price series by allowing variance to evolve through time. This model is widely used in the financial market since it considers the heteroskedasticity properties of return series. The P-GARCH model, which was proposed by Bollerslev and Ghysels (1996)[23], accounts for the seasonal changes or the period cycle; at first, it was developed for the high frequency financial data, the opening and closing time will make the price has seasonal change property. Later, this model was introduced to estimate and forecast the prices of the commodities, such as those of energy, agricultural commodities, and so on. The variance function of modified P-GARCH model is as follows:

$$h_t = \omega_{s(t)} + \sum_{i=1}^{q} \alpha_{i,s(t)} \epsilon_{t-1}^2 + \sum_{i=1}^{p} \beta_{i,s(t)} h_{t-1} \tag{20}$$

where $s(t)$ denotes the stage of the seasonal cycle at time t, implying that different GARCH parameters will be estimated in different seasons. In our study, we will define $s(t) = k$ $(k = 1, 2, 3, 4)$, k represents different quarters, that is $k=1$ denotes the quarter 1, $k=2$ denotes the quarter 2, and so on. Here, we only consider the simple version of the P-GARCH model as there are seasonal changes in $\omega_{s(t)}$ and $\alpha_{i,s(t)}$, but $\beta_{i,s(t)}$ are constant. This is because in this model, α measures the immediate impact of the news on volatility, while β measures the smooth, long-term change on volatility. Second, the empirical study shows that the estimations are always not converged, therefore it is difficult to estimate β.

$$Model \quad One : h_t = \omega_{s(t)} + \sum_{i=1}^{q} \alpha_i \epsilon_{t-1}^2 + \sum_{i=1}^{p} \beta_i h_{t-1} \tag{21}$$

$$Model \quad Two : h_t = \omega_{s(t)} + \sum_{i=1}^{q} \alpha_{i,s(t)} \epsilon_{t-1}^2 + \sum_{i=1}^{p} \beta_i h_{t-1} \tag{22}$$

For the first step of DCC model, the margins can be estimated by the different GARCH models, however, in our analysis, since the target futures are commodity futures and the seasonal trends in volatility are pervasive in commodities futures market. It is necessary to capture this characteristic to improve the estimation. The effect of period GARCH (P-GARCH) is to capture the seasonal volatility is well-known. The most advantage of P-GARCH is to eliminate the skewness and kurtosis of the residuals (Xue and Sriboonchitta, 2014) [24].

4.2 Conditional Covariance Matrix Estimates

To obtain an optimal portfolio, we used the dynamic models we introduced in section 2 to estimate the conditional covariance matrix. We divided procedures into two parts corresponding to the two steps in the DCC and PGC estimation. We do the first step: the marginal estimation. Then, these standardized residuals series were brought into the second stage for dynamic conditional correlation estimating. Table 3 shows the estimated parameters of DCC and Copula-GARCH under the quasi-maximum likelihood estimation (QMLE).

Table 2. Volatility Estimation of Periodic GARCH model with skewness t

	Corn	Oil
μ	0.127	0.228
	-1.071	-1.558
α	0.126	0.085
	(3.514)***	(4.112)***
β	0.748	0.879
	(11.007)***	(29.014)***
ω_1	1.585	0.993
	(2.649)**	(2.144)*
ω_2	1.669	-0.332
	(2.233)*	(-0.708)
ω_3	3.037	-
	(1.766)*	-
ω_4	-1.195	-
	(-2.818)***	-
ξ	8.006	8.42
	(3.866)***	(4.135)***
γ	-0.067	-0.201
	(-1.471)	(-4.248)***
Standardized Residuals		
Skewness	-0.292	-0.443
Kurtosis	1.134	1.854
Loglikelihood	2597.375	2769.382

Table 3. Correlation Estimation of DCC, Gaussian and T Copula

	DCC	Gaussian Copula	t copula
α	0.031	0.096	0.305
	(1.060)**	(1.976)**	(4.049)***
β	0.853	0.771	-0.372
	(1.060)**	(5.306)***	(-1.097)
γ	-	-0.2	-0.307
		(-1.982)*	(-2.152)*
Dof	-	-	242.41
			(-1.089)
Joint llh	-4579.804	-4577.95	-4695.417

Tables 2 shows the GARCH margin estimation results. It can infer that the Periodic GARCH model with skewness t achieves good statistic performance. Both AIC and BIC are lower than the GARCH(1,1) model, and after filtering the skewness and kurtosis of residuals are almost zero.

We compare different models by using two copulas: Gaussian and t copula and DCC model with normal distribution marginal within the whole data span. The advantage of the copula can be shown: it can link any margins to get a multivariate distribution. The results are present in Table 4. In t copula estimation, the degree of freedom is 242, which converges to normal copula. And also t copula has much smaller joint likelihood value; therefore we just compare the results and DCC and Gaussian Copula here and in the next section.

4.3   The Out-of-Sample Comparisons

The weights of the portfolio are computed by the expected return and the conditional covariance, which are estimated by DCC and Gaussian Copula models. We construct a nonparametric benchmark portfolio, since we would like to construct a baseline one-period variance matrix to compare with parametric methods.

There are many ways to construct the nonparametric approach.

$$\hat{\sigma}_{ij,t} = \sum_{l=-t+1}^{T-t} \omega_{ij,t+l}(r_{i,t+l} - \mu_i)(r_{j,t+1} - \mu_j) \tag{23}$$

$$\omega_{ij,t+l} = e^{t+l} / \sum_{l=-t+1}^{T-t} e^{t+l}$$

where r_{it} and r_{jt} denote the returns on assets i and j, respectively; T is the number of observations in the sample. However, Fleming et $al.$, (2001)[8] find that volatility timing is more effective using a smaller decay rate than implied by the minimum MSE criterion. We construct the benchmark portfolio by following decay rate weights to the variance matrix. This approach will be easy to conduct and fit for our need.

After building the covariance matrix estimates, we form the dynamic portfolios by our parametric methods and evaluate their performance. The economic value of dynamic model is the estimated fee that a risk-averse investor would be willing to pay to switch from the ex ante optimal benchmark portfolio to the dynamic portfolio.

Each forecasting value was estimated by 581 sample observations over 10 years. Therefore we forecast from 8/1/2007 to 5/25/2011, totally 200 out-of-samples forecasting.

Table 4. Means and Volatilities of One-period Ahead Optimal Portfolios

Target(%)	DCC			Normal Copula			Nonparamatric		
	μ	σ	SR	μ	σ	SR	μ	σ	SR
5	5.24	5.52	0.94	6.36	5.39	1.17	5.55	7.08	0.77
6	6.37	8.46	0.75	8.09	8.26	0.97	6.84	10.85	0.62
7	7.50	11.40	0.65	9.81	11.13	0.88	8.13	14.62	0.55
8	8.63	14.33	0.60	11.54	14.00	0.82	9.43	18.39	0.51
9	9.75	17.27	0.56	13.26	16.87	0.78	10.72	22.16	0.48
10	10.88	20.21	0.54	14.99	19.74	0.76	12.01	25.93	0.46
11	12.01	23.15	0.52	16.71	22.61	0.74	13.3	29.69	0.45
12	13.14	26.08	0.50	18.44	25.48	0.72	14.60	33.46	0.43
13	14.27	29.02	0.49	20.16	28.34	0.71	15.89	37.23	0.43
14	15.39	31.96	0.48	21.89	31.21	0.70	17.18	41.00	0.42
15	16.52	34.9	0.47	23.62	34.08	0.69	18.47	44.77	0.41

In Table 4, we evaluate the out-of-sample performance of three models with different target returns and different risk aversions. And we also draw the optimal portfolio weights in Figure 1 using the rolling covariance estimates by three methods. The annualized Sharpe ratios for PGC (0.81) is higher than the non-parametric model (0.50) and even DCC (0.59). That implies the PGC model much more awardable.

In Table 5, we give out the average switching fees from one adopted model to another. We set relative risk aversion as 1, 5, and 10. As for the performance fees with different relative risk aversions, in general, an investor with a higher risk aversion should be willing to pay more to switch from the benchmark portfolio by nonparametric method to the dynamic ones. With higher target returns, the performance fees increasingly grow. Moreover, Table 5 also reports the performance fees for switching from DCC to Periodic GARCH Copula. The results show that the Periodic GARCH Copula model can give more significant

Table 5. Switching Fees with Different Relative Risk Aversions

Target(%)	Nonparametric to DCC			Nonparametric to NC			DCC to NC		
	$\triangle_1$	$\triangle_5$	$\triangle_{10}$	$\triangle_1$	$\triangle_5$	$\triangle_{10}$	$\triangle_1$	$\triangle_5$	$\triangle_{10}$
5	4.68	7.80	8.58	6.08	9.20	10.14	1.71	1.86	1.98
6	10.92	17.68	18.72	13.26	19.97	21.32	2.34	2.29	2.60
7	19.76	29.12	31.72	22.88	33.28	35.36	3.12	4.16	3.64
8	30.68	43.68	46.28	34.32	48.36	50.96	3.64	4.68	4.68
9	42.12	59.28	61.88	47.32	65.00	67.60	5.2	5.72	5.72
10	55.12	74.88	77.90	61.88	81.64	84.76	6.76	6.76	6.86
11	69.16	91.00	94.64	76.96	99.32	102.60	7.8	8.32	7.96
12	83.2	107.64	112.32	92.56	117.26	120.64	9.36	9.62	8.32
13	98.28	124.80	128.96	109.2	135.20	139.36	10.92	10.40	10.40
14	114.92	142.48	146.12	125.84	153.92	157.56	10.92	11.44	11.44
15	131.56	159.64	163.80	143.52	172.12	176.28	11.96	12.48	12.48

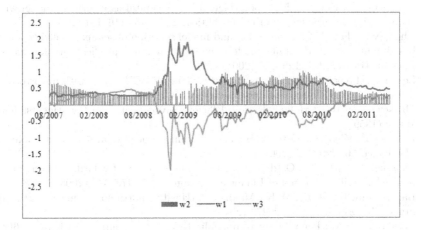

Fig. 1. The Optimal Weights by Period-GARCH-Copula

economic value in forecasting variance and covariance matrices than DCC, when applying to the commodity futures return.

5 Conclusion

In this paper, we answered the question that whether the periodic GARCH copula on forecasting the volatility of commodities futures has economic value when compared with the nonparametric method and traditional DCC models. After our analysis, we can conclude that the model shows better performance than others. This result is so important since the Periodic GARCH Copula has not only statistically improved the traditional method, but has economic value in its application. The in-sample and out-of-sample results both show that a risk-averse investor should be willing to switch from the non-parametric method, DCC model to a Copula based Model.

Acknowledgement. We thank famous Prof. Hung T. Nguyen for sharing the knowledge of Copula, and we thank two anonymous reviewers for their great insights.

References

1. Engle, R.F.: Autoregressive conditional heteroscedasticity with estimates of the variance of United Kingdom inflation. Econometrica: Journal of the Econometric Society, 987–1007 (1982)
2. Bollerslev, T.: Generalized autoregressive conditional heteroskedasticity. Journal of econometrics **31**(3), 307–327 (1986)
3. Bollerslev, T.: Modelling the coherence in short-run nominal exchange rates: a multivariate generalized ARCH model. The Review of Economics and Statistics, 498–505 (1990)
4. Bollerslev, T., Engle, R.F., Wooldridge, J.M.: A capital asset pricing model with time-varying covariances. Journal of Political Economy, 116–131 (1988)
5. Chen, X.H., Fan, Y.Q.: Estimation and model selection of semiparametric copula-based multivariate dynamic models under copula misspecification. Journal of Econometrics **135**(1), 125–154 (2006)
6. Lee, T.H., Long, X.: Copula-based multivariate GARCH model with uncorrelated dependent errors. Journal of Econometrics **150**(2), 207–218 (2009)
7. Markowitz, H.M.: Porfolio Selection: Efficient Diversification of Investments. John Wiley (1959)
8. Fleming, J., Kirby, C., Ostdiek, B.: The economic value of volatility timing. Journal of Finance **56**, 329–352 (2001)
9. Fleming, J., Kirby, C., Ostdiek, B.: The economic value of volatility timing using realized volatility. Journal of Financial Economics **67**, 473–509 (2003)
10. Bai, Z., Liu, H., Wong, W.K.: Making Markowitz's portfolio optimization theory practically useful. SSRN 900972 (2010)
11. Markowitz, H.M.: Foundations of portfolio theory. The Journal of Finance **46**(2), 469–477 (1991)

12. Du, X., Yu, C.L., Hayes, D.J.: Speculation and volatility spillover in the crude oil and agricultural commodity markets: A Bayesian analysis. Energy Economics **33**(3), 497–503 (2011)
13. Nazlioglu, S., Erdem, C., Soytas, U.: Volatility spillover between oil and agricultural commodity markets. Energy Economics **36**, 658–665 (2013)
14. Engle, R.: Dynamic conditional correlation: A simple class of multivariate generalized autoregressive conditional heteroskedasticity models. Journal of Business and Economic Statistics **20**(3), 339–350 (2002)
15. Engle, R.F., Sheppard, K.: Theoretical and empirical properties of dynamic conditional correlation multivariate GARCH (No. w8554). National Bureau of Economic Research (2001)
16. Embrechts, P., McNeil, A., Straumann, D.: Correlation and dependence in risk management: properties and pitfalls. Risk management: Value at Risk and Beyond, 176–223 (2002)
17. Chinnakum, W., Sriboonchitta, S., Pastpipatkul, P.: Factors affecting economic output in developed countries: A copula approach to sample selection with panel data. International Journal of Approximate Reasoning **54**(6), 809–824 (2013)
18. Puarattanaarunkorn, O., Sriboonchitta, S.: Copula based GARCH dependence model of chinese and korean tourist arrivals to thailand: implications for risk management. In: Huynh, V.-N., Kreinovich, V., Sriboonchitta, S. (eds.) Modeling Dependence in Econometrics. AISC, vol. 251, pp. 395–416. Springer, Heidelberg (2014)
19. Wichian, A., Sriboonchitta, S.: Econometric Analysis of Private and Public Wage Determination for Older Workers Using A Copula and Switching Regression. Thai Journal of Mathematics, 111–128 (2014a)
20. Wichian, A., Sriboonchitta, S.: Econometric Analysis of Older Workers' Hours of Work Using A Copula and Sample Selection Approach. Thai Journal of Mathematics, 91–110 (2014b)
21. Joe, H.: Multivariate models and multivariate dependence concepts, vol. 73. CRC Press (1997)
22. Rakonczai, P., Tajvidi, N.: On Prediction of Bivariate Extremes. International Journal of Intelligent Technologies and Applied Statistics **3**(2) (2010)
23. Bollerslev, T., Ghysels, E.: Periodic autoregressive conditional heteroscedasticity. Journal of Business and Economic Statistics **14**(2), 139–151 (1996)
24. Xue, G., Sriboonchitta, S.: Co-movement of prices of energy and agricultural commodities in biofuel Era: a period-GARCH copula approach. In: Modeling Dependence in Econometrics, pp. 505–519. Springer International Publishing (2014)

On the Estimation of Western Countries' Tourism Demand for Thailand Taking into Account of Possible Structural Changes Leading to a Better Prediction

Nyo Min[✉] and Songsak Sriboonchitta

Department of Economics, Chiang Mai University, Chiang Mai, Thailand
nyo.min@gmail.com

Abstract. Forecasting tourist arrivals is an essential feature in tourism demand prediction. This paper applies Self Exciting Threshold Autoregressive (SETAR) models. The SETAR takes into account of possible structural changes leading to a better prediction of western tourist arrivals to Thailand. The finding reveals that although the forecasting method such as SARIMA GARCH is the state of art model in econometrics, forecasting tourism demand for some specific destinations without consideration of the potential structural changes means ignoring the long persistence of some shocks to volatility and the conditional mean values leading to less efficient forecast results than SETAR model. The findings show that SETAR model outperforms SARIMA GARCH model. Then this study based on the SETAR model uses the Bayesian analysis of Threshold Autoregressive (BAYSTAR) method to make one step ahead forecasting. This study contributes that SETAR overtakes SARIMA GARCH as it takes into account of the nonlinear features of the data via structural changes resulting in the better forecasting of Western Countries tourism demand for Thailand.

Keywords: SARIMA GARCH · Structural change · SETAR · BAYSTAR · Volatility · Forecasting tourist arrivals

1 Introduction

For countries around the world, travel and tourism are some of the most important economic activities because over one billion people travel around the world every year. While the tourism industry is popular among Asian countries at the present, it plays a major role in the growth of Thailands industry to become developed with direct and indirect benefits to the economic growth.

Forecasting the future trend and potential of tourism demand essentially help policy makers in both the private and public sectors make strategic planning and policy for tourism development of Thailand. It is obvious that forecasting assists not only for success in tourism, but also avoids the management failure to meet tourism demand. In terms of management, this supporting information is

© Springer International Publishing Switzerland 2015
V.-N. Huynh et al. (Eds.): IUKM 2015, LNAI 9376, pp. 452–463, 2015.
DOI: 10.1007/978-3-319-25135-6_41

essential for all parties in tourism sector to have efficient and effective planning for investment and all kinds of related destination infrastructures development for short, medium, and long term horizons.

In every tourism industry, there must be seasonality effects and impacts. The seasonality can be seen as a pattern in all trends whether or not there is any structural change in the trends, but it can also be seen changing itself.

This study aimed to apply the time series model called the SARIMA - GARCH model to successfully analyze and forecast the demand for tourism in Taiwan after the Chinese authorities allowed Mainland China tourists to visit Taiwan [20]. Liang applied SARIMA - GARCH model with small sample size and without considering structural change. In general, time series analysis, especially tourism demands, have been done by many researchers including Songsak Sriboonchitta, Jianxu Liu, Jirakom Sirisrisakulchai [32], Kanjanatarakul et al. [17], [18], & [19], Xue Gong, Songsak Sriboonchitta [14], Autchariyapanitkul et al. [1], Apiwat Ayusuk, and Songsak Sriboonchitta [3], Phochanachan et al. [27], Vladik et al [35], Nyo Min et al. [25], Wichian et al. [37], Liu and Sriboonchitta [23], Tang el at. [33], Piamsuwannakit and Sriboonchitta [28], Liu et al [22], Puarattanaarunkorn et al [29], Liu and Sriboonchitta [21], Chia-Lin Changa et al. [6], [7], Maureen Ayikoru [2], Gunter, and Önder [15], Melda Akin [24] , and Morleya et al. [26], but none of them considered structural change in their works. That might lead to model misspecification.

Structural change in sample size can also provide bias if the researchers neglect to handle this feature in data analysis. This paper applies SETAR model which allows to provide some break points and structural change.

In this paper, data analysis forecasts western tourist arrivals to Thailand for the following month by using BAYSTAR as well as SARIMA - GARCH.

The paper is organized as follows: in the next section, we provide literature reviews, concepts of the paper, and a brief review on SARIMA and SETAR models, the literature review related to these models and concepts of the paper, and methodology. Section 3 gives details of data we used in this study, Section 4 presents empirical results of SARIMA GARCH and SETAR models, and Section 5 forecasts for one period ahead for the tourist arrival by using BAYSTAR and discusses some findings. Our conclusions are drawn in Section 6.

2   Literature Reviews

Chia-Lin Chang et al estimated SARIMA model in 2009 [7]. In 2014, Nyo Min, Jirakom Sirisrisakulchai and Songsak Sriboonchitta forecasted tourist arrivals to Thailand by using belief function, and in the paper they used SARIMA model as statistical model to estimate. [25].A.K. Diongue et al investigated conditional mean and conditional variance forecasts on electricity price reaction using a dynamic model following a k-factor GIGARCH process, and compared their forecast performance with a SARIMAGARCH benchmark model using the year 2003 as the out-of-sample [11]. C. Sigauke, and D. Chikobvu applied a seasonal autoregressive integrated moving average (SARIMA) model, a SARIMA model with generalized autoregressive conditional heteroskedastic (SARIMA-GARCH) errors, and a regression-SARIMAGARCH (Reg-SARIMAGARCH) model to forecast daily peak electricity demand in South Africa. They used these models for out of sample prediction of daily peak demand and found that the Reg-SARIMAGARCH model generated better forecast accuracy in terms of mean absolute percent error (MAPE) [31]. J. Guo et al stated that a stochastic seasonal autoregressive integrated moving average plus generalized autoregressive conditional heteroscedasticity (SARIMA + GARCH) process has the ability to jointly generate traffic flow level prediction and associated prediction interval [16]. Yi-Hui Liang proposed the SARIMAGARCH model to analyze and forecast tourism demand in Taiwan, and compared the predictive power of this model and other forecasting models. Liang said that the result of his study could provide favorable on the model [20]. Up to now, only few researchers apply SARIMA-GARCH model to forecast tourism demands including tourist arrivals in Thailand Tourism Sector. The Generalized Autoregressive Conditional Heteroskedastic (GARCH) model is another extension of univariate time-series analysis in tourism demand.

In 1977, Howell Tong introduced Self Exciting Threshold Autoregressive (SETAR) model, and in 1980 the model was fully developed by H. Tong and K. S. Lim [34]. Edward P. Campbell presented Bayesian Selection of Threshold Autregressive Models to recommend model selection based on SETAR model [5]. In 2007, Apostolos Serletis and Asghar Shahmoradi conducted an analysis of purchasing power parity between Canada and the United States, and in their paper they applied SETAR to nonlinearities in the real exchange rate [30]. Since 2001, Michael P. Clements and Jeremy Smith have researched the ability of two SETAR exchange rate models in their forecasting performance [8]. Gilles Dufrenot et al presented SETAR for volatility identifying [12] and Pedro Galeano and Daniel Pea worked model selection criteria for SETAR [13]. Generally, some researchers, including Michael E Clements, Jeremy Smith [8], Jan G. De Gooijer, and Paul T. De Bruin [9] tried to apply SETAR model as forecasting tool as soon as the model was introduced.

3   Methodology

This paper used the self-exciting threshold autoregressive (SETAR) model and then applies the Bayesian analysis of threshold autoregressive (BAYSTAR) model based on some information from SETAR to forecast one-step-ahead on the two regimes.

Self-Exciting Threshold Autoregressive (SETAR) Model

Suppose the conditioning variable is a dependent variable itself after some delay, then this model can be called self-exciting threshold autoregressive (SETAR) model.

The SETAR model can be generalized to the pth order autoregressive as follows;

$$regime\ 1\ \ y_t = \beta_0^{(1)} + \beta_1^{(1)} y_{t-1} + \cdots - \beta_p^{(1)} y_{t-p} + \varepsilon_t\ \ if\ y_{t-d} \leq r_1,$$

$$regime\ 2\ \ y_t = \beta_0^{(2)} + \beta_1^{(2)} y_{t-1} + \cdots - \beta_p^{(2)} y_{t-p} + \varepsilon_t\ \ if\ r_1 \leq y_{t-d} \leq r_2,$$

$$\vdots$$

$$regime\ k\ \ y_t = \beta_0^{(k)} + \beta_1^{(k)} y_{t-1} + \cdots - \beta_p^{(k)} y_{t-p} + \varepsilon_t\ \ if\ r_{k-1} \leq y_{t-d} \leq r_k. \quad (1)$$

The above mentioned model is defined as self-exciting model because the regimes are the functions of the past realizations of y_t sequence itself.In the SETAR model, positive integer d is delay parameter and $r_i, i = 1, \cdots, k$ are threshold variables. The delay parameter is lag of the process to identify the threshold variables, and the thresholds show which values of the threshold variablse separate the data points into the regimes. Identification of threshold value and delay parameters are complicated and the Akaike information criterion is one of the methods used to select delay and threshold values in practical researches [4].

4 Data

This paper used the monthly western tourist arrivals to Thailand for model estimation and evaluation. The data series extended from January 1997 until October 2014 to apply in the study and consisted of 214 observations. All data that we used for the paper were obtained from EcoWin data base and the data showed seasonality.

After testing Augmented Dickey-Fuller test the data showed whether there is unit root with ADF statistic 0.199 and probability 0.972. In addition, the data clearly showed whether there is seasonality. Therefore, the study had to handle the data to remove unit root, and the type of data control are taking log return, taking first difference after obtaining log return, and taking twelve differences.

As per visual figure of the data, there were some breaks in the tourists arrivals. Then, the study tested for potential break points with multiple break point tests; results showed that there were three break points on Nov. 2000, Nov. 2006, and Nov. 2010. These results were confirmed with Chaw break point test and Quandt-Andrews unknown breakpoint test whether there were breakpoints. When the study tested the data with recursive residuals and CUSUM test, the graphical results showed that there is structural change.

5 Empirical Results

Self-Exciting Threshold Autoregressive (SETAR) Model

The self-exciting threshold autoregressive (SETAR) process is the relevant model for the discontinuities due to internal changes. In the SETAR process, the value

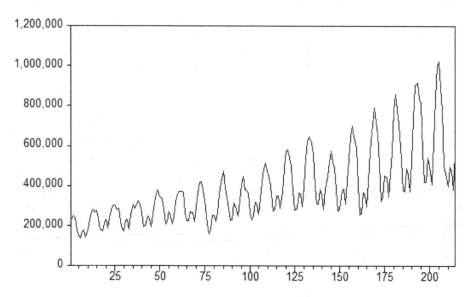

Fig. 1. Western tourists arrival to Thailand between January 1997 and October 2014

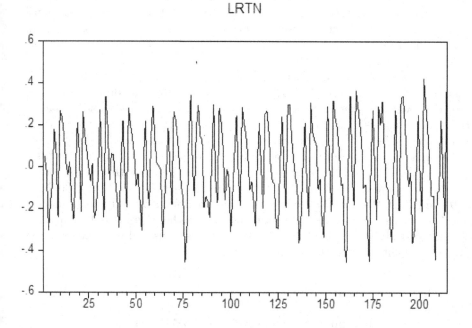

Fig. 2. Western tourists arrival to Thailand after taking log-return

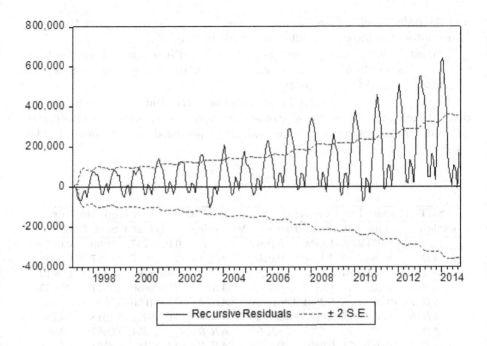

Fig. 3. Recursive test result on OLS of the western tourists arrival

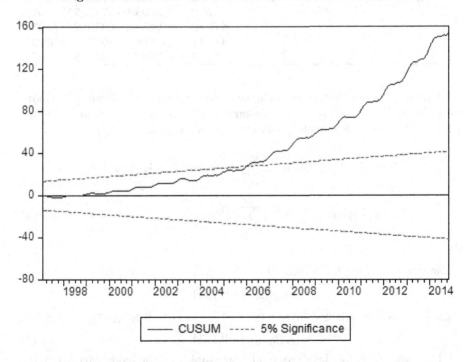

Fig. 4. CUSUM test result on OLS of the western tourists arrival

p of the order of autoregression was set equal for all the regimes, but the order of the autoregression can be different in each regime [36].

The study found out the best fit AR model with p value 12 for the high regime and 8 for the low regime among other AR models. The model had AIC (-2.353645), and SC (-2.139999).

By depending on the SETAR model selection, the study went on the prediction with maximum autoregressive order for low regime 8, maximum autoregressive order for high regime 12, threshold values not fixed, and the delay 1. The results showed as follows;

Table 1. SETAR(2,8,12) with d=1, r $= -0.123360$

SETAR model in Low Regime				SETAR model in High Regime			
Variable	Coefficient	t-Stat	Prob	Variable	Coefficient	t-Stat	Prob
C	-0.0715623	-1.8866	0.0607	C	0.0259767	2.2666	0.0245
AR 1	-0.2855323	-2.4150	0.0167	AR 1	-0.0608667	-0.6857	0.4938
AR 2	-0.1113070	-0.9434	0.3467	AR 2	-0.4658366	-6.7441	1.801e-10
AR 3	0.3094146	1.6581	0.0989	AR 3	-0.3036690	-4.4537	1.438e-05
AR 4	-0.6422825	-5.2831	3.461e-07	AR 4	-0.3591600	-5.5324	1.036e-07
AR 5	-0.3678217	-4.0955	6.229e-05	AR 5	-0.0390225	-0.4918	0.6234
AR 6	-0.4928772	-5.8267	2.382e-08	AR 6	-0.2193351	-3.0067	0.003
AR 7	-0.0876771	-0.9691	0.3337	AR 7	-0.1557164	-1.8521	0.0656
AR 8	0.1922540	3.3463	0.0010	AR 8	-0.2012147	-2.2862	0.0233
				AR 9	-0.4049866	-4.7575	3.870e-06
				AR 10	-0.3335471	-4.3730	2.017e-05
				AR 11	-0.0010045	-0.0135	0.9892
				AR 12	0.4054097	5.0961	8.333e-07

Although some estimates for some parameters were not significant, all parameters are stable as $|\beta_j| < 1$ is necessary condition for the stability, where j $=1, 2, ...$ The SETAR models we selected can be written as;

$$low\ regime\ y_t = \beta_0^{(1)} + \sum_{i=1}^{8} \beta_i^{(1)} y_{t-i} + \varepsilon_t^{(1)} \ if\ y_{t-d} \leq r,$$

$$high\ regime\ y_t = \beta_0^{(2)} + \sum_{i=1}^{12} \beta_i^{(2)} y_{t-i} + \varepsilon_t^{(2)} \ if\ y_{t-d} > r. \tag{2}$$

Or

$$low\ regime\ y_t = -0.0715623 - 0.2855323 y_{t-i} - \cdots + 0.1922540 y_{t-8} + \varepsilon_t^{(1)}$$

$$if\ y_{t-0} \leq -0.1234,$$

$$high\ regime\ y_t = 0.0259767 - 0.0608667 y_{t-i} - \cdots + 0.4054097 y_{t-12} + \varepsilon_t^{(2)}$$

$$if\ y_{t-0} > -0.1234. \tag{3}$$

The parameters for the above model are stable and the model has AIC value (-1150), BIC value (-1073.172) and MAPE (101.2).

6   Foresting and Discussion

6.1   Forecasting

All models used in this study and analysis are feasible for forecasting with out of sample forecast method; it is also called ex-ante forecasts. Indeed, these models can be also applied for recursive (Window expanding) out of sample forecasting. Generally, the forecasting in nonlinear autoregressive models, can utilize the following equation;

$$\hat{y}t + h = E[y_{t+h}|y_t, Y_{t-1}, \dots], \tag{4}$$

where $h = 1, 2, \cdots$.

Forecasting with SETAR model can be done with the following equation:

$$low \ regime \ y_{t+1} = \beta_0^{(1)} + \sum_{i=1}^{p} \beta_i^{(1)} y_{t-i+1} + \varepsilon_{t+1}^{(1)} \ if \ y_{t-d} \leq r,$$

$$high \ regime \ y_{t+1} = \beta_0^{(2)} + \sum_{i=1}^{p} \beta_i^{(2)} y_{t-i+1} + \varepsilon_{t+1}^{(2)} \ if \ y_{t-d} > r, \tag{5}$$

where $d = $ delay, and $r = $ threshold value.

In this study, BAYSTAR method was used to forecast for one step ahead. The BAYSTAR is Threshold Autoregressive model: Bayesian approach. To get the best possible forecast on the parameters, the BAYSTAR applies Markov Chain Monte Carlo (MCMC) method for ten thousand times. Since some of the coefficients from the full mmodel in BAYSTAR are not significant, it is recommended to drop those related variables out of the mmodel to be parsimonous. The corrected BAYSTAR model is then presented in Table 2.

Table 2. The results of BAYSTAR for $10,000$ MCMC reiterations

Parameter	Mean	Median	s.d.	Lower	Upper
$\beta_{1.4}$	-0.5845	-0.5839	0.1232	-0.8240	-0.3492
$\beta_{1.5}$	-0.4872	-0.4846	0.0906	-0.6693	-0.3128
$\beta_{1.6}$	-0.5547	-0.5558	0.0740	-0.6979	-0.4055
$\beta_{1.8}$	0.2255	0.2251	0.0550	0.1190	0.3356
$\beta_{2.2}$	-0.3251	-0.3246	0.0588	-0.4403	-0.2117
$\beta_{2.3}$	-0.2061	-0.2063	0.0542	-0.3132	-0.0998
$\beta_{2.4}$	-0.2635	-0.2623	0.0551	-0.3747	-0.1586
$\beta_{2.6}$	-0.2514	-0.2514	0.0603	-0.3713	-0.1340
$\beta_{2.8}$	-0.2779	-0.2777	0.0705	-0.4164	-0.1395
$\beta_{2.9}$	-0.1791	-0.1797	0.0604	-0.2961	-0.0613
$\beta_{2.10}$	-0.2898	-0.2892	0.0643	-0.4175	-0.1656
$\beta_{2.12}$	0.5160	0.5166	0.0774	0.3626	0.6657
σ_1^2	0.0063	0.0062	0.0013	0.0042	0.0091
σ_2^2	0.0046	0.0045	0.0006	0.0036	0.0058
γ	-0.1167	-0.1160	0.0121	-0.1442	-0.0925

The results came with residuals for each regime of the model estimated, posterior, coefficients, threshold values, the deviance information criterion (DIC) = -853.551; a Bayesian method for model comparison and the mean forecast error (MFE) is -0.007.

Based on the results, the forecasting equation for one step ahead can be written as following;

$low\ regime\ y_{t+1} = -0.0715623 - 0.5845y_{t-3} + \cdots - 0.5547y_{t-5} + 0.2255y_{t-7} + \varepsilon_{t+1},$

$high\ regime\ y_{t+1} = 0.0259767 - 0.3251y_{t-1} \cdots - 0.2635y_{t-3} - 0.2514y_{t-5}$

$$-0.2779y_{t-7} - 0.1791y_{t-8} - 0.2898y_{t-9} + 0.5160y_{t-11} + \varepsilon_{t+1}. \tag{6}$$

The above forecasting model had RMSE = 0.0692, and MAPE = 101.2%, whereas SARIMA-GARCH forecast had RMSE = 0.0752, and MAPE = 115.3786%. The actual western tourists arrival to Thailand in November 2014 was $722,391$ [10] and for the log return value > -0.1167, the forecasted value was $784,883$. Therefore the study indicated that only 0.9% overforecasted from the actual value.

6.2   Discussion

In this study, we searched the efficiency and effectiveness of SARIMA GARCH in forecasting as the combination of notorious forecasting methods, SARIMA and GARCH family, have proven that this combination is reliable for forecasting, especially in forecasting for tourists arrival. This method was applicable for Chinese mainlanders arrival to Taiwan, but the researchers for this study did not consider the structural changes and breakpoints in the sample. Normally, the GARCH family members need to be predicted with the sample size with a thousand data set, at least. The breakpoints, structural change, and small sample size made the model weak. In this study the result of SARIMA GARCH had had unstable parameters and unrealistic estimate with negative sign.

In the second part of the study, SETAR process was focused with the confidence that SETAR model may be the answer for the study, and revealed that it was applicable and could forecast tourists arrival nearly correct. When the study compares SARIMA GARCH and SETAR, the findings favored the SETAR. For forecasting, the study forecasted on only one period ahead with out of sample forecasting method although the model allowed for multi steps ahead forecasting. To forecast, the study used BATSTAR forecasting method by applying MCMC 1000 times.

When the study applied BAYSTAR to the model with 8 autoregressive lags for lower regime and 12 autoregressive lags for higher regime, the process dropped some lags for better forecasts. The findings of the study claim that SETAR model was better than SARIMA GARCH at the presence of break points and structural change in the sample, especially for small samples.

7   Conclusion

The study contributed examination on SARIMA - GARCH, a tourism demand forecasting model proposed by Yi-Hui Liang to determine some facts to be aware of when researchers apply the model. Although the SARIMA GARCH is a state of the art model for tourism forecasting, model specification is crucial for the SARIMA part whereas taking a big enough sample size, and considering breaks and structural change are also important features for the GARCH parts. The study also proved predictive power of SETAR model and a reliable forecasting method BAYSTAR. The model matching between SARIMA - GARCH and SETAR showed that SETAR outperforms SARIMA - GARCH at the presence of structural change in the sample size. The SETAR model confirmed that the model can predict western tourists arrival to Thailand effectively and forecast with BAYSTAR acceptably. The prediction value and forecasted values came for each threshold. It is flexibility for forecasting and it can be used an interval for more confidence in the forecasting. As the study tried to handle structural change, the study suggests the researchers work on other possible answers, such as (Double Threshold GARCH) DTGARCH and smooth transition autoregressive (STAR) models for the data set in which there are breaks and structural.

Acknowledgment. The authrs are very thankful to Professor Hung T. Nguyen and Professor Cathy W.S. Chen for the valuable comments. The authors extend their thanks to Puay Ungphakorn Centre of Excellence in Econometrics, Faculty of Economics, Chiang Mai University for financial support.

References

1. Autchariyapanitkul, K., Chanaim, S., Sriboonchitta, S.: Evaluation of portfolio returns in fama-french model using quantile regression under asymmetric laplace distribution. Studies in Computational Intelligence **583**, 233–244 (2014)
2. Ayikoru, M.: Destination competitiveness challenges: A Ugandan perspective. Tourism Management **50**, 142–158 (2015)
3. Ayusuk, A., Sriboonchitta, S.: Risk, return and international portfolio analysis: Entropy and linear belief functions. Studies in Computational Intelligence **583**, 319–328 (2014)
4. Bratcikoviene, N.: Adapted SETAR model for Lithuanian HCPI time series. Nonlinear Analysis: Modelling and Control **17**(1), 2746 (2012)
5. Campbell, E.P.: Bayesian Selection of Threshold Autoregressive Models. Journal of Time Series Analysis 467–482 (2004)
6. Chang, C.-L., McAleer, M., Slottje, D.J.: Modelling international tourist arrivals and volatility: An application to Taiwan. Contributions to Economic Analysis **288**, 299–315 (2009)
7. Chang, C.-L., Sriboonchitta, S., Wiboonpongse, A.: Modelling and forecasting tourism from East Asia to Thailand under temporal and spatial aggregation. Mathematics and Computers in Simulation **79**(5), 1730–1744 (2009)

8. Clements, M.P., Smith, J.: The performance of alternative forecasting methods for SETAR models. International Journal of Forecasting **13**(4), 463–475 (1997)

9. De Gooijer, J.G., De Bruin, P.T.: On forecasting SETAR processes. Statistics & Probability Letters **37**(1), 7–14 (1998)

10. Department of Tourism: International Tourist Arrivals to Thailand 2014 (BY NATIONALITY) (November 2014). http://www.tourism.go.th/home/details/11/221/23044

11. Dionguea, A.K., Guganb, D., Vignalc, B.: Forecasting electricity spot market prices with a k-factor GIGARCH process. Applied Energy **86**(4), 505–510 (2009)

12. Dufrenota, G., Gueganb, D., Peguin-Feissolle, A.: Economics Letters **86**(2), 237–243 (2005)

13. Galeana, P., Pea, D.: Improved model selection criteria for SETAR time series models. Journal of Statistical Planning and Inference **137**(9), 2802–2814 (2007)

14. Gong, X., Sriboonchitta, S.: Optimal Portfolio Selection Using Maximum Entropy Estimation Accounting For The Firm Specific Characteristics. Studies in Computational Intelligence **583**, 305–318 (2015)

15. Gunter, U., Önder, I.: Forecasting international city tourism demand for Paris: Accuracy of uni- and multivariate models employing monthly data. Tourism Management **46**, 123–135 (2015)

16. Guoa, J., Huanga, W., Williams, B.M.: Adaptive Kalman filter approach for stochastic short-term traffic flow rate prediction and uncertainty quantification. Transportation Research Part C: Emerging Technologies **43**, 50–64 (2014)

17. Kanjanatarakul, O., Lertpongpiroon, P., Singkharat, S., Sriboonchitta, S.: Econometric forecasting using linear regression and belief functions. In: Cuzzolin, F. (ed.) BELIEF 2014. LNCS, vol. 8764, pp. 304–312. Springer, Heidelberg (2014)

18. Kanjanatarakul, O., Kaewsompong, N., Sriboonchitta, S., Denoeux, T.: Forecasting using belief functions: An application to marketing econometrics. International Journal of Approximate Reasoning **55**(5), 1113–1128 (2014)

19. Kanjanatarakul, O., Kaewsompong, N., Sriboonchitta, S., Denux, T.: Estimation and Prediction Using Belief Functions: Application to Stochastic Frontier Analysis. Studies in Computational Intelligence **583**, 171–184 (2015)

20. Liang, Y.H.: Forecasting models for Taiwanese tourism demand after allowance for Mainland China tourists visiting Taiwan. Computers & Industrial Engineering 111–119 (2014)

21. Liu, J., Sriboonchitta, S.: Analysis of volatility and dependence between the tourist arrivals from China to Thailand and Singapore: A copula-based GARCH approach. Advances in Intelligent Systems and Computing 283–294 (2013)

22. Liu, J., Sriboonchitta, S., Nguyen, H.T., Kreinovich, V.: Studying Volatility and Dependency of Chinese Outbound Tourism Demand in Singapore, Malaysia, and Thailand: A Vine Copula Approach. Advances in Intelligent Systems and Computing **251**, 259–274 (2014)

23. Liu, J., Sriboonchitta, S.: Empirical evidence linking futures price movements of biofuel crops and conventional energy fuel. Studies in Computational Intelligence 287–303 (2015)

24. Melda, A.: A novel approach to model selection in tourism demand modeling. Tourism Management **48**, 64–72

25. Min, N., Sirisrisakulchai, J., Sriboonchitta, S.: Forecasting tourist arrivals to Thailand using belief functions. Studies in Computational Intelligence **583**, 343–357 (2015)

26. Morleya, C., Rossellb, J., Santana-Gallegob, M.: Gravity models for tourism demand: theory and use. Annals of Tourism Research **48**, 1–10 (2014)

27. Phochanachan, P., Sirisrisakulchai, J., Sriboonchitta, S.: Estimating oil price value at risk using belief functions. Studies in Computational Intelligence 377–389 (2015)

28. Piamsuwannakit, S., Sriboonchitta, S.: Forecasting risk and returns: CAPM model with belief functions. Studies in Computational Intelligence 259–271 (2015)

29. Puarattanaarunkorn, O., Kiatmanaroch, T., Sriboonchitta, S.: Dependence structure between TOURISM and TRANS sector indices of the stock exchange of Thailand. Thai Journal of Mathematics **2014**, 199–210 (2014)

30. Serletisa, A., Shahmoradib, A.: Chaos, self-organized criticality, and SETAR nonlinearity: An analysis of purchasing power parity between Canada and the United States. Chaos, Solitons & Fractals **33**(5), 1437–1444 (2007)

31. Sigaukea, C., Chikobvub, D.: Prediction of daily peak electricity demand in South Africa using volatility forecasting models. Energy Economics **33**(5), 882–888 (2011)

32. Sriboonchitta, S., Liu, J., Sirisrisakulchai, J.: Willingness-to-pay estimation using generalized maximum-entropy: A case study. International Journal of Approximate Reasoning **60**, 1–7 (2015)

33. Tang, J., Sriboonchitta, S., Yuan, X.: Forecasting Inbound Tourism Demand to China Using Time Series Models and Belief Functions. Studies in Computational Intelligence **583**, 329–341 (2015)

34. Pemberton, J., Tong, H.: Threshold Autoregression, Limit Cycles and Cyclical Data. Journal of the Royal Statistical Society 245–292 (1980)

35. Kreinovich, V., Nguyen, H.T., Sriboonchitta, S.: What if we only have approximate stochastic dominance? Studies in Computational Intelligence 53–61 (2015)

36. Watier, L., Richardson, S.: Modelling of an epidemiological time series by a threshold autoregressive model. The Statistician 353–364 (1995)

37. Wichian, A., Sirisrisakulchai, J., Sriboonchitta, S.: Copula based polychotomous choice selectivity model: Application to occupational choice andwage determination of older workers. Studies in Computational Intelligence 359–375 (2015)

Welfare Measurement on Thai Rice Market: A Markov Switching Bayesian Seemingly Unrelated Regression

Pathairat Pastpipatkul[(✉)], Paravee Maneejuk, and Songsak Sriboonchitta

Department of Economics, Chiang Mai University, Chiang Mai, Thailand
ppthairat@hotmail.com, mparavee@gmail.com

Abstract. This paper aimed to measure the welfare of the Thai rice market and provided a new estimation in welfare measurement. We applied the Markov Switching approach to the Seemingly Unrelated Regression model and adopted the Bayesian approach as an estimator for our model. Thus, we have the MS-BSUR model as an innovative tool to measure the welfare. The results showed that the model performed very well in estimating the demand and supply equations of two different regimes; namely, high growth and low growth. The equations were extended to compute the total welfare. Then, the expected welfare during the studied period was determined. We found that a mortgage scheme may lead the market to gain a high level of welfare. Eventually, the forecasts of demand and supply were estimated for 10 months, and we found demand and supply would tend to increase in the next few months before dropping around March, 2015.

Keywords: Welfare · Thai rice · Markov switching · Bayesian seemingly unrelated regression

1   Introducion

Agriculture has played a key role in the Thai economy where rice is one of the most important agricultural commodities that create enormous economic value for exports. In spite of the strong capacity ranked of Thailand competitive countries for producing rice, i.e. India and Vietnam, Thailand still has ranked in the top three largest rice exporters (countries) in the world. Agriculture accounts for 7.91% of Thailand's total exports in which rice takes 30.25% of primary agricultural exports (Bank of Thailand, 2014), and it still employs over half of the total labor force.

Thai government considers rice as a major agricultural crop of Thailand for both its economic value and strong socio-cultural root. Several policies have been proposed by the Thai government in order to boost up the rice market and to support Thai rice farmers. It is normally known that Thai rice farmers have faced many problems while taking this career. For instance, some of them do not own the land he/she works and most of them are still in debt even they have worked in this career for a long time. Also the fluctuation of international rice prices in the market creates a great impact on the domestic rice price. The Thai government realizes these problems and tries to resolve them by proposing policies such as the rice price guarantee scheme and the

© Springer International Publishing Switzerland 2015
V.-N. Huynh et al. (Eds.): IUKM 2015, LNAI 9376, pp. 464–477, 2015.
DOI: 10.1007/978-3-319-25135-64_42

rice mortgage scheme. But the solutions seem like making the rice to become a political crop. Authority is vested to the government to control either the domestic rice prices or how much rice the farmers should produce. This situation leads to the questions that, is the political crop system efficient?

To answer these questions, it might be good to look at some economic theorems and techniques to evaluate the policies in terms of the effects on the well-being of society. Therefore, now, we are taking into account welfare economics which is a branch of economics that can be used for appraising the efficiency of the policy. In this paper, we intend to measure the welfare of the rice market during a studied period, 2006 to 2014, when several policies were used to support the market. Then, we aim to capture what types of policy could make the market better off by seeking for the policy, which creates the highest level of welfare.

As welfare is our key to answer these questions, some effective tools for measuring welfare were substantially sought. Many papers have been reviewed, and we found that the classical method has been used very often including the seemingly unrelated regression model (SUR), e.g., Baltagi and Pirotte [3], Phitthayaphinant, Somboonsuke and Eksomtramage [14], Kuson, Sriboonchitta and Calkins [12], and Lar, Calkins and Sriboonchitta [13]. SUR is still taken into account since the demand and supply equations of the market are linear system equations, and the welfare is technically measured by both demand and supply curves. Not only using SUR model, we also applied a regime switching approach in the study. As we know several policies have been used to support and boost up the rice market, and some are not likely to work but rather harm the market. Thus, the market movement might not be rectilinear; it ought to be either an upturn or downturn. For this reason, we provided SUR model with regime switching as an innovative tool to determine the welfare of Thai rice market.

To improve on the estimation, the Bayesian approach was implemented for estimation. Even though the classical estimator, i.e., the maximum likelihood, has been used very often in econometric fields, e.g., Ayusuk and Sriboonchitta [2], Boonyanuphong and Sriboonchitta [4], Chinnakum, Sriboonchitta, and Pastpipatkul [6], and Sriboonchitta, Liu, Kreinovich, and Nguyen [16], we still believe that an estimator with Bayesian statistic might be better than the classical one. Therefore, in this study, SUR model with the regime switching based Bayesian estimation was provided as the innovative tools to estimate the welfare of Thai rice market.

2   Welfare Economics

Welfare economics –or social welfare- is a branch of economics that evaluates the well-being of the community. One of the main reasons that economic policies are used is to make people happier. Welfare economics is like a conceptual tool for judging an achievement of the policies. To understand the analysis of welfare, a partial equilibrium of a competitive market is desirable since the market is efficient in terms of allocating resources. We used the area between the equilibrium price (P*) and the demand and supply curves to measure the welfare.

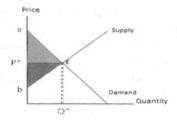

Fig. 1. Consumer Surplus and Producer Surplus

Consider the market demand and supply curves shown in Figure1. The market equilibrium is located at point E. We can use the demand and supply curves to measure the net benefit for the market. Consumer surplus (CS) is the value of the difference between the price that consumers are willing to pay for a product on the demand curve and the price actually paid. Figure 1 shows that the green shaded triangle, the area under the demand curve and above the market price, represents CS measured in dollars. Similar to the concept of CS, Producer surplus (PS) is the value of the difference between the actual selling price of the product and the price that producers are willing to sell it for on the supply curve. Figure 1 shows that PS measured in dollars is represented by the red shaded triangle. It is the area above the supply curve and under the market price. The sum of CS and PS is given by the distance ab for the first unit produced. The total welfare continues to increase as additional output is produced up to the equilibrium level. This level of production can be achieved only if the price is at the competitive equilibrium. Therefore, the total dollar value of the total welfare of society is the whole triangle area consisting of both the consumer and producer surplus triangles shown in Figure 1. Clearly, for the total social welfare, it also can be written as Total Social Welfare (TS) = CS + PS.

3   Methodology

After the first introduction of Markov switching autoregressive model of Hamilton [7], it became very popular in several fields used for applications since the model is able to capture a different pattern of the time-series data. However, the model seems to lack efficiency in estimate the system equation. Thus, in this study, the Markov switching approach has been applied to seemingly unrelated regression (SUR) model as proposed by Zellner [17] in order to estimate the system linear regression equation for each regime. In addition, to make our study better, the Bayesian approach is implemented to estimate the model. So, we try to propose a Markov Switching, Bayesian, Seemingly unrelated regression (MSB-SUR) in this paper and a brief introduction of SUR and MS-BSUR model can be described as follows.

3.1   Seemingly Unrelated Regression (SUR) Model

The system of equations, SUR, was introduced by Zellner [17]. The model was proposed to estimate the multiple correlated equations and to improve estimation efficiency. In considering the structure of SUR model, we found that it consists several equations where each equation is allowed to have a different both dependent and independent variables. We assumed that each equation has the same right hand side number of independent variables: say, N independent variables. Thus, M equations can be written as follows:

$$y_{t,1} = \underline{x}_{t,1}\beta_1' + \varepsilon_{t,1}$$
$$\vdots$$
$$y_{t,M} = \underline{x}_{t,M}\beta_M' + \varepsilon_{t,M}$$

We can also write the model as a vectorial form:

$$Y_t = X_t\beta + U_t$$

Where

$$Y_t = \begin{bmatrix} y_{t,1} \\ \vdots \\ y_{t,M} \end{bmatrix}, \quad X_t = \begin{bmatrix} x_{t,11} & \cdots & x_{t,1N} \\ \vdots & \ddots & \vdots \\ x_{t,M1} & \cdots & x_{t,MN} \end{bmatrix}, \quad \beta = \begin{bmatrix} \beta_{11} & \cdots & \beta_{1N} \\ \vdots & \ddots & \vdots \\ \beta_{M1} & \cdots & \beta_{MN} \end{bmatrix}'$$

Y_t is a vector of dependent variables $y_{t,i}, i = 1,...,M$ and X_t is a matrix of independent variables (regressors) $x_{t,ij}, i = 1,...,M, j = 1,...,N$, it does not include an intercept. β is a matrix of an unknown parameters (regression coefficients). U_t is a vector of errors, $U_t = \begin{bmatrix} \varepsilon_{t,1} & \varepsilon_{t,2} & \cdots & \varepsilon_{t,M} \end{bmatrix}'$ where $\varepsilon_{t,i} \sim N(0, \sigma_i^2), i = 1,...,M$. The errors, U, are assumed to correlate across equations. Thus, we can estimate the M equations jointly. Then, a variance-covariance matrix can be given as:

$$\Gamma = \Sigma \otimes I = \begin{bmatrix} \sigma_{11}I & \sigma_{12}I & \cdots & \sigma_{1M}I \\ \sigma_{21}I & \sigma_{22}I & \cdots & \sigma_{2M}I \\ \vdots & \vdots & \ddots & \vdots \\ \sigma_{M1}I & \sigma_{M2}I & \cdots & \sigma_{MM}I \end{bmatrix}, \text{ where } \Sigma = \begin{bmatrix} \sigma_{11} & \sigma_{12} & \cdots & \sigma_{1M} \\ \sigma_{21} & \sigma_{22} & \cdots & \sigma_{2M} \\ \vdots & \vdots & \ddots & \vdots \\ \sigma_{M1} & \sigma_{M1} & \cdots & \sigma_{MM} \end{bmatrix}$$

And I is an identity matrix .Thus, we can write the estimated parameters as

$$\beta_{sure} = (X_t'\Gamma^{-1}X_t)^{-1}X_t'\Gamma^{-1}Y_t$$

3.2   Markov Switching Seemingly Unrelated Regression (MS-SUR)

The Markov Switching approach was first introduced by Hamilton in 1989 as the regime switching model. The main idea of the Markov switching is that there exists a switching in the model structures consisting of an intercept, coefficient of regressors, and covariance. The switching is controlled by an unobserved variable [8]. For the Markov switching seemingly unrelated regression (MS-SUR) model, let's consider the following general form of M equations model.

$$Y_t = \alpha(s_t) + X_t \beta(s_t) + U_t(s_t) \tag{1}$$

Y_t is the vector of the dependent variables $y_{t,i}, i = 1,...,M$ and X_t is the matrix of the independent variables (regressors) $x_{t,ij}, i = 1,...,M, j=1,...,N$. The intercepts, regression coefficients, and the covariance all depend on the state or regime (s_t), where $s_t = 1,...,k$. For example, if we consider two regimes which are high and low, then $s_t - 1, 2$. Furthermore, α is the vector of the intercepts $\alpha_i, i = 1,...,M$ and it is regime-dependent as well as β, which is the matrix of the regression coefficients with regime-dependent.

The errors $U_t \sim i.i.d.N(0, \Gamma(s_t))$ and $\alpha_1(s_t),...,\alpha_M(s_t), \beta_{ij}(s_t),....,\beta_{MN}(s_t)$, $\Sigma(s_t) = \gamma$ are regime-dependent parameters. The structure errors are correlated across equations since Zellner [17] suggested that the model can gain more efficiency if the error term of each equation is related. Moreover, it is also assumed to have a nonsingular normal distribution. Thus, the structure error term for regime k can be written as

$$\Sigma(s_t) = \begin{bmatrix} \sigma_{11}(s_t) & \sigma_{12}(s_t) & \cdots & \sigma_{1M}(s_t) \\ \sigma_{21}(s_t) & \sigma_{22}(s_t) & \cdots & \sigma_{2M}(s_t) \\ \vdots & \vdots & \ddots & \vdots \\ \sigma_{M1}(s_t) & \sigma_{M1}(s_t) & \cdots & \sigma_{MM}(s_t) \end{bmatrix} \tag{2}$$

Recall, the regime (s_t) which is unobserved, is governed by first order Markov process. It is defined by the transition probabilities matrix P. The regime variable s_t is finite and does not change over time t. We get filtered regime probabilities $\Pr(s_t = j | s_{t-1})$ by using the Filter and Smoothing algorithm as proposed in Brandt [5] and Kim and Nelson [10].

The estimation technique implemented for MS-SUR models is called Expectation Maximum (EM) algorithm which is consist of 2 steps. In the first step is to compute the expected likelihood by including latent variables as if they were observed, and second step is to compute the maximum likelihood estimates of the parameters by maximizing the expected likelihood found in the first step. The EM algorithm can estimate the unique optimal solutions for estimated parameters (γ) which are guaranteed by the maximum properties.

The joint likelihood for observation in each regime is defined as $\Pr(y_t, s_t = i)$, $\forall i = 1, 2, \ldots k$. In this study, we assume a Markov switching with 2 regimes, thus from (1) we obtain the likelihood function of MS-SUR with 2 regimes as follows:

$$Log\,L = \prod_{t=1}^{T} \left[\sum_{s_t=1,2} p(Y_t | Y_{t-1}, X_t, \gamma, Q) p(s_t | Y_{t-1}, X_{t-1}, \gamma, Q) \right] \tag{5}$$

The EM iterations are employed to maximize the likelihood function (5). The likelihood function will increase at each step and reach a maximum correspondence to convergence [14].

3.3   Prior and Posterior

The prior of the model has a form as follows:

$$\Pr(\gamma, Q, s_t) = \Pr(\gamma) \Pr(Q) \Pr(s_0 | \gamma, Q) \prod_{t=1}^{T} \Pr(s_t | \gamma, Q, s_{t-1}) \Pr(s_t) \tag{6}$$

The prior is separated into three parts, which are the estimated parameter (γ), transition matrix (Q), and the Markov process (s_t). In this study, the prior for parameter coefficient, $\Pr(\gamma)$, is Normal-wishart. The prior for the Transition matrix, $\Pr(Q)$, is assumed to be Dirichlet. The prior for Markov process, $\Pr(s_t)$, is the ergodic distribution. Combining these priors (6) with the likelihood function (5) using Bayes theorem to compute the posterior, thus the posterior estimation can be formed as follows:

$$\Pr(\gamma, Q | Y_t, X_t) = \Pr(\gamma, Q) \Pr(Y_t, X_t | \gamma, Q) \tag{7}$$

However, the posterior density $p(\gamma | Y_t, X_t)$ is unknown thus the Gibb sampler is used for sampling the following block conditional posterior distributions:

1) The conditional posterior for regression coefficients for each regime is ($p(s_t | Y_t, X_t, \gamma, Q)$). In this block, the cross-product matrices of each equation i in regime $k \in \{1, 2\}$ can be computed from

$$\Sigma_{XX,i,k} = \sum_{t \in s_t = k} X_t X_t' \quad , \quad \Sigma_{XY,i,k} = \sum_{t \in s_t = k} X_t Y_t'$$

Then, the conditional posterior precision is computed by

$$\Sigma_{i,k} \otimes \Sigma_{XX,i,k} + diag((\lambda_0 \times \lambda_1)^2 \otimes \tilde{\beta}_{i,k}) \tag{8}$$

where λ_0 is the overall tightness of the prior, λ_1 is proportional standard deviations around the coefficients, and $\tilde{\beta}_{i,k}$ is the initial estimated coefficient. Next, inverting the positive definite square matrix of Cholesky decomposition in (8) to obtain the standard error of the estimated coefficient ($\tilde{\beta}_{se}$). Thus $p(s_t | Y_t, X_t, \gamma, Q)$ can be estimated by

$$\hat{\beta}_{i,k} = (\beta_{i,k} + \tilde{\beta}_{se,i,k})$$

2) The conditional posterior for transition matrix is ($p(Q | Y_t, X_t, \gamma, s_t)$). This block is to draw $p(Q | Y_t, X_t, \gamma, s_t)$ where the estimated draws can be written as

$$\prod_{i=1}^{k=2} p_{i,j}^{n_{i,j} + \alpha_{i,j}} \tag{9}$$

where $n_{i,j}$ are the number of times that regime i changes to regime j; $\alpha_{i,j}$ are the Dirichlet prior element for the initial transitions

3) The conditional posterior for the Markov process is ($p(s_t | Y_t, X_t, \gamma, Q)$). To estimate $p(s_t | Y_t, X_t, \gamma, Q)$ in order to obtain the regime probabilities, we employ the Baum-Hamilton-Lee-Kim (BHLK) and smoother to estimate the filter probabilities and use the standard forward-filter-backward-sample algorithm to draw the matrix of regimes.

3.4   Forecast MSB-SUR

To forecast the MS-BSUR, we follow the estimation procedure of Krolzig [11], Fruwirt-Schnatter [9], and Brandt [5]. First of all, the formula of MS-BSUR forecast for period T+f can be written as:

$$Y_{t+f} = Y_t + SS(Y_{t+f-1})(\gamma + \delta_i)$$

Where SS is State-space initialization produced from sampling posterior of an MS-BUR model in previous step, and γ is the sum of estimated intercept and coefficient term from MS-BSUR model. δ_i is the structural shock of the model. Then, generating the burn-in and final posterior are drawn by 1,000 and 2,000 draws, respectively.

4   Estimation

In this study, we use data set related to Thai rice consisting of the demand for rice (Q_t^d), output of grain rice (Q_t^s), export price of Thai rice (P_t^{exp}), Pakistan's exported rice price (P_t^{Paki}), Vietnam's exported rice price (P_t^{Viet}), export price of India's rice

(P_t^{Ind}), producer price of rice (P_t^{Farm}), rainfall (R_t), and water storage (W_t). The data set are monthly frequency collected from M1/2006 to M9/2014, covering 105 observations. All series are transformed into the log-log before we start the estimation.

The Model Specification

In this study, we attempt to measure the social welfare based on supply and demand functions, thus, the log-log supply and demand functions are preferred for estimating MS-BSUR model and the model can be specified as follows:

$$Q_t^d = \alpha_1(s_t) + \beta_1(s_t) P_t^{export} + \beta_2(s_t) P_t^{viet} + \beta_3(s_t) P_t^{paki} + \beta_4(s_t) P_t^{india} + U_{1,t}(s_t)$$
$$Q_t^s = \alpha_2(s_t) + \delta_1(s_t) P_t^{export} + \delta_2(s_t) W_t + \delta_3(s_t) R_t + \delta_4(s_t) P_t^{farm} + U_{2,t}(s_t)$$

(10)

The estimation of MS-BSUR with 2 regimes comprises 3 main steps. In the First step, the sequence estimation based the linear optimization algorithm has been conducted to estimate the initial or starting parameter values, Υ, as a baseline for MS-SUR model. Sets an initial value for transition matrix as $p_{11} = 0.9$ and $p_{22} = 0.9$.

In the second step, the MS-SUR log-likelihood function (5) is estimated recursively in order to get the estimated parameters. Following Sims, Waggoner, and Zha [15] and Brandt [5], as they proposed an algorithm that is designed for multiple equations called block-wise maximum likelihood algorithm. This method can update initial parameters , $\hat{\gamma}$, where the first block is MS-SUR intercept coefficients , the second block is regressor or instrument coefficients, the third block is error covariance, and the last block is the transition matrix. We maximize each block separately, while holding the other block fix. These blocks are iterated by 1,000 iterations for fitting our initial model from the sequence estimation step.

The unknown parameters, γ, can be estimated from filtering and smoothing the observed process for Y_t and X_t to find the $\Pr(s_t | Y_{t-1}, \gamma, P)$ as proposed in Brandt [5] and Sims, Waggoner, and Zha [15]. To derive the filter probability in MS-SUR model, the dynamic of transition probability which controls the probabilities of switching between the regimes is computed by using the Hamilton filter. In the third step, the sequence estimated parameter (γ_0), transition matrix (Q_0), and covariance matrix (Σ_0) based on the likelihood of MS-SUR are used as a starting value to draw using Gibbs sampling. In this step, the Gibbs sampling is used to draw first 4,000 rounds as burn-in draws and draw those updated parameters which are obtained from the first draws again by 20,000 rounds[1]. In addition, we forecast the demand and supply functions with 10 step-ahead forecasts.

---

[1] Brandt [5] explained the basic algorithm for Gibbs sampling comprising three estimated blocks which are filter probabilities sampling, MS process sampling, and parameter sampling.

Finally, we extend the estimated parameters which are obtained from the MS-BSUR posterior to compute the following equations considering intercept and price elasticity of demand and supply, namely, $\hat{\alpha}_1, \hat{\alpha}_2, \hat{\beta}_1$ and $\hat{\delta}_1$, as follows:

$$\begin{bmatrix} \hat{\alpha}_1(s_t) \\ \hat{\alpha}_2(s_t) \end{bmatrix} \begin{bmatrix} -\hat{\beta}_1(s_t) & 1 \\ -\hat{\delta}_1(s_t) & 1 \end{bmatrix}^{-1} = \begin{bmatrix} \kappa_1(s_t) \\ \kappa_2(s_t) \end{bmatrix}$$

$$CS(s_t) = \frac{\kappa_1(s_t) \times (\hat{\beta}_1(s_t) - \kappa_2(s_t))}{2}$$

$$PS(s_t) = \frac{\kappa_1(s_t) \times (\kappa_2(s_t) - \hat{\delta}_1(s_t))}{2}$$

Therefore, according to the above equations, in this paper the total social welfare (TS) can be simply measured as $TS(s_t) = CS(s_t) + PS(s_t)$, where CS and PS are the consumer surplus and the producer surplus, respectively.

5   Empirical Results

Before estimating results, we begin with checking whether the data we used are stationary. Augmented Dickey Fuller (ADF) unit root test has been used as a tool. The results show that all variables are stationary; they passed the test at level with probability equals to zero. In addition, by the Bayesian Approach, the results of Gibbs sampler draws which produce trace plots and density plots of intercepts and coefficients of the model. From the results of trace plot, the Markov chain has reached stationarity because the mean and variance of the trace plots of these intercepts and coefficients are constant over time. Moreover, the results of the density plots also confirmed that the distribution of intercepts, coefficients are converging to the normal distribution after a burn-in 4,000 steps. These results confirm that 4,000 steps of the burn-in and 20,000 final posterior draws are sufficient condition for convergence. Therefore, these parameters can be estimated under Normal-wishart prior.

5.1   Estimation Results for Demand and Supply

The first part of estimated results of this study are displayed as in the following table. We have estimated the system of demand and supply equation of Thai rice using the MS-BSUR. Since we aim to measure the welfare of the Thai rice market, the demand and supply equations have to be constructed and, here, Table 1 shows the obtained results. It shows the estimated results for demand and supply of each regime where the intercepts and the coefficients in each regime are different. Despite the fact that there is not much difference in numbers between regime 1 and regime 2, we still consider that the demand and supply equations in each regime are separate. The result allows us to define regime 1 as the "high growth regime" and regime 2 as the "low growth regime" because the intercepts of both demand and supply of regime 1 are higher than regime 2 [1].

The model performed very well, as we can see obviously that the results are almost significant at the level of 0.01. The coefficients of all variables estimated from this model seem to make sense of the basic demand and supply theory. The coefficients of Thai rice export prices are statistically significant equal to -0.1804 and -0.1773 for the demand equations in regime 1 and regime 2 respectively. For the supply equation, the coefficients of Thai rice export prices are statistically significant equal to 3.0284 and 2.8392 in regime 1 and 2, respectively. In addition, Pakistan rice price also has a negative sign implying that it might be complementary for Thai rice. On the other hand, export rice of Vietnam and India are considered to be substitution goods since they have a positive sign in their rice prices. Furthermore, the transition matrix shown in Table 1 allows us to notice an asymmetry in the rice market by looking at the duration of each regime where the high growth regime has the duration of approximately 3.875 periods and 3.676 periods for the low growth regime.

Table 1. Estimates of MS(2)-BSUR for demand and supply

Regime-dependent intercepts

	Demand	Supply
Regime 1	0.5349***	-4.0834***
Regime 2	0.4612***	-4.1172***

Regime-dependent Autoregressive parameters at lag 2

		Regime1	Regime2
Demand	P_t^{export}	-0.1804***	-0.1773***
	P_t^{viet}	0.3567***	0.3516***
	P_t^{paki}	-0.5138***	-0.5172***
	P_t^{india}	0.0332	0.0124***
Supply	P_t^{export}	3.0284***	2.8392***
	W_t	-0.0013	-0.0284
	R_t	-0.3389**	-0.3370***
	P_t^{farm}	0.0122***	0.017

	p_{1t}	p_{2t}	**Duration**	**Observations**
Regime 1	0.724	0.271	3.875	53
Regime 2	0.276	0.728	3.676	57

Source: Calculation.
Note: "*," "**," and "***" denote rejections of the null hypothesis at the 10%, 5%, and 1% significance levels, respectively.

5.2   Welfare Measurement

Welfare, which is our concern, can be measured by its definition: summing up consumer surplus (pink shaded triangle) and producer surplus (blue shaded triangle). Figure 2 shows that PS is higher than CS for both regimes since the supply curve is steep. Corresponding with a numerical result shown in Table 2, the value of PS is greater than CS for both regimes as well. There is a possible main reason for the steep supply. That is -or should be- due to production capacity of the rice be almost full. Considering the fact that of rice arable area almost 90% is already used. Furthermore, the government subsidy policy kept the Thai rice production high despite the decreasing world rice price. Therefore, the supply of rice would not be decreased. Figure 2 also shows that the total welfare is estimated to be 3.3235 units in regime 1 (high growth) and approximately 3.4747 units for regime 2 (low growth).

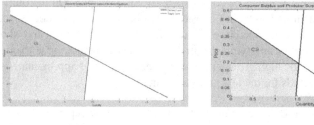

(a) Regime 1   (b) Regime 2

Fig. 2. Welfare measurement (Source: Calculation)

Table 2. Welfare measurement

	Regime 1	Regime 2
Consumer surplus	0.1868	0.2042
Producer surplus	3.1366	3.2704
Total Welfare	**3.3235**	**3.4747**

Source: Calculation.

Expected Welfare

Here is the key point of our study. After getting the welfare values and the probability from the MS-BSUR, we are able to calculate the expected welfare at every time point during the studied period to capture the efficiency of the policies. The result can be displayed as Figure 3. It shows that there is the highest level of expected welfare in

Fig. 3. Expected Welfare (Source: Calculation)

Q1/2006. The government provided the rice mortgage scheme, and the policies established price[2] was slightly close to the market price (Office of Agricultural Economics, Thailand, 2015).

We might not say that this scheme is the best fit for a system of Thai rice over time, but this scheme can create the highest level of welfare during our studied period. Despite the mortgage scheme creating the highest level of welfare, the result also shows that there are low levels of welfare explicitly in Q2/2008 and Q4/2011 to Q1/2012 when the mortgage scheme was used. To clarify this converse result, we found that if there is much difference between the policies established price and the market price, the welfare will be at a low level. On the contrary, a similarity between the policies established price and the market price may create a high level of welfare to the market[3].

5.3   Forecasting Demand and Supply

In this section, we forecast the demand and supply of the rice for 10 steps ahead using the estimated coefficients from the previous result. Using the previous monthly samples, we computed the forecast series from the obtained MS-BSUR model and found that the model perform well in forecasting the demand and supply of Thai rice. The forecast results of supply and demand are illustrated in Figure 4. It provides a comparison between the estimated forecast and the actual number of demand and supply during the period from M1/2006 to M9/2014. The forecast period chosen for prediction is November 2014 to August 2015, being 10 months. According to the above results, it is evident that the estimated forecast series of demand and supply are close to the actual values, indicating that the model performs very well. Moreover, the Mean absolute percentage error (MAPE) is employed to measure the accuracy of the MS-BSUR forecast and the result confirms a high performance of the model with the value of MAPE less than 0.1%.

There is a tendency for the rice market to exhibit a great swing of the demand and supply in the next 10 months. For the supply of Thai rice, the model predicts that supply's growth will increase and reach the highest growth in December, 2015 before dropping around 0.9% in March, 2015. The possible reason behind the swing of the supply forecast is the harvest seasons of Thai rice. Thai farmers normally grow rice two times a year: one is in rainy season and the other off-season. Thus Thailand usually has a large volume of production and exports large amounts of rice to many countries after the harvest seasons. Different results are obtained for the demand of

---

[2] A policies established price, here, means a price that is set or established by the government in order to prevent the market from slumping rice prices

[3] In 2006, the policies established price was 7,000 THB per ton while the market price was 6,784 THB per ton. Comparing with the prices in around Q4/2011 to Q1/2012 where the policies established price was 15,000 THB per ton and the market price on average was 10,026 THB per ton, there is a big gap between the policies established price and the market price. According to the results we got, it is noticeable that if the policies established price is quite similar to the price that market actually pays, the rice welfare will be at the high level. But if there is a big gap, the welfare turns to be in the low level.

Thai rice. The growth rate increases around 0.45% in October, 2014 before dropping to 0.02% in February, 2015. As the demand forecast goes up obviously, we sought for the reason and found that the government may need to release a batch of the rice stocks which is holding around 16 million metric tons. To get rid of the over stock problem the government has to sell that rice at a price substantially below the market prices. This will lead Thai rice price to be cheaper than those of other export countries.

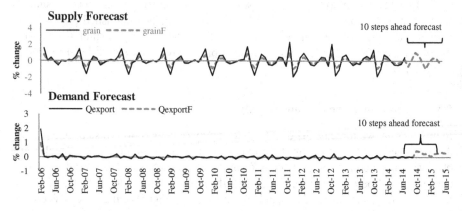

Fig. 4. Demand and Supply forecasting (Source: Calculation)

6   Conclusions

We measured the welfare of Thai rice market using our innovative tool, the MS-BSUR model. The model performed very well in estimating demand and supply equations since the results were statistically significant and the numbers seemed to make sense. The welfare can be estimated nicely for two different regimes. Then, we created the expected welfare at every time point during the studied period to capture the efficiency of the policies. We found that the mortgage scheme could create the highest level of the welfare for Thai rice market if it is conditioned on a similarity between the policies established price and the market price. This empirical result should be useful for the policy makers since it demonstrated that the policy could bring the highest welfare to the society. This study also provided new information to the policy makers about the gap between the policies established price and the market price. It suggested that even if the government provides a high price for the market like a populist price, the result may turn to be bad in terms of the low welfare resulted from the big gap between the realistic price and the populist price. That would not help the rice farmers but rather harm them. Eventually, the forecasting of demand and supply were also estimated for 10 months, and we found that demand and supply tend to increase swing forwards positive growth.

Acknowledgments. This paper would not be successful without support from Ms. Doungtip Sirikarnjanaruk, senior economist at Bank of Thailand, who provided us with exclusive data and very useful comments. Also, many thanks to Mr. Woraphon Yamaka, who always supports us and gives much helpful suggestions, especially in methodology and estimation - we really would like to thank him for this kind support. In addition we are grateful for financial support from Puay Ungpakorn Centre of Excellence in Econometrics, Faculty of Economics, Chiang Mai University.

References

1. Artis, M., Krolzig, H.M., Toro, J.: The European business cycle. Oxford Economic Papers **56**(1), 1–44 (2004)
2. Ayusuk, A., Sriboonchitta, S.: Risk Analysis in Asian Emerging Markets using Canonical Vine Copula and Extreme Value Theory. Thai Journal of Mathematics 59–72 (2014)
3. Baltagi, B.H., Pirotte, A.: Seemingly unrelated regressions with spatial error components. Empirical Economics **40**(1), 5–49 (2011)
4. Boonyanuphong, P., Sriboonchitta, S.: The Impact of Trading Activity on Volatility Transmission and Interdependence among Agricultural Commodity Markets. Thai Journal of Mathematics 211–227 (2014)
5. Brandt, P.T.: Empirical Regime Specific Models of International, Inter-group Conflict, and Politics (2009)
6. Chinnakum, W., Sriboonchitta, S., Pastpipatkul, P.: Factors affecting economic output in developed countries: A copula approach to sample selection with panel data. International Journal of Approximate Reasoning **54**(6), 809–824 (2013)
7. Engel, C., Hamilton, J.D.: Long swings in the exchange rate: Are they in the data and do markets knowing it? (No. w3165). National Bureau of Economic Research (1989)
8. Frühwirth-Schnatter, S.: Finite Mixture and Markov Switching Models: Modeling and Applications to Random Processes. Springer (2006)
9. Hamilton, J.D.: A new approach to the economic analysis of nonstationary time series and the business cycle. Econometrica: Journal of the Econometric Society 357–384 (1989)
10. Kim, C.J., Nelson, C.R.: State-space models with regime switching (1999)
11. Krolzig, H.M.: Markov switching vector autoregressions: modelling, statistical inference and application to business cycle analysis. Springer, Berlin (1997)
12. Kuson, S., Sriboonchitta, S., Calkins, P.: The determinants of household expenditures in Savannakhet, Lao PDR: A Seemingly Unrelated Regression analysis. The Empirical Econometrics and Quantitative Economics Letters **1**(4), 39–60 (2012)
13. Lar, N., Calkins, P., Sriboonchitta, S., Leeahtam, P.: Policy-based analysis of the intensity, causes and effects of poverty: the case of Mawlamyine, Myanmar. Canadian Journal of Development Studies/Revue canadienne d'études du développement **33**(1), 58–76 (2012)
14. Phitthayaphinant, P., Somboonsuke, B., Eksomtramage, T.: Supply response function of oil palm in Thailand. Journal of Agricultural Technology **9**(4), 727–747 (2013)
15. Sims, C.A., Waggoner, D.F., Zha, T.: Methods for inference in large multiple-equation Markov-switching models. Journal of Econometrics **146**(2), 255–274 (2008)
16. Sriboonchitta, S., Liu, J., Kreinovich, V., Nguyen, H.T.: A vine copula approach for analyzing financial risk and co-movement of the Indonesian, Philippine and Thailand stock markets. In: Huynh, V.-N., Kreinovich, V., Sriboonchitta, S. (eds.) Modeling Dependence in Econometrics. AISC, vol. 251, pp. 281–294. Springer, Heidelberg (2014)
17. Zellner, A.: An efficient method of estimating seemingly unrelated regressions and tests for aggregation bias. Journal of the American statistical Association **57**(298), 348–368 (1962)

Modeling Daily Peak Electricity Demand in Thailand

Jirakom Sirisrisakulchai$^{(\boxtimes)}$ and Songsak Sriboonchitta

Faculty of Economics, Chiang Mai University, Chiang Mai, Thailand
sirisrisakulchai@hotmail.com, songsakecon@gmail.com

Abstract. Modeling of daily peak electricity demand is very crucial
for reliability and security assessments of electricity suppliers as well as
of electricity regulators. The aim of this paper is to model the peak
electricity demand using the dynamic Peak-Over-Threshold approach.
This approach uses the vector of covariates including time variable for
modeling extremes. The effect of temperature and time dependence on
shape and scale parameters of Generalized Pareto distribution for peak
electricity demand is investigated and discussed in this article. Finally,
the conditional return levels are computed for risk management.

Keywords: Dynamic POT approach · Daily peak electricity demand ·
Risk management · Extreme value theory

1 Introduction

Electricity is the backbone of each industrialized economy. Reliability of electricity supply is crucial for the well-functioning modern economies. Digital technologies, communication infrastructures and industrial processes all depend on
reliable and efficient electricity systems. The optimal day-to-day operation of
a power plant requires a forecasting of the electricity demand. This forms the
basis for power system planning, security, and reliability. The major forecasting methods include time series such as exponential smoothing, autoregressive
integrated moving average (ARIMA), regression, and transfer function [1]. These
methods only focus on the mean forecasting of demand and sometimes ignore the
occurrence of extreme events. It is very essential for the electricity suppliers to
know if there are any extreme events that can not be explained by their demand
forecasting models. Therefore, to match up the electricity demand and supply,
electricity suppliers need not only to forecast the electricity demand but also
have to understand the behavior of any rare extreme events. The unexpected
extreme electricity demand may increase the risk of power blackout, which in
turn leading to the large amount of economic loss.

In Thailand, there are growing concerns about the probability of blackouts in
summer weather when residential air-conditioning demand adds proportionately
to peak loads. Time-of-use pricing, as an example of demand management strategy, have emerged in 1991 to give the incentive to customers for reducing peak

© Springer International Publishing Switzerland 2015
V.-N. Huynh et al. (Eds.): IUKM 2015, LNAI 9376, pp. 478–487, 2015.
DOI: 10.1007/978-3-319-25135-6_43

demand. A range of demand management strategies will be expected to launch in the near future in order to either redistribute or reduce electricity consumption during peak periods [7]. Peak electricity demand has many factors involved, for example, population growth, economic conditions, weather conditions as well as the randomness in individual usage [8].

Extreme value theory (EVT) is widely used as a tool for risk management [3]. Local maxima which exceed a high threshold are modeled based on the asymptotic theory of extremes for stationary series (specifically, independent and identically distributed random variables. In many cases, such as daily peak electricity demand, there is clear non-stationarity in the series. The requirement of stationary assumption is not satisfied. However, several researchers applied this theory to assess load forecast uncertainty by preprocessing the series before performing the standard EVT models (see [9],[10]). In this paper, we apply the dynamic EVT approach, as proposed by [2], to model the daily peak electricity demand. [2] treated the non-stationary in the series by introducing the covariates to the Generalized Pareto distribution parameters.

The aim of this paper is to explain the seasonal pattern of daily peak electricity demand with the temperature variation and time inhomogeneity dependence. The results from this paper can be used as an additional information in reliability assessment and risk management for power development planning.

2   Standard EVT Approach for Modeling Extremes

The standard approaches for modeling the extreme events of a stationary time series are the Block Maxima approach and the Peak-Over-Threshold (POT) approach. The former approach is to model the maxima for each set of blocks dividing the time series. The POT approach will focus on modeling the exceedances over a fixed or varied high thresholds. The latter has an advantage of being more flexible in modeling data in the sense that more data points can be incorporated in the models in comparison with the former approach. For the standard EVT approaches for modeling extremes, the readers are referred to the standard textbooks such as [4]. The method used in this paper is the extension of the POT approach to a non-stationary data as proposed by [2], which will be discussed in the next sections.

The standard POT approach for stationary time series can be briefly summarized as follows. Let $\{t_1, ..., t_n\} \subseteq \{t'_1, ..., t'_{n'}\}$ be the time points, in increasing order, for which $X_{t'_1}, ..., X_{t'_{n'}}$ exceed the threshold u. Thus, $X_{t'_1}, ..., X_{t'_{n'}}$ represent the exceedances over threshold u with the corresponding excesses $Y_{t_i} = X_{t_i} - u, i \in 1, ..., n$. It can be found in [3] that 1) the number of exceedances N_t will approximately follow a Poisson process with intensity parameter λ. We denote $N_t \sim Poi(\Lambda(t))$ with the integrated rate function $\Lambda(t) = \lambda t$, and 2) the excesses $Y_{t_i} = X_{t_i} - u, i \in 1, ..., n$ will approximately follow a generalized Pareto distribution (GPD). We denote $GPD(\xi, \sigma)$, for $\xi \in \mathbb{R}$, and $\sigma > 0$. ξ and σ are called shape and scale parameters, respectively. The cumulative distribution function for GPD can be expressed as

$$G_{\xi,\sigma}(x) = \begin{cases} 1 - (1 + \xi\frac{x}{\sigma})^{-1/\xi}, & \xi \neq 0, \\ 1 - \exp(\frac{-x}{\sigma}), & \xi = 0, \end{cases} \tag{1}$$

for $x \geq 0$, if $\xi \geq 0$, and $x \in [0, -\sigma/\xi]$, if $\xi < 0$. Finally the quantile at level $\alpha \in (0,1)$, q_α can be estimated by

$$q_\alpha = u + \frac{\hat{\sigma}}{\hat{\xi}}[(\frac{n}{N_u})(1-\alpha)^{-\hat{\xi}} - 1] \tag{2}$$

where n is the number of time points, and N_u is the number of time points exceed the threshold u

3   Dynamic Peak-Over-Threshold Approach

In practice, the stationary assumptions for time series sometimes are violated. [2] proposed to model the non-stationary time series by letting the model parameters to depend on covariates. These models are considered as dynamic models in the sense that we can also include the function of time t in the covariates. For convergence of the simultaneous fitting procedure for ξ and σ, [6] suggested that these parameters should be orthogonal with respect to the Fisher information metric. Thus we reparameterized the GPD parameter σ by $\nu = \log((1 + \xi)\sigma)$. Then we modeled the distribution of excesses by a GPD$\{\xi(\mathbf{x}, t), \nu(\mathbf{x}, t)\}$. $\xi(\mathbf{x}, t)$ and $\nu(\mathbf{x}, t)$ can then be modeled as

$$\xi(\mathbf{x}, t) = f_\xi(\mathbf{x}) + h_\xi(t), \tag{3}$$
$$\nu(\mathbf{x}, t) = f_\nu(\mathbf{x}) + h_\nu(t), \tag{4}$$

where f_ξ, f_ν are functions in the factor levels of the covariate $\mathbf{x}$, and $h_\xi(t), h_\nu(t)$ can be either a parametric function or, a non-parametric function of t. Then $\sigma(\mathbf{x}, t)$ can be recovered by

$$\sigma(\mathbf{x}, t) = \frac{\exp(\nu(\mathbf{x}, t))}{1 + \xi(\mathbf{x}, t)} \tag{5}$$

By assuming function h in (3) and (4) to be sufficiently smooth functions, in [2], it is shown that the parameter vector of the dynamic POT model can then be estimated by maximizing the penalized log likelihood function. Let $\theta \in \Theta$ be the vector of all parameters to be estimated in the above models. The penalized log-likelihood function is

$$l(\theta; .) - \gamma_\xi \int \left(h_\xi''(t)^2\right) dt - \gamma_\nu \int \left(h_\nu''(t)^2\right) dt, \tag{6}$$

where $l(\theta; .)$ is the log-likelihood function for Generalized Pareto Distribution. The penalty terms are introduced to avoid the overfitting of the smooth function h [5]. The parameters γ are then chosen to regulate the smoothness of the curves. The larger values of parameter γ indicate the smoother curves in comparison with the smaller one.

4 Application to Daily Peak Electricity Demand

In this section, we apply the dynamic POT approach to modeling daily peak electricity demand in Thailand, during the period from 1998 to 2013. The dataset consists of 5,574 time points (Days) of 15-minute daily peak electricity demand. The data is provided by the Electricity Generating Authority of Thailand (EGAT). The covariates used in these dynamic POT models are the temperatures and time. The temperature data is provided by the Thai Meteorological Department (TMD). We used the maximum temperature (Celsius degree) for all regions (Northern, Northeastern, Central, Eastern, Southern East Coast, and Southern West Coast) in Thailand. Then we computed the average of the maximum temperature to be used as a covariate in the models.

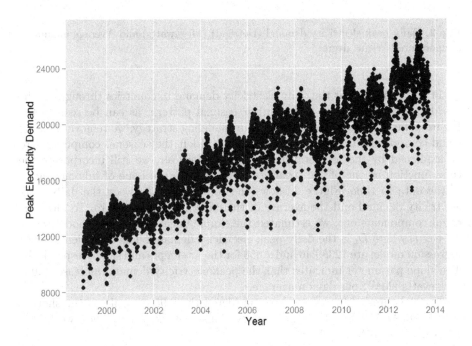

Fig. 1. Daily peak electricity demand in Thailand (Megawatts)

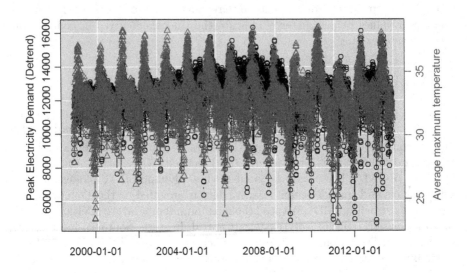

Fig. 2. Daily peak electricity demand (Detrend) (Megawatts) and Average maximum temperature (Celsius degree)

From the facts that the daily electricity demand in countries throughout the world clearly shows time trend and a seasonal pattern, as can be seen in the Figure 1, for the case of Thailand. In our modeling strategy, we remove the time trend components from the series ,but we still left the seasonal components to be explained by the temperature variations. However, we still incorporate the time function as one of the covariate to investigate the issue of inhomogeneity of the data in time. Figure 2 shows the comovement between the daily peak electricity demand and the average of the maximum temperature. To filter the trend components out, we estimated the simple linear time trend model, $D_t = \alpha_0 + \alpha_1 t$, where D_t is the daily peak electricity demand. The coefficients of the regression model are 12389.46 and 2.058 for the intercepts and slope, respectively. The slope parameter indicates that the peak electricity demand increases 2.058 Megawatts (MW) per day on average.

For the threshold selection, we considered the 0.95-quantile as threshold u (i.e. 14343.92 MW) and the 205 observations that exceed u. Then we fit the following models for (ξ, ν):

$$\xi(x,t) = c_\xi, \qquad\qquad \nu(x,t) = c_\nu, \qquad\qquad (7)$$

$$\xi(x,t) = x\beta_\xi, \qquad\qquad\qquad \nu(x,t) = c_\nu, \qquad\qquad (8)$$

$$\xi(x,t) = x\beta_\xi + c_\xi t, \qquad\qquad \nu(x,t) = c_\nu, \qquad\qquad (9)$$

$$\xi(x,t) = x\beta_\xi + h_\xi(t), \qquad\qquad \nu(x,t) = c_\nu, \qquad\qquad (10)$$

$$\xi(x,t) = x\beta_\xi + h_\xi(t), \qquad\qquad \nu(x,t) = x\beta_\nu, \qquad\qquad (11)$$

$$\xi(x,t) = x\beta_\xi + h_\xi(t), \qquad\qquad \nu(x,t) = x\beta_\nu + c_\nu t, \qquad\qquad (12)$$

$$\xi(x,t) = x\beta_\xi + h_\xi(t), \qquad\qquad \nu(x,t) = x\beta_\nu + h_\nu(t), \qquad\qquad (13)$$

where $h_\nu(t)$ is a smooth function of time with variable degree of freedom.

We selected the best fitted model based on the likelihood ratio test. The likelihood ratio test based on Models (7) and (8) shown that temperature has a significant effect on ξ. By adding time variable to Models (9) and (10), the results from the likelihood ratio test reveal that the linear time trend does not have a significant effect on ξ but the smooth function of time does. We therefore use the smooth function of time and temperature as covariates for ξ. As the test shows for Models

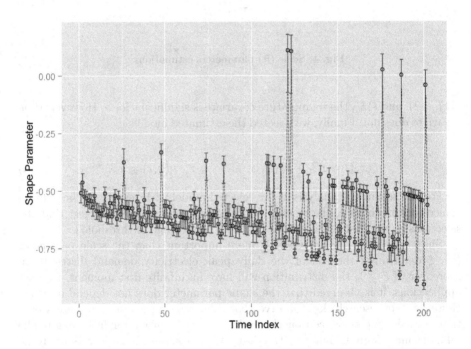

Fig. 3. Shape $(\hat{\xi})$ parameters estimations

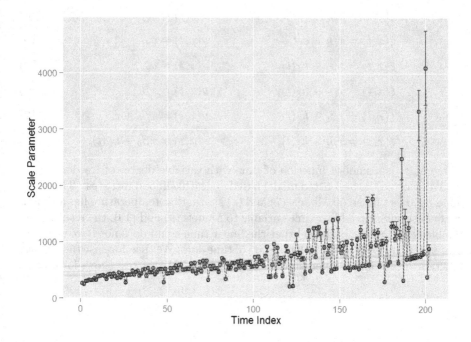

Fig. 4. Scale ($\hat{\sigma}$) parameters estimations

(11), (12), and (13), the temperature covariate is significant for ν. However, time covariate does not. Finally, we selected the estimated models

$$\xi(x,t) = x\hat{\beta}_\xi + \hat{h}_\xi(t), \qquad \nu(x,t) = x\hat{\beta}_\nu. \qquad (14)$$

For the sensitivity analysis of the selected model with respect to the varying choices of threshold u, the results (which are not shown here) suggest that the selections of the best fitted Model (14) hold for a range of thresholds.

Shape parameter (ξ) of GPD is more important than the scale parameter (σ). ξ determines the tail of the daily peak electricity demand distribution. Moreover, if $\xi \geq 1$, the distribution will have an infinite first moment. In most applications, it is observed that the shape parameter does not depend on time. However, our research does observe that. The varying shape parameters over time provides the evidence that the underlying distribution might belong to the different maximum domain of attractions. This issue need to be further analyzed in the future research.

Figure 3 and Figure 4 shows the estimates $\hat{\xi}$ and $\hat{\sigma}$ over time, respectively. We found that temperature has negative effect on shape parameter but has positive effect on scale parameter.

Graphical check for goodness-of-fit test for the selected model was performed by Quantile and Quantile plot (QQ-plot). A QQ-plot is used to compare the residuals computed from the fitted model to a theoretical distribution. Figure 5 shows a Q-Q plot of the residuals (r_i) computed from the best fitted model. The residuals were computed by $r_i = -\log(1-G_{\hat{\xi}_i,\hat{\sigma}_i}(y_{t_i}))$, $i \in \{1, ..., n\}$. The straight line is the reference line when the theoretical quantiles equal quantiles of the fitted model's residuals. The nearby dashed-lines are confidence intervals for QQ-plot. These residuals are approximately distributed as the standard exponential distribution. The plot indicates well fitted in lower quantiles. In our justification, the selected model is fairly accepted for the given data.

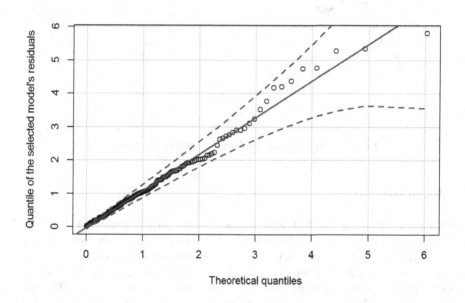

Fig. 5. QQ-plot for the selected model

Finally, the conditional 10-year return levels were computed using Equation (2) for the estimates $\hat{\xi}$ and $\hat{\sigma}$. Figure 6 shows the estimates 10-year conditional return level for peak electricity demand over time. The return level is defined as a quantile of the excesses, Y_{t_i}, conditionally on the covariates. The meaning of 10-year return level is that the probability of the daily peak electricity demand at the given time t_i exceed this level will be once in 10 years.

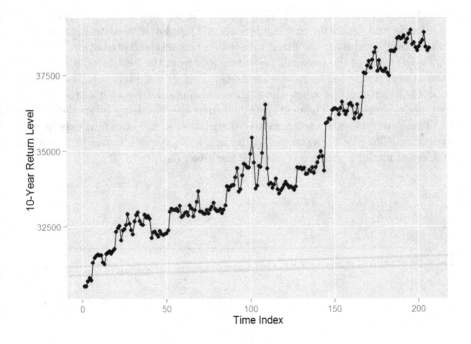

Fig. 6. 10-year conditional return level estimations

5   Conclusions

In this paper, daily peak electricity demands were modeled using a dynamic POT approach. This approach models the extremes depending on covariates in order to capture the non-stationary patterns in the series. The covariates used in our models consisted of temperature and time variables. In the best fitted model for daily peak electricity demand, temperature has a negative effect on shape parameter but has a positive effect on scale parameter. Moreover, the shape parameter are varying over time through the smooth function.

However, there are some further areas have to be investigated in the future. Firstly, non-linear dependence of temperature on parameters have to be investigated in the future work. Several studies on the dependence of temperature on electricity demand have shown non-linear patterns [11] [12]. Secondly, another approach for modeling non-stationary extremes using varying threshold should be performed in comparison with the dynamic POT approach. Finally, we remark here that the use of dynamic POT approach for prediction is questionable. Clearly, the dynamic approach has the superior ability relative to the standard method in capturing the underlying data-generating process and is likely to improve the return level estimation under a greater range of covariate scenarios. However, the smooth function of time covariate is difficult to justify for the future covariate scenarios. Practitioners have to be aware of these issues in their works.

Acknowledgments. The authors are very thankful to the anonymous referees for their valuable suggestions.

References

1. Almeshaieia, E., Soltan, H.: A methodology for Electric Power Load Forecasting. Alexandria Engineering Journal **50**(2), 137–144 (2011)
2. Chavez-Demoulin, V., Embrechts, P., Hofert, M.: An Extreme Value Approach for Modeling Operational Risk Losses Depending on Covariates. Journal of Risk and Insurance (2015). doi:10.1111/jori.12059
3. Embrechts, P., Kluppelburg, C., Mikosch, T.: Modelling Extremal Events for Insurance and Finance. Springer-Verlag, Berlin (1997)
4. Coles, S.G.: An Introduction to Statistical Modelling of Extreme Values. Springer, London (2001)
5. Green, P.J., Silverman, B.W.: Nonparametric Regression and Generalized Linear Models: A Roughness Penalty Approach. Chapman & Hall/CRC, London (2000)
6. Chavez-Demoulin, V.: Two Problems in Environmental Statistics: Capture-Recapture Analysis and Smooth Ecternal Models. Ph.D. Thesis, EPF Lausanne (1999)
7. Faculty of Economics, Chiang Mai University: Development of Demand Response Programs in Thailand (in Thai), Unpublished Technical Report (2014)
8. Hyndman, R.J., Fan, S.: Density Forecasting for Long-term Peak Electricity Demand. IEEE Transactions on Power systems **25**(2), 1142–1153 (2010)
9. Hor, C., Watson, S., Infield, D., Majithia, S.: Assessing load forecast uncertainty using extreme value theory. In: 16th PSCC, Glasgow, Scotland, July 14–18, 2008
10. Sigauke, C., Verster, A., Chikobvu, D.: Extreme Daily Increases in Peak Electricity Demand: Tail-quantile estimation. Energy Policy **53**, 90–96 (2013)
11. Apadula, F., Bassini, A., Elli, A., Scapin, S.: Relationships between Meteorological Variables and Monthly Electricity Demand. Applied Energy **98**, 346–356 (2012)
12. Moral-Carcedo, J., Vicens-Otero, J.: Modelling the Non-linear Response of Spanish Electricity Demand to Temperature Variations. Energy Economics **27**, 477–494 (2005)

Author Index

Autchariyapanitkul, Kittawit   151, 163, 428

Chaniam, Somsak   151

Duong, Phuc H.   303

Entani, Tomoe   65

Fermüller, Christian G.   19

Gong, Xue   440

Hattori, Yuichi   214
He, Yanyan   336
Honda, Katsuhiro   102, 204, 214, 226, 247
Huynh, Van-Nam   290

Innan, Shigeaki   77
Inuiguchi, Masahiro   65, 77, 89, 247

Jiroušek, Radim   254

Kartbayev, Amandyk   326
Kawaguchi, Mayuka F.   43
Kiatmanaroch, Teera   428
Kirby, Robert M.   336
Kondo, Michiro   43
Kosheleva, Olga   112, 138
Kreinovich, Vladik   112, 138
Krejčová, Iva   254
Kudo, Yasuo   279

Le, Anh Ngoc   315
Le, Cuong Anh   315
Le, Tuan Q.   303
Liu, Jianxu   403, 440
Ly, Sel   126

Ma, Tieju   350
Maneejuk, Paravee   464
Min, Nyo   452
Mirzargar, Mahsa   336
Miyamoto, Sadaaki   11

Miyazaki, Taro   247
Murai, Tetsuya   279

Nakano, Takaya   226
Nguyen, Hien T.   303
Nguyen, Hung T.   1, 112, 138
Nguyen, Nhat Bich Thi   315
Nguyen, Van-Doan   290
Niwitpong, Sa-Aat   171, 193
Niwitpong, Suparat   171, 193
Nohara, Yuhumi   102
Notsu, Akira   102, 204, 214, 226, 247
Novák, Vilém   15

Oshio, Shunnya   204
Ouncharoen, Rujira   163

Panichkitkosolkul, Wararit   183
Pastpipatkul, Pathairat   362, 374, 389,
    415, 464
Perfilieva, Irina   32
Pham, Tai Dinh   315
Pham, Uyen H.   126
Phochanachan, Panisara   403
Piamsuwannakit, Sutthiporn   163
Puarattanaarunkorn, Ornanong   428

Qin, Zengchang   236

Ruankham, Warawut   389

Saito, Koki   102
Sangnawakij, Patarawan   193
Sariddichainunta, Puchit   89
Sasaki, Yasuo   54
Seki, Hirosato   267
Sirisrisakulchai, Jirakom   478
Sriboonchitta, Songsak   112, 138, 151, 163,
    362, 374, 389, 403, 415, 428, 440, 452,
    464, 478

Tang, Jiechen   403
Tibprasorn, Phachongchit   151

Tran, Hien D.   126
Tran, Thanh Trung   315

Ubukata, Seiki   102, 204, 214, 226, 247
Ueno, Takanori   214

Van Bui, Vuong   315
Vo-Duy, T.   126

Wiboonpongse, Aree   374, 389

Yamaka, Woraphon   362, 374, 415
Yan, Hong-Bin   350
Yang, Yanping   350

Zhang, Zhipeng   279
Zhao, Hanqing   236

Printed in the United States
By Bookmasters